CREATIONAL THEOLOGY AND THE HISTORY OF PHYSICAL SCIENCE

THE CREATIONIST TRADITION FROM BASIL TO BOHR

BY

CHRISTOPHER B. KAISER

BRILL
LEIDEN · NEW YORK · KÖLN
1997

This book is printed on acid-free paper.

Library of Congress Cataloging-in-Publication Data

Kaiser, Christopher B.
  Creational theology and the history of physical science : the creationist tradition from Basil to Bohr / by Christopher B. Kaiser.
    p. cm. — (Studies in the history of Christian thought, ISSN 0081-8607 ; v. 78)
  Includes bibliographical references and indexes.
  ISBN 9004106693 (cloth : alk. paper)
  1. Creation—History of doctrines. 2. Religion and science—History. 3. Physical sciences—History. I. Title. II. Series.
BT695.K355   1997
261.5'5'09—dc21   97-12491
                                                                  CIP

Die Deutsche Bibliothek - CIP-Einheitsaufnahme

Kaiser, Christopher B.:
Creational theology and the history of physical science : the creationist tradition from Basil to Bohr / by Christopher B. Kaiser. – Leiden ; New York ; Köln : Brill, 1997
  (Studies in the history of Christian thought ; Vol. 78)
  ISBN 90–04–10669–3 Gewebe

ISSN 0081-8607
ISBN 90 04 10669 3

© *Copyright 1997 by Koninklijke Brill, Leiden, The Netherlands*

All rights reserved.
No part of this publication may be reproduced,
translated, stored in a retrieval system, or transmitted in any form
or by any means, electronic, mechanical, photocopying,
recording or otherwise, without prior written permission
from the publisher.

PRINTED IN THE NETHERLANDS

CONTENTS

Acknowledgements	vii
Abbreviations	ix
Introduction	1
Chapter One: The Early Church and Greco-Roman Science (Through the Twelfth Century A.D.)	13
A. Background: From Intertestamental Judaism to Basil of Caesarea	13
B. The Creationist Tradition: Comprehensibility of the World	21
C. The Creationist Tradition: Unity of Heaven and Earth	27
D. The Creationist Tradition: Relative Autonomy of Nature	32
E. The Creationist Tradition: Ministry of Healing and Restoration	60
Chapter Two: The Medieval Church and Aristotelian Science (Thirteenth Century to the Fifteenth Century)	84
A. The Reception of Aristotelian Science	84
B. The Impact of Aristotelian Science on Scholastic Theology	88
C. The Influence of Medieval Theology on Natural Science	112
Chapter Three: Renaissance, Reformation, and Early Modern Science (Fifteenth Century through the Seventeenth Century)	134
A. Renaissance Science through Copernicus and Paracelsus	134
B. Renaissance Science and Reformation Theology through Kepler and Bacon	162
C. The Seventeenth Century: Spiritualist, Mechanist, and Platonic Traditions through Leibniz, Boyle, and Newton	199
Chapter Four: The Heritage of Isaac Newton: From Natural Theology to Naturalism (The Eighteenth Century)	252
A. The Newtonian Tradition from Newton to Hutton	252
B. Post-Newtonian Materialists	295
C. British and American Anti-Newtonians	319
D. Post-Newtonian Cosmogonists and Neo-Mechanists	335

Chapter Five: The Creationist Tradition and the Emergence of
Post-Newtonian Mechanics (Nineteenth and Early Twentieth
Centuries) 352

A. The Nineteenth-Century Context 352
B. The Mechanical Philosophy Challenged (Oersted, Davy,
   and Faraday) 357
C. The Mechanical Philosophy Restated and Formalised
   (Whewell, Joule and Kelvin) 366
D. The Mechanical Philosophy Generalised (Maxwell) 379
E. Conclusion: The Contribution of the Creationist Tradition to
   Twentieth-Century Physics (Einstein and Bohr) 388

Retrospect and Prospect 400

Bibliography 408

Indices 440
   Names 440
   Places 446
   Subjects 447

ACKNOWLEDGEMENTS

I wish to thank Professor Heiko Oberman, editor-in-chief of Studies in the History of Christian Thought, for graciously accepting a work that explores the theological dimensions of the history of science and stretches the usual limits of historical theology.

Parts of the book (chapters 3A and 4) were written while I was a resident member of the Center for Theological Inquiry at Princeton, New Jersey. I greatly benefitted from the office space, library privileges, and excellent housing they provided for my myself and family. The two semesters I spent at the center were also supported by sabbatical leaves granted by Western Theological Seminary and a grant for theological scholarship and research from the Association of Theological Schools.

An abridged version of this book was published in 1991 as volume 3 in the Marshall-Pickering History of Christian Theology series, edited by Paul Avis. I have received helpful comments and criticisms on the manuscript from many professional colleagues, including Robert Bast, Robert Brungs, Dan Deffenbaugh, Eugene Klaaren, Colin Russell, Robert Snow, Jeff Tyler, Wentzel van Huyssteen, and Howard Van Till. I am also indebted to Laurie Baron for the final copy editing and to Russell Gasero for preparing the camera-ready copy. I wish to express my deep gratitude to all of these good friends.

In view of the fact that the two of us managed to work full time and raise three children during these fifteen years of research and writing, my penultimate word of thanks must go to my wife, Martha Winnifred Mercaldi Kaiser.

Sola Deo Gloria.

Christopher B. Kaiser
Holland, Michigan, Easter, 1997

ABBREVIATIONS

ACW	*Ancient Christian Writers.* 56 vols. to date, Westminster, Md. and Mahwah, N.J., 1949-.
AHES	*Archive for History of Exact Sciences*
AJP	*American Journal of Physics*
ANF	*Ante-Nicene Fathers.* Edited by Alexander Roberts and James Donaldson. 8 vols., Buffalo, 1885-86; 2 supplementary vols., Buffalo and New York, 1887, 1896.
AV	*Authorized Version of the King James Bible*
BHM	*Bulletin of the (Institute of the) History of Medicine*
BJHS	*British Journal for the History of Science*
CCL	*Corpus Christianorum: Series Latina.* Turnholt: Brepolis, 1954-.
CH	*Church History*
CNTC	*Calvin's New Testament Commentaries.* Edited by David W. and Thomas F. Torrance. 12 vols. Edinburgh: Oliver and Boyd, 1959-73.
CO	*Ioannis Calvini Opera.* Edited by William Baum et al., 59 vols. (Corpus Reformatorum, vols. 29-87), Brunswick and Berlin, 1863-1900.
DHI	*Dictionary of the History of Ideas.* Edited by Philip P. Weiner. 4 vols., New York: Scribner s, 1968, 1973.
DNB	*Dictionary of National Biography.* Edited by Leslie Stephen and Sidney Lee. 22 vols., London: Oxford Univ. Press, 1921-22.
DSB	*Dictionary of Scientific Biography.* Edited by Charles Coulston Gillispie. 16 vols., New York: Scribner s, 1970-80.
ECS	*Eighteenth-Century Studies*
FCPS	*Fathers of the Church, Patristic Series.* 82 vols. to date, Washington, D.C.: Catholic Univ. of America Press, 1947-.
HSPS	*Historical Studies in the Physical Sciences.* Ed. Russell McCormmach. Princeton: Princeton Univ. Press, 1970-.
HTR	*Harvard Theological Review*
JHI	*Journal of the History of Ideas*
JHMAS	*Journal of the History of Medicine and Allied Sciences*
JHP	*Journal of the History of Philosophy*
JIH	*Journal of Interdisciplinary History*
KGW	Johannes Kepler: *Gesammelte Werke.* Edited by Max Caspar et al. Munich: Beck, 1938-.

LCC	*Library of Christian Classics*. Edited by John Ballie, John T. McNeill, and Henry P. Van Dusen. 26 vols., London: SCM, 1953-70.
LW	*Luther's Works*, American Edition. Edited by Jaroslav Pelikan and Helmut T. Lehmann. 55 vols., St. Louis and Philadelphia: Concordia and Fortress, 1955-76.
MPG	*Patrologiae Cursus Completus... Series Graeca*. Edited by J. P. Migne. 162 vols., Paris, 1857-66.
MPL	*Patrologiae Cursus Completus... Series Latina*. Edited by J. P. Migne. 221 vols., Paris, 1844-79.
NPNF	*Nicene and Post-Nicene Fathers, First Series*. Edited by Philip Schaff. 14 vols., Buffalo and New York, 1886-90.
NPNF II	*Nicene and Post-Nicene Fathers, Second Series*. Edited by Philip Schaff and Henry Wace. 14 vols., Buffalo and New York, 1890-1900.
OTP	*Old Testament Pseudepigrapha*. Edited by J. H. Charlesworth. 2 vols., Garden City: Doubleday, 1983, 1985.
PAPS	*Proceedings of the American Philosophical Society*
SC	*Sources Chrétiennes*. Paris: Cerf, 1940-.
SHPMP	*Studies in the History and Philosophy of Modern Physics*
SHPS	*Studies in the History and Philosophy of Science*
TAPS	*Transactions of the American Philosophical Society*
WA	*Martin Luthers Werke... Weimarer Ausgabe*. 90 vols., Weimar: Böhlaus, 1883-1968 (repr. Graz: Akademische Druck, 1966-70).

INTRODUCTION

A. Bridges Between Science and Theology

The subject of this book crosses three different fields: the history of science, science and religion, and historical theology. The plurality of fields may make the book difficult to classify, but the overlap is intentional. The underlying vision that has motivated my work over the last thirty years has been one of bridge-building between science and theology, or more generally between the secular and the spiritual aspects of our culture.

There are many ways of building such bridges between science and faith. One could link them through the philosophy of science and philosophical theology, for example, or through the subjects of science and society and theological ethics. The methods and procedures used here are strictly historical ones, but philosophical and ethical issues are frequently involved. Historical work can never be impersonal, since every historian works out of a context and is affected by the conflicts and dilemmas that motivate historical research within that context. I should say something about my own philosophical and ethical interests here so that the reader can discern where they may impinge on the historical narrative.

Philosophically, there have been two burning issues for me. The first is what I shall call the mystery of science: Einstein described it as the mystery of the 'comprehensibility of the world'.[1]

Of course, science is not generally thought to be a mystery. In fact, for many people, science dispels all the mystery in nature. To be sure, there are apparent boundaries to science like the beginning of time, the fine-tuning of the parameters of cosmology and nuclear physics, and the failure of determinism at the level of quantum uncertainty. But some of these mysteries might be dispelled by further advances in physics. In any case, they are all due to the disproportion between the realms of deep science and the conditions of everyday life.

The mystery of science is neither located in nature itself nor in science as a body of knowledge, but in the fact that ordinary human beings are the ones who investigate nature and pursue the science.[2]

---

[1] Albert Einstein, *Ideas and Opinions* (London: Alvin Redman, 1954), p. 46.
[2] It is not necessary to be an epistemological realist in order to appreciate the mystery of science. Even instrumentalists and social constructivists, who view scientific theories merely as human artifacts like tools or works of art, should recognise the wonder of the fact that solutions to new scientific problems continue to be found. Only in the case of an epistemology

I recall my own experience as a university student majoring in physics in the early 1960s. As the mathematical elegance of classical mechanics, electricity and magnetism, and quantum theory was unfolded in the lecture hall, I felt as though there was something mystical about our common task. It was as if our minds were entering another dimension of human creativity. Deep science develops cognitive tools that are as far removed from everyday logic as the realms it explores are remote from the conditions of everyday life.[3]

The mystery of science can be treated as a special case of another mystery, that of the human intelligence required to pursue modern science. Early humans were probably selected for a pragmatic understanding of nature by the course of natural evolution. Physical anthropologists argue that there was survival value in the developing cognitive skills of early humans like *homo erectus* and early *homo sapiens*. Hence they gradually developed the ability to recognise patterns in nature like the seasons and to make predictions concerning things like the behaviour of animals. These skills are shared to some extent by other hominoids like chimpanzees, but they are far more developed in humans than in any other species. Presumably the harsh conditions of life in the Paleolithic era provided the environmental stimulus for such advances in intelligence.

Naturalistic accounts of human evolution are helpful for developing an outline of our prehistory, but they do not account for the kind of intelligence and creativity required for the emergence of modern science. The phenomenal patterns that modern scientists explore are entirely different from those of everyday human life in Paleolithic times or even today. Granted that we are better at investigating novel features of nature than chimpanzees are, why should we be that much better?

Consider an analogy with human physiology. Our bodies are adapted for walking upright. Our hominid ancestors have been walking for millions of years. As it turns out, our arms and legs are also adapted for swimming, even though we may not have needed this skill in early hominid days. In this case we are said to be 'preadapted' for swimming. It is a kind of bonus paid in the evolutionary economy. What about flying? No such luck there. Flying requires special machines and the fortuitous provision of fossil fuels. Arms and legs are not enough. Similar balance sheets could be drawn up for

---

that recognised no criteria for distinguishing good theories from bad ones would there be no mystery at all.

[3] When personal experience is introduced into the discussion of history, many of us worry about the dangers of arbitrary subjectivity. My own view is that all historical research is an attempt to answer the question of where we and our world come from. So all history is internalised and filtered through subjective experience. But there is also a reciprocal movement. Subjectivity is externalised and given structure through the investigation of history. My historical investigation in this work attempts to explore the deep history of science and subsequent developments leading up to the twentieth-century and the background of my own experience as a student.

other aspects of human physiology like manual dexterity and digestion. In each case, it will be found that, when an evolutionary development occurs to meet one environmental condition, it may by chance also meet another, but the range of conditions addressed is very limited.

In the parallel case of human intelligence, we have long since excelled the likely benefits of preadaptation. Evolutionary adaptation does not even come close to accounting for it. So the question arises: is there a sufficient condition for human intelligence and creativity as exhibited in modern science? How can we account for the progress of science in areas that lie beyond the conditions under which we evolved?[4] Is it just a case of absurdly good luck? Certainly luck is a factor in any enterprise, but it can hardly be the only basis for our belief in the long-term prospects for science. Lottery tickets would offer better chances!

One of the great paradoxes of twentieth-century thought is that there has been so little discussion of this question. Some of the great scientists of our time have raised it, but for the most part it goes unnoticed, even in educated circles.[5] Perhaps, as moderns, we have become desensitised to the mystery of science by its pervasiveness in our world, just as St Augustine said that people were oblivious to the miraculous element in familiar phenomena of

---

[4] I am not concerned here with the 'anthropic principle'. The anthropic principle states that the conditions of the universe must be compatible with the evolution of living creatures capable of observing the universe and asking simple questions about it. My question here is rather different. It has to do with the conditions necessary for the emergence of creatures intelligent enough to probe phenomena completely beyond the natural conditions under which their intelligence evolved. Amazing as it is, the fact that the laws governing nuclear reactions must have the parameters they do in order for carbon-based life to exist is not a sufficient condition for the evolution of creatures capable of understanding nuclear physics. For a nontechnical introduction to the anthropic principle, see John Polkinghorne, 'A Potent Universe', in John Marks Templeton, ed., *Evidence of Purpose: Scientists Discover the Creator* (New York: Continuum, 1994), pp. 105-115.

[5] The first writer, to my knowledge, to treat the issue within the context of Darwinian evolution was Arthur James Balfour, *The Foundations of Belief* (London, 1895), pp. 300-302. Current interest in the subject dates from Eugene Wigner's essay, 'The Unreasonable Effectivenesss of Mathematics in the Natural Sciences'; originally published in *Communications in Pure and Applied Mathematics*, vol. xiii (Feb. 1960); reprinted in *idem*, *Symmetries and Reflections* (Bloomington: Indiana Univ. Press, 1979), pp. 222-37. See also R.W. Hamming, 'The Unreasonable Effectiveness of Mathematics', *American Mathematical Monthly*, vol. lxxxvii (Feb. 1980), pp. 81-90; Mark Steiner, 'The Appplication of Mathematics to Natural Science', *Journal of Philosophy*, vol. lxxxvi (Sept. 1989), pp. 449-80. The most wide ranging discussions to date have been those of Paul C.W. Davies, 'The Intelligibility of Nature', in Robert John Russell et al., eds., *Quantum Cosmology and the Laws of Nature: Scientific Perspectives on Divine Action* (Vatican City: Vatican Observatory, 1993), pp. 145-61; 'The Unreasonable Effectiveness of Science', in J.M. Templeton, ed., *Evidence of Purpose*, pp. 44-56; 'The Mind of God', in Jan Hilgevoord, ed., *Physics and Our View of the World* (Cambridge: Cambridge Univ. Press, 1994), pp. 230-38. See also the mind-clearing criticisms made by Willem B. Drees, *Beyond the Big Bang: Quantum Cosmologies and God* (La Salle, Ill.: Open Court, 1990), pp. 107-110; 'Problems in Debates about Physics and Religion', in Hilgevoord, ed., *Physics and Our View of the World*, pp. 201-3.

their time.⁶ If so, it makes sense to consider the thinking of past generations on the subject. Premoderns were enthralled by the wonder of the human intellect, even though it had accomplished only a fraction of what it has today. At one level, *Creational Theology and the History of Physical Science* is my personal review of their way of accounting for human intelligence in terms of the theology of creation.

A second philosophical issue that concerns me is whether the creative activity of God is conceivable in a world governed by natural laws. Or, vice versa, can the integrity of science and the autonomy of nature be established within the framework of a creational theology. I call this the problem of the 'relative autonomy of nature'.

I was raised in a liberal, secular milieu, in which such compatibility was unthinkable. As an aspiring young physicist, I received quite a shock when I first met Christians who took the biblical narratives seriously. But, the more I learn to question my inherited standards of plausibility, the more I wonder how and why we ever reached such a juncture in history. Did earlier generations find the relation of divine activity and natural law to be as much of a dilemma as we do today? If not, how did they avoid it? And when and why did the dilemma arise?

At yet another level, *Creational Theology and the History of Physical Science* is a personal quest for evidence of ethical criteria in the history of science.

Ethically, my sensitivities have been shaped by the counter-cultural critique of the 1960s and '70s. I wonder whether science and technology have been, on balance, beneficial or harmful for life on earth.⁷ Or, under what conditions could they be beneficial?

As I write, the Cold War is over, and the threat of a nuclear holocaust seems to have receded. But the continued depletion of ozone in the stratosphere causes genetic damage on land and in the seas, while excessive levels of ozone at ground levels aggravate respiratory disease. Natural resources are consumed at an accelerating rate, and life expectancy is increased at the expense of a population explosion which heightens ethnic conflict around the world. Traditional structures of communal existence and spiritual awareness atrophy as networks of modern communication and commerce are developed.

In light of such developments, I see no apologetic value in showing the

---

⁶ Augustine, *City of God* XXI.4. Examples of overlooked miracles cited by Augustine include the properties of fire, charcoal, lime, diamonds, and the loadstone.

⁷ The issue of the social impact of science and technology was highlighted at the International Conference on Faith, Science, and the Future, sponsored by the World Council of Churches at the Massachusetts Institute of Technology, July 12-24, 1979. For my observations on this historic meeting, see 'Faith and Science and the W.C.C.', *Reformed World*, vol. xxxv (1979), pp. 330-36.

compatibility of modern science with theology unless ethical criteria can be established that can govern the uses of science. Techniques of exploiting nature must be directed toward the relief of suffering, not its aggravation. Advances in medicine must be compatible with the vitality of communities and the dignity of all people, not just a privileged few. So the aim of this book is not to defend tradition, but rather to recover its meaning for life and work in society. Throughout the book, therefore, I search for evidence of what I call the 'ministry of healing and restoration'.

These philosophical and ethical commitments might be suspected of distorting the historical record if it were supposed that modern philosophy and ethics are fundamentally sound. At root, I am motivated by the deep sense that we have lost too much in our push toward modernity. If I tend to exaggerate certain aspects of the past as the result of my commitments, I do it in the spirit of a physician prescribing what the patient seems to lack. Distortion can be justified, if it is wholesome.

B. PARAMATERS AND CAVEATS

Although this work straddles the fields of history of science, science and religion, and historical theology, there are definite limitations as stated in the full title.

First, the 'science' discussed here is primarily physical science. Very little is said about the life sciences in these pages, particularly following coverage of the eighteenth century, when the life sciences first became clearly differentiated from physical science. In this sense, this work is more narrowly defined than other historical treatments of the science-theology interaction which include Darwinian evolution and the complex controversy it raised.[8]

This Darwinian deficit is hopefully compensated for in other ways. For one thing, the scope of physical science is broadened to include applied sciences like medicine, alchemy, and geology as well as the more theoretical ones like physics and cosmology. For another, greater attention is given to

---

[8] On the Darwinian controversy, see Meal C. Gillispie, *Charles Darwin and the Problem of Creation* (Chicago: Univ. of Chicago Press, 1979); James R. Moore, *The Post-Darwinian Controversies: A Study of the Protestant Struggle to Come to Terms with Darwin in Great Britain and America, 1870-1900* (New Cambridge: Cambridge Univ. Press, 1979); David N. Livingstone, *Darwin's Forgotten Defenders: The Encounter between Evangelical Theology and Evolutionary Thought* (Grand Rapids: Eerdmans, 1987); Jon H. Roberts, *Darwinism and the Divine in America: Protestant Intellectuals and Organic Evolution, 1859-1900* (Madison: Univ. of Wisconsin Press, 1988).

the intertestamental, patristic, and medieval background of modern science than is usually the case in Protestant treatments of the subject.[9]

As I see it, the Christian faith has a continuous history from its background in Second Temple Judaism and the early church to modern times, and the Protestant Reformation was basically just an attempt to recover the teachings and practises of the early Christians. I find many historic differences among Christians with regard to creation and science, but none of them correlate with traditional differences between Roman Catholics and Protestants, or between medieval and Reformation theology.

If the point of origin is pushed back into biblical times, the point of closure in the early twentieth century constitutes a second limitation of *Creational Theology and the History of Physical Science*. Why stop with the work of Einstein and Bohr (chap. 5, sec. E)?

The simplest explanation is that I am tracing a historical tradition, and like all traditions it has a beginning and an end. The Jewish-Christian belief in creation was instrumental in the investigation of nature for a period of over 2,000 years—beginning with the Hellenistic period a few centuries before Christ, and extending through the nineteenth century AD. James Clarke Maxwell was probably the last major proponent of the continuous tradition.

I conclude with the work of Einstein and Bohr, even though neither of them professed a traditonal creational theology, because they were deeply influenced by Maxwell and because they are the two pillars of twentieth-century physics. In this way, I try to show that the values of the historic creationist tradition were assimilated by Einstein and Bohr and passed on to the twentieth century in spite of its evident secularism. The legacy of the creationist tradition is still with us in a practical way, even

---

[9] In this sense, my work parallels that of Stanley L. Jaki, particularly his *Science and Creation: From Eternal Cycles to an Osciallating Universe* (Edinburgh: Scottish Academic Press, 1974). But I view the creationist tradition in more positive and eclectic terms than Jaki does. Three of the best historical treatments of science and theology to date have focused largely on developments since the sixteenth century: John Dillenberger, *Protestant Thought and Natural Science: A Historical Interpretation* (Garden City, N.Y.: Doubleday, 1960); Colin A. Russell, *Cross-Currents: Interactions between Science and Faith* (Grand Rapids, Mich.: Eerdmans, 1985); John Hedley Brooke, *Science and Religion: Some Historical Perspectives* (Cambridge: Cambridge Univ. Press, 1991). As if to counterbalance the usual Protestant bias, Harold Nebelsick, with whose work I feel the closest affinity, has emphasised the period up to the seventeenth century: *Circles of God: Theology and Science from the Greeks to Copernicus* (Edinburgh: Scottish Academic Press, 1985); *Renaissance and Reformation and the Rise of Science* (Edinburgh: Scottish Academic Press, 1992). A historian of science who has studied the impact of the early and medieval church is David C. Lindberg: 'Science and the Early Christian Church', *Isis*, vol. lxxiv (Dec. 1983), pp. 509-530, reprinted in D.C. Lindberg and R. L. Numbers, eds., *God and Nature: Historical Essays on the Encounter between Christianity and Science* (Berkeley: Univ. of California Press, 1986); idem, *The Beginnings of Western Science: The European Scientific Tradition in Philosophical, Religious, and Institutional Context* (Chicago: Univ. of Chicago Press, 1992).

though the faith that sustained it has been relegated to the (rather limited) sphere of religion.

The point is that secularism is an illusion. It can only be maintained at the expense of deep cultural contradictions like those explored by Daniel Bell and Bruno Latour.[10] Some of these contradictions were noted by the philosopher, Immanuel Kant, in the mid-eighteenth century. Kant managed to define them out of existence (exorcise them might be a better description) by making careful incisions between the noumenal and the phenomenal and the theoretical and the practical. But the contradictions of secularism are still with us: persistent belief in scientific progress in spite of evidence for genetically determined limits to human abilities; heroic efforts toward education and moral improvement in the midst of recalcitrant patterns of social behaviour. Perfect consistency is impossible, even in a rigorous discipline like pure mathematics. But one task of any discipline must be to recognise and articulate the inconsistencies it cannot do without.

C. Some Theses and Terms Defined

Whenever I have discussed my writing with academic colleagues, I have been asked, 'What is your thesis?' And somehow I felt unworthy of the high art of scholarship if I could not come up with a proposition that seemed both stunning and defensible. I honestly did not know what I would find as I began my project in the early 1980s. As indicated above, my motivation had more to do with a quest for wisdom than with the defense of a proposition. I have also been inspired by the words of the Jewish sage, Jesus ben Sirach, who said, 'I have not laboured for myself alone, but for all who seek wisdom' (Ecclus. 24:34; 33:18).

Nonetheless, I do have beliefs that will be tested throughout the work. Since those beliefs run counter to the received wisdom of our time, I should also offer some account of the differences. For those who like theses, therefore, I offer the following four. All four are to be tested by exploring the views of historical figures, both scientists and theologians, whose thinking on the subject appears to be at least as coherent as our own in the twentieth century.

Thesis 1. Scientific work is entirely compatible with a shared commitment to a creational theology of nature and human life.

By 'scientific work', I mean the investigation of nature based on the

---

[10] Daniel Bell, *The Cultural Contradictions of Capitalism* (Basic Books, 1978); Bruno Latour, *We Have Never Been Modern* (Cambridge, Mass.: Harvard Univ. Press, 1993).

assumption of universal, natural laws and the use of analytical reason (in theoretical sciences).[11] Under the rubric of 'scientific work', I include also the effort to redirect nature in order to improve the human condition (in applied sciences) as explained above. By 'a creational theology', I mean a historic worldview and moral stance based on the biblical belief that nature and humanity are created by a wise, powerful God who intends good for them.[12] A creational theology, as I understand it, has little or nothing to do with 'creation science', which is predicated on a literal reading of the Bible and a presumed *incompatibility* between scientific work (as defined here) and creationist belief. By 'a shared commitment', I mean a vision of the world and an agenda for the future that is shared by most members of a profession or community. Many in the scientific community today would accept the thesis, if only as a gesture of tolerance, provided that creationist belief were limited to personal convictions. But the history reviewed here indicates a corporate commitment sustained by institutions like the church. In fact, a stronger version of Thesis 1 can be defended as follows.

Thesis 2. Scientific work can be engendered and sustained by a shared commitment to a creational theology of nature and human life.

The historical influence of theology on science is more interesting than mere compatibility, but it is also more difficult to demonstrate. Adequate

---

[11] The term 'scientist' was not coined until the mid-nineteenth century (by William Whewell), and the medieval notion of science (*scientia*) was radically different from the modern one, particularly with respect to the agency of God. However, there is a continuous commitment to rationality and universality which justifies the use of the terms, 'science', 'scientist', and 'scientific', throughout the work.

Andrew Cunningham argues that the term, 'science', should not be used to describe the 'history of science' prior to the late-eighteenth century; 'How the *Principia* Got Its Name; or, Taking Natural Philosophy Seriously', *History of Science* 29 (1991), 381-89. He argues instead for use of the term, 'natural philosophy', as describing a pre-Enlightenment discipline that treated God and his creation.

As to the issue of terminology, medieval philosophers did, in fact, use the term *scientia* (as Cunningham acknowledges on p. 387). But for them *scientia* was a broader study (including metaphysics) than natural philosophy, not a narrower one as Cunningham supposes (see chapter 2, note 1, below).

More importantly, it will be shown below that the modern separation of science and theology has roots in the Middle Ages, even though the separation was not complete until the late-nineteenth century. Cunningham neglects the medieval tradition that stressed the autonomy of nature and made a clear distinction between matters of science and matters of faith (e.g., Adelard of Bath, William of Conches, Albertus Magnus, and Jean Buridan). On the other hand, I agree with Cunningham's basic argument that premodern natural philosophy presupposed the work of God in creation; cf. his 'Getting the Game Right: Some Plain Words on the Identity and Invention of Science', *Studies in the History and Philosophy of Science*, vol. xix (1988), pp. 384-85.

[12] I have taken the phrase 'creationist tradition' from the work of Richard C. Dales on twelfth-century natural philosophy and have adapted it to the earlier Jewish and patristic context. For citations of Dales's work and a comparison of our definitions, see chapter 1, section D, below. 'Creational theology' is just one aspect of the creationist tradition, since the latter includes the historical impact of that theology in secular disciplines. Roughly, the creationist tradition is to creational theology what the history of science is to scientific work.

evidence will be presented in the following pages, however, to illustrate the point. However, Thesis 2 does not entail either a necessary or a sufficient condition for scientific work. For the sake of clarity, then, this thesis should be differentiated from even stronger versions like the following:

(2a) Scientific work readily results from a shared commitment to a creational theology of nature and human life. In other words, creational theology is a sufficient condition for the rise of modern science. Such a thesis appears to be false. It ignores the existence of contingencies that may be independent corporate commitments. And it could be sustained only at the expense of overlooking the creational theology of cultures that have not developed a natural science on the basis of universal laws and analytical reason. The Eastern Orthodox and Oriental Orthodox traditions, for example, share the creational theology of the early church, yet their circumstances and development have been quite different from that of Western Christendom. I have no desire to argue the thesis of a creational thelogy as sufficient condition. Nor do I wish to argue the following.

(2b) The long-term sustainability of scientific work requires a shared commitment to a creational theology of nature and human life. This version is more interesting than 2a. Strictly speaking, it could never be historically verified, since more episodes of scientific work are possible than will ever be observed. Nor could it be falsified in the near term. Much scientific work today is carried on without the stipulated commitment (although many individual scientists are still very religious). And, even if current scientific work should turn out not to be sustainable in the long term, economic and technological factors would be as determinative as ideological ones. It is best to leave such a thesis to historians of the long-term future and return to the lessons of the observable past.

Thesis 3. A shared commitment to a creational theology of nature and human life can be reinforced (if not engendered) by scientific work. In other words, the historical influence between faith and science is a reciprocal one. In fact, there is ample evidence of a fairly continuous tradition (I call it the 'creationist tradition'), lasting over two millennia, in which creational theology and scientific work did mutually reinforce each other. Again, neither a necessary nor a sufficient condition for the effect on theology is entailed.

But, if these three theses can be sustained from history, why is it so often presumed that the progress of science detracts from biblical faith? Or why is it held that a consistent commitment to faith undermines science? In order to address this question, we must consider something besides the relation between science and theology. We need also to consider the historical shift that has occurred between different ways of relating the two.

Thesis 4. The common presumption that scientific work and creational theology are incompatible is the historical result of the reconstrual of plausible options in terms of mutually exclusive alternatives.

Abstractly, it is possible to view scientific work and creational theology as either mutually reinforcing or mutually incompatible. In practise, however, it is very difficult for an individual to shift from one perspective to the other. The shift primarily occurs at the societal level, and such a shift has occurred in Western culture.

In the chapters that follow, we shall carefully trace the steps through which that shift occurred. In order to avoid the idea of simple continuity in the creationist tradition, four of the most important steps should be noted in advance. First, the application of Aristotelian dialectic in the early Middle Ages led to a semantic differentiation between the natural and the supernatural (discussed in chap. 1, sec. E). Second, the assimilation of Aristotelian cosmology in the high Middle Ages restructured the God-world relation in terms of divine causation (efficient cause and effect) rather than divine decree (word and response, chap. 2, sec. B). Third, the rise of the mechanical philosophy in the early modern period led some to search for gaps in the natural order as evidence for God's continued activity in the world. The mechanical model led others to view any success in eliminating such gaps as evidence against the activity of God. In either case, we have lost sight of the creational image of God operating in and through the laws and forces of nature (chap. 3, sec. C, and chap. 4). Finally, the interiorisation of religious faith and the mechanisation of industry and commerce in the nineteenth century made the dichotomy of spirit and nature a social reality as well as a philosophic one (chap. 5, sec. A).[13]

The book, therefore, closes with a paradox: modern science is moving far beyond the mechanical philosophy at the very same time that modern culture is becoming increasingly mechanised.[14] The major advances of twentieth-century physics in general relativity, quantum physics, and chaos theory, important as they are for our philosophical understanding of God, have done little to alter the basic dichotomy that has been institutionalised through industrialisation and the privatisation of theology.[15] As

---

[13] On the way in which industrialisation alters the implicit worldview and contributes to the privatisation of religion, see Peter L. Berger, Brigitte Berger, and Hansfried Kellner, *The Homeless Mind: Modernization and Consciousness* (New York: Random House, 1973). On the social construction of modern dichotomies as a historical process, see my article, 'From Biblical Secularity to Modern Secularism: Historical Aspects and Stages', in S. Marianne Postiglione and Robert Brungs, eds., *Secularism versus Biblical Secularity* (St. Louis, Mo.: ITEST Faith/Science Press, 1994), pp. 1-43.

[14] The paradox was brilliantly highlighted by James Burke in the TV series, 'The Day the Universe Changed'. See the book by the same title (London: London Writers, 1985), pp. 284-85.

[15] I argue that the new quantum technologies effectively disguise the supramechanical features of quantum physics in my article, 'The Laws of Nature and the Nature of God', in

scientifically trained individuals, many of us know that there should be no conflict between faith and science, but as social beings embodied in a technological culture, we can not completely avoid being embroiled in the old dichotomies.

What would it take for us to break out of the modern faith-science dichotomy? If the history presented here is any guide, it would take a social transformation comparable to the one that institutionalised that dichotomy in the first place. Just as the mechanical philosophy was enshrined in the market and industrial revolutions of the eighteenth and nineteenth centuries, the new physics would have to be coupled to a new popular ideology and a new social structure. At this point, no one can say for sure what that social order might look like, but it would likely involve a significant change in our relationships and lifestyles.

D. HISTORICAL METHODOLOGY

A strict methodology can suffocate the spirit of open inquiry. For the most part, I have simply tried to be thorough in reviewing a wide range of historical material and considering all the differing interpretations available. But, in investigating historical developments as long range and as complex as those treated here, issues of methodology can be determinative. And in writing the material I have been guided by certain principles.

In searching and selecting historical material, my procedures have been decidedly externalist. In dealing with scientists and physicians, I have always looked for evidence of their theological convictions and religious motivation. Conversely, in dealing with theologians and literary figures, I have tried to ascertain their understanding of contemporary issues of natural philosophy. I have been open to the possibility of ironies in history: those who criticise or oppose a particular form of science have sometimes been among the greatest contributors to scientific progress; conversely, critics of organised religion have often been profoundly theological. I have also tried to consider the relevant social and political issues of the time without fear of being reductionistic. Since so much of theology is concerned with issues of peace and justice, social and political factors are not at all extraneous.

In developing outlines for organising the material, I have attempted to use categories that are contextual. While the questions I ask of the material are much the same throughout the work, the organisation of the material varies according to the recognised issues of each historical period. In treating

Jitse Van der Meer, ed., *Facets of Faith and Science*, vol. 4: *Interpreting God's Action in the World* (Lanham, Md.: University Press of America, 1995).

the theological material of the patristic period (chap. 1) and the Protestant Reformation (chap. 3, sec. B), I have organised the material according to the basic outline of the creationist tradition. In this way, I have been able to assess the degree of continuity in the creationist tradition from biblical into early modern times.

But I found it best to organise the material for other periods in terms of varying responses to the natural philosophy that was dominant at the time. In the Middle Ages, for example, everything revolved around the assimilation, influence, and criticism of Aristotelian natural philosophy, and I have organised the material accordingly (chap. 2). Eighteenth-century science revolves around the natural philosophy of Newton (chap. 4), and nineteenth-century physics is best understood in terms of varying stances with respect to the mechanical philosophy (chap. 5). However, the early modern science of the seventeenth century is best subdivided according to the major styles of science that arose before the hegemony of the mechanical philosophy was established (chap. 3, sec. C).

Of course, none of these outlines is perfect, and, if taken too literally, they would distort the individuality of the scientists and theologians concerned.[16] I hope the number of intermediate cases and cross influences will alert the reader to the flexibility of the categories used. However, I find that contextually appropriate categories are helpful for mapping the historical terrain and for locating the various figures in ways that would have made sense to them and their colleagues.

Finally, I have tried to discern the changing meaning of words and ideas over time. When individuals or schools of thought oppose each other, they often place different values on the terms of the debate and therefore use the terms differently from earlier generations. So words like 'science', 'natural law', and 'miracle' have dramatically shifted in meaning over the centuries. Common modern distinctions, like those of natural versus supernatural and nature versus miracle, were not systematically made in biblical and patristic times. Conversely, traditional distinctions, like that between the orders of creation and the requirements for salvation, are nowhere near as influential in public discourse today as they were in premodern times. Even a basic term like 'reason' has shifted in meaning from something akin to divine illumination to (Plato, Augustine, Aquinas, and Bonaventure) to its modern sense of detached objectivity. So, while it is fair to speak of a continuous tradition of creational theology, it must be noted that significant discontinuities have occurred when new issues arose and semantic displacement occurred.

---

[16] The difficulty of categorising Robert Boyle, for example, has been illustrated by Michael Hunter in his introduction to M. Hunter, ed., *Robert Boyle Reconsidered* (Cambridge, Cambridge Univ. Press, 1994), pp. 1-18.

CHAPTER ONE

THE EARLY CHURCH AND GRECO-ROMAN SCIENCE
(Through the Twelfth Century AD)

A. BACKGROUND: FROM INTERTESTAMENTAL JUDAISM TO BASIL OF CAESAREA

The emergence of the early Christian church must be understood in continuity with earlier developments in intertestamental Judaism. The military conquests of Alexander the Great (late fourth century BC) led to the formation of an international Hellenistic culture which drew from the traditions of various Near Eastern populations: Egyptian, Phoenician, Babylonian, and Persian as well as Greek. The forceful inclusion of the Hebrews in this ecumenical world brought about the first real contact between the faith of the Jews and the philosophy of the Greeks. It also brought about a new appreciation for the indigenous cultures of the Near East, particularly the Egyptian and Babylonian. Jewish dialogue with this international Hellenistic culture probably began in the early third century BC,[1] but the first records we have of it date from the late third and early second centuries. Already in them, some characteristic features of what was to become an ongoing contest between progressives and conservatives within the community of faith are readily apparent.

1 Early Jewish Responses to Greek Science

On the one hand, there were those who adopted a receptive attitude toward the dominant Greek culture and made an attempt to promote it amongst Jewish youth in Jerusalem. On the other, there was a conservative reaction against the process of Hellenisation based on the quite legitimate fear that it would undermine the distinctive values of the Jewish law (1 Macc. 1:11-15; 2 Macc. 4:4-17). Consequently, the claims of Greek science and technology were perceived as part of a broad cultural challenge affecting all aspects of life, much as Western European science and technology are perceived in many parts of the third world today.

In the context of this intense cultural interaction, we find the first instances of what we might call Jewish apologetists in the second century BC. Their intent was partly to gain respect for the Jewish faith in the eyes of non-Jews and partly to reassure those Jews who were properly impressed

---

[1] Michael E. Stone, *Scriptures, Sects, and Visions: A Profile of Judaism from Ezra to the Jewish Revolts* (Philadelphia: Fortress Press, 1980), pp. 27f.

with the accomplishments of Greek culture that their own tradition was equally good or better.² It was argued by Artapanus, for example, that the Egyptians learned their science and technology from Moses.³ Eupolemus claimed that astronomy (or astrology—the two were not distinct) had been invented by Enoch and that it was Abraham who later taught it to the Phoenicians and Egyptians, from whom the Greeks were in turn supposed to have learned it.⁴ Even Pythagoras and Plato borrowed philosophical ideas from Moses according to Aristobulus.⁵ All of these sources date from the late third to mid-second century BC.

What we find in these early sources is not only a courageous affirmation of the value of Greek science (or natural philosophy) and technology, but an underlying belief in the essential unity of all knowledge. The Jewish patriarchs who had such insight into the laws of God and, in apocalyptic literature, into the composition of the spirit world were believed also to have complete understanding of the laws of nature.

Side by side with the above, there was a more negative assessment of foreign wisdom. First (Coptic) Enoch 6-11 associated pharmacology, metallurgy, and Babylonian astronomy (astrology) with the fall of the angels and their illicit intercourse with humans. And the apocryphal Book of Baruch flatly denied that there is any wisdom in foreign cultures and criticised the younger generation for departing from the Mosaic law (Baruch 3:9-4:4). The concern in both these cases was with the adverse effects of unbridled social, cultural, and technological change brought about by the emergence of new local elites patronised by foreign powers.

A more discriminating attitude was taken by Jesus ben Sirach (early second century BC), who discouraged unguarded philosophical (or theosophical) speculation (Ecclus. 3:21-24) while sanctioning the use of secular medicine as a gift from God (38:1-15, esp. the Hebrew text). As one recent study has put it, ben Sirach was 'entirely open to Hellenic thought *as*

---

² On the following, see Martin Hengel, *Judaism and Hellenism* (London: SCM Press, 1974), vol. I, pp. 88-95, 163-69.

³ Fragment preserved in Eusebius, *Preparation for the Gospel*, trans. Edwin Hamilton Gifford. (Oxford: Clarendon Press, 1903; Grand Rapids: Baker Book House, 1981), IX.xviii.420a-b; xxvii.431d-432b. J.J. Collins cites Cerfaux's persuasive arguments for a date in the late third century BC (*OTP* 2:890f.).

⁴ Ibid., IX.xvii.2-9.418c-419d; xxvi.431c; cf. the anonymous Samaritan text in ibid., IX.xviii.2.420b-c; Clement of Alexandria, *Miscellanies* I.23 (*ANF*, vol. II, p. 335b). According to F. Fallon, the date of Eupolemus is 158-57 BC (*OTP* 2:863). On the authenticity of the first fragment of Eupolemus and the 'anonymous Samaritan', see R. Doran, OTP 2:878.

⁵ Ibid., VIII.x.376c; XIII.xii.663d-664b; cf. Clement of Alexandria, *Miscellanies* I.22 (*ANF*, vol. II, p. 334b). Aristobulus wrote 160-125 BC. A. Yarbro Collins places it 155-145 BC (*OTP* 2:833).

long as it could be Judaized'.⁶ This attitude was not unlike that of third-world nations today who seek the benefits of Western science and technology while insisting on retaining their traditional values and beliefs.

2 Early Christian Attitudes to Greek Science (second to third century)

The first comparable interaction of Christian faith with Greco-Roman science took place in the second and third centuries, when Christians suffered persecution much as the Jews had earlier. As in Jewish apologetics, there were those who claimed all truth to be inspired by God and hence suitable material for Christian scholarship. The first clear statement of this viewpoint was made by Justin Martyr (*c.* AD 165). Justin borrowed the Stoic idea of a seminal Word (*logos spermatikos*) implanted by God in all humans and maintained that this seed inspired the best philosophy of the Greeks as well as the prophecies of the Old Testament. Hence, 'Whatever things were rightly said among all men, are the property of us Christians'.⁷ In the same breath, however, Justin noted that the various schools of Greek philosophy contradicted each other and concluded that they knew only that part of the Logos that was distributed to them and not the fullness of the Word which was embodied in Christ.⁸ In another context, Justin recounted the opinion of his own teacher that the Greek philosophers were motivated by a desire for personal fame and only taught a select few, while the Hebrew prophets were inspired by God's Spirit and 'saw and announced the truth to all'.⁹

Such a positive attitude towards the arts and sciences was taken also by Clement of Alexandria, Origen, and Pseudo-Clement (purportedly Clement of Rome) in the third century. All three were concerned with communicating the gospel to pagan inquirers and advocated the study of what later became known as the *quadrivium* (geometry, arithmetic, astronomy, and music) as a prerequisite for a proper understanding of Christian theology.[10]

---

[6] Jack T. Sanders, *Ben Sira and Demotic Wisdom* (Chico, Calif.: Scholars Press, 1983), p. 58. Sirach's opposition to Hellenism was stressed by Victor Tcherikover, *Hellenistic Civilization and the Jews* (Philadelphia: Jewish Publication Society of America, 1959; New York: Atheneum, 1970), pp. 143f. Hengel points out both Sirach's polemic against an uncritical acceptance of Hellenism (esp. Epicureanism) and his affinity with Hellenism (esp. Stoicism) in *Judaism and Hellenism*, pp. 138-53; cf. George W.E. Nickelsburg, *Jewish Literature Between the Bible and the Mishnah* (Philadelphia: Fortress Press, 1981), p. 64.

[7] Justin, *Second Apology* 13 (*ANF*, vol. I, p. 193a).

[8] *Ibid.*, 8, 10, 13.

[9] *Dialogue with Trypho* 7.1 (translating *mê êttêmenoi doxês* as 'not influenced by glory', rather than 'not influenced by opinion'); cf. *Second Apology* 10.6ff. The Jewish roots of the critique are best seen in Josephus, *Against Apion* II. 16, 158f., 168 (labelled II.17 in *The Works of Flavius Josephus*, trans. William Whiston (Philadelphia, 1833; Grand Rapids: Baker Book House, 1974, vol. IV, pp. 217-20).

[10] Clement of Alexandria, *Miscellanies* I.5; VI.10f.; Origen, *To Gregory Thaumaturgus* 1f.; Pseudo-Clement, *Recognitions* VIII.8-57; X.42.

On the other side, Irenaeus and Tertullian (late second to early third century) were more critical of Greek philosophy, primarily because they had to deal with the rise of numerous heresies within the ranks of the church. Irenaeus made a sweeping condemnation of the natural philosophers (Thales, Anaximander, Anaximenes, Pythagoras, Empedocles, et al.), calling their teachings 'a heap of miserable rags' from which the Valentinian Gnostics had sewed together a cloak to cover their own deviations from orthodoxy.[11] Natural mysteries like the rising of the Nile and the dwelling place of birds, he argued, were far beyond the reach of human knowledge, and, while much could be said concerning their causes if they were properly searched into, 'God alone who made them can declare the truth regarding them'.[12] Christians should confine their studies to the scriptures and the apostolic rule of faith (an early form of the Apostles' Creed). If they were foolishly to inquire into the wonders of nature they would develop conflicting schools of thought, like those of the Greeks, and undermine the God-given unity of the church.[13] Irenaeus's attitude towards pagan learning was clearly coloured by his experience of it in the teachings of the Gnostics.

Tertullian was even more vehement in his condemnation of the natural philosophers, calling their teachings 'uncertain speculations', 'worthless fables', and 'promiscuous conceits'. The philosophers, he complained, 'indulge a stupid curiosity on natural objects, which they ought rather (intelligently to direct) to their Creator and Governor'.[14] Like Irenaeus, Tertullian associated the influence of Greek cosmological speculation (Platonist, Stoic, and Epicurean) and dialectics ('unhappy Aristotle') with the Gnostic heresies of Valentinus and Marcion and the impending dissolution of the church into opposing sects. So when he exclaimed, 'What indeed has Athens to do with Jerusalem?' he went on to say, 'What concord is there between the Academy and the Church? What between heretics and Christians?'[15] His overriding concern was with the unity of the church and with the purity of its doctrine.

No one with any appreciation for Greek philosophy could fail to be offended by Tertullian's tirade. Modern critics of Christianity have frequently cited his words as evidence of anti-intellectualism in the early church. Once one allows for the vituperativeness of Tertullian's style, however, there is really nothing to which an informed pagan philosopher of the second or

---

[11] Irenaeus, *Against Heresies* II.14.1-6.
[12] *Ibid.*, II.28.2 (*ANF*, vol. I, p. 399a). The Empiricists and Methodists were two schools of medicine in the first and second centuries AD that stressed the futility of seeking hidden causes of mysterious events; G.E.R. Lloyd, *The Revolutions of Wisdom* (Berkeley: Univ. of California Press, 1987), pp. 159-62.
[13] *Ibid.*, II.27.1.
[14] Tertullian, *To the Nations* II.1, 4 (*ANF*, vol. III, pp. 130a, 133b).
[15] *Prescription Against Heretics* 7 (*ANF*, vol. III, p. 246b).

third century would take exception in the substance of his comments. As modern scholars like Edgar Zilsel and Ludwig Edelstein have pointed out, the fatal flaw of Greek science was its division into a multiplicity of schools and its lack of any means of accountability that would allow the resolution of disputes. In fact, this failure was already appreciated by the leading thinkers of the second century, Diodorus, Galen, and Ptolemy, to name but a few.[16]

The long-range welfare of natural science depended on the development of an ecumenical community of scholars dedicated to the pursuit of truth. This ideal was appreciated by leading thinkers of late antiquity, but the needed substructure was not available. As we shall see in section E, the ecumenical foundation of modern science was to be provided by the monastic movement of the Middle Ages, a movement based on the very discipline that was advocated by Irenaeus and Tertullian. Such are the ironies of history!

3 Basil of Caesarea and the Hexaemeral Tradition

The next major phase of the interaction of Christian faith with Greek science began with the recognition of Christianity as a legal religion in the early fourth century and its progressive assumption of the responsibilities of an established religion through the sixth century. The principal figure of the fourth century was Basil, who was ordained bishop of Caesarea (in Cappadocia) in AD 370.

Basil established what was to become a long-standing tradition in the church known as the *Hexaemeron* ('Work of Six Days'), a popular series of sermons or lectures on the work of God during the first six days of creation. Now that the churches were attracting members of the middle class who had heard popular expositions of Greek science, there was a need for an explanation and defence of the biblical account of creation that would stand up to criticism.[17]

Basil could not dismiss the philosophers wholesale as Irenaeus and Tertullian had done. Yet he could not accept philosophical ideas quite as uncritically as Clement of Alexandria and Origen had done either. He advocated study of the *quadrivium* but pointed out that all such endeavours were futile if the student fell into the trap of believing that the world was coeternal with the Creator as Aristotle had taught. The astronomers have

---

[16] Edgar Zilsel, 'The Genesis of the Concept of Scientific Progress', *JHI*, vol. vi (June 1945), p. 327; Ludwig Edelstein, 'Recent Trends in the Interpretation of Ancient Science', *JHI*, vol. xii (Oct. 1952), pp. 597-603.

[17] *Hexaemeron* III.5: 'None of you assuredly will attack our opinion; not even those who have the most cultivated minds, and whose piercing eye can penetrate this perishable and fleeting nature' (NPNF II, vol. VII, p. 68a). On the origins of the hexaemeral tradition, see Frank Egleston Robbins, *The Hexaemeral Literature* (Univ. of Chicago Press, 1912), esp. chap. 3, 'Early Christian Hexaemera Before Basil'.

measured the distances to the stars, he said, yet they have not realised that God is their Creator and Judge.[18]

Basil made free use of the Aristotelian theory of the four elements to explain the appearance of heaven and earth and the separation of the dry land from the seas. He also accounted ingeniously for the predominance of the element water through the Stoic idea that water was gradually consumed by celestial fire, leading to the eventual destruction of the world.[19] Yet he rejected the Aristotelian idea that each element has a natural place in the cosmos (earth at the centre, water next to the earth, then air and fire) and attributed the support of the earth in space and the gathering of the waters to their proper place (Gen. 1:9) to the ordaining and sustaining work of God.[20]

Basil's *Hexaemeron* was one of the first in a series of criticisms of Aristotle, a series that was to last for more than 1,200 years and give rise at last to modern (post-Aristotelian) science in the seventeenth century. Some of the key points of this critique were:

(1) that the behaviour of the elements must be understood in terms of laws ordained by God rather than in terms of their essences;[21]

(2) that the heavens are corruptible like the earth so that the same laws of physics should apply to both;[22]

(3) that nature, once created and put in motion, evolves in accordance with the laws assigned to it without interruption or diminishment of energy.[23]

The importance of these ideas in the development of science has been recognised by a number of historians, though the insight of Basil and the influence of his commentary have not always been properly credited.[24] They are the foundation of what Richard C. Dales has termed 'the creationist tradition of Christianity',[25] a tradition that was to last for 1,600 years and give birth to modern Western science and technology before it degenerated into pure naturalism in the eighteenth and nineteenth centuries.

4 The Roots of the Historic Creationist Tradition

The historic creationist tradition is not to be confused with modern-day

---

[18] *Ibid.*, I.3, 4.
[19] *Ibid.*, I.7; III.5; IV.5.
[20] *Ibid.*, I.8ff.; IV.3.
[21] *Ibid.*, I.8; III.5; IV.5; V.10; VIII.1.
[22] *Ibid.*, I.3 (heavens were created and will pass away); II.2 ('one universal sympathy'); III.5ff. (denies fifth element, affirms heat of the sun); III.9 (heavens are not alive).
[23] *Ibid.*, V.16; IX.2.
[24] E.g., Shmuel Sambursky, *The Physical World of Late Antiquity* (London: Routledge and Kegan Paul, 1962), pp. 4f., 157-73; 'John Philoponus', *DSB*, vol. VII, pp. 134f.
[25] Richard C. Dales, 'The De-Animation of the Heavens in the Middle Ages', *JHI*, vol. xli (Oct. 1980), p. 533.

'creation science' or 'creationism'. It does share with creation science a critique of naturalism, but evolutionary science can be viewed as a truncated version of the historic creationist tradition as much as creation science can. We shall offer some thoughts on the origins of 'flood geology' in the late eighteenth and early nineteenth centuries in chapter 4 section A. Aside from that, our treatment of the creationist tradition has nothing to do with modern creationism.

We shall trace the historic creationist tradition through the twelfth century in the remainder of this chapter and shall refer back to it in later chapters. In order to place the development in perspective, it is appropriate at this point to say something about its sources prior to Basil. At the end of the chapter we shall note the ways in which it differed in its early stages from what it was to become in the later Middle Ages.

The fundamental idea in the creationist tradition is that the entire universe is subject to a single code of law, which was established along with the universe at the beginning of time. The origin of the universe is beyond human understanding, depending as it does on the wisdom and will of God, but its subsequent operation can be understood due to the fact that human reason is in some way a reflection or image of that same lawfulness or reason that governs the world. In the hexaemeral tradition of commentary on Genesis 1, a distinction was often made between the way things happened during the 'six days' of creation (usually figuratively understood) and the way things happen after the sixth day. During the first 'six days' all depended directly on God's immediate activity. As of the seventh (sabbath) day, however, God rested and nature could operate in accordance with the laws already established.[26]

As far as we know, the roots of this tradition go back to the early stages of Mesopotamian civilisation in the fourth and third millennia BC.[27] The Mesopotamians viewed the universe as a cosmic state in which the wills of the various gods, like the wills of humans, were bound by common law. In a second-millennium revival of these ideas (the *Enûma elish*), the Babylonian god Marduk was credited with having ordained laws for the stars, which were identified with the lesser gods, just as the kings of Babylon had given laws to their subjects.[28] The writers of the Old Testament,

---

[26] The LXX of Genesis 2:2 specifies that God finished his work on the sixth day, but the Hebrew text is ambiguous. Jerome and the Latin Vulgate followed the Hebrew text; Daniel Nodes, 'The Work of the Seventh Day: The Exegetical Tradition from Philo to the Twelfth Century', paper delivered at the 21st International Congress on Medieval Studies, Kalamazoo, Michigan, Session 87, 9 May 1986.

[27] See also Leo G. Perdue, *Wisdom and Cult* (Missoula, Mont: Scholars Press, 1977), pp. 85f., 94, for the Mesopotamian (Sumerian-Akkadian-Babylonian-Assyrian) background.

[28] *Enûma elish* V; A. Heidel, *The Babylonian Genesis* (Chicago: Univ. of Chicago Press, 1942, 1951), pp. 44f.; Henri Frankfort et al., *The Intellectual Adventure of Ancient Man*

particularly those associated with the Israelite monarchy, developed this tradition stressing the unique sovereignty of Yahweh, the God of Israel, and the complete subservience of all nature, both in heaven and on earth, to his command.[29]

Beginning with the sixth century BC, a parallel, though divergent, development took place among the early Greek natural philosophers. Anaximander, Pythagoras, Heraclitus, and others developed the ancient Near Eastern idea of divine laws into a more secular concept of laws of nature.[30] The seemingly naturalistic implications of this early Greek science were modified by Plato and the Stoics (fourth and third centuries BC) who developed the notion of a universal logos related to the operation of a divine world soul.[31]

In continuity with the Old Testament tradition, and later influenced to some degree by popular schools of Greek thought like Platonism and Stoicism, intertestamental Judaism developed the concept of Wisdom as an intelligence responsible for the orderly behaviour of the world and for the

---

(Univ. of Chicago Press, 1946), 1949), chap. 5; Jonathan Z. Smith, 'Wisdom and Apocalyptic' in *Visionaries and their Apocalypses* (Paul D. Hanson, ed., London: SPCK, 1983), pp. 105f.

The persistence of the henotheistic model of natural law is evidenced in Porphyry's citation of an oracle of Apollo, according to which the God of the Hebrews gave laws to the lesser deities in charge of the heaven, earth, and sea; *apud* Augustine, *City of God* XIX.23.

[29] E.g., Gen. 1:1-25; Job 28:25f., 38:4-11; Pss. 19:4ff., 104:9; Prov. 8:29; Jer. 5:22, 31:35f. On the ancient Near Eastern view of the order of creation as the horizon for OT theology, see H.H. Schmid, 'Creation, Righteousness, and Salvation: "Creation Theology" as the Broad Horizon of Biblical Theology', in B.H. Anderson, ed., *Creation in the Old Testament* (Philadelphia: Fortress Press, 1984), chap. 6.

[30] According to Peter Gorman, Pythagoras interpreted the Phoenician tradition in more mystical, panpsychic terms, but he still viewed the gods as representing impersonal forces or numbers; *Pythagoras* (London: RKP, 1979), pp. 24, 26, 33f., 35 f. 53f., 105ff. According to Charles H. Kahn, the cosmology of Empedocles, even in his early treatise, *On Nature*, is 'religious in the same sense that this term applies to the thought of Parmenides or Plato'; 'Religion and Natural Philosophy in Empedocles' doctrine of the Soul', reprinted in Alexander P.D. Mourelatos, *The Pre-Socratics* (Garden City, N.Y.: Doubleday/Anchor, 1974), p. 433.

[31] E.R. Goodenough, *By Light, Light: The Mystic Gospel of Hellenistic Judaism* (New Haven: Yale Univ. Press, 1935), pp. 54ff; Edgar Zilsel, 'The Genesis of the Concept of Physical Law', *Philosophical Review*, vol. iii (May 1942), pp. 249-52; S. Sambursky, *The Physical World of the Greeks* (London: Routledge and Kegan Paul, 1956, 1960), pp. 81ff., 101ff.; Charles H. Kahn, *Anaximander and the Origins of Greek Cosmology* (New York: Columbia Univ. Press, 1960), pp. 183-93, 206f., 222f., 238f.

Joan R. Kung has challenged the notion of a pre-Stoic belief in universal law in Greek philosophy in 'Review Essay on *Magic, Reason and Experience*, by G.E.R. Lloyd', *Nature and System*, vol. iv (1982), pp. 101-5. Helmut Koester stresses the discontinuity between the Stoic and Hebrew concepts in '*NOMOS PHUSEOS*: The Concept of Natural Law in Greek Thought', in J. Neusner, ed., *Religions of Antiquity* (Leiden: Brill, 1968), p. 521-41. John R. Milton dismisses most references to the idea of laws of nature in the Greeks because they are small in number and are not attributed to the lawgiving of a transcendent God in 'The Origin and Development of the Concept of "Laws of Nature"', *Archive for European Sociology*, vol. xxii (1981), pp. 173ff., 186f., 189f.

reasoning faculty in humans, as well.[32] As a result, the processes of nature were believed to be governed by laws and hence open to human comprehension.[33] The fully developed creationist tradition can thus be dated from about the second or first century BC, the period immediately prior to formation of the New Testament.

We have already noted at least three distinct ideas in the creationist tradition: the comprehensibility of the world, the unity of heaven and earth, and the relative autonomy of nature. To these three ideas we shall add a fourth: the ministry of healing and restoration, which is a practical program as much as an idea. At this point in our treatment, it will be convenient to pursue the history of these four themes of the tradition separately.

B. THE CREATIONIST TRADITION: COMPREHENSIBILITY OF THE WORLD

The idea that human reason is an image of the same Logos that is implanted in all the world was a recurring theme in early Christian writings. We find it, for example, in Apostolic Fathers like Clement of Rome in the late first century AD.[34] It recurred in Alexandrian writers like Origen and pseudo-Silvanus, as well as in Latin writers like Tertullian and Lactantius. In the fourth century it was articulated by Athanasius and the two Gregories, as well as by Basil.[35]

1 Agnosticism Concerning Causes Beyond Human Experience

The openness of the world to human comprehension was counterbalanced by the view that many things in the cosmos transcended human understanding inasmuch as they lay beyond the reach of human experience.

In the biblical tradition, one finds a certain agnosticism about hidden mysteries, particularly in the wisdom literature. For example, the later chapters of the book of Job stress the fact that the erratic behaviour of wind, rain, and lightning; the foundations of the earth; and other wonders of creation are understood by God alone.[36] The reason given for human

---

[32] In the case of Jesus ben Sirach, continuity with the OT wisdom tradition has been stressed by Sanders, *Ben Sira*, pp. 26, 50ff. And, as Dieter Georgi has pointed out, the influence was mutual: popular Hellenistic schools like Neopythagoreanism, Middle and Neo-Platonism, and Stoicism were influenced by Hellenistic Jewish Apologetics; *The Opponents of Paul in Second Corinthians* (Philadelphia: Fortress Press, 1986), p. 370, n. 54.

[33] Job 28:26; 38:33; Jer. 31:35f., 33:25; Wisd. 7:15-28; 9:16f.; Ecclus. 1:9f.; 1 Enoch 33; 41; 79; Aristobulus *apud* Eusebius, *Preparation* XIII.12, 667; Philo, *apud* Eusebius, *Preparation* VII.13, 323.

[34] 1 Clement 33.3ff.; cf. John 1:1-10.

[35] E.g., Gregory of Nazianzus, *Oration* XXVIII.16.

[36] Job 28:23-27; 37:5-17; 38:1-38; cf. Prov. 30:2-4; Eccl. 7:24; 11:5; Jer. 31:37; Wisd. 9:16; Ecclus. 1:2-3; 4 Ezra 5:34-40; 13:52.

ignorance in each case was that mortals did not have ready access to the regions of the cosmos described or were not present at their creation. Many of these texts seem deliberately ambiguous: human ignorance persists in the face of comprehensive divine rationality (weight, measure, number, time, place, and order).[37] The possibility of long-term progress is left open provided that humans fear God and adhere to God's laws.[38]

Similar limitations of human knowledge were pointed out by early Christian writers like Irenaeus, Tertullian, Lactantius, and Basil. Examples they cited include the reasons (aside from God's decree) for the ebb and flow of the tides; the causes of rain, thunder, and lightning; differences in properties among various metals and stones (Irenaeus); the causes of celestial phenomena like the phases of the moon (Irenaeus, Lactantius); and the mode of the earth's support in space (Basil). Note that the facts themselves were not in doubt, only the reasons for them.[39]

This agnosticism with regard to causes and the consequent restriction of natural philosophy to the knowledge of patterns in nature was quite in keeping with general trends in the science of late antiquity. In the first century BC, Posidonius and Geminus had thus distinguished astronomy, which was concerned with the modeling of phenomena, from physics, which attempted to discover the underlying causes. Aristotle himself was principally a physicist, but the progress of astronomy from the second century AD through the Middle Ages was to depend more on the work of Ptolemy, who was concerned primarily with 'saving the phenomena'.[40] Some historians have regarded this pragmatic tendency to be harmful.[41] Others, however, have seen it as necessary, at least for that particular period.[42]

The agnosticism of the Old Testament and the early church was not antiscientific; it assumed that deeper understanding of nature would be possible if circumstances allowed a wider range of human experience.[43]

---

[37] Job 28:25-26; Ps. 147:4; 148:5-8; Prov. 8:27-30; Isa. 40:12, 26; Jer. 31:35-36; Wisd. 1:20b; Ecclus. 16:26-28; 43:1-10; 4 Ezra 4:36-37; 6:4-5; 1 Enoch 2:1-5:3; 69:16-26; 93:13-14; 2 Apoc. Baruch 21:8; 48:4-10; Pss. Sol. 18:10-12; 1QH 1:9-13.

[38] Job 28:28; 42:1-6; Wisd. 7:1-22; 8:9-9:12; Ecclus. 42:19; 43:33; Bar. 3:36-4:4.

[39] Lactantius, *Divine Institutes* III.3-6; cf. Philo of Alexandria, *On the Creation* XIX.13.61. Minucius Felix attributed the same sense of limits to his pagan friend Caecilius in his *Octavius* 5.5f. (FCC, vol. X, pp. 326f.).

[40] Sambursky, *Physical World of Late Antiquity*, pp. 133-45.

[41] Stephen Toulmin and June Goodfield, *The Fabric of the Heavens* (New York: Harper and Row, 1961), pp. 145-49.

[42] Edelstein, 'Recent Interpretations', pp. 576ff.; Sambursky, *Physical World of the Greeks*, pp. 83f., 224f.

[43] So Augustine in criticism of sceptics: 'But because these small things [sparrows and grass; Matt. 10:29; 6:30] are before us and are perceived by our senses and because we can easily search into them, the plan of creation shines forth [to human eyes] in them. But the things whose plan we cannot see are judged to be unplanned by those who think that nothing

Jeremiah was promised knowledge of hidden mysteries; the context includes the ordinances of heaven and earth as well as God's plan for the restoration of Israel (Jer. 33:2-3, 25; cf. 31:35-37). Based on the prophetic and wisdom traditions of Israel, Jewish apocalyptic literature also developed an interest in cosmic secrets. In this case, the limits of ancient technology were overcome by invoking the assistance of angels and thereby opening the cosmos to spiritual ascent and exploration. Ancient patriarchs like Enoch and Moses were allowed to examine the workings of the heavens, even study the behaviour of the wind, rain, and lightning, and found them to be quite comprehensible.[44] Similar knowledge was sometimes promised to all the elect during the end times.[45]

Clearly, we can not easily systematise the 'teachings' of the Old Testament or the early church on this topic. Therefore, we must think of the idea of the comprehensibility of the world in terms of the issues that are probed and tested, rather than as fixed dogma. As we shall see, all the themes of the creationist tradition allow for a variety of articulations, some more optimistic and others more pessimistic, depending on the historical situations to which they respond. What the various articulations share is a common ideal and a common vocabulary. They differ in the extent to which they forsee the realisation of that ideal in the near-term future.

2 Finite or Infinite?

Frequently associated with the idea that the world is accessible to human understanding was the belief that it is encompassed by God, hence that it is finite in both size and duration. In contrast to the unbounded or infinite, which was deemed incomprehensible by many of the ancients, the physical world was believed to be literally comprehended by God and hence

exists unless they can see it...'; *On Genesis Word for Word* V.xxii.43.

Similarly, Augustine cited Psalm 148:7f. as evidence of God's plan even in the chaotic elements of wind and water: '...the Psalmist...made it quite clear that the plan in these phenomena subject to God's command is hidden from us rather than that it is lacking to universal nature' (ibid., V.xxi.42, p. 173).

[44] 1 Enoch 41:3-7; 60:10-22; 72:1-37; 79:1-6; 2 Enoch 23:1; 40:1-12; 2 Apoc. Baruch 59:5-11.

[45] Jer. 30:24d; Dan. 11:33a; 12:4b; 1 Enoch 93:10-14; 4 Ezra 6:1 (Armenian version). On the Armenian version of 4 Ezra 6:1, see L. Joseph Kreitzer, *Jesus and God in Paul's Eschatology* (Sheffield: JSOT Press, 1987), p. 68.

The advance of apocalyptic (and Hekhalot) literature beyond the agnosticism of Old Testament texts like Job is stressed by Ithamar Gruenwald, 'Knowledge and Vision: Towards a Clarification of Two "Gnostic" Concepts', *Israel Oriental Studies*, vol. iii (1973), pp. 69-76; *idem. Apocalyptic and Merkavah Mysticism* (Leiden: Brill, 1980), pp. 7-16; *idem, From Apocalypticism to Gnosticism* (Frankfurt am Main: Peter Lang, 1988), pp. 73-83; 127-29.

Unfortunately, Gruenwald overlooks the potential implications of human access to the heavenly council in Old Testament texts like Num. 24:15-17; 1 Kgs. 22:19-22; Pss. 110:1; 138:1; Isa. 6:1-8; Jer. 23:18-22; Zech. 3:7; cf. Guy Couturier, 'La vision du conseil divin: étude d'une forme commune au prophétisme et à l'apocalyptique', *Science et Esprit*, vol. xxxvi (1984), pp. 5-43.

comprehensible in the objective sense. Aristotle and the Stoics both believed the world to be finite in spatial extent and in some sense encompassed by God. The influence of the Stoic and Neopythagorean idea of God containing and giving coherence to the world may be detected in the biblical period in such writers as Jesus ben Sirach ('by his word all things hold together', Ecclus. 43:26), Aristobulus (the light of wisdom 'in which all things are comprehended'),[46] the Wisdom of Solomon ('the Spirit of the Lord...holds all things together', Wisd. 1:7), Philo ('God contains all things and is contained by none'),[47] and Paul (Acts 17:28; Col. 1:17).

Some scholars have argued, however, that there was also an independent source for the idea of a finite world in the early rabbinic description of God as the 'place' (Hebrew: *maqôm*) of the world.[48] As early as the late third century BC, the Egyptian historian Hecataeus described the God of the Jews as one who surrounded the world as its heaven (cf. Gen. 24:3; Neh. 1:4; Ps. 136:26, passim).[49]

The belief that God is the place of the world, containing it by his word or by his power, is repeated in numerous early Christian writings: The Preaching of Peter (*Kerygma Petrou*), The Shepherd of Hermas, Theophilus, and Irenaeus being among the earliest (all second century).[50] Many early Christian writers also argued against the idea of the eternity of the world—an idea which was held in one form or another by almost all of the Greek philosophers.[51] The arguments varied, but for the most part they were based on the belief that God was the sole origin of all things so that the world, including the matter of which it is formed, must be limited in

---

[46] Aristobulus, *apud* Eusebius, *Preparation* XIII.xiii.667a-d (following the translation of *suntheoreitai* given by Hengel, *Judaism and Hellenism*, vol. I., pp. 166f.; cf. *OTP* 2:841: which translates it as 'contemplated'). See note 32 above on the reciprocal influence of Jewish thought on later Neopythagoreanism and Stoicism.

[47] Philo, *On Sobriety* 63, ed. Loeb, vol. III, p. 477; cf. *Allegorical Interpretation* III.6, 51; *On Dreams* I.63; William R. Schoedel, '"Topological" Theology and Some Monistic Tendencies in Gnosticism', in Martin Krause, ed., *Nag Hammadi Studies*, vol. III (Leiden: Brill, 1972), pp. 92-99.

[48] E.g., Genesis Rabbah 68:9; see A. Marmorstein, *The Old Rabbinic Doctrine of God* (New York: Ktav Publishing House, 1968), pp. 92f.; Max Kadushin, *The Rabbinic Mind* 3rd ed. (New York: Bloch, 1972), pp. 256f.; Ephraim E. Urbach, *The Sages—Their Concepts and Beliefs* (Jerusalem: Magnes Press, 1975), chap. 4.

[49] *ton periechonta ten gen ouranon* (Hengel, *Judaism and Hellenism*, vol. I, p. 256). Cf. the statement now attributed to Posidonius: *to periechon hemas apantas kai gen kai thalattan* [Attic for *thalassan*], *on kaloumen ouranon* (*ibid.*, pp. 147, 259).

[50] Preaching of Peter, frag. 2a, 2b, *apud* Clement of Alexandria, *Miscellanies* VI.v.39 (*ANF*, vol. II, pp. 489a); Hermas, Mandate I.1; Theophilus, *To Autolycus* II.3; Irenaeus, *Against Heresies* I.xv.5; II.i.1f.; xxx.9; IV.xx.2; cf. Schoedel, 'Topological Theology', p. 90.

[51] The primary exception being Eudorus of Alexandria (1st century BC), who taught that all things, including matter itself, originated in time from God: John Dillon, *The Middle Platonists* (Ithaca, N.Y.: Cornell Univ. Press, 1977), pp. 121, 126ff.; Christopher Stead, review of *Schöpfung aus dem Nichts*, by Gerhard May, *Journal of Theological Studies*, vol. xxx (Oct. 1979), p. 548.

duration as well as in spatial extent.[52]

An emphasis on the sovereignty of God could also lead to an emphasis on the inexhaustibility of the world from the perspective of fallible, finite minds (e.g., Job 38:16-38; Eccl. 3:21ff., 8:16f., 11:5; Jer. 31:37; 4 Ezra 5:38; Wisd. 9:16; Ecclus. 1:2ff.). The inexhaustibility of the the divine order did not contradict its comprehensibility, however. The two ideas frequently appeared side by side in Scripture and were tied together by the belief that God's wisdom was available to humans who sought it (Job 28:12-28; Jer. 31:31-37; Wisd. 9:13-18; Ecclus. 1:1-10, 19; 17:1-12; 24:19-34).

A similar juxtaposition of teachings was offered by Origen of Alexandria in the first half of the third century. On the one hand, the eternity and omnipotence of God suggested to Origen that there must always have been a world in which the Deity could exercise his power.[53] On the other hand, Origen clearly denied the coeternity of the visible world with God and argued that, since God comprehends all things, the world must have both a beginning and an end.[54] Henry Chadwick has resolved the apparent contradiction by making a distinction as follows: the eternal object of God's power for Origen was the spiritual world of angels and human souls, whereas the world with a beginning and an end was a world of material bodies.[55]

Basil provided an interesting interpretation of Origen's speculations in his *Hexaemeron*. Before the creation of this present, finite world, he said, there may well have been a spiritual world for the angels to live in. Their world would have been eternal and infinite, since purely intellectual creatures are not confined by bodies and hence can comprehend the infinite! So a world could be infinite and still be orderly, according to Basil, but we ourselves can say nothing about such a world since it would transcend *our* comprehension.[56]

On balance, then, we may say that the creationist tradition required the finitude of the present, visible world in both spatial extent and duration, though it could allow the existence of other worlds, beyond our comprehension, that could be infinite and even eternal. The two principal Christian contributors to scientific development in the sixth century stressed the basic idea of finitude. Boethius (d c. 525), writing in Italy under the Ostrogoths,

---

[52] E.g., Theophilus, *To Autolycus* II.4; Tertullian, *Against Hermogenes* 8, 33, 39.

[53] Origen, *On First Principles* I.2.10; 4.3f., III.5.3; cf. Clement of Alexandria, *Miscellanies* V.14.262.

[54] *Ibid.*, I.3.3; II.9.1; III.5.1ff.; *Commentary on Matthew* XIII.1. See Thomas F. Torrance, *Space, Time and Incarnation* (London: Oxford Univ. Press, 1969), p. 12.

[55] Henry Chadwick, *Early Christian Thought and the Classical Tradition* (New York: Oxford Univ. Press, 1966), p. 117.

[56] Basil, *Hexaemeron* I.5; II.5.

wrote an influential treatise on arithemetic in which he stipulated that 'nothing which is infinite can be found in science nor can be comprehended in science'.[57] In Alexandria, John Philoponus (d c. 565) wrote two treatises refuting the idea of the eternity of the world, one against Proclus and the other against Aristotle. Among the arguments presented in the latter was the statement that the present motions of the heavens would be inexplicable if they had no beginning since, in that case, there would be an infinite series which would defy human comprehension.[58] In this, as in other arguments, Philoponus turned Aristotle's principles against Aristotle's conclusions with devastating effect. The inspiration for his critique may, however, be credited to the creationist tradition he inherited from Athanasius and Basil.[59]

The issue of the temporal duration of the cosmos was never quite settled. Neoplatonist Christians continued to adhere to the eternity of the world in spite of Philoponus's refutation,[60] and some later Arab philosophers, like al-Farabi and Ibn Sina (known in the Latin West as Avicenna), were ambiguous, at best, on the issue. On the other hand, al-Kindi and al-Biruni followed the reasoning of Philoponus, al-Biruni arguing against Ibn Sina on this and many other issues.[61] The issue was to become a prominent one again in the scholasticism of thirteenth-century Europe, as we shall see in chapter 2.

The principal idea we are tracing, however, is that the natural world is comprehensible to humans because it is circumscribed and because the same Logos that is responsible for its ordering is also reflected in human reason. Enough has been said to show that this idea became deeply ingrained in the Christian creationist tradition, particularly where it was reinforced either by Neoplatonism or, as in the later Middle Ages, by Arab philosophy. Thus, for example, Adelard of Bath (d c. 1150), often regarded as the first truly scientific thinker of Western Europe,[62] reflected this tradition on both the world and the human mind. According to Adelard, the visible universe was

---

[57] Marshall Clagett, *Greek Science in Antiquity*, 2nd ed. (New York: Collier Books, 1963), pp. 185f.

[58] John Philoponus, *apud* Simplicius, *Commentary on Aristotle's 'Physics'* VIII.1; see Gerard Verbeke, 'Some Later Neoplatonic Views on Divine Creation and the Eternity of the World' in Dominic J. O'Meara, ed., *Neoplatonism and Christian Thought*, Studies in Neoplatonism, vol. III (Albany: SUNY Press, 1982), p. 48.

[59] Verbeke, 'Some Later Neoplatonic Views', pp. 49, 52f.

[60] *Ibid.*, p. 46 and note 3.

[61] Herbert A. Davidson, 'John Philoponus as a Source of Medieval Islamic and Jewish Proofs of Creation', *Journal of the American Oriental Society*, vol. lxxxix (1969), pp. 357-91; Seyyed Hossein Nasr, *An Introduction to Islamic Cosmological Doctrines*, rev. ed. (Boulder, Colo.: Shambhala Publications, 1978), p. 167.

[62] Adelard is the first figure treated in two standard texts of medieval science: A.C. Crombie, *Augustine to Galileo*, vol. I, 2nd ed. (Harmondsworth: Penguin Books, 1952, 1959); Richard C. Dales, *The Scientific Achievement of the Middle Ages* (Philadelphia: Univ. of Pennsylvania Press, 1973).

subject to quantification (and hence to scientific analysis) because it was limited by its very nature and there were only a finite number of individuals in any given species.[63] For its part, the human soul could understand things and investigate their causes because God had endowed it with sufficient mental power as part of the divine image.[64] Adelard also extolled the arts for their ability to teach the human soul to intuit the divine pattern of things based on her God-given affinity with the divine *rationes*, or seminal reasons implanted within them.[65]

On the point at issue, then, there is demonstrable continuity in the creationist tradition through the twelfth century and, as we shall see, at least through the seventeenth. There was also considerable dissension within the creationist tradition, but that occurred more in relation to other points we shall consider.

C. THE CREATIONIST TRADITION: UNITY OF HEAVEN AND EARTH

Basil believed the heavenly bodies were hot just like terrestrial fire and denied that they were intelligent like angels or humans.[66] His principal target was Aristotle, who had taught that the heavens were composed of a fifth element, that they were divine, and that the stars and planets moved along with them by virtue of their being alive and having eternal souls.[67] The Pythagoreans and Platonists also believed the stars to be divine intelligences due to the regularity of their motion. The Epicureans denied that the stars were living, as Anaxagoras had before them, but they did so by denying the apparent regularity of their motion. The Stoics treated both heaven and earth as being permeated by the divine world soul, practically

---

[63] Tina Stiefel, 'The Heresy of Science: A Twelfth-Century Conceptual Revolution', *Isis*, vol. lxviii (Sept. 1977), p. 351; *idem, The Intellectual Revolution in Twelfth-Century Europe* (New York: St Martin's Press, 1985), pp. 40ff.

[64] According to Adelard:
The Creator of things, supremely good, drawing all creatures into his own likeness so far as their nature allows, has endowed the soul with that mental power which the Greeks call *nous*....She examines not only things in themselves but their causes as well, and the principles of their causes, and from things present has a knowledge of the distant future....Once bound by the earthly and vile fetters of the body, she loses no small portion of her understanding, but that elemental dross cannot wholly obliterate this splendour (*De eodem et diverso*, c. 1105-10; Winthrop Wetherbee, 'Philosophy, Cosmology, and the Twelfth-Century Renaissance', in P. Dronke, ed., *A History of Twelfth-Century Western Philosophy*

[65] W. Wetherbee, in the introduction to *The 'Cosmographia' of Bernardus Silvestris* (New York: Columbia Univ. Press, 1973), p. 9.

[66] See note 21.

[67] Principally the *De caelo* ('On the heavens'); see John Herman Randall, Jr., *Aristotle* (New York: Columbia Univ. Press, 1960), chap. 7.

identifying God with the cosmos. They did succeed, however, in eliminating the Aristotelian dualism of heaven and earth, and they regarded the substance of the heavenly bodies to be fire like that found on earth.[68]

1 Old and New Testaments

In the biblical tradition, the sun, moon, and stars were believed to move at the command of God,[69] or in accordance with his laws.[70] Occasionally they were personified[71] or associated with angels,[72] but terrestrial elements were also personified and associated with angels,[73] so there was no difference between heaven and earth in this respect. The biblical teaching was not a 'de-animation of the heavens' so much as a nonduality of heaven and earth.

The rending of the heavens and the alteration of its luminaries was a prominent feature in biblical theophanies.[74] The complete destruction of the heavens and the extinction of its luminaries was expected on the great and terrible day of the Lord.[75] This was to be followed by the creation of a new heaven and a new earth in which there would no longer be any need for luminaries.[76] Another significant feature of the biblical cosmology was that there were waters (supposedly just like those on earth) above the heavens.[77] In other words, the waters of the cosmos gathered wherever God commanded them to.

The Aristotelian dualism of heaven and earth was countered by Stoic writers like Cleanthes who held that universal reason penetrated all things and gave them unity.[78] Jewish wisdom literature of the intertestamental period adopted this language and applied it to God's Wisdom, Word, and Spirit as the ground of cosmic unity.[79]

---

[68] For a survey of these viewpoints, see Sambursky, *Physical World of the Greeks*, pp. 53ff., 103, 129, 206.

[69] Josh. 10:12ff.; Job 9:7; Ps. 147:4; Isa. 40:26; 45:12; Hab. 3:11; Ecclus. 43:5, 10.

[70] Gen. 1:14-19; Ps. 148:3-6; Jer. 31:35f.; 1 Enoch 2:1, 33:4, 41:5ff., 79:1ff, 83:11; T. Naph. 3:2; Pss. Sol. 18:10ff.; 1QH 1.11f.; 12:4-11.

[71] Gen. 37:9; Judg. 5:20; Neh. 9:6; Job 38:7; Pss. 19:1-6; 89:9; 148:3f.; Isa. 14:12; Ecclus. 16:27f.; Baruch 3:34; Ezekiel the Tragedian, *Exagoge* 79ff.

[72] Job 38:7; Dan. 8:10; 12:3; 1 Enoch 18:13-16; 21:1-6; 41:5ff.; 43; 72-82; 86:1-6; 2 Enoch 4; 11-16; 19; 2 Apoc. Baruch 51:10; 3 Baruch 6-9; Rev. 1:20; 16:8; 19:17.

[73] Job 38:36; Ps. 104:4, 7ff.; 114:3-7; 148:7ff.; Jubilees 2:2; 1 Enoch 60:14-22; 61:10; 66; 69:22ff.; 2 Enoch 5-6; 11-16; 19:4; Rev. 7:1ff.; 14:18; 16:1-5, 17.

[74] Judg. 5:4; 2 Sam 22:10; Pss. 18:9; 144:5; Isa. 24:23; 64:1; Hab. 3:11.

[75] Isa. 13:9-13; 34:4; 51:6; Joel 2:1-11, 30f.; 3:14f.; Mark 13:25, 31; 2 Pet. 3:5-12.

[76] Isa. 65:17; 66:22f.; 2 Pet. 3:13; Rev. 21:1, 23f.; 22:5.

[77] Gen. 1:6ff.; Ps. 148:4; Dan 3:60; 1 Enoch 14:11; 2 Enoch (J version) 3:3; A 4:2; T. Levi 2:7; 4 Ezra 4:7.

[78] *Apud* Cicero, *On the Nature of the Gods* I.39. Cleanthes was head of the Stoa from 262 to 232 BC.

[79] Cf. Ecclus. 1:9; 43:26; Wisd. 1:7-8; 7:24; 8:1; Philo, *On the Confusion of Tongues* 136. On Sirach's affinity with Stoicism, see Hengel, *Judaism and Hellenism*, 1:147ff.

The writers of the New Testament claimed a lordship for Jesus Christ which was coextensive with that of God the Father. Jesus was thus Lord over all things and (as divine Wisdom) permeated all things in both heaven and earth.[80] As the writer of Colossians stated: 'in him all things in heaven and on earth were created...and in him all things hold together' (Col. 1:15-17). In all of these ways, Scripture made it clear that the heavens were not to be accorded any special status and that they were subject to the same laws as the earth and its inhabitants.

2 The Heavens: Animate or Inanimate?

In the second century AD, Tatian and Athenagoras both criticised Aristotle for limiting providence to the heavens.[81] 'God's eternal providence', said the latter, 'is equally over us all'.[82] Athenagoras also rejected Aristotle's notion that the heavenly substance was divine, though he allowed the angels a role in the ordering of both heaven and earth.[83]

The early Christian appropriation and critique of Greek science continued with some regularity until the time of Basil. Here we may note some of its most distinctive contributions.

Origen allowed that the sun, moon, and stars were endowed with life and intelligence. Significantly, he based his conclusion on the biblical facts that the luminaries received commands from God in Scripture and that they were subject to change just like earthlings.[84] Here again, Origen shows us the variation that could occur within the basic outline of the creationist tradition. The notion that mute creatures exhibited a form of intelligence was later revived by the hermetic and alchemical traditions in the Renaissance (chap. 3, sec. A) and by the spiritualist tradition in the seventeenth century (chap. 3, sec. C). Although this idea poses the danger of disregarding the uniqueness of humans, its abandonment by modern science has posed the opposite danger of disregarding the integrity of all God's creatures. The variety of interpretations allowed by the historic creationist tradition thus had some value.

Tertullian and Lactantius adopted the Stoic view that the Spirit of God is diffused through all things (cf. Wisd. 1:7; 7:24), though they rejected the Stoic identification of the world with God.[85] Tertullian pointed out (AD

---

[80] Matt. 28:18; John 1:3-4; 1 Cor. 8:6; 15:24-28; Eph. 1:9-10, 20-23; 4:8ff.; Phil. 2:9ff.; Col. 1:15-20; Heb. 1:2f.

[81] Tatian, *Address to the Greeks* 2; Athenagoras, *Plea on Behalf of Christians* 25.2.

[82] Plea 25.2, trans. Cyril C. Richardson, *Early Christian Fathers*, LCC, vol. I (London: SCM Press, 1953), p. 328.

[83] Athenagoras, *Plea* 6.3.; 10.5; 24.2f.

[84] Origen, *On First Principles* I.7.2f.

[85] Tertullian, *To the Nations* II.2f.; *Against Marcion* I.xi.3; Lactantius, *Divine Institutes* VIII.3.

197) that the sun and moon could not be gods since they undergo change, for example, in eclipses.[86] Lactantius (early fourth century) argued that the stars could not be animate because their motions showed no variation, thus turning the Platonic argument for their vitality on its head.[87]

In the early third century, Eusebius of Caesarea argued that the uniqueness of the divine Word was reflected in the unity of the cosmos. Citing some of the biblical motifs described above, he concluded that there is a single life-force in all things, the heavens and the stars as well as the earth and the sea.[88]

Athanasius, in the fourth century, picked up on the biblical idea that the Word of God holds all things together and argued that the same act of divine will was responsible for the straight-line motions we observe on earth as for the circular motions we find in heaven.[89] In citing examples of the upholding work of the divine Logos, Athanasius treated celestial phenomena side by side with terrestrial ones.[90]

Basil, then, stood within a well-established tradition in denying any special status to the heavens in his *Hexaemeron*.[91] He seems, however, to have been the first Christian writer to follow the Platonists and Stoics in explicitly denying the existence of a fifth element peculiar to the heavens.[92]

In the late fourth and early fifth centuries, the issue of the animation of the celestial bodies was still an open one. Theodore of Mopsuestia (in Cilicia) allowed the angels a role in moving the stars. Jerome denied that the sun, moon, or stars were alive, as Basil had. Augustine was undecided.[93]

In the sixth century, John Philoponus attacked Theodore's speculations on the role of angels and argued that what was visible must also be tangible. Hence the stars could not be angels. Philoponus also followed Basil in regarding the heavenly bodies as fire, pointing out that differences in colour and magnitude indicated differences of composition, just as with terrestrial fires. Indeed, he went so far as to compare the radiation of stars with that of animals like glowworms and luminescent fish, thus evoking a charge of philosophical heresy from his Neoplatonist contemporary Simplicius. And, again using Aristotelian principles against Aristotelian conclusions,

---

[86] Tertullian, *To the Nations* II.6.

[87] Lactantius, *Divine Institutes* II.5; see Sambursky, *Physical World of the Greeks*, pp. 53ff.

[88] Eusebius, *Proof of the Gospel* IV.5. Texts cited include Wis. 1:7; 7:22a, 24; 8:1; Sir. 43:26b; John 1:3-4; Rom. 11:36; Col. 1:17.

[89] Athanasius, *Against the Gentiles*, 42.

[90] *Ibid.*, 35ff.

[91] Cf. note 22 above.

[92] Origen (*On First Principles* III.6.6) rejects the notion that the resurrection body will consist of a fifth element, and Eusebius (*Preparation* XV.7) cites the opinion of Atticus against Aristotle on the fifth element.

[93] See Dales, 'De-Animation', pp. 533f., for references.

Philoponus argued that celestial bodies must have both form and substance, hence they must be composite and perishable like all other bodies.[94] Thus, on both observational and theoretical grounds, Philoponus established the creationist position of the unity of heaven and earth over against Aristotle. In the estimate of one modern historian (I. P. Sheldon-Williams), Philoponus thereby put the Christian doctrine of creation on a scientific basis.[95]

3 Islamic and Medieval European Discussions

On the whole, the idea of the unity of heaven and earth was not challenged after the work of Basil and Philoponus, though its radical implications for physics were not to be fully realised until the seventeenth century. The issue of the animation of the heavens was still an open one, however. As we have seen in the cases of Origen and Theodore of Mopsuestia, the creationist tradition did not rule out this idea absolutely.

In the East, John of Damascus (d c. 750), citing 'the divine Basil' as an authority, regarded the heavens to be corruptible, like all things, 'according to the law of their nature' and declared that the luminaries were inanimate and insensible.[96]

Islamic philosophers and scientists from the ninth through the twelfth centuries were divided on the issue, however. Although the Qur'an protrayed the sun, moon, and heavens as directly subject to the ordinances of God,[97] the rediscovery of Aristotle and Neoplatonism in the ninth century made such an impact that leading thinkers like al-Farabi (d AD 950) and Ibn Sina (Avicenna, d 1037) attempted to synthesise them with the teachings of the Qur'an and postulated a hierarchy of intelligences, corresponding to the heavens, as a bridge between the unity of God and the multiplicity of the terrestrial world.[98] Influenced by Philoponus and taking a stricter reading of the Quran, al-Biruni (d 1048) rejected the vitality of the heavenly bodies, and in the twelfth century, al-Bitruji (Alpetragius) wrote a treatise which attempted to explain the motion of the heavens without recourse to celestial intelligences altogether.[99]

Al-Bitruji's treatise was translated into Latin by Michael Scot in 1217, at the height of the influx of Greco-Arabic learning into Western Europe. In

---

[94] Sambursky, *Physical World of Late Antiquity*, pp. 158ff., 164f.; Richard Sorabji, ed., *Philoponus and the Rejection of Aristotelian Science* (Ithaca, N.Y.: Cornell Univ. Press, 1987), pp. 25f., 51f., 117.
[95] In A.H. Armstrong, ed., *The Cambridge History of Later Greek and Early Medieval Philosophy* (Cambridge: Cambridge Univ. Press, 1967), p. 478.
[96] John of Damascus, *On the Orthodox Faith* II.6f.
[97] Surahs XIII.2; XXXVI.37ff.; XLI.12.
[98] S.H. Nasr, 'Islamic Conceptions of Intellectual Life', DHI, vol. II, pp. 644f.
[99] Dales, 'De-Animation', pp. 538f.

the meantime, the Western tradition had vascillated on the issue as much as the Syrian-Arab world had. Of the two principal scientific writers of the early Middle Ages, Isidore of Seville (d 636) had allowed for the possibility of the stars having souls, while the Venerable Bede (d 735), perhaps the first critical mind in Western European thought, omitted any reference to these speculations in his revision of Isidore's work.[100]

In the twelfth century the Neoplatonists who were associated with the cathedral school of Chartres freely speculated on the role of intelligences among the celectial spheres. William of Conches (d *c*. 1150), in particular, held that terrestrial events like the formation of Adam's body (as distinct from his soul) were governed by the stars and the spirits associated with them, rather than directly by God. His synthesis of Scripture with Neoplatonism was attacked by the more fundamentalist William of St Thierry, and he subsequently retracted a few of his more unguarded statements such as the one noted above.[101]

The process of the 'de-animation of the heavens', as Richard Dales calls it,[102] was not completed until the idea of the unity of heaven and earth could be given a mathematical form in which the need for celestial intelligences was eliminated. For such a mathematisation of the idea, we have to wait for the work of Isaac Newton. But not even Newton saw the universe as inanimate in the mechanical sense (chap. 3, sec. C).

D. THE CREATIONIST TRADITION: RELATIVE AUTONOMY OF NATURE

The ideas we have traced so far, the comprehensibility of the world and the unity of heaven and earth, have been widely accepted in Western thought since the time of Basil, despite uncertainty over related issues like the eternity of the world and the animation of the heavens. Our third theme, the relative autonomy of nature, has been the cause of far more misunderstanding and dispute. Our discussion, then, will provide an introduction to the contest between science and religion that had its roots in the West in the twelfth century, a contest that we shall be surveying in the following chapters of this volume.

By the 'relative autonomy' of nature, we mean the self-sufficiency nature possesses by virtue of the fact that God has granted it laws of operation. Like all laws, the laws of nature may come to be viewed as enslaving and

---

[100] *Ibid.*, p. 534; Thomas R. Eckenrode, 'The Growth of a Scientific Mind: Bede's Early and Late Scientific Writings', *Downside Review*, vol. xciv (July 1976), pp. 208f.

[101] E.J. Dijksterhuis, *The Mechanization of the World Picture* (Oxford: Clarendon Press, 1961), p. 120ff.

[102] See note 25 above.

inflexible, but, in their original sense, at least, they were viewed as liberating (from chaos) and life-giving. The autonomy of nature is thus 'relative' in the sense of being relational (to God), as well as in the sense of not being self-originated or entirely self-determined.[103]

1 Old Testament and Intertestamental Judaism

Among the texts of the Old Testament contributing to the idea, Genesis 1 is best known. Day and night follow each other automatically once their alternation is established (Gen. 1:5), and new generations of plants and animals succeed each other without interference through the normal processes of reproduction (Gen. 1:11f., 21f., 24f.). Elsewhere in the Old Testament, lawfulness is attributed to the courses of the sun, moon, and stars (Job 38:33; Pss. 19:4ff.; 148:3ff.; Jer. 31:35f.; 33:25); the ebb and flow of the tides (Job 38:8-11; Ps. 104:9; Prov. 8:29; Jer. 5:22); the alternation of seasons (Gen. 8:22); and even to meteorological phenomena like wind, rain, and lightning (Job 28:25ff.; 38:24f.; Ps. 148:8).

Within the Old Testament understanding of time, however, wherever the beneficient effects of God's mighty deeds were seen to continue, God's foundational work was also viewed as continuing. Creation once and for all was also continual creation (*creatio continua* or *creatio continuata*).[104]

In other words, the order of nature is a dependent order and, like an executive decree, is subject to the regular ratification of God. It is not rigid but flexible, and it can be altered when its fulfillment in good is at stake (Gen. 6:5-9:17). The natural order is not separate from history and its denouement. It is neither impersonal nor amoral; hence it is not to be set

---

[103]The construction of the phrase 'relative autonomy' is an example of what Ian Ramsey has called a model with a qualifier [*Religious Language* (London: SCM Press, 1957), chap. 2]. On any straightforward reading, the ideas of 'relative' (or 'relational') and 'autonomy' are contradictory. The phrase therefore functions as a 'disclosure model' rather than a definition. Ramsey's strategy is particularly helpful when mapping a position in a semantic field where none of the options allowed is suitable by itself. In the semantic field of modern Western discourse, the options are conceived in terms of causation. The standard model for the God-world relationship in this field is the autonomy of nature (sometimes referred to as deism). The polar opposite is the direct control of nature by God. A third possibility is the autonomy of nature with gaps—a combination of the two major options. None of these models is adequate to express the God-world relationship as conceived in the historic creationist tradition. In accordance with Ramsey's treatment of religious language, therefore, I take the standard model (here 'autonomy') and qualify it so radically ('relative' or 'relational') that the semantic field itself is called into question. Hopefully, the 'penny will drop'.

[104]Hans-Jürgen Hermisson, 'Observations on the Creation Theology in Wisdom' in *Israelite Wisdom*, John G. Gammie et al., eds., (Missoula, Mont.: Scholars Press, 1978), pp. 43-57. John H. Stek gives an excellent treatment of the biblical concept in Howard J. Van Till et al., *Portraits of Creation* (Grand Rapids, Mich.: Eerdmans, 1990), pp. 246ff. Unfortunately, however, Stek equates the Reformed doctrines he critiques with the Christian doctrine as a whole (pp. 211f., 247).

over against the freedom and responsibility humans experience in everyday life (Pss. 19; 93; 104).[105] Any supposed order that might ultimately lead to chaos, anarchy, or injustice would not, in the biblical view, be true order. Hence, the upholding of natural order not only allows but requires its emendation at points where irreversible damage may occur. We shall return to this idea in our discussion of resurrection and healing in the following section.

During the intertestamental period, the Jews developed the idea of the relative autonomy of nature considerably, partly as the result of their dialogue with Greek natural philosophy. In the early second century BC, Jesus ben Sirach, writing probably in Jerusalem, gave a stunning description of the ceaseless regularity of natural rhythms:

> The works of the Lord have existed from the beginning by his creation, and when he made them, he determined their divisions. He arranged his works in an eternal order, and their dominion for all generations: they neither hunger nor grow weary, and they do not cease from their labours. They do not crowd one another aside, and they will never disobey his word (Ecclus. 16:26ff.).

The stress here on nature's obedience to God's word was intended as a contrast to the foolishness of humans who disregard God's (moral) law, as the context makes abundantly clear (Ecclus. 16; 17). The contrast between the obedience of the luminaries and the rebelliousness of humans was made even more explicit by a near contemporary of ben Sirach in an early segment of 1 Enoch (2:1-5:5), and it reappeared in the following century in the Testament of Naphtali, the Psalms of Solomon, and the Dead Sea Scrolls.[106]

Already in the second century BC, however, an indication was given of the notion of mechanical inexorability that could develop within the creationist tradition. This first occurred with Aristobulus of Alexandria (mid-second century BC), sometimes called the first Jewish 'philosopher' of the Hellenistic period. Aristobulus argued that God had arranged the order of creation (Genesis 1) in such a way that it would last for all time. God would keep it as is and would not make any changes. In spite of his belief in the invariability of the order of nature, Aristobulus did not regard God as being absent or inactive. Indeed the very regularity of natural law was for him a sign of God's presence and activity.[107] This is not deism as we have know

---

[105] Gerhard Von Rad, *Old Testament Theology*, vol. I (London: Oliver and Boyd, 1962), pp. 424-28.
[106] 1 Enoch 2:1-5:5; T. Naph. 3:2ff.; Pss. Sol. 18:10ff. (12ff.); 1QS 3:15ff. (stressing predestination); 1QH 1:9-20; Prayer for the Feast of Weeks 2:1ff.
[107] According to Aristobulus:
   But what is clearly stated by the Law, that God rested on the seventh day, means

it in the modern Western world.

2 Early Christian Texts

The transmission of the idea of nature's relative autonomy into Christian circles is best illustrated by the Odes of Solomon, composed in Syria or Palestine, probably during the early second century AD. Like Aristobulus, the author reflected on the seven days of Genesis 1-2:[108]

> And he [the Lord] set the creation and aroused it,
> then he rested from his works.
> And created things run according to their courses,
> and work their works,
> and they are not able to cease and be idle.
> And the hosts are subject to his word.

A comparison with the passage from Jesus ben Sirach quoted earlier shows a certain degree of dependence. The idea of God's resting from his works, however, must have come from the kind of philosophical interpretation of the Sabbath we find in Aristobulus and Philo.[109]

Moving on to the fourth century, we have several witnesses to belief in the relative autonomy of nature just prior to the time of Basil. Arnobius of Sicca (north Africa), writing during the persecution under Diocletian in the early fourth century, argued that accusations that Christians had disrupted the world were unfounded. Natural events were still in accord with the 'laws established in the beginning', and the 'fabric of this machine and mass [of the universe], by which we are all covered and in which we are held enclosed' remained intact even after the advent of Christianity![110] There are several allusions here to the work of Lucretius, the Latin Epicurean, but the ideas of laws of nature being established by God at the beginning of time and the finitude of the cosmos are clearly biblical and, if anything, aimed against the physics of the Epicureans, who taught the randomness and unboundedness of nature. The comparison of the universe to a machine was to become a popular theme in later medieval and early modern European thought.

In the East the sufficiency of natural law was taught by Eusebius of

---

not, as some suppose, that God henceforth ceases to do anything, but it refers to the fact that, after he has brought the arrangement of his works to completion, he has arranged them thus for all time. For it points out that in six days he made the heaven and the earth and all things that are therein, to distinguish the times and predict the order in which one thing comes before another: for after arranging their order, he keeps them so, and makes no change (*apud* Eusebius, *Preparation* XIII.12, 667b-c; trans. E.H. Gifford, p. 721).

[108] Odes of Solomon 16:12ff.; *OTP* 2:749.
[109] Philo, *On the Decalogue* XX.99ff.; cf. *On Creation* XIX.13.61.
[110] Arnobius, *Against the Heathen* I.2f. (*ANF*, vol. VI, pp. 413f.); cf. Lucretius, *On the Nature of Things* V. 91-6; 1 Clement 60.1.

Caesarea (in Palestine, c. 314), who benefited directly from the writings of Aristobulus. One of the first church fathers to use the Greek term *physis* for universal nature, he described Moses as attributing to God the framing of the 'laws of universal nature'[111] as well as those of the nation of Israel.[112] The stability of the firmament; the suspension of the earth; the orbits of the sun, moon and stars; and the alternation of the seasons were all established in the beginning by God's word and law.[113]

Dating from around the middle of the fourth century is a remarkable chapter of Christian cosmic speculation in the Pseudo-Clementine *Recognitions*, which were originally written in Greek and later translated into Syriac and Latin so as to have an extensive influence on subsequent

---

[111]The expressions used by Eusebius are: 'laws of universal nature' (or 'laws of the nature of the universe', *nomôn... en tê physei tôn holôn*); 'universal nature the all-mother' (*tên pammêtora tôn holôn physin*); 'laws which concern the nature of the universe' (*tous peri tês tôn holôn physeôs...nomous*); *Preparation* VII.9, 10 (MPG 21:532f.; trans. Gifford, 314a, 314d, 315a). To my knowledge, this is the first clear instance in Greek patristic literature of the use of the term 'nature' (*physis*) in the sense of 'universal nature' or 'nature of the universe'. Surprisingly, it is not discussed by Robert M. Grant, *Miracle and Natural Law in Graeco-Roman and Early Christian Thought* (Amsterdam: North-Holland, 1952), p. 26.

Other Greek Christian texts from the first and second centuries (Barnabas, Aristides, Athenagoras, Acts of John) spoke of the 'nature' (*physis*) of individual creatures or of 'natural law' (*nomos physeôs*) in the moral sense, but not of 'universal nature' in the physical sense, i.e., the 'rational principle of order in the universe' (Grant, *Miracle and Natural Law*, p. 24). Among the early Latin fathers (Tertullian, Minucius Felix, and Arnobius), the idea of nature was slightly more developed, though often in reference to the ideas held by pagan philosophers.

In Hellenistic Jewish literature, the term *physikos* ('according to nature') was used in the fragments of Aristobulus cited by Eusebius (frags. 2 and 5; *apud* Eusebius, *Preparation* VIII.10.2; XIII.12.9). The concept of 'laws of nature' (*physeôs nomoi*) was applied by Philo to the principles of Pythagorean mathematics (*On Creation* 13) and divine providence (*ibid.* 172; cf. *On Special Laws* IV.232f.). Note also the striking phrase 'universal mother' (*pammêtora*), taken from the Greek poets (*On Creation* 133). These Jewish writers were probably Eusebius's sources (hence also Basil's; see below) for the idea of laws of universal nature; cf. note 112 below.

The idea of an 'order of nature' (i.e., the order of creation) was clearly present in Qumran-related literature: e.g., T. Naph. iii.4f. (*taxis physeôs*); CD iv.21. The biblical basis is found in the Priestly tradition (Gen. 1; 8); Psalms (e.g., Pss. 19; 104; 147; 148); Jeremiah's 'Book of Consolation' (Jer. 31; 33); and ben Sirach (Ecclus. 16). A possible Christian antecedent for Eusebius is found in Athenagoras, *Plea* 22.8, where the phrase *peri tês physeôs* may allude to the idea of univeral nature. William R. Schoedel suggests that Athenagoras took the idea from Empedocles via Aetius; *Athenagoras* (Oxford: Clarendon Press, 1972), p. 53n. Eusebius, however, never once mentioned Athenagoras; cf. C.C. Richardson, *Early Christian Fathers* (London: SCM Press, 1953), p. 290.

[112]*Preparation* VII.9f. (trans. Gifford, 313d-315d). The idea of Moses combining moral and physical law probably reflects Aristobulus (frag. 2, *apud* Eusebius, *Preparation* VIII.10.3) and/or Philo (*On Creation* 3). Aristobulus, himself, seems to have taken the treatment of the Sabbath rest in Exod. 20:11 as his model (frag. 5; *apud* Eusebius, *Preparation* XIII.12.11f.).

[113]'...for by his decree and power all things have received their being, and by his laws [*nomois*] and limitations again the whole duration of time is directed in its course and order.' Eusebius, *Preparation* VII.10, trans. Gifford, 314b.

Christian thought in both East and West.[114] As in Arnobius we find phrases like 'machine of the world' and 'fabric of the world' which were current in popular expositions of Epicurean science and well suited to the expression of biblical faith.[115] The paths of the stars, we are told, are governed by 'fixed laws and periods' and are evidence of divine creation.[116] The reproduction of animals is also regular even though God has ordained a few special cases (like those of the crow, which conceives through the mouth, and the weasel, which brings forth through the ear!) to remind us that the order of the world is due to his appointment rather than to nature itself.[117]

Thus the power of effective causation, for Pseudo-Clement, was present in nature only by virtue of God's creative decree. For example, the germination of seeds and the growth of plants was due to the 'power of the spirit' (i.e., moisture) which God had implanted in water at the moment of its creation.[118] The manner of the germination of seeds, furthermore, was open to rational investigation, at least by the 'worthy and faithful' who understood something of the mysteries of God's ways. Thus the author himself was able to 'prove by fact and example' that the process depends entirely on the power of water by means of a thought experiment in which the amount of earth used is weighed both before planting and after the harvest and is found to be exactly the same.[119] This was perhaps the greatest encouragement any ancient Christian writer was to give to rational scientific investigation.

3 Basil of Caesarea

We come back then to Basil in the seventh decade of the fourth century AD. In spite of the precedents we have studied, Basil is usually cited as the first major Christian contributor in the creationist tradition, and not without reason.[120] In view of the influence he was to have in both East and West, several passages in his *Hexaemeron* take on great significance for the

---

[114]J. Irmscher, 'The Pseudo-Clementines', in Wilhelm Schneemelcher, ed., *New Testament Apocrypha*, vol. II (London: Lutterworth Press, 1965), pp. 534f.

[115]E.g., *Recognitions* VIII.15, 21. The speaker is a former Epicurean; cf. VIII.7.

[116]*Ibid.*, 20 (*ANF*, vol. VIII, p. 171a).

[117]*Ibid.*, 25. The text has 'order of nature' meaning 'order...assigned by nature'; 26; cf. 20, 34 (pp. 172a, 170b, 174b).

[118]'For there is in water a certain power of the spirit given by God from the beginning, by whose operation the structure of the body that is to be begins to be formed in the seed itself, and to be developed by means of the blade and the ear [cf. Mark 4:28].' *Ibid.*, 26 (*ANF*, vol. VIII, p. 172b).

[119]*Ibid.* 27; cf. 34, 42. According to A.C. Crombie, this kind of thought experiment implies the idea of the conservation of matter; *Augustine to Galileo*, vol. II, p. 111; cf. Herbert M. Howe, 'A Root of Van Helmont's Tree', *Isis*, vol. lvi (winter 1965), pp. 415ff.

[120]R.C. Dales, 'A Twelfth-Century Concept of the Natural Order', *Viator*, vol. ix (1978), p. 179; *idem*, 'De-Animation', pp. 532f.

evolution of our belief in the relative autonomy of nature. His comments on Genesis 1:11 ('Let the earth put forth vegetation...') and 1:24 ('Let the earth bring forth living creatures...') are particularly striking in this connection. Genesis 1:11 mentions 'vegetation, plants yielding seed, and fruit trees bearing fruit in which is their seed'.

Basil first notes the wisdom of the order: first vegetation (LXX: 'grass'); then trees. This order, he notes, is followed by the earth to this day and will continue for all time:[121]

> For the voice that was then heard and this command were as a natural and permanent law [*nomos physeôs*] for it; it gave fertility and the power to produce fruit for all ages to come.

As in the Pseudo-Clementine *Recognitions,* the power to cause germination is seen to be present in nature by virtue of creation, although that power appears to reside in the element earth, for Basil, rather than in water.

In concluding his homily on this text, Basil returns to the theme of the relative autonomy God grants to nature by his command and, in so doing, gives us the first of two classic examples of what later became known as the concept of impetus or momentum:[122]

> It is this command which, still at this day, is imposed on the earth....Like tops [*strobiloi*],[123] which after the first impulse, continue their revolutions, turning upon themselves when once fixed in their centre; thus nature, receiving the impulse of this first command, follows without interruption the course of ages, until the consummation of all things.

Spinning tops were a phenomenon known to every child that strained the basic principles of Aristotelian physics. Belonging to the terrestrial world, they yet moved in circular fashion like the celestial spheres, thus demonstrating once again the non-duality of heaven and earth. Moreover, the relatively stable state of spinning was regarded as 'unnatural' in the Aristotelian view and required the *ad hoc* supposition of a thin layer of air whirling around the top to keep it going. For Basil, the motion of the spinning top was perfectly 'natural', however. It was just like the regular cycle of seedtime and harvest (Gen. 8:22) that one observes in terrestrial nature. In either case there is an initial impulse (the twist of fingers, in one

---

[121]Basil, *Hexaemeron* V.1 (*NPNF II*, vol. VIII, p. 76a; cf. SC 26:278f.: 'comme une loi de la nature'); cf. Eusebius, *Preparation* VII.10, trans. Gifford, 315ab.

[122]*Ibid.,* V.10 (ibid., p. 81b). The concept has Stoic roots; cf. the fragment of Chrysippus in Cicero, *On Fate* 42f.

[123]The Greek term, *strobilos*, can mean either a spinning top or a millwheel. Eustathius (*c.* 400) translated it as 'a top made of pine wood, having been set in motion by the prior pull of a cord', Dales, 'De-Animation', pp. 532f. Later Syrian and Latin texts use the example of a millwheel.

case; the command of God, in the other) the effect of which continues indefinitely even after the original action has ceased. In modern science, the principle exhibited in the case of the spinning top is called the law of the conservation of momentum (in this case, angular momentum) or the principle of inertia.[124] For Basil, not only tops but all of nature, organic as well as inorganic, moved in regular intervals in accordance with the command of God.

Basil's second example was that of a rolling ball. Here Basil describes the spontaneous generation of life from the earth in response to the command of God in Genesis 1:24 ('Let the earth bring forth living creatures'). He compares it to the way a ball, once set in motion, rolls down an inclined plane without further assistance.[125] The details are not entirely clear, but the original impulse is apparently a small push which brings the ball to the edge of the downward slope. Acceleration occurs as the ball begins to roll down the decline and continues until the ball reaches level ground again and eventually rolls to a stop.

Aristotle's physics could account for the acceleration downwards but had difficulty with the continuation along the horizontal.[126] As in the case of the spinning top, however, the motion described was perfectly natural according to Basil. As the nature of soil and seminal potency in seeds produces the cycle of birth and death once the latter is set in motion by God's word, so the contour of the ground and the spherical shape of the ball produce the familiar pattern once the ball is set in motion. From a modern perspective, the physics is slightly more complicated than that of the spinning top, involving in modern terms the conservation of energy as well as momentum, but the basic reasoning is the same, and each case reinforces the point of the other.

It would be more than 1,200 years before Galileo, Descartes, and Newton would formulate a principle of inertia in mathematical terms that could be used in calculations. However, the idea of relative autonomy that lay behind

---

[124] Sambursky (*Physical World of Late Antiquity*, pp. 70ff.) credits the first enunciation of the principle of direct momentum transfer after Hipparchus to John Philoponus, thus ignoring Basil's contribution.

[125] According to Basil:
Behold the word of God pervading creation, beginning even then the efficacy which is seen displayed today, and will be displayed to the end of the world! As a ball, which one pushes, if it meet a declivity, descends, carried by its form and the nature of the ground, and does not stop until it has reached a level surface; so nature, once put in motion by the divine command, traverses creation with an equal step, through birth and death, and keeps up the succession of kinds through resemblance, to the last (*Hexaemeron* IX.2; *NPNF II*, vol. VIII, p. 102a).

Note that Basil has substituted the divine command for the Stoic fate; cf. the fragment of Chrysippus as preserved in Aulus Gellius, *Attic Nights* VII.2.11.

[126] Sambursky, *Physical World of the Greeks*, pp. 92-99.

it was clearly fixed by the time of Basil. Indeed it was deeply embedded in the Hellenistic-Jewish-Christian tradition that Basil inherited, as we have seen. Basil merely gave practical examples from everyday experience to illustrate the principle of the relative autonomy of nature as it had been understood since the time of Jesus ben Sirach and Aristobulus.

4 Transition to the Middle Ages Surveyed

The idea that motion is conserved and that its quantity depends only on the magnitude of the intitial impulse was developed in the sixth century by John Philoponus as part of his programmatic attack on the physics of Aristotle.[127] Through the writings of Philoponus, and also through the Syriac hexaemeral tradition,[128] it was passed on to Arab philosophers of the eleventh and twelfth centuries like Ibn Sina (Avicenna), Ibn Bajjah (Avempace), al-Baghdadi, and al-Bitruji (Alpetragius). The idea recurred, with significant alterations, in the Western scholastics of the threenth and fourteenth centuries like Thomas Aquinas, Peter John Olivi, and Francis of Marchia (chapter 2, section C).[129] The degree to which this Western development was stimulated by ideas transmitted by the Arabs is difficult to determine. It may have been an independent development based on the fundamental idea of the autonomy of nature embedded in the hexaemeral tradition.

[127]Clagett, *Greek Science*, pp. 213ff.; Sambursky, *Physical World of Late Antiquity*, pp. 74ff.; Sorabji, ed., *Philoponus*, pp. 9, 97. Michael Wolff's claim that Philoponus's idea of impetus was not indebted to his Christian faith does not take the hexaemeral tradition into account; 'Philoponus and the Rise of Preclassical Dynamics', Sorabji, ed., *Philoponus*, pp. 107f.

[128]Basil's *Hexaemeron* was translated into Syriac in the fifth century and into Arabic probably by the eighth or ninth century (Paul J. Fedwick, 'The Translations of the Works of Basil Before 1400', *Basil of Caesarea*, [Toronto: Pontifical Inst. of Medieval Studies, 1981], pp. 449, 486). Syriac *hexaemera* were written by James of Edessa (d 708) and Moses bar Kepha (d 903); see F.E. Peters, *Aristotle and the Arabs* (New York: New York Univ. Press, 1968), pp. 116, 132f. The illustration of a spinning wheel was used by Job of Edessa (d c. 835) in his *Book of Treasures* V.12 (A. Mingana, trans., *Encyclopaedia of Philosophical and Natural Sciences as Taught in Baghdad about AD 817* [Cambridge: W. Heffer and Sons, 1935] p. 224).

[129]Nasr, 'Islamic Conception', pp. 645-49; Ernest A. Moody, 'Galileo and Avempace: The Dynamics of the Leaning Tower Experiment', *JHI*, vol. xii (1955), pp. 385-94 (reads Avempace's alterations back into Philoponus); Shlomo Pines, *Studies in Abu-l-Barakat al-Baghdadi* (Collected Works, vol. I, Jerusalem: Magnes Press, Hebrew University, 1979), pp. 31-83 (in French). Some historians have argued that Western scholastics must have developed their concept of momentum independently of the Arabs since they did not have access to the relevant Arabic texts; e.g., Anneliese Maier, *Zwei Grundprobleme der scholastischen Naturphilosophie*, 2nd ed. (Rome: Edizioni di Storia e Letteratura, 1951), pp. 127-33; A.C. Crombie, *Augustine to Galileo*, vol. II, pp. 66-74. However, the Neoplatonic Islamic idea of a continuously impressed force, at least, was available through a late twelfth-century translation of al-Ghazali's summary of Avicenna in his *Intentions of the Philosophers*; Fritz Zimmermann, 'Philoponus's Impetus Theory in the Arabic Tradition', in Sorabji, ed., *Philoponus*, pp. 122ff., 129.

The alterations in the idea were largely due to the influence of Neoplatonism. Ibn Sina and Ibn Bajjah had reinterpreted the impartation of impetus as a continuously impressed force, thus weakening the basic idea of the autonomy of nature and ruling out the possibility of conservation of momentum in the absence of a continuous force. In this altered form, the idea of an impressed force continued until the time of John Buridan (mid-fourteenth century), who revived the idea of a conserved impetus by appealing to the efficacy of God's original act of creation, as Basil had done almost a thousand years earlier.[130]

In the meantime, the idea of the relative autonomy of nature had been transmitted to the Latin West and had prepared the way for the medieval developments we just noted. Basil's *Hexaemeron* was paraphrased by Ambrose in 389, and an elegant Latin translation by Eustathius appeared around the turn of the fifth century.[131] Basil's work was also known and used by Augustine, Cassiodorus, and the Venerable Bede. The Pseudo-Clementine *Recognitions* were translated by Rufinus (d 410) about the same time and influenced both Isidore and Bede.

5 Augustine of Hippo

Augustine (writing 386-430) developed the idea of the autonomy of nature to an unprecedented degree by stressing the transcendence of God, for whom there was no time as we know it,[132] and explaining the unfolding of nature (and history) in terms of seminal causes that God implanted at creation so as to have their effects in a predetermined sequence.[133] Thus, in his earlier

---

[130]Nasr, *Islamic Cosmological Doctrines*, pp. 228f.; Moody, 'Galileo and Avempace', pp. 395, 404-9. According to Buridan:
> If you cause a large and very heavy smith's mill [wheel] to rotate and you then cease to move it, it will still move a while longer by the impetus it has acquired....And thus one could imagine that it is unnecessary to posit intelligences as the movers of celestial bodies....For it could be said that, when God created the celestial spheres, he began to move each of them as he wished, and they are still moved by the impetus which he gave to them because, there being no resistance [as there is for millwheels on earth], the impetus is neither corrupted nor diminished (*Questions on the Heavens and the Earth* II.xii.6f. as translated in Marshall Clagett, *The Science of Mechanics in the Middle Ages* [Madison: Univ. of Wisconsin Press, 1959], p. 561; cf. Moody, 'Galileo and Avempace', p. 409; Dales, *Scientific Achievement*, pp. 116f.)

[131]Fedwick, ed., *Basil of Caesarea*, pp. 459f.

[132]Augustine, *On Genesis Word for Word* I.ix.15ff.; *City of God* XI.21; cf. Gregory of Nazianzus, *Oration* XXX.18; see C.B. Kaiser, *The Doctrine of God* (London: Marshall, Morgan and Scott, 1982), pp. 79ff.

[133]*Confessions* I.vi.9; *On Genesis, Incomplete* 51; *On Genesis Word for Word* IV.xxxiii.51f.; V.xxiii.44f.; VIII.iii.7; IX.xvii.32; *On the Trinity* III.viii.13-ix.16; *City of God* V.9. On the Genesis commmentaries, see the valuable discussion in James Y. Simpson, *Landmarks in the Struggle Between Science and Religion* (London: Hodder and Stoughton, 1925), pp. 114ff. The connection between the timelessness of God's decree and the (partial)

writings, Augustine accounted for the fact that heaven and earth were created in the beginning of God's work (Gen. 1:1), yet did not take form as firmament and dry land, respectively, until the second and third 'days' (Gen. 1:6-10). He solved the problem by referring the initial act of creation (Gen. 1:1) to the seeds (as well as the unformed material) of heaven and earth.[134]

Likewise, in a later commentary, *On Genesis Word for Word* (401-15), Augustine explained the two different accounts of the creation of the first human pair (Gen. 1 and 2) as describing successive stages: the initial creation of their seminal principles in the earth and the later emergence of their concrete forms from the earth.[135] In effect, the seminal causes of all things were created at once in the first instant of time, and the work of the 'six days' in Genesis 1 was really an inventory of the potencies contained in those principles.[136]

As a result of Augustine's interpretation, the seminal causes, which had been so closely related to the divine world soul in Stoicism and Neoplatonism and to the Wisdom or Word of God in intertestamental Judaism and early Christianity, could now be regarded as distinct from God's (transcendent) essence.[137] Moreover, the beginning of God's rest and of nature's relative autonomy were pushed back from the first Sabbath (Gen.

---

predetermination of nature by seminal causes is illustrated in *On Genesis Word for Word* VIII.iii.7; xx.39. Once temporal succession was detached from its roots in the eternal decree (via the 'seven days' work'), Augustine's foreknowing God could be replaced by a Laplacean intelligence who knew the future by knowledge of the present state of the world and simple calculation based on mathematical laws.

[134]*On Genesis Against the Manichaeans* I.11f.; *On Genesis, Incomplete* 10 (Simpson, *Landmarks*, pp. 115ff.).

[135]*On Genesis Word for Word* VI.v.8ff. On this later reading, the creation of the seminal causes of birds and fish and their implantation in the watery realms are treated under Day 5. The seminal causes of land animals and the first human pair are treated under Day 6 (*On Genesis Word for Word* IV.xxxiii.51f.; V.iv.10f.; v.14; VI.v.8). Seminal causes here are immaterial causal principles, not material seeds (*ibid.*, V.xxxiii.52; VI.vi.10f.; X.xx.35) and are associated with numbers (*ibid.*, IV.xxxiii.52; V.v.14; vii.20; VI.xiii.23; xvi.27; X.xxi.37; cf. A's use of the trinity of 'measure, number and weight' [Wisd. 11:20b] in IV.iii.7-vi.13). The six 'days' of Gen. 1 were simultaneous stages in which the angels knew all these creatures (first in the Word of God, then in their own concrete natures), or, in the case of living creatures, the potencies that later gave rise to them (I.ix.19; II.viii.16f.; IV.xxxiii.52-xxxiv.55).

[136]*On Genesis Word for Word* III.xiv.23; IV.xviii.33; V.v.12f., 15; VI.vi.11 (for living beings only); cf. *Confessions* XII.xii.15; *City of God* XI.33.

The influence of Origen and Gregory of Nyssa should be noted here: cf. Origen, *On First Principles* I.2.2; *Against Celsus* VI.60; Gregory of Nyssa, *Hexaemeron* 72b-77d (Robbins, *Hexaemeral Literature*, pp. 15-18, 67f.). Even further back, of course, is the influence of Plato, the Stoics, and Philo: *On Creation* 13-16, 25, 28, 43, 67f.; *Allegorical Interpretation* I.19ff.; *Questions on Exodus* I.1.

[137]On the other hand, from the transcendent (angelic) viewpoint, the sempiternal reasons of events were eternally in and with God; *Confessons* I.vi.9; *On Genesis Word for Word* II.vi.12ff.; IV.xxiv.41; V.xvi.34; *passim*. Augustine was agnostic regarding the existence of the world soul but emphatically denied its deity; *On Genesis Word for Word* VII.ii.3-iv.6; see Vernon J. Bourke, 'St Augustine and the Cosmic Soul', *Giornale di metafisica*, vol. x (1954), pp. 431-40.

2:2) to the very first instant of time. This is, therefore, the first clear indication we have of the concept of autonomous nature that was to prevail in the West after the seventeenth century.[138]

Augustine was not a deist in the modern sense, however, for he regarded God's eternal will and power as terminating in time. God's eternal decree functioned as a continuously creative activity by virtue of which seminal causes could produce their respective effects.[139] Still, given the fact of that continuous activity, the inevitability and predictability of cause-effect sequences seemed to follow.[140]

6 Boethius and Cassiodorus

The sixth century was a pivotal one for the development of Western civilisation. Two Italian Christian leaders of that time were instrumental in the formation of medieval European scholarship. Boethius (d *c.* 524) established the format of the medieval curriculum (the three literary arts, or *trivium*, and the four mathematical sciences, or *quadrivium*)[141] and wrote texts on the four sciences (arithemetic, music, geometry, and astronomy). He thus made available the rudiments of Euclid, Nichomachus, and Ptolemy to later generations, though most of this was lost and not rediscovered until the late tenth century.[142] He also began a translation program that was to include the major works of Plato and Aristotle and anticipated the later, more comprehensive, translation program that began in the twelfth century.[143]

---

[138]For an alternative reading of Augustine as the source of the modern Western view of autonomous nature (stressing the extrinsic character of divine grace and the divine ideas in Augustine's thought), see Philip Sherrard, 'The Desanctification of Nature', in Derek Baker, ed., *Sanctity and Secularity* (Oxford: Basil Blackwell, 1973), pp. 13ff. Sherrard points to an important problem but removes the ambiguity of Augustine's (and all Neoplatonic) thought and ignores the critical shifts that occurred between Augustine and Aquinas in the twelfth century.

[139]*On Genesis Word for Word* IV.xii.22f.; V.iv.11; xx.40f.; xxiii.45; *City of God* XII.26; XXII.24. See Ernan McMullin, ed., *Evolution and Creation* (Notre Dame, Ind.: Univ. of Notre Dame Press, 1985), pp. 11-15. For Augustine, the passage of time from one moment to the next was made possible by the fact that each moment passed through (was contained by) the eternal present of God; *Confessions* I.vi.10.

[140]An example of the idea of inevitability is the following quote:
   ...since whatever comes forth to our sight by being born, receives the first beginnings of its course from hidden seeds, and takes the successive increments of its proper size and its distinctive forms from these as it were original rules (*On the Trinity* III.8.13; *NPNF*, vol. III, p. 61a).

[141]The format of the curriculum had its roots in Plato, Varro, and Capella, but Boethius was the first to use the term *quadrivium* ('fourfold way'); William Harris Stahl, *Roman Science* (Madison, Wis.: Univ. of Wisconsin Press, 1962), p. 199.

[142]Clagett, *Greek Science*, pp. 185ff. For a summary of the contents of these texts, see Henry Chadwick, *Boethius* (Oxford: Clarendon Press, 1981), chap. 2.

[143]For an accurate list of his translations, see David C. Lindberg, 'The Transmission of Greek and Arabic Learning to the West', *Science in the Middle Ages* (Univ. of Chicago

Even if Boethius had lived to complete his great work, it is doubtful whether there was a sufficient market for scientific ideas in his time to have allowed a scientific renaissance of any magnitude. The scientific development of the Greco-Roman era had reached a point of diminishing returns for lack of an adequate social and technological base,[144] and the Christian influence that was to catalyse the next major advance was only beginning to make itself felt.

However, the assimilation of Boethius's works in the tenth and eleventh centuries helped prepare the West for the rapid assimilation of Greco-Arabic science in the twelfth and thirteenth centuries.[145] As Richard Dales has put it: 'From Boethius's textbooks, the Middle Ages...learned to conceive of the world of nature as an ordered whole and to deal with it rationally.'[146] Though Boethius's philosophical ideas came principally from Neopythagorean and Neoplatonic sources (the compatibility of which with biblical thought at some points, at least, has already been noted), it was his creationist faith and his Christian altruism that motivated his efforts and sustained him through periods of doubt right up until his untimely death.[147]

Boethius's contemporary Cassiodorus (d c. 575?) was less of a theoretician and more of an organiser. Like Boethius, he left his stamp on the scholarship of medieval Europe during its formative stages. The monastic community Cassiodorus founded in Calabria was not influential in itself, but the library he assembled and the advice he gave his monks concerning the importance of the arts and sciences were to encourage the love of learning in the rapidly spreading Benedictine monastic order which inherited them.[148] By playing down the mystical aspects of Neopythagorean mathematics and showing the usefulness of the arts and sciences for an understanding of Scripture, Cassiodorus helped allay the fears that many

---

Press, 1978), p. 53.

[144]Edelstein, 'Recent Trends', pp. 580-85, 597-601; 'Motives and Incentives for Science in Antiquity', in A.C. Crombie ed., *Scientific Change* (London: Heinemann, 1963), pp. 23ff., 40; Sambursky, *The Physical World of the Greeks*, pp. 73, 223-30. On the extent and level of Boethius's audience, see Helen Kirby, 'The Scholar and His Public' in Margaret Gibson, ed., *Boethius: His Life, Thought and Influence* (Oxford: Basil Blackwell, 1981), chap. 2.

[145]H. Liebenschultz, 'Boethius and the Legacy of Antiquity', in *Cambridge History of Later Greek and Early Medieval Philosophy*, p. 543,; Chadwick, *Boethius*, pp. 252f.

[146]Dales, *Scientific Achievement*, p. 30.

[147]On the issue of Boethius's Christian assumptions, see E.K. Rand, *Founders of the Middle Ages* (Cambridge, Mass.: Harvard Univ. Press, 1928; pp. 257ff., 177f.; Chadwick, *Boethius*, pp. 237f., 248-52.

[148]So most authorities: e.g., Dom Cuthbert Butler, *Benedictine Monachism: Studies in Benedictine Life and Rule*, 2nd ed. (London: Longmans, Green and Co., 1924), p. 332; contrariwise, James J. O'Donnell, *Cassiodorus* (Berkeley: Univ. of California Press, 1979), pp. 219ff., 239, 251f. Cassiodorus's library included the *Hexaemera* of Basil and Ambrose and the two principal commentaries on Genesis by Augustine as well as medical works of Dioscorides, Hippocrates, Galen, and Caelius Aurelianus; Leslie Webber Jones, trans., *Introduction to Divine and Human Readings* I.1, 31 (New York: Columbia Univ. Press, 1946), pp. 74f., 135.

devout monks would naturally have about the contamination of pure doctrine with pagan ideas.[149]

The impact of monasticism on European culture was to be felt as much on the social and technological levels as it was on pure scholarship. One aspect of this impact which relates to the idea of the autonomy of nature was the concern for cosmic time and its measurement for liturgical purposes. The determination of the time of day, the month, and the year had already been an important task of astrologer-priests in the ancient Near East. Some Jewish sects believed that the worship of God was carried on through the night by the hosts of angels in heaven, and the early Christians developed their own prayer cycles as a means of participating in this heavenly cosmic liturgy.[150] The Qumran community (second century BC to first century AD) had developed a common life in which work and study were coordinated with the cycles of worship: human life was thus constructed in the image of the lawful mechanism of the cosmos.[151] Beginning with the fourth century AD, the common or cenobite life was adopted and adapted by Christians: Pachomius in Upper Egypt; Eustathius of Sebaste and Basil in Asia Minor; Augustine and Cassian in the West. The sixth century saw the founding of the Benedictine order which was to dominate Western European history for 700 years and lay the foundations of Western science and technology. The rules developed by the monastic fathers specified a certain number of hours (usually six) for work each day in addition to the regular duties of worship.

It is in connection with this more practical side of the creationist tradition that Cassiodorus made his most influential contributions. He provided his monastery with both a sundial and a water clock: the one for sunny days, the other for cloudy days and for nights, '...in order that the soldiers of Christ, warned by the most definite signs, may be summoned to the carrying out of their divine tasks as if by sounding trumpets.'[152] This is the first indication we have of the importance time-keeping devices were to assume in the monastic life of the Middle Ages.

Also significant was the composition (whether by Cassiodorus or by someone commissioned by him) of a revised *computus* (computational tables), keyed to the year 562, for determining the dates of successive Easters. This document was innovative in that it used the nineteen-year (235

---

[149]Stahl, *Roman Science*, pp. 207ff.

[150]Luke 2:13; Heb. 12:22ff.; Rev. 4:8; 7:15; 1 Enoch 39:12; 2 Enoch 22:2. Cf. Hippolytus, *Apostolic Tradition* 36.12; Basil, Letter 207.3. For the Jewish background see G.H. Box, ed., *The Apocalypse of Abraham* (London: SPCK, 1919), p. 47 n7; John J. Gunther, *St Paul's Opponents and Their Background* (Leiden: Brill, 1975), pp. 194, 202ff.

[151]See references in note 106 above. Cf. Josephus, *Jewish War* II.8 128-36; 1QS 1:11-15; 3:9ff; 10:1-8.

[152]*Introduction to Divine and Human Readings* I.30 (trans. Jones, p. 135).

month) cycle of Dionysius Exiguus (AD 525), rather than the less accurate eight-year cycle popularised by Pliny in the first century AD.[153]

7 The Computus Tradition through Bede

From the sixth century to the twelfth, progress in calendrics and astronomy was slow, but the motivation was strong. In the late sixth century, Gregory of Tours (d 594) wrote a treatise on the courses of the stars to enable monks to determine the proper hours for night offices from the positions of the constellations at different times of the year.[154] Isidore of Seville (d 636), writing in what is generally regarded to be the depth of the dark ages, made a clear distinction between astronomy and astrology: the former, which he called 'natural astrology', dealt with the lawful courses of the heavenly bodies (*lex astrorum*) and played an imporant role in meteorology, agriculture, and medicine, while the latter dealt with personal horoscopes.[155]

At this point, the scene of activity shifted to the British Isles, where a *computus* tradition was already flourishing under the aegis of Celtic monasticism at Iona and Lindisfarne. In the year 668, Theodore of Tarsus was appointed Archbishop of Canterbury, and under his supervision, Hadrian of Africa established a comprehensive curriculum, including astronomy and computation, at the monastery school of Canterbury which became the model for monastic schools founded by Benedict Biscop at Wearmouth (674) and Jarrow (*c.* 681).[156]

It was at Jarrow that the Venerable Bede (*c.* 672-735) grew up and received his training. Bede is recognised as the first indigenous scientist of the West. In an autobiographical note he appended to his *History of the*

---

[153]The nineteen-year cycle was discovered by Meton (*c.* 432 BC) and Christianised by Theophilus of Alexandria, in the late fourth century, and by Cyril of Alexandria, in the early fifth; Charles W. Jones, 'Development of the Latin Ecclesiastical Calendar', *Bedae: Opera de Temporibus* (Cambridge, Mass.: Mediaeval Academy of America, 1943), pp. 29ff., 68ff.; O. Neugebauer, 'On the Computus Paschalis of "Cassiodorus"', *Centaurus*, vol. xxv (1982), pp. 292-302; Elias J. Bickerman, 'Calendars and Chronology', in W.D. Davies and L. Finkelstein, eds., *Cambridge History of Judaism*, vol. I (Cambridge: Cambridge Univ. Press, 1984), pp. 62f.; G.J. Whitrow, *Time in History* (Oxford: Oxford Univ. Press, 1988), p. 189.

[154]*De cursu stellarum* (On the Sourse of the Stars), written after Gregory's consecration as bishop of Tours in 573; Clagett, *Greek Science*, p. 195; Stephen C. McCluskey, 'Gregory of Tours, Monastic Timekeeping, and Early Christian Attitudes to Astronomy', *Isis*, vol. lxxxi (March 1990), pp. 10ff.

[155]Isidore, *Of Diversities* II.152; after Clement of Alexandria, *Exhortation to the Greeks* VI.58; Pseudo-Clement, *Recognitions* VIII.20ff.; Augustine, *On Christian Doctrine* II.29.46; Cassiodorus, *Introduction* II.7; cf. Richard Kieckhefer, *Magic in the Middle Ages* (Cambridge: Cambridge Univ. Press, 1990), p. 127.

Isidore has been credited as the pivotal turning point in the Western evaluation of the mechanical arts by two recent studies: Elspeth Whitney, *Paradise Restored* (Philadelphia: American Philosophical Society, 1990), 59, 61, 63; Valerie I. J. Flint, *The Rise of Magic in Early Medieval Europe* (Princeton: Princeton Univ. Press, 1991), pp. 98, 128, 146.

[156]Bede, *History of the English Church* IV.1f., 18; V.21.

English Church, he described himself as having been devoted all his life (since the age of seven) to the monastic discipline at Jarrow. But, he said, he found his chief delight in study, teaching, and writing.[157] Among other things, Bede's 'study, teaching, and writing' included the organisation of a cooperative programme for monitoring the tides up and down the English coast ('the establishment of a port') and the writing of two major works on time measurement and chronology, one of which became a standard text for centuries to come.

While basing much of his work on the texts of Pliny and Isidore, Bede was more willing to check inherited wisdom against his own observations than Isidore, and he gave the first clear statement of the sphericity of the earth in medieval times. (The idea was thus well established by Columbus's time.)

In general, Bede presented the following generation of scholars with an ordered universe of cause and effect in which as many phenomena as possible were reduced to general laws. Bede was venerated by the leaders of the Carolingian renaissance half a century later, and his scientific works were still influential at the cathedral school of Chartres in the twelfth century.[158]

The eighth century thus marked a turning point at which the creationist tradition, together with the disciplined monastic life it fostered, gave rise to the earliest stages of Western scientific thought. The turning point was also marked by the fact that, for the first time in the West, belief in the relative autonomy of nature was sensed to be a threat to ecclesiastical authority, and a mild conservative reaction occurred.

8 Boniface's Reaction to Virgil of Salzburg

Around the year 748, Boniface complained to Pope Zachary that Virgil (later to be bishop of Salzburg) held the view that the opposite side of the world (the antipodes) might be inhabited by humans, a teaching generally associated with writings of the classical Epicurean, Lucretius![159]

We know little of Virgil's background other than that he was born in Ireland or Scotland. It has been suggested that his speculations about inhabitants of the remoter portions of the globe were rooted in Celtic

---

[157]Bede, *History*, 'Autobiographical Note' following V.24; cf. Butler, *Benedictine Monachism*, p. 337; Eckenrode, 'Growth of a Scientific Mind', p. 210.

[158]Crombie, *Augustine to Galileo*, vol. I, pp. 38-41; Clagett, *Greek Science*, pp. 197-202; T.R. Eckenrode, 'Venerable Bede as Scientist', *American Benedictine Review*, vol. xxii (Dec. 1971), pp. 486-507; idem, 'Growth of a Scientific Mind', pp. 197-212.

[159]The classical notion that the earth had separate bands of human habitation was also known in the West through Macrobius's Commentary on [Cicero's] *Dream of Scipio* V, ed. William Harris Stahl (New York: Columbia Univ. Press, 1952), pp.200-206.

traditions concerning a race of immortals that dwelt underground.[160] If so, his citation of the classical idea of the antipodes was more indebted to folk religion than to empirical science in the modern sense. It was an early attempt of synthesis between science and faith rather than an instance of opposition between the two.

Boniface was a product of Anglo-Saxon Benedictine education and something of a classical scholar himself, though of more a literary than a scientific bent. A humanist by inclination, he worked hard at organisation and discipline, helping establish more than sixty monasteries in Germany (the 'Apostle of Germany') and attempting, almost single-handedly, an extensive reform of the church in France.[161] So we can readily understand that the pronouncements of Virgil on an issue that had been regarded as highly speculative by Basil, Ambrose, and Augustine[162] and was associated with classical Epicureanism and Celtic mythology might elicit his concerns about the reemergence of paganism. Boniface is still known today for his symbolic act of felling the Oak of Thor at Geismar.

In response to Boniface's complaint, Pope Zachary denounced the idea that the other side of the earth might be inhabited, apparently on the grounds that it implied the existence of a race of beings not descended from Adam and Eve (the equatorial zone was believed to be impassable). Such an idea Zachary declared to be 'in opposition to God' (i.e., contrary to divine revelation; cf. Gen. 10:32; 11:8; Matt. 24:14) and detrimental to Virgil's own soul. The pope accordingly instructed Boniface to excommunicate and defrock Virgil on the supposition that his report of Virgil's words was accurate.[163] Apparently no action was taken, however. In fact, Virgil became bishop of Salzburg in 767 (or 755?) and was canonised in 1233.

The brief altercation between Boniface and Virgil in the mid-eighth century was one of the first in what was to become a long series of conflicts between the interests of intellectual inquiry and the moral and spiritual concerns of church order in Western Christendom. We recall that church leaders of the second, third, and fourth centuries (e.g., Irenaeus, Tertullian, and Lactantius) had been overtly critical of much of Greek science, largely for ecclesiastical or pastoral reasons. In the fourth century, Pseudo-Clement

---

[160] John Carey, 'Ireland and the Antipodes: The Heterodoxy of Virgil of Salzburg', *Speculum*, vol. lxiv (1989), pp. 5f., 9, 10.

[161] Jean Decarreaux, *Monks and Civilization* (London: Allen and Unwin, 1964), pp. 284f., 300-311.

[162] Basil, *Hexaemeron* I.8f.; Ambrose, *Hexaemeron* I.6; Augustine, *City of God* XVI.9. Isidore (*Etymologiae* IX.ii.133) and Bede (*De temporum ratione* 34) followed Augustine; Carey, 'Ireland', note 6.

[163] A translation of Pope Zachary's words is given in Carey, 'Ireland', p. 1. Cf. the words of Lucretius: 'There are unknown races of men, and unvisited realms, drawing a shared light from a single sun', *De rerum natura* I.374f.

(of the *Recognitions*) and Basil took a more positive approach to pagan learning, and this new departure combined with the emergence of monastic discipline to produce the interest in natural phenomena we found in the Venerable Bede.

So the issue between Virgil and Boniface was *not* a conflict between science and religion.[164] Both protagonists were loyal churchmen. Indeed, both were missionaries for their faith. Both of their respective viewpoints—belief in the comprehensibility and unity of nature and concern for corporate responsibility—were born and nurtured within the creationist tradition. They may be viewed as representing two wings of that tradition, the one more progressive and the other more conservative. Both viewpoints would prove to be instrumental in the development of modern science. What we have, then, is a tension between two ideals or goods—both sanctioned by the church and both conducive to scientific progress—not a conflict between science and religion as such.

9 Rabanus, Gerbert, and Fulbert

With the work of Boniface and Virgil, the focus of our story shifts from Britain back to France and Germany, where the Carolingian Renaissance was engineered by Alcuin of York (d 804), an educator trained by a former student of Bede. The work of building an indigenous northern European intelligentsia was continued by Alcuin's student Rabanus Maurus (d 856), who was archbishop of Mainz and later known as *primus praeceptor Germaniae* ('the foremost teacher of Germany').[165]

As if it had not already been proven that an interest in natural science was no bar to promotion in the church, the last year of the millennium saw Gerbert of Aurillac, the leading European mathematician of the day, become

---

[164]Among earlier historians who cited Virgil's case as a prototypical instance of the conflict of science and the church are Andrew Dickson White, *A History of the Warfare of Science with Theology in Christendom*, vol. I (New York: D. Appleton, 1896), pp. 105f.; J.L.E. Dreyer, *A History of the Planetary Systems from Thales to Kepler* (Cambridge: Cambridge Univ. Press, 1906), pp. 224f. According to Carey ('Ireland', note 9), Michael Maestlin compared Copernicus to Virgil (letter to Kepler, 1605), and Johannes Kepler compared himself to Virgil.

[165]Dijksterhuis, *Mechanization*, p. 101; Stahl, *Roman Science*, pp. 233ff. Two recent studies point to Rabanus's development of Isidore's ideas as pivotal in the Western evaluation of the mechanical arts; Whitney, *Paradise Restored*, 63, 68; Flint, *Rise of Magic*, 98 (on astrology, following Isidore) and 370f. (on magic, independently of Isidore).

Rabanus also illustrates the difficulty in establishing consistent polarities in Western thought prior to the twelfth century. He was closely associated with the establishmentarian reaction of Hincmar of Reims against Gottschalk's (Augustinian) emphasis on the sovereignty and 'trine deity' of God. On the other hand, Hincmar was opposed to Ratramnus, who on the matter of the Virgin Birth and the eucharist (in opposition to Radbertus) placed the emphasis (also Augustinian) on the order of nature and empirical reality; D.E. Nineham, 'Gottschalk of Orbais', *Journal of Ecclesiastical History*, vol. xl (Jan. 1989), pp. 2f., 6, 9, 10, 12f., 15ff.; Jaroslav Pelikan, *The Christian Tradition*, vol. III, pp. 73f., 76f.

(by appointment of the German emperor, Otto III) the first French pope, Sylvester II. Gerbert, trained in the classics by the Benedictines at Aurillac, had traveled to Spain to study Arabic science and had taught astronomy at the cathedral school of Rheims. Using Arabic models, he devised astronomical instruments with which to illustrate his lectures and was perhaps the first Christian in the West to use Hindu-Arabic numerals instead of the more cumbersome Roman ones.[166] Gerbert thus gives us the first indication of the way in which the creationist tradition of the West produced the kind of demand for scientific knowledge that would pave the way for the major influx of Greco-Arabic learning in the following three centuries.[167]

Gerbert's pupil, Fulbert, was responsible for building the cathedral school at Chartres into a centre of scientific studies in the early eleventh century. It is among twelfth-century Chartrians, stimulated in part by a renewed interest in Neoplatonism, that many historians of science today see the beginnings of the modern scientific outlook. Thierry of Chartres (d after 1156) and his students developed the idea of relative autonomy to the point where natural processes became almost mechanical in the modern sense. For Thierry, even the six days' work of God described in Genesis 1 could be explained *secundum phisicam* (according to the principles of physics), in other words, in terms of the natural properties of the four material elements (earth, water, air, and fire). Only when describing creation *ad litteram* (according to the letter of Genesis 1) did Thierry find it appropriate to refer to the formative work of God's Spirit. [168]

The Chartrian Platonists were not the only exponents of the idea of autonomy in the twelfth century: there were also Peter Abelard (d 1142), Adelard of Bath (d *c.* 1150), William of Conches (d *c.* 1150), and Honorius of Autun (d 1152)—all testimony to the fact that the idea had become deeply ingrained in European thought well before the twelfth century, as our survey has shown.[169]

---

[166]Dijksterhuis, *Mechanization*, pp. 103f.; Dales, *Scientific Achievement*, p. 35.

[167]D.J. Struik, 'Gerbert', *DSB*, vol.V., p. 365a. Already in the 950s, Abbot John of Gorze had studied Arabic astronomy in Spain—apparently he was the first Western European to do so; McCluskey, 'Gregory of Tours', 21.

[168]Thierry of Chartres, *Tractatus de sex dierum operibus* (1130s or 1140s); see the discussions by Charlott Gross, 'Twelfth-Century Concepts of Time: Three Reinterpretations of Augustine's Doctrine of Creation *Simul*', *JHP*, vol. xxiii (July 1985), pp. 328-31; Willemien Otten, 'Nature and Scripture: Demise of a Medieval Analogy', *HTR*, vol. lxxxviii (April 1995), pp. 272-76. As Otten points out, Thierry's intention was pious. He wanted to exegete Scripture in harmony with natural philosophy in order to bring the Christian reader closer to God.

[169]M.D. Chenu, *Nature, Man, and Society in the Twelfth Century* (Chicago: Univ. of Chicago Press, 1968), pp. 1-48; Tina Stiefel, 'Science, Reason and Faith in the Twelfth Century', *Journal of European Studies*, vol. vi (March 1976), pp. 1-16; R.C. Dales, 'A Twelfth-Century Concept of the Natural Order', pp. 179-92; *idem, The Intellectual Life of Western Europe in the Middle Ages* (Washington, D.C.: University Press of America, 1980),

10 Academic and Ecclesiastical Dimensions of Autonomy

The revival of natural philosophy in the eleventh and twelfth centuries was closely associated with a renewed interest (stimulated by the rediscovery of Boethius's works) in the use of Aristotelian dialectic (logic) to interpret Scripture and organise theological discourse generally. This was the first wave of Aristotelian influence in the Latin Middle Ages, often referred to the 'old logic'. The 'old logic' of Aristotle did not include the major corpus of Aristotelian natural philosophy, but it prepared the way for the reception of the latter in the late twelfth and early thirteenth centuries.

Also in the eleventh and twelfth centuries, there was a conservative reaction to both Neoplatonic natural philosophy and Aristotelian dialectic. This reaction was similar to that of Boniface in the eighth century but far more extensive in scope. The use of the word 'conservative' in this context is traditional, but it may be misleading, since many of the anti-dialecticians also advocated reforming the church and ending abuses that stemmed from secular interference in ecclesiastical affairs.

An important alignment of social and theological interests thus came about that was to persist in varying forms for more than seven centuries. The freedom of the church from secular control was associated with the freedom of theological discourse from the constraints of secular (Aristotelian) reason.[170] It was a pattern that was to repeat itself in various ways in the fourteenth, seventeenth, and nineteenth centuries. Both sides sought autonomy in their own terms: autonomy of the natural sphere was mirrored in academic privileges, on the one hand; autonomy of the spiritual, sacramental, and moral spheres was mirrored in clerical privileges (and the restriction of dialectic to its place as just one of the arts), on the other. What God had joined together was beginning to come apart, at least, as far as Western society was concerned.

11 God's Absolute Power: Damian, Manegold, and William of St Thierry

Peter Damian (d 1072), first and foremost among these critics of dialectic, had himself taught grammar and rhetoric (the other two arts in the *trivium*) at Ravenna before being drawn to monastic life. His skill in dialectic was also amply demonstrated by his devastating use of the art in arguing against its more zealous advocates.

Damian's basic concern, however, was that the extensive use of reason to

---

pp. 265-69. Adelard may have been a pupil of Thierry; William, of Bernard of Chartres; A. Clerval, *Les écoles de Chartres au moyen âge* (Chartres, 1895), pp. 181ff., 188ff.; cited by Richard Lemay, *Abu Ma'shar and Latin Aristotelianism in the Twelfth Century* (Beirut: American University, 1962), note on p. 157.

[170]Gordon Leff, *Medieval Thought* (Harmondsworth: Penguin Books, 1958), pp. 90ff. It should be noted, however, that advocates of reform like Damian and Humbert were skilled dialecticians.

draw inferences in theological matters neglected the sovereign freedom of God, or, equivalently, that an increasing emphasis on the autonomy of the natural order overlooked the biblical basis of natural law in the divine decree. Consequently, he stressed what later became known as the *potentia absoluta* ('absolute power') of God to alter the course, or even the existing state, of nature.[171] To take an extreme case: God could not only reverse the effects of a past injustice, but even cause it not to have happened in the first place![172] Such were the lengths some conservatives felt compelled to go in order to counter the threat of an emerging naturalism and rationalism.

Writing a little later than Damian, Manegold of Lautenbach (*c.* 1080) shared many of his concerns. The natural philosophers, he complained, were so concerned with the physical nature of things that they could no longer conceive of a substantial being existing beyond the natural order.[173]

William of St Thierry (d 1148), one of the greatest medieval mystics, was a friend of Bernard of Clairvaux. He wrote to Bernard against both Peter Abelard and William of Conches, succeeding in the former case in inducing the more famous churchman to take up his own pen. The grounds in that instance seem to have been purely theological, resting, of course, on the underlying issue of the value of dialectic. His case against William of Conches (1141), however, concerned the latter's radically naturalistic

---

[171]Damian enunciated the conceptual distinction between the two powers of God even though the actual terminology was not developed until the early thirteenth century; see William J. Courtenay, *Capacity and Volition: A History of the Distinction of Absolute and Ordained Power* (Bergamo: Pierluigi Lubrina, 1990), p. 68.

On the scholarly debate over the meanings of *potentia absoluta* and *potentia ordinata* during the Middle Ages, see Francis Oakley, *The Western Church in the Later Middle Ages* (Ithaca, N.Y.: Cornell Univ. Press, 1979), pp. 143ff. We shall use the term *potentia ordinata* to refer to the normal course of nature without prejudice to the degree to which that course is altered by God in actual fact or could be altered in principle. The difference between various medieval writers has more to do with their assessment of the degree of alteration and its significance than with the meaning of the terms and the ideas they represent.

[172]For instance, in restoring the virginity of a woman who had had premarital sex; *On Divine Omnipotence* (John F. Wippel and Allan B. Wolter, eds., *Medieval Philosophy* [New York: Free Press, 1969], pp. 143-52). Damian was entirely consistent with the feudal idea of discretionary rule, which was to be displaced by the idea of natural law beginning in the twelfth century; Geoffrey Koziol, 'Lord's Law and Natural law', in Harold J. Johnson, ed., *The Medievel Tradition of Natural Law*, Studies in Medieval Culture, XXII, (Kalamazoo, Mich.: Medieval Institute Publications, 1987), pp. 103-117.

Augustine had stressed God's power to alter the course of nature in *On Genesis Word for Word* IX.17; *City of God* XXI.8, but he played down its importance for rational inquiry in *ibid.*, II.1.

[173]*Opusculum contra Wolfelmum* 22 (*MPL* 155:170; Wetherbee, *Cosmographia*, p. 7). Manegold criticised Wolfelm for following Macrobius's teaching that the four separate quarters of the earth were inhabited by different branches of the human race (separated by impassable oceans). This conflicted with the church's teaching that salvation in Christ was available to all humans. On Manegold's opposition to Wolfelm and his denial of the antipodes, see A.J. Macdonald, *Authority Reason in the Early Middle Ages* (London: Oxford Univ. Press, 1933), pp. 103f.

interpretation of Genesis, as well as theology.[174] It would not be fair to say that William of St Thierry was opposed to science itself, however, for he wrote a treatise of his own, *On the Nature of the Body and the Soul*, which was based on some of the latest medical ideas introduced into Western Europe by Constantine the African (chap. 1, sec. E).[175] Indeed, one of William of St Thierry's principle concerns was to affirm the unity of body and soul against the more dualistic tendencies of William of Conches and other Platonists of the twelfth century.[176]

Such was what is sometimes called the 'conflict between religion and science' in the eleventh and twelfth centuries. The episode has been noted by recent historians as marking the first time since antiquity that the problem of scientific versus religious thinking received serious attention.[177]

However, the conflict at this stage was not one of religion and science, so much as one of two different emphases within the creationist tradition.[178] As we shall see in later chapters, the left wing of the creationist tradition gradually abandoned its theological orientation and the right wing eventually lost its interest in science. But the process of divorce was not completed until the specialised professions emerged in the nineteenth century. In the meantime, we are dealing with two opposing interpretations of creational theology, not with an opposition between science and theology as such.

Still, those historians who date the problem from the eleventh or twelfth century are partly right, for it was during that period that the dichotomy between the natural and the supernatural, so ingrained in modern Western thought, had its origin. The scholastics (after Anselm and Peter Lombard) began to make a systematic distinction between the regular power (*potentia*

---

[174]For instance, William of Conches's view that Eve was formed by natural processes rather than supernaturally from the rib of Adam; William of Conches, *De philosophia mundi* I.23 (MPL 172:55f.; Tullio Gregory, 'La nouvelle idée de nature et de savoir scientifique au XIIe siècle', in J.E. Murdoch and E.D. Sylla, eds., *The Cultural Context of Medieval Learning*, [Dordrecht: Reidel, 1975], pp. 195f.). William of St Thierry attacked this view in *De erroribus de Guillaume de Conches*, AD 1141; MPL 180:339f.; Chenu, *Nature, Man, and Society*, p. 16; Stiefel, 'Science, Reason, and Faith', p. 10.

Augustine had argued that the supernatural was required in the creation of Eve by the fact that new substance was created both to build Eve out of Adam's rib and to replace rib with new flesh; *On Genesis Word for Word* IX.xv.26, 28.

[175]Jean Jolivet, 'The Arabic Inheritance', in P. Dronke, ed., *History of Twelfth-Century Western Philosophy*, p. 129.

[176]Richard McKeon, 'Medicine and Philosophy in the Eleventh and Twelfth Centuries: The Problem of Elements', *Thomist*, vol. xxiv (1961), pp. 252f. Cf. Augustine's critique of Neoplatonism in *City of God* XII.25ff., and McMullin, ed., *Evolution and Creation*, pp. 14f. and note 49.

[177]Stiefel, 'Science, Reason and Faith', pp. 1-5.

[178]'What may look in retrospect like an irreconcilable dichotomy between "science" and "faith" (or "reason" and "religion") began as a mere conflict between two ways of understanding the religious universe.' Thomas Goldstein, *Dawn of Modern Science* (Boston: Houghton Mifflin, 1980), pp. 69f.

ordinata) of God, reflected in the normal sequences of cause and effect, and his absolute power (*potentia absoluta*) at any time to suspend or alter those sequences—a distinction that was quite useful in the interpretation of Scripture.[179] However, already in the eleventh and twelfth centuries, the normal sequences of nature were viewed as due to a power delegated to nature by God, and the distinction became an opposition that was quite foreign to the sense of Scripture. In place of a *relative autonomy* of nature based on the efficacy of God's creative Word, one was forced to make a choice: either an autonomous world, created by God but virtually independent of God's continued presence and power; or else a world so utterly dependent on God's will moment by moment that all rational, scientific investigation became impossible. In effect, we have the beginning of the dissolution of the creationist tradition itself, even though the complete demise of the tradition took seven centuries to complete.

12 The Order of Nature: Adelard and William of Conches

Two examples of the problem will suffice: first, Adelard of Bath, who wrote in the early decades of the twelfth century. In his *Natural Questions*, Adelard tried to defend his interest in Arabic science against an attack by conservatives by equating the work of God with the strictly miraculous and contrasting it to the work of nature. The wording of his defense shows how the separation of God and nature came about:[180]

> I take nothing away from God, for whatever exists is from him and because of him. But the natural order does not exist confusedly and without natural arrangement, and human reason should be listened to concerning those things it treats of. But when it completely fails, then the matter should be referred to God.

---

[179]On the biblical view of miracle in relation to natural law, see H. Wheeler Robinson, *Inspiration and Revelation in the Old Testament* (Oxford: Clarendon Press, 1946), chap. 3; Von Rad, *Theology of the Old Testament*, vol. I, pp. 360f.

Augustine differentiated two types of formative principles (*rationes*) in God: those that had been planted in nature as seminal causes and acting in accordance with the ordinary course of nature (e.g., the emergence of vegetation and animals from the earth) and those that God reserved in the secrecy of the divine will so as to act contrary to the ordinary course of nature, though not contrary to nature itself (e.g., the formation of Eve from the rib of Adam); *On Genesis Word for Word* IX.xvii.32.

[180]*Quaestiones naturales* IV (Dales, *Scientific Achievement*, p. 40). Dales places this work in the 1120s (*ibid.*, p. 38). Others date it as early as 1107: C.H. Haskins, *Studies in the History of Medieval Science* (Cambridge, Mass.: Harvard Univ. Press, 1924, 1927), p. 27; Brian Lawn, *The Salernitan Questions* (Oxford: Clarendon Press, 1963), pp. 28f. Cf. William of Conches, *De philosophia mundi* I.23 (from the 1120s).

Compare also Augustine's criticism of Christians who invoked God's *potentia absoluta* in support of the idea of supra-celestial waters against the arguments of philosophers of their time in *On Genesis Word for Word* II.i.2. Note, however, that Augustine did not make reason and revelation, or the work of God and the events of nature, mutually exclusive; cf. *ibid.*, IV.xxxiii.51 (suggesting a sharper dichotomy); V.iv.11; vi.18; *passim*.

The underlying ideas cited in this passage—the creation of all things by God, the consequent order and rationality of the cosmos, and the power of human reason—all stem from the Judeo-Christian creationist tradition, dating back at least to the second century BC. What was new was that Adelard set the natural order and the work of God, rational investigation and Christian faith, over against each other as alternatives ('...when human reason fails, *then* the matter should be referrd to God').[181]

The consequence of this polarisation was that, for Adelard, God was removed from the natural order in such a way that natural law became inflexible and impersonal:[182]

> Truly, whoever thinks to abolish the innate order within nature is mad....For he who disposes is most wise and, consequently, is least of all either willing or even able to abolish the fundamental order in nature...and among [natural] philosophers it is agreed that any upsetting of this order is least likely to occur.

The belief that God does not normally alter his established order (*potentia ordinata*) had been an essential part of the creationist tradition since Genesis, but, for earlier theologians like Aristobulus and Augustine, the natural order itself was upheld by God (through his word, will, or power). For Adelard, on the other hand, the only sort of properly divine action envisioned was God's abolition or upsetting of that order (*de potentia absoluta*), and even this was deemed to be unlikely. The order of nature was so fixed that God

---

[181]In his *De sacra coena* (1068-69), Berengar of Tours had laid down the hermeneutical principle that reason should be followed in preference to the authority of the fathers whenever the two were found to be in conflict. Significantly, Berengar legitimised his trust in reason by appealing to the biblical teaching of the creation of humanity in the image of God. However, Berengar accepted the virgin birth and the miracles of the Old Testament on the basis of authority; Macdonald, *Authority and Reason*, pp. 87f., 96.

In Adelard's more immediate background was the figure of Petrus Alfonsi, who immigrated from Aragon to serve in the court of Henry I sometime between 1106 and 1116. Petrus may have been Adelard's teacher in astronomy; see John Tolan, *Petrus Alfonsi and His Medieval Readers* (Gainesville: Univ. of Florida Press, 1993), pp. 10f., 42, 60 n53. In his *Dialogi contra Iudaeos*, written in 1108 or 1110, Petrus clearly differentiated healing accomplished through *physica* (medicine) from that performed through the power of God. But Petrus appealed to rational explanations of biblical miracles only in order to defend the New Testament against the criticisms of Aristotelians and Jews (*MPL* 157:647; Tolan, *Petrus Alfonsi*, pp. 10, 45, 47f., 49).

[182]*Quaestiones naturales* IV (Stiefel, 'Heresy of Science', p. 351); cf. Dales, 'Twelfth-Century Concept', pp. 182f. The emphasis on reason and nature had been made (with respect to the eucharist) by Berengar of Tours (where Adelard studied; Lawn, *Salernitan Questions*, p. 26, n. 8) and had been answered by an appeal to the absolute power of God over against the power of nature by Eusebius Bruno, *Epistola ad Berengarium magistrum* (*MPL* 147:1203a-b); cf. Irven M. Resnick, 'Peter Damian on the Restoration of Virginity', *Journal of Theological Studies*, vol. xxxix (April 1988), pp. 131, 133. Note also the similarity of Adelard's argument to that of Celsus *apud* Origen, *Contra Celsum* V.14.

was neither 'willing nor even able' to alter it![183]

William of Conches, our second example, began teaching in the 1120s at Paris (and possibly later at Chartres)[184] and was the most influential master of the time until he was forced to retire around 1144. In one of his later works, his gloss on Plato's *Timaeus* (1145-9), William made a differentiation between the work of God and the work of nature very similar to that of Adelard:[185]

> It must be recognized that every work is the work of the Creator or of nature, or the work of a human artisan imitating nature. The work of the Creator is the first creation without pre-existing material, for example the

---

[183]Augustine had argued that God exercised power not arbitrarily, but 'by the strength of wisdom', hence that he would not act contrary to the principles he had implanted in nature; *On Genesis Word for Word* IX.xvii.32 ('since he himself is not more powerful than himself'); xviii.34 ('God would do nothing contrary to the causes which his will created'). Adelard could have appealed to the text of Augustine in support of his views, but in doing so he would clearly have had to distort its meaning. For Augustine, the causal principles often only determined the potentialities of creatures and left the actual course and speed of their realisation to the secret will of God; ibid., VI.xv.26-xviii.29; IX.avii.31f. Moreover, Augustine allowed that God could contravene the natural properties of substances (e.g., oil floating on water, *ibid.*, II.i.2), even though he would not contravene the causal principles he had planted within them.

[184]As to whether William taught at Chartres in the late 1130s as his pupil John of Salisbury claims (*Metalogicon* I.5; II.10), see R.W. Southern, 'The Schools of Paris and of Chartres', in Robert L. Benson and Giles Constable, eds., *Renaissance and Renewal in the Twelfth Century* (Cambridge Mass.: Harvard Univ. Press, 1982), pp. 129ff.; T. Stiefel, '"Impious Men": Twelfth-Century Attempts to Apply Dialectic to the World of Nature', in Pamela O. Long, ed., *Science and Technology in Medieval Society* (New York: New York Academy of Sciences, 1985), p. 190 with notes.

[185]*Glosses on Plato's Timaeus* 28a (Chenu, *Nature, Man, and Society*, p. 41; cf. Stiefel, 'Science, Reason and Faith', p. 9). For the Latin texts of this passage and a parallel one in William's *Glosses on Boethius's Consolation of Philosophy*, see Joseph-Marie Parent, *La doctrine de la création dans l'école de Chartres* (Paris: J. Vrin, 1965), pp. 127f., 147; Edouard Jeauneau, ed., *Glosae super Platonem* (Paris: Libr. Philosophique J. Vrin, 1965), pp. 104f. The threefold distinction between the works of God, nature, and humanity appears also in Hugh of St Victor, *Didascalicon* I.9 (written in late 1120s; appeared *c*. 1130); cf. Jerome Taylor, trans. (New York: Columbia Univ. Press, 1961, p. 55. The distinction is based on Chalcidius, *On Plato's Timaeus* I.23.73.10ff. On the influences of Chalcidius's commentary on William, see Benedicta Ward, *Miracles and the Medieval Mind* (Philadelphia: Univ. of Pennsylvania Press, 1982), pp. 5f.

In his earlier (1120s) *On the Philosophy of the World* (*De philosophia mundi* I.23), William had tried to protect himself against his critics by referring the power of nature back to God (Tina Stiefel, *The Intellectual Revolution in Twelfth-Century Europe* [New York: St Martin's Press, 1985], pp. 82, 85).

By way of contrast, Augustine had clearly stated the unity of the three forms of action (divine, angelic/human, and natural) and stoutly refused to limit God's works to creation and miracle: e.g., *On Genesis Word for Word* VIII.viii.15-ix.18; xii.25. In fact, Augustine (*ibid.* VIII.xv.27) virtually equated the work of nature with that of God in contrast to that of a human agent like a farmer or physician (although both natural and human were then classified as two aspects of providence). In general, Augustine's view of the creative forces in nature was more positive than his view of the creative arts of humanity. Humanity's freedom with respect to nature was more one of understanding and cooperating than one of restructuring; cf. *ibid.*, VIII.viii.16.

creation of the elements or of spirits, or it is the things we see happen contrary to the accustomed course of nature, as the virgin birth and the like. The work of nature is to bring forth like things from like through seeds or offshoots, for nature is an energy inherent in things and making like from like.

This statement is, if anything, more extreme than Adelard's: Adelard had at least attributed the ordering, as well as the creation, of the elements to God. For William, however, the ordering of nature was due to the inherent properties of the elements, and only the creation of the elements (and spirits) and their properties was due to God's agency.[186] Consequently, the ordering described in Genesis was itself subject to rational scrutiny and could not be accepted as authoritative.

William went so far as to reject the literal truth of a biblical statement about the existence of waters above the visible heaven (Gen. 1:7; cf. Ps. 148:4). Such an elevation of a heavier element would be unnatural and contrary to reason, he argued. At best, the biblical text could be interpreted as a loosely worded reference to the clouds in the atmosphere.[187] The difference here from earlier, allegorical interpretations of Scripture was not just a matter of degree. Basil and other early representatives of the creationist tradition believed in the order of nature *as* the work of God, whereas William believed in nature as self-ordering and self-perpetuating. The work of nature was autonomously 'to bring forth like things from like', and any act of God subsequent to the first moment of creation (e.g., the virgin birth) was 'contrary to the accustomed order of nature'.[188] Here we have the idea of 'miracle' as a violation of the laws of nature that was to become popular in Western philosophy of the seventeenth and eighteenth centuries (e.g., David Hume).

---

[186] Petrus Alfonsi had also attributed the sequence of creation to the God-given powers of the elements and the world soul in his *Dialogi* (*MPL* 157:555, 562; Tolan, *Petrus Alfonsi*, pp. 50f.).

[187] *De philosophia mundi* II.1ff. (*MPL* 172:57f.; Stiefel, 'Science, Reason and Faith', pp. 7f.); cf. Helen Rodnite Lemay, 'Science and Theology at Chartres: The Case of the Supracelestial Waters', *BJHS*, vol. x (1977), p. 231. William based his understanding of the properties of the elements on Aristotle's teaching; John Kirkland Wright, *The Geographical Lore of the Time of The Crusades* (Washington, D.C.: American Geographical Society, 1925), pp. 151, 183. Needless to say, the issue we are concerned with here is not the waters themselves, but the reasoning about their existence exhibited by William of Conches.

[188] For earlier Christian writers, miracles were also modifications of the 'natural order' and 'contrary to the accustomed course of nature'. E.g., Radbertus: '...note all the miracles of the Old and New Testaments which...were accomplished by God contrary to natural order' (*On the Lord's Body and Blood* I.2; *LCC* 9:94). But, prior to the twelfth century, miracles were not so exclusive of natural processes and human art as they were for William. Compare, e.g., Cassiodorus: 'Mechanics is...almost Nature's comrade, opening her secrets, changing her manifestations, sporting with miracles, feigning so beautifully that what we know to be an illusion is accepted by us as a truth'; *Variae* I.45, Thomas Hodgkin, trans., *The Letters of Cassiodorus* (London, 1886), p. 170; Whitney, *Paradise Restored*, pp. 67f.

We should note the effect the differentiation between the natural and divine orders had on William of Conches's epistemology. The consistent testimony of the church fathers up to and including the Venerable Bede, whom William and others of his time greatly admired, had been in favour of the existence of supracelestial waters almost without exception.[189] So William was forced to limit the authority of the fathers (and, by implication, that of Scripture) to matters of religious faith and morals.[190] This resulted in a clear dichotomy in modes of knowing correlated with the ontological dichotomy between God and nature. In matters concerning the work of God, now limited to the initial impetus of creation and occasional miracles, the authority of Scripture and the church was to be taken on faith. With regard to the workings of nature, however, reason alone was to be followed. Moreover, these two procedures were mutually exclusive and antithetical. Given this view of the alternatives, already expressed in the mid-twelfth century, it is no wonder that many creative modern thinkers have become sceptical about theology and about matters of faith generally!

Finally, we should note the dualism that emerges in William of Conches between two spheres of human existence: (1) the moral and spiritual, now relegated to the jurisdiction of the church; and (2) the technological and natural, based on human art and science. On one hand, William's dichotomy between faith and reason, noted above, left the realm of human traditions and values ('the establishment of custom') on the side of pure faith. On the other hand, his dichotomy between the work of God and the work of nature left human technology ('the work of an artisan imitating nature') on the side of nature, based on human bodily needs and a knowledge of the structures of nature.[191]

---

[189] Interestingly, the Wisdom of Solomon omitted any reference to the firmament and the waters above it (James M. Reese, *Hellenistic Influence on the Book of Wisdom and its Consequences* [*Analecta Biblica*, vol. xli, Rome: Biblical Institute Press, 1970], p. 88). Early Christian texts like the Ascension of Isaiah (vii.9-13) and the Pseudo-Clementine *Recognitions* (IX.3) regarded the firmament as separating the visible from the angelic world and made no mention of the waters. This alternative interpretation was transmitted to the Middle Ages by Augustine (*Confessions* XIII.7.8; 15.18; 32.47; cf. *City of God* XI.34), who interpreted the waters allegorically as referring to the angelic world, and by John Scotus Erigena (*On the Division of Nature* III.26), in the ninth century, who interpreted the waters as referring to the intellectual world of primordial causes.

[190] According to William of Conches:
> In those things which pertain to the Catholic faith or to the establishment of custom, it is not right to contradict Bede or any other of the Holy Fathers (against the authority of Scripture). But, in those things which relate to [natural] philosophy, if they err in any way, it is permissible to assert something different (Gratarol, ed., *Dragmaticon*, pp. 65f., c. 1145; Stiefel, 'Science, Reason and Faith', p. 6; cf. Lemay, 'Science and Theology', pp. 231f.; Gregory, 'La nouvelle idée', p. 197, gives the Latin text).

[191] See note 185 above. On Gilbert of Poitiers's more unitive, theistic treatment, see Chenu, *Nature, Man, and Society*, pp. 39f. The idea that human *technê* either imitates nature or assists it goes back to Plato (*Laws* X.889a-d) and Aristotle (*Physics* II.viii.199a.16ff.; J. H.

The resulting antithesis of ethical mores and technological skills was probably not intentional in William of Conches, but it was an immediate implication of his thought. It also reflected and was reinforced sociopolitically by the increasing autonomy of the secular order from the church as witnessed by the rapid development of lay guilds during this period and the growth of mercantile capitalism.

Still, while we have used terms like natural as opposed to supernatural, reason as opposed to faith, secular and technological as opposed to moral and ecclesiastical, it should be kept in mind that we are really talking about two strands of the same creationist tradition, two strands that were beginning to unravel in the twelfth century. If the two seem so separate to us today, we must remember that we in the West are the products of the history here described, and that 'from the beginning it was not so' (Matt. 19:8). Our thesis will receive further support as we consider the rationale for Western medicine and technology that developed out of Christian moral and spiritual values prior to the split in the creationist tradition in the twelfth century.

13 Our Definition of the Creationist Tradition Defended

At this point, we are in a position to note the differences between the definition of the 'creationist tradition' we have offered, going back to the biblical and patristic periods, and the definition Richard Dales offers in his studies of developments in the twelfth century. The features that characterised the 'creationist' (or 'transcendent-creationist') tradition according to Dales are a gulf between God and the world; the de-animation of nature; creation *ex nihilo*; the consequent goodness, self-sufficiency, and rationality of nature; and a tendency to view nature in mechanistic terms. Dales's conclusion is that the creationist tradition was inherently unstable and tended toward modern atheism.[192]

For comparison, our own definition has consisted thus far of the rationality of nature (with or without the stipulation that it be finite), the commensurable rationality of the human mind (hence the comprehensibility of nature to humans), the unity of heaven and earth (with or without the association of the heavens with ruling angels), and the relative autonomy of nature based on God's creative word and power.

So the creationist tradition does not entail (*pace* Dales) a gulf between God and the world or a de-animation or mechanisation of nature in the

Randall, *Aristotle* [New York: Columbia Univ. Press, 1960], pp. 275f.). John Scotus Erigena described the mechanical arts as being based on 'some imitation or human devising' in his commentary on Martianus Capella's *Marriage of Philology and Mercury* (Whitney, *Paradise Restored*, pp. 70f.). Note that this classical view, while allowing a high degree of autonomy for nature (*natura naturans*), maintained a sense of humility on the part of human artisans.

[192]Dales, 'Twelfth-Century Concept', pp. 191f.; cf. *idem*, 'De-Animation', pp. 533f.

modern sense. Such an emphasis did begin to enter the tradition with Augustine's separation of the seminal causes from God's consubstantial Word and Spirit, but it was not essential to the tradition itself. The idea of the complete autonomy, or even mechanicity, of nature did not enter until the gulf opened by Augustine widened to the point of suggesting a dichotomy between God's ordering of nature and his absolute power, or even between nature itself and God. This did not happen until the eleventh or twelfth century. In our view, then, it was not the original biblical and patristic tradition, but a distortion of it, that tended toward the determinism, reductionism, and atheism that characterises so much of modern Western thought.

E. The Creationist Tradition: Ministry of Healing and Restoration

Thus far we have concentrated on the theoretical aspects of the historic creationist tradition, which were given coherence by Basil in his sermons on Genesis 1. Three principal themes of this tradition—the comprehensibility of the world, the unity of heaven and earth, and the relative autonomy of nature—can be traced through the centuries in both East and West, as we have found, until they are seen to converge in Western Europe during the twelfth century and find their places at the foundation of modern Western science.

But the creationist tradition and Basil's contribution, in particular, were not just theoretical in nature. They had strong practical components that were closely related to the theoretical but took on a life of their own and influenced the history of science just as much, if not more, than the theoretical. We have already discussed the importance of the liturgical concern for time and the regulation of monastic life as vehicles for the sense of regularity in the rhythms of the cosmos. In this section we turn to the healing and helping ministries of the early church, which were rooted in the biblical beliefs of creation, resurrection, and the possibility of miracle and, through the work of Basil and his contemporaries, gave rise to the Christian traditions of medical science and technology in the Middle Ages. Much of what we discuss here relates to the histories of medicine and technology.

1 Jewish and Greek Assessments of Technology

What evidence we have for the Old Testament and intertestamental attitudes towards technology is rather paradoxical. The Israelite monarchy, purportedly dating back to the time of David and Solomon (tenth century BC), was made possible in part by the development of an iron-based technology competitive with that of the Philistines and Canaanites.[193] The early monarchy was thus a time of major growth for all the arts and crafts needed to maintain a small oriental dynasty.[194] On the other hand, the association of human fabrication with the manufacture of idols and with political repression gave rise to a lasting suspicion of the manual arts, particularly among the working classes, from the time of Solomon onwards.[195]

This reaction intensified during the Babylonian exile when the Jews were confronted with a superior technology in alliance with a polytheistic religion (Isa. 40:19f.; 41:6f.; 44:9-20; *passim*). Babylonian, and later Greek, technology were thus associated, in certain strata of the apocalyptic tradition, with the antediluvian fall of the angels and the beginnings of warfare and immorality (e.g., 1 Enoch 6-11, of the late third century BC or earlier). This occasionally negative attitude in the post-exilic period must be interpreted in light of the clear association of technology with oppression and persecution by foreign powers, however (cf. Isa. 54:16f.).

In other words, the Jewish position was neither for nor against technology as such: it was for relief from suffering and oppression, and all technology was assessed in terms of that criterion (Isa. 25:1-5; 61:1-7; cf. Wisd. 7:16-22; Ecclus. 38:32a, 24a; Jubilees 4:15-19; 1 Enoch 52). What we are about to trace is a thousand-year development through which

---

[193]Josh. 17:16, 18; Judg. 1:19, 4:2f., 13ff.; 1 Sam. 13:19-22. Cf. Paula M. McNutt, *The Forging of Israel* (Sheffield: Almond Press, 1990), pp. 209ff., 224. Even if the specific uses of iron described date from later in the monarchy (*c*. 700), they reflect an understanding of the role of earlier technologies in establishing that monarchy. For a critical review of the evidence for a kingdom under David and Solomon, see Thomas L. Thompson, *Early History of the Israelite People* (Leiden: Brill, 1992), pp. 312-16, 409-12.

[194]1 Kgs. 5-10; cf. Exod. 31:1-11 [P]; 1 Chron. 22; 2 Chron. 2-9. Cf. G. Ernest Wright, *Biblical Archeology* (London: Duckworth, 1957), pp. 90-93, 129-42. For the Mycenaean and Philistine background, see James D. Muhly, 'How Iron Technology Changed the Ancient World and Gave the Philistines a Military Edge', *Biblical Archaeology Review*, vol. viii (Nov. 1982), pp. 48-54.

[195]Deut. 27:15; 1 Kgs. 5:13-18; 9:15-23; 11:1-8; 12:25-33; 16:31-34; Isa. 2:7f.; 17:7f.; Jer. 10:9; 22:13-17; Hos. 8:4ff.; 13:2. Artisans and smiths were listed together with the elites who were carried off into exile by Nebuchadnezzar (the king, court officials, warriors, and 'leaders of Judah and Jerusalem': 2 Kgs. 24:14, 16; Jer. 24:1; 29:2). As McNutt states, this fact 'suggests that they were highly regarded for their social contributions' (*Forging*, 236), but the roles of all such elites were ambiguous, particularly in the deuteronomistic history and Jeremiah (cf. *Forging*, 237). In the Priestly text, dating from just before or during the Exile, crafts were portrayed as the gift of God (Exod. 31:1-11), but this was presumably in anticipation of the Restoration and the rebuilding of the Temple.

technology once again (as in the days of David and Solomon) became associated with the relief of human misery, a development which contributed to the social basis needed for the growth of modern Western science.

The fundamental belief of the Jews was in God as the creator of all things (Gen. 1; Pss. 33:6-9; 139:13-16; 148:5f.; Prov. 8:22-31; Jer. 10:12; 51:15). In dialogue with schools of Greek philosophy that posited an pre-existent substratum of unformed matter, the Jews affirmed creation 'out of nothing': God was the creator of all things, even of the matter of which all things consist.[196] The corollary of this belief was that things can be changed. Things do not have to continue as they now are because their existence depends on a God who created them beginning with nothing, who can therefore transform them as he will, and who has promised that such a transformation will take place through the power of his Spirit (Pss. 104, 146; Isa. 40:27-31; Jer. 33; Ezek. 37; Wisd. 19:6-12; 1 Macc. 7:28).

A comparison with the classical Greek outlook is useful at this point. The Greeks were able to maintain a tradition of political independence through much of their history even during the period of Persian hegemony. Even under later Roman rule (after 146 BC), Greeks enjoyed the privilege and status of providing the dominant cultural ambience.[197] So it is not surprising to find that technology, especially the technology of warfare, was widely praised and patronised among the Greeks.[198]

On the other hand, the Greek philosophers, almost without exception,

---

[196]Ep. Arist. 136; 2 Macc. 7:28; Rom. 4:17; 2 Enoch 24:2; 2 Apoc. Baruch 21:4f.; 48:8; Gen. Rab. 1:9. As Gerhard May has argued, the philosophical articulation of the doctrine of *creatio ex nihilo* took place in the context of the Gnostic crisis of the late-second century AD. But the substance of the doctrine was already present in the Old Testament teaching of total creation; *Creatio Ex Nihilo* (Edinburgh: T. & T. Clark, 1994), pp. xiif. However, May evaluates Jewish texts in terms of the standards of self-contained philosophical disquisition and fails to interpret them in terms of their own intertextual concerns (*ibid.*, pp. 6ff., 11f., *passim*).

The statement in Wisd. 11:17 that God formed the world 'out of formless matter' (*ex amorphou hylês*) should be read in the context of the Septuagint of Gen. 1:1-2: God created the world in a formless (or invisible) state and then gave it visible form. In other words, creation took place in two stages, the first of which was the creation of primordial matter (out of nothing; cf. Wisd. 1:14; 9:1). The Letter of Aristeas 136; Heb. 11:3; 4 Ezra 6:38-40; and 2 Enoch 24:2 make much the same point. 2 Macc. 7:28 is perhaps unique in that it appears to argue against this two-stage interpretation on Gen. 1. But creation *ex nihilo* is implied rather than argued in the canonical and deuterocanonical texts. A phrase like 'total creation' would describe biblical teachings better than 'creation out of nothing'.

[197]Except in Rome itself where Latinised forms could compete with the Greek; Moses Hadas, *Hellenistic Culture* (New York: Columbia Univ. Press, 1959), pp. 90f.

[198]Compare Aeschylus, *Prometheus Bound* 436-506 with 1 Enoch 7-9. The often cited (and overstated) disdain that Greek philosophers had for manual arts should be characterised as 'postmaterialist', not antimaterialist; Harold Dorn, *The Geography of Science* (Baltimore: Johns Hopkins Univ. Press, 1991), pp. 77f., 84. See Whitney, *Paradise Restored*, pp. 23-26, 43-50, for a review of the evidence. In any case, the claim we make here is that technology was patronised, not that it was idealised.

viewed matter as eternal and uncreated. If there was a God, he had to do the best he could with the prescribed properties of matter.[199] In other words, the Greek philosophers did not believe in the *possibility* of radical change largely because they did not experience the *need* for radical change. As Ludwig Edelstein has put it: 'The world was there to live in, not to be used or to be made over.'[200]

So there was a paradox at the heart of both Jewish and Greek thought. The Greeks, following the Babylonians and Egyptians, had developed the rudimentary techniques of engineering, but their outlook precluded them from sensing the full value of those techniques. The Jews, on the other hand, due to their faith and their social condition, understood the need for physical redemption, even though they failed (after David and Solomon) to see the potential of technology in the fulfillment of that need. The development of Western technology required both of these contributions and, therefore, required an outlook on life that could bring them together. This integrated outlook was provided by the early Christian church.

2 The New Testament Church: Healing Power and Social Benefit

The distinctive contribution of the New Testament was the belief that the power and love of the Creator had been poured out on humanity through the ministry of Jesus and his disciples. The outpouring of God's Spirit had empirical correlates in the relief of the oppressed, the healing of the sick, and the raising of the dead (Matt. 8:1-10:8; Mark 1:9-45; Luke 4:14-19; 6:17ff.; Acts 3:1-16; 5:29ff.; *passim*). Indeed, these works of the Spirit were regarded as the principal signs or evidences that promises of the Old Testament had been fulfilled and that the message of the disciples of Jesus was trustworthy (Matt. 11:2-24; 12:15-32; Luke 7:21f.; 13:32; Acts 3:17-4:22).[201]

The display of healing power was regarded as evidence of the Spirit's

---

[199]Most notably Plato, *Timaeus* 30a, 37c-d, 47e-48a, 53a-b, 69b; *Thaetetus* 176a. Compare Prometheus's conclusion to the enumeration of his technical innovations: 'Craft (*technê*) is much weaker than necessity' (*Prometheus Bound* 514). The only major exception in Greek thought was Eudorus of Alexandria; cf. note 51 above.

[200]Edelstein, 'Recent Trends', p. 584; cf. *idem*, 'Motives and Incentives', p. 25: 'Men no more claimed than did their gods to be creators out of nothing, to act with a free will that imposes its law on things that have no [a priori] nature of their own. Rather did they feel called upon to shape matter that was given and, here below at any rate, refractory to reason.' The upper-class orientation of Greek science has been pointed out by Zilsel, 'Genesis of the Concept of Scientific Progress'; pp. 328f.; Sambursky, *Physical World of the Greeks*, pp. 226ff.; Peter Green, *Alexander to Actium* (Berkeley: Univ. of California Press, 1990), pp. 454-59.

[201]According to Dieter Georgi, the Hellenistic Jewish-Christian wing of the early church (including the opponents of Paul in 2 Corinthians, Mark, and Luke but not Matthew) stressed the continuity of Jesus' (and their own) ministry with that of the patriarchs and prophets on the basis of pneumatic displays and miracles; *Opponents*, pp. 167-74, 252f., 264, 275f.

presence, but it was not sufficient evidence in itself. No people of antiquity were more aware of the possible misuses of power than the Jews. Even the Jewish apologists who portrayed Moses as a great king in Hellenistic terms stressed the idea that he did not use his power for personal advantage, but rather for the benefit of others.[202] The New Testament writers insisted on the same criterion for the assessment of spiritual gifts. The powers of prophecy and exorcism, even accompanied by the name of Jesus, were not to be trusted if used for evil or selfish ends (Matt. 7:21ff.).[203] Christians whose gifts were recognised were not to use them for their own advantage, nor to regard themselves as superior to others. Everything was to be done as Christ had done it, for the edification or benefit of others (Mark 10:42-45; Rom. 12:16; 14:13-15:6; 1 Cor. 10:23-11:1; 14:1-33; Eph. 5:1f., 25; Phil. 2:3-7; 1 Thess. 2:1-8; 1 Pet. 4:10f.).[204]

To sum up: the early Christians believed in the possibility of healing and restoration that would truly benefit the needy. Underlying this belief was faith in a God who had created and could restore, a Messiah who had initiated God's final rule over both the forces of nature and the structures of society, and a Spirit who had been poured out on the believers, enabling them to carry on the work of Jesus and to extend it to all nations.

3 Two Streams in the Second Century: Apocryphal Acts and Apologists

During the second century, the ideas we have documented from the New Testament divided temporarily into two streams and merged again in the

---

[202]Philo, *On the Migration of Abraham* 121; *On the Life of Moses* I.148-53, 328; cf. *Special Laws* I.97; Josephus, *Against Apion* II.16.158f. (II.17 in William Whiston translation); cf. Origen, *Against Celsus* II.50ff. Cf. Georgi, *Opponents*, p. 258.

This emphasis on Moses' concern for the benefit of others (*philanthropia*) was partly occasioned by Greek charges that Jews were misanthropes (*misanthropoi*); J.N. Sevenster, *The Roots of Pagan Anti-Semitism in the Ancient World* (Leiden: Brill, 1975), pp. 89-94. Note particularly the sequence in Josephus from Apollonius Molon's accusation of misanthropy (*Against Apion* II.14.147-50) to the depiction of Moses as altruistic (II.16.158f.).

[203]Eduard Schweizer, *The Good News According to Matthew* (Atlanta: John Knox Press, 1975), pp. 179-84, is an excellent discussion, though it overstates the differences between Matthean and Pauline communities. Both reacted against the pneumatic emphasis of Hellenistic Jewish Christianity in similar ways; cf. Georgi, *Opponents*, pp. 165, 171 (on Matthew), 273, 276, 316 (on Paul).

Another factor in the background was Roman suspicion of the occult arts. As early as the fifth century BC, laws had been passed that labeled all harmful acts of magic criminal; Alan F. Segal, 'Hellenistic Magic', in R. van den Broek and M.J. Vermaseren, eds., *Studies in Gnosticism and Hellenistic Religions* (Leiden: Brill, 1981), pp. 356ff. Both Jews and Christians had to differentiate their healing arts (esp. those of Moses and Jesus) from those of the magicians.

[204]Thus the biblical critique of pagan magic; Clinton E. Arnold, *Ephesians: Power and Magic* (Cambridge: Cambridge Univ. Press, 1989), pp. 99f. Contrast the attitude of Aristotle: 'if [a man] does anything for his own sake or for the sake of his friends or with a view to [personal] excellence, the action will not appear illiberal; but if done for the sake of others, the very same action will be thought menial and servile'; *Politics* VIII.2.1337b.18ff.

writings of Irenaeus and Pseudo-Clement.

The first stream, that of the apocryphal books of acts, stressed the healing power of the apostles and related it to their faith in a God who created all things. Of particular interest is a document known as the Third Epistle to the Corinthians (Acts of Paul 8:1-3), which dates from the late second century and was later included in the Syriac and Armenian canons of the New Testament.[205] In the context, Paul was preaching against Gnostics, who denied that the body was God's creation and that there would be a resurrection. Claiming that God had created the body, together with all things, and would not forsake his creation when it was lost, Paul explained that Christ had come in the flesh, had died, and had been raised in the flesh in order to secure the salvation of all flesh.

In the sequel (preserved in the Coptic Heidelberg Papyrus), Paul and a young woman named Frontina were thrown down into a pit and Frontina was killed. When Paul saw the grief of Frontina's mother, he prayed to the Lord Jesus Christ for deliverance. Thereupon, Frontina was restored to life, and the crowd cried out with one voice: 'One is God, who has made heaven and earth, who has given life to the daughter....'[206]

The story is, of course, apocryphal, but it gives us an idea of the healing ministry of the church as it was understood in the second century and shows the close connection between Christian faith and practise. Belief in creation meant the possibility of healing and restoration to life. Similar incidents can be found in the Acts of Peter (late second century), the Acts of John, and the Acts of Thomas (both of the third century), though the connection of healing power with the doctrine of creation is not made so explicit.

The second stream of thought in the second century is found in the major Christian apologists: Justin, Athenagoras, and Theophilus. Since these writings are apologetic rather than narrative in character, we would not expect to encounter the same emphasis on the mighty deeds of the apostles that we found in the Apocryphal Acts. We do find arguments for the future resurrection based on a belief in God as creator (cf. 2 Macc. 7:28; Rom. 4:17; Heb. 11:3, 19), but these are not clearly related to the present healing ministry of the church.[207]

More significantly, we find polemical arguments against Greek philosophy and science to the effect that the philosophers were primarily motivated by a desire for personal fame, in contrast to the prophets and

---

[205] W. Schneemelcher, 'Acts of Paul', *New Testament Apocrypha*, vol. II, pp. 326, 351.
[206] Acts of Paul viii, ad fin. (R. McL. Wilson, trans., *New Testament Apocrypha*, vol. II, p. 378).
[207] E.g., Justin, *First Apology* 10, 18f.; *Dialogue with Trypho* 69; Athenagoras (attr.), *On the Resurrection* 3; Theophilus, *To Autolycus* I.8.

apostles who served the people without any expectation of reward.[208] Here, for the first time, the Judeo-Christian concern about the misuse of power was developed into an effective critique of the privileged status of intellectuals in the classical world. There was no question of Justin or the other apologists being anti-intellectual themselves, for they regarded Greek philosophy highly and borrowed from it freely. Yet they were also concerned about the uses to which philosophy was put, and thus they introduced an element of social assessment that has played an important role in Western thought ever since.[209]

4 Two Streams Rejoined: Irenaeus and Pseudo-Clement

The two streams came back together in the late second century in the work of Irenaeus of Lyons. In his treatise *Against Heresies*, directed primarily against the Gnostics, Irenaeus argued that the possibility of resurrection followed from the fact of creation, and also that the possibility of healing followed from the fact that the Creator had become flesh and performed healings when his handiwork had become impaired. Christian ministers continued to perform healings by calling on the name of the Lord Jesus Christ who had made all things in the beginning. Indeed, the reason the Gnostics were unable to raise the dead was that they did not believe in either the creation or the resurrection of the body.[210] Irenaeus's thought here is identical to that of the Aprocryphal Acts noted above.

But Irenaeus did not reduce the controversy with the Gnostics to a contest of power as the Apocryphal Acts tended to do: like the Apologists, he also dealt with the matter of motivation and its effect on the persons being healed. Simon Magus and his followers, he argued, would perform miracles

---

[208]E.g., Justin, *Dialogue* 7.1 (cf. note 9 above on the translation); Theophilus, *To Autolycus* III.2; Tertullian, *Apology* 36.

[209]On the importance of unselfish ends in the progress of science in the West, see Zilsel, 'Genesis of the Concept of Scientific Progress', pp. 325-49.

It is true that Hellenistic schools like the Stoics and Cynics had championed the ideal of taking their philosophy to the people, and this shift from an aristocratic to a more democratic style of philosophy may well have influenced Jewish and Christian practise; Robert L. Wilken, 'Toward a Social Interpretation of Early Christian Apologetics', *The Charismatic Figure as Miracle Worker* (Missoula: Univ. of Montana Press, 1972), pp. 73-77. However, the criticisms of these schools by writers of the second to fourth centuries, pagan and Christian alike, suggest that many of those who were in need of philosophical guidance received little or no help from the popular healers and teachers of their day: Pliny, *Natural History* XXVI.9; XXIX.5, 7 (note the anti-Greek bias and antiprofiteering rhetoric here; Howard Clark Kee, *Medicine, Miracle and Magic in New Testament Times* [Cambridge: Cambridge Univ. Press, 1986], pp. 5ff.); Lucian, *Menippus* 4ff.; Justin, *Dialogue* 2; Pseudo-Clement, *Homilies* I.3; IV.9; *Recognitions* I.3ff.; Porphyry, *Life of Plotinus* 3; Lactantius, *Divine Institutes* III. 15, 25ff.; Libanius, *Letter to Basil* (among the *Letters* of Basil, no. 338).

[210]Irenaeus, *Against Heresies* V.3.2f.; 12.6; II.31.2; 32.5.

for personal reward, not for the well-being of those whom they treated. The Christians, on the other hand, took no payment for their services but gave what little they had to the needy and, like Jesus, were exclusively concerned with the welfare of others:[211]

> It is not possible to name the number of the gifts which the Church throughout the whole world has received from God in the name of Jesus Christ...and which she exerts day by day for the benefit of the Gentiles, neither practicing deception upon any, nor taking any reward....Nor does she perform anything by means of angelic invocations... but, directing her prayers to the Lord, who made all things, in a pure, sincere, and straightforward spirit...she has been accustomed to work miracles for the advantage of mankind, and not to lead them into error.

Whether the church of the second century entirely lived up to Irenaeus's standard or not, the ideal itself was clear, and, given the recurrence of popular mistrust toward science and technology through the centuries, it was an ideal without which the social support needed for the development of modern science in the West could not have emerged. However, Irenaeus could not have claimed the ideal with any credibility unless the church of his time at least approximated its realisation.[212] Both ideal and example, then, were to have an important effect on the development of Western civilisation.

The combination of the two streams, emphasis on the possibiliy of healing and the criterion of human benefit, is also found in writings of the third century, like those of Pseudo-Clement.[213] Being a dramatic narrative with extensive dialogue based on the alleged teachings of Peter, the Pseudo-Clementine *Homilies* could bring together the narrative style of the Apocryphal Acts and the theological arguments of the Apologists even more forcefully than Irenaeus did. Thus we find Peter prefacing the healing of a crippled old woman with the words: 'If I be a herald of truth, in order to [confirm] the faith of the bystanders, that they may know that there is one God, who made the world, let her straightway rise whole.'[214] Conversely, when he was told that Simon could perform prodigies like making statues walk and flying through the air (both associated with technological feats in

---

[211] *Ibid.*, II.31f.; quote from 32.4f. (*ANF*, vol. I, p. 409). The apologetic value of Christian concern for the poor was also pressed in the Apocryphal Acts; e.g., Acts of Peter ii (*Actus Vercellenses*) 4.8; 6.17; 8.28; 9.30; cf. the canonical Acts 2:45; 3:6; 4:34f.; 8:18ff. The ideal of performing services without expectation of remuneration may reflect the Hippocratic policy of public physicians: A.R. Hands, *Charities and Social Aid in Greece and Rome* (Ithaca, N.Y.: Cornell Univ. Press, 1968), pp. 131ff., 204 (Document 67).

[212] See, e.g., 2 Clement 13 for the ideal and an exhortation to live up to it.

[213] Taking the first half of the third century as the probable date of the basic writing ('G'): after J. Irmscher, 'The Pseudo-Clementines', p. 533.

[214] Pseudo-Clement, *Homilies* XII.23 (*ANF*, vol. VIII, p. 297a); cf. *Recognitions* VII.23; *Homilies* VII.2; XIV.5.

the ancient world as today), Peter pointed out that such wonders were designed to astonish and deceive rather than to heal and to save. In contrast, '...the miracles of compassionate truth are philanthropic, such as you have heard that the Lord [Jesus] did, and that I after him accomplish by my prayers....'[215]

5 Early Monastic Medicine

Enough has been said to show how the New Testament beliefs and values were translated during the second and third centuries into an effective ministry of healing and restoration that was seen as a viable alternative to the claims of Greek science and technology. All of this took place while Christians were a persecuted minority in a pagan empire. The following centuries saw a distinct shift in outlook, though not in the underlying faith, as a new generation of Christian leaders attempted to meet the challenges that came with religious recognition and responsibility for secular affairs. Here again the key figure was Basil of Caesarea, but the larger movement in which Basil participated was the rise of cenobite (communal) monasticism.

The fourth century saw the rise of the monastic movement and the beginnings of public hospital care, two movements which were to be closely related throughout the Middle Ages, even though we do not generally associate them together today. In fact, it was the early cenobite communities that were responsible for converting the miracle-based healing ministry of the post-apostolic period into a systematic program of health care that could be made available on a regular basis. Some modern writers like Leslie Weatherhead have argued that this transition entailed a weakening of Christian faith and a capitulation to Greek science.[216] Undoubtedly, a new source of tension was introduced into the Church, and, of course, the development of modern medicine does suggest a gradual dilution of the specifically religious element, but the transition of the fourth century itself was a legitimate extension of early Christian faith and practice into new areas of responsibility, as we shall see.

The problems of maintaining a minimal degree of health care came to the fore in the first cenobite communities organised by Pachomius in Upper Egypt, where the rigours of asceticism were monitored and kept within strict

---

[215]*Homilies* II.32ff.; quote from 34 (*ANF*, vol. VIII, p. 235b); *Recognitions* III.60; X.66; cf. Origen, *Against Celsus* I.68; II.50f. The charge of being more interested in personal gain than in either finding the truth or in benefitting others is also leveled against the Greek philosophers in *Homilies* I.3; IV.9; cf. *Recognitions* I.2; X.50. About the same time (260s), Plotinus charged that the Gnostics only boasted of being able to expel spirits associated with diseases in order to 'enhance their [own] importance with the crowd'; *Enneads* II.ix.14.

[216]Leslie D. Weatherhead, *Psychology, Religion and Healing* (London: Hodder and Stoughton, 1951), pp. 78-83.

limits. Illness was a frequent occurrence and required special allowances for diet and for rest.[217] Then, too, people would come from miles around to be healed of their various infirmities. In such cases, neither Pachomius nor his chief disciple, Theodore, claimed any special powers, but they affirmed their faith in the possibility of healing, as the early Christians had, on the basis of the fact that God was the creator of all things.[218] This monastic ministry was an important component in the background of Basil.

6 Basil's Ministry of Medical Care

Basil was uniquely suited to promote a synthesis of this simple Christian faith with Greek medicine at this juncture in history for three reasons. First, he had received his early training under Libanius (at Antioch or Constantinople, c. 349).[219] Libanius was a classical rhetorician and undoubtedly helped imbue Basil with the late Stoic ideal of philanthropy as an ordinance of God's providence or natural law, essential to the welfare of civilisation.[220] The Stoic concept of philanthropy tied in nicely with the Judeo-Christian ideal of selfless service and provided Basil with a simple conceptual scheme for applying the latter to the needs of fourth-century Greco-Roman society.[221]

Secondly, Basil studied the classical sciences (astronomy, geometry, and arithemtic) at Athens (c. 350-55) and took a special interest in the art of medicine. According to his companion, Gregory of Nazianzus, he had a sufficiently good grasp of the mathematical disciplines 'not to be baffled by those who are clever in such sciences', yet he refused to devote himself to them entirely since they did not contribute to spiritual development. Medicine, on the other hand, was useful to Basil in dealing with his own delicate condition, and in serving the monks under his care in later years. Accordingly, 'he attained to a mastery of the art, not only in its empirical

---

[217]*Life of Pachomius,* trans. Apostolos N. Athanassakis (First Greek) 24, 28, 51, 53, 64, 95 (Missoula, Mont.: Scholars Press, 1975). There may have been precedent for monastic medicine in the Jewish communities of the Essenes and Therapeutae; Josephus, *Jewish Wars* II.vii.6; Philo, *That Every Good Man is Free* 87 (on the Essenes); *On the Contemplative Life* 2; *Hypothetica* II.11ff. (Therapeutae); Kee, *Medicine,* pp. 23, 25.

[218]*Ibid.*, 43ff., 133; cf. 4, on Pachomius's own first experience of Christian charity, and 48, 96, on his belief in the creation and possible resurrection of the human body.

[219]Fedwick, *Basil,* pp. 5, 180. Gregory of Nazianzus and John Chrysostom also studied rhetoric under Libanius, the latter at Antioch.

[220]Hubert Martin, Jr., 'The Concept of *Philanthropia* in Plutarch's *Lives*', *American Journal of Philology,* vol. lxxxii (1961), pp. 174f.; Jacob Viner, *The Role of Providence in the Social Order* (Princeton: Princeton Univ. Press, 1972), pp. 36f.; M. Hengel, *Property and Riches in the Early Church* (London: SCM Press, 1974), chap. 1.

[221]E.g., in his appeal to the merchants of Caesarea to open their stores of grain to the poor during the great famine of 369; see *Homily* VI.3, 6; cf. Gregory of Nazianzus, *Oration* XLIII.34ff., 63.

and practical branches, but also in its theory and principles'.²²²

Thirdly, Basil had gained firsthand acquaintance with the monastic way of life during a tour he made (356-57) of Syria, Mesopotamia, Palestine, and Egypt following in the steps of Eustathius of Sebaste.²²³ Here he came into contact with the cenobite ideal of mutual service and the ministry of helping and healing in the surrounding community. These ideals were later reflected in the monastic rules Basil drew up for the communities under his own supervision in the Pontus and at Caesarea.²²⁴ Eustathius had founded a hostel for the poor in his diocese of Sebaste in Armenia, and Basil himself organised a small network of hospitals under his care in the vicinity of Caesarea.²²⁵

Our principal source of information about the work of Basil at Caesarea is Gregory of Nazianzus's panegyric (*Oration* XLIII) written about two years after Basil's death (AD 379). Gregory's testimony is invaluable to us not only because it tells us what Basil did, but because it gives us the shared understanding of the theological meaning of those actions. Since Basil and Gregory worked so closely together, we may take Gregory's comments as if they were Basil's own: as Gregory puts it, their friendship was so close that they were of one life and one nature like 'one soul, inhabiting two bodies'.²²⁶ The consistency of Gregory's comments with Basil's thinking is substantiated at various points by the text of Basil's recorded sermons.

There were two particularly notable instances, according to Gregory, in which Basil worked to 'benefit the people' of Caesarea: during the famine of 369 and in the construction of a public hospital a few years later.

6.1 Basil's relief work during the famine of 369

During the famine of 369 AD, Basil (then still a presbyter) appealed to the merchants of Caesarea to donate some of their stores of grain to feed the poor. Then he gathered as many of the victims of the famine as he could and, together with his monks, fed and cared for them 'imitating the ministry

---

²²²Gregory of Nazianzus, *Oration* XLIII.23, 61f.
²²³*Letters* 1 (to Eustathius); 223.2. On the possibility that Basil studied in one of Eustathius's schools as a child, see P.J. Fedwick, *The Church and the Charisma of Leadership in Basil of Caesarea* (Toronto: Pontifical Inst. of Medieval Studies, 1979), pp. 159f.
²²⁴*Longer Rules* VII; cf. Gregory of Nazianzus, *Oration* XLIII.34; E.F. Morrison, *St Basil and His Rule* (London: Oxford Univ. Press, 1912), chap. 13. On the hospice for cripples at Alexandria, see Palladius, *Lausiac History* VI.5; Cassian, *Conferences* XIV.4. Basil may well have visited this exemplary monastic institution.
²²⁵Epiphanius, *Against Heresies* III.1, 55 (75); Basil, *Letters* 142-44. On the possible influence of Eustathius on Basil, see Timothy S. Miller, *The Birth of the Hospital in the Byzantine Empire* (Baltimore: Johns Hopkins Univ. Press, 1985), pp. 78f., 85-88.
²²⁶Gregory of Nazianzus, *Oration* XLIII.14, 20. On Basil and Gregory's collaboration in theological matters, see *ibid.*, 69.

of Christ'.²²⁷

The theological rationale for this public ministry had two aspects, according to Gregory, corresponding to the two streams of thought we have traced from intertestamental Judaism to the fourth century. First, Basil believed in the possibility of radical change, a belief which, as we have seen, was based on the idea of God as creator of all things. But, whereas previous instances concentrated on the miraculous element in God's action of healing, Basil related this faith to the practise of charity he had learned from the monastic tradition. Gregory explained this development by noting three ways in which Basil's work could be compared to that of Moses, Elijah, and Jesus: (a) it was based on the same power of God (available through Christ); (b) it was inspired by the same faith; and (c) it had the same result of feeding the hungry.²²⁸ The only difference was that Moses, Elijah, and Jesus altered the nature of the elements themselves, as was proper for 'their time and its circumstances', whereas Basil had to sway the hearts and minds of the rich—no small feat in itself!

The second aspect of Basil's theological rationale was the Judeo-Christian ideal of selfless service. Gregory explains that unlike Joseph, who made grain available to the poor in Egypt in the hope of personal advancement (Gen. 41), Basil neither gained nor expected any personal profit from the venture: his services were entirely gratuitous.²²⁹ The point we wish to make here is that there was a continuity of ideals between the biblical ministry of miracles and Basil's ministry of social service in spite of the changes in circumstance and strategy. Those, like Basil, who ministered to the needy saw their service as a direct expression of their Christian faith.²³⁰

---

²²⁷*Ibid.*, 35f.
²²⁸Gregory of Nazianzus on Basil:
> He indeed could neither rain bread from heaven by prayer, to nourish an escaped people in the wilderness, nor supply fountains of food without cost from the depth of vessels which are filled by being emptied...nor feed thousands of men with five loaves....These were the works of Moses, and Elijah, and my God [Christ], from whom they too derived their power. Perhaps also they were characteristic of their time and its circumstances: since signs are for unbelievers not for those who believe. But he did devise and execute with the same faith things which correspond to them, and tend in the same direction. For by his word and advice he opened the stores of those who possessed them, and so, according to the Scripture dealt food to the hungry... (Oration XLIII.35; *NPNF II*, vol. VII, p. 407).

The examples of Moses and Elijah had been cited by Basil, *Homily* VIII.5f. (Delivered During the Time of Famine and Drought).

²²⁹*Ibid.*, 36. The example of Joseph had also been used by Basil, *Homily* VI.2.
²³⁰See, e.g., John Chrysostom, *Homily* XIV.11 (*NPNF*, vol. XI, p. 451), who draws a striking parallel between the Christian's 'tending [the poor] and setting them upright' and the creative work of God who 'brought the poor from not being into being' in the first place. According to Palladius, Chrysostom contributed to the founding of two hospitals in Constantinople (*Life of Chrysostom* V).

6.2 Basil's public hospital.

A few years later (c. 372), after he had been appointed bishop of Caesarea, Basil sensed the need for a more permanent form of ministry to the needy. Again using funds solicited from the wealthy, he founded what modern scholars regard as history's first hospital (or infirmary) open to the public on a regular basis.[231] The significance of this event for the historical development of medicine has been pointed out by others.[232] What needs to be stressed here is that the same twofold faith we have been tracing as part of the creationist tradition was again cited by Gregory as the theological basis for Basil's action: (a) as in the case of the famine of 369, he viewed his medical care as the functional equivalent of the healing miracles Jesus had been able to perform by a mere word; and (b) he founded the hospital for the support of the poor rather than for personal fame.[233] And again the shift from miracle to method in medicine was viewed as an appropriate response to changing conditions and new responsibilities. Basil still believed in the possibility of miraculous cures, and he was credited with having, on more than one occasion, been granted such a miracle in response to his prayers. Whether healing came with or without visible means, God was the author of the healing power, so faith and prayer were essential to the process in any case.[234]

---

[231] For a description of Basil's 'hospital' and rehabilitation program, cf. his *Letter* 94. According to Timothy S. Miller, Basil's institution was a legitimate hospital, even though it treated only the very poor (including refugees from the countryside) and was staffed by monks rather than professional physicians. Basil also hired professional physicians and nurses on a semipermanent basis; 'Byzantine Hospitals', *Dumbarton Oaks Papers*, no. 38 (1984), pp. 54ff., and note the shift from the less committal evaluation of Basil's hospital in Miller, 'The Knights of St John and the Hospitals of the Latin West', *Speculum*, vol. liii (1978), pp. 723f.

The Romans had private infirmaries for their slaves and military hospitals for their soldiers, but...'For the origins of care for the sick in public institutions, one looks to the newly arrived Christianity'; John Scarborough, *Roman Medicine* (Ithaca, N.Y.: Cornell Univ. Press, 1969), p. 77; cf. E.D. Phillips, *Aspects of Greek Medicine* (London: Thames and Hudson, 1973), pp. 193f.; George E. Gask and John Todd, 'The Origin of Hospitals,' in E. Ashworth Underwood, ed., *Science, Medicine and History*, vol. I (London: Oxford Univ. Press, 1953), pp. 122-28; Timothy S. Miller, *Birth*, chap. 3.

There is also evidence for the existence of hospitals or infirmaries in India that are roughly contemporary with that of Basil (fourth-fifth cent. AD in Pataliputra/Patna). In this case, the socioreligious base was Buddhist monasticism; Kenneth G. Zysk, *Asceticism and Healing in Ancient India* (New York: Oxford Univ. Press, 1991), pp. 44f.

[232] '...if Medicine, like the Church, had any way of canonizing its most worthy sons, it would surely number Basil among its elect'; Gask and Todd, 'The Origin of Hospitals', p. 128.

[233] Gregory of Nazianzus, *Oration* XLIII.63. Basil regarded all of the productive arts as participating in the creative power of God (*Hexaemeron* I.7).

[234] *Ibid.*, 54f.; cf. Basil, *Longer Rules* LV.2,5. The recognition that God's creative power could work in conjunction with the use of physical therapy had ample precedent in intertestamental literature (Wisd. 16:12; Ecclus. 38:1-15) and in the church's practise of anointing with oil (*Apostolic Tradition of Hippolytus* 5; *Sacramentary of Serapion* 5, 15, 17; *Apostolic Constitutions* VIII.29).

We have followed the progress of the New Testament belief in the possibility of radical change and the ideal of selfless service in the ministry of healing and restoration up to the pivotal work of Basil of Caesarea. We must now briefly trace the tradition of medical care up to the twelfth century and then say a few words about the cultivation and theological interpretation of the other manual arts and the significance of all this for the development of modern Western science.

7 Early Medieval Medicine: Three Lines of Transmission

There is evidence for a growing tradition of medical care in Syria and Mesopotamia following the time of Basil.[235] During a famine in Edessa in 372-73, Ephraim of Syria appealed to the rich for funds and, following Basil's example, established an infirmary with some 300 beds which was made open to the public, native and foreigner alike.[236] Subsequently, in the fifth century, Edessa became a respected medical centre with some teaching facilities. When the Byzantine emperor Zeno closed the school of Edessa in 489, many of its faculty moved to Nisibis in Sassanid Persia and helped establish that city as a major centre of learning. In the mid-sixth century, the Sassanid ruler, Khusro I (r 531-79), tried to attract Christian scholars to the capital city of Gunde-Shahpur (Arabic, Jundi-Shapur). Later records indicate that there was a hospital in or near Gunde-Shahpur, and that medical training was also available. In the late eighth and early ninth centuries, Syrian Christian physicians were employed in turn by the Abbasid Moslem caliphs for the translation of Greek and Syriac scientific treatises into Arabic. Christian physicians also assisted in the founding of a hospital and two medical schools in Baghdad. Thus the Syrian medical tradition, dating back to the examples of Basil and Ephraim, was one of the foundations of Islamic medical science and thereby contributed to the beginnings of medieval Western medicine through the translations of Constantine the African (*c.* 1065 to *c.* 1087) and the school of Salerno in southern Italy in the eleventh century.[237]

---

[235]See, e.g., J.B. Segal, *Edessa, 'The Blessed City'* (Oxford: Clarendon Press, 1970), pp. 71, 148; W. Stewart McCullough, *A Short History of Syriac Christianity to the Rise of Islam* (Chico: Scholars Press, 1982), pp. 168f.

[236]Sozomen, Ecclesiastical History III.16. On the merit of the tradition that Ephraim had visited Basil at Caesarea (*c.* 371), see Segal, *Edessa*, p. 71n.; McCullough, *Short History*, p. 58.

[237]See, e.g., De Lacy O'Leary, *How Greek Science Passed to the Arabs* (London: Routledge and Kegan Paul, 1949; 1979); Peters, *Aristotle and the Arabs*; S. H. Nasr, *Islamic Science* (London: World of Islam Festival, 1976), chap. 8 and further references cited there, pp. 155n., 172n; and, for a more critical review of the evidence, Michael W. Dols, 'The Origins of the Islamic Hospital: Myth and Reality', *BHM*, vol. lxvi (1987), pp. 367-90.

The Salerno tradition of medicine was introduced into France by William of Dijon, and from there into England by the Benedictine monks who followed in the wake of the Norman

Before considering the theological aspects of this development, we should note two other lines of transmission from Basil to the medieval West. First, in the eastern Roman (Byzantine) empire there was a succession of Christian physicians, beginning with Caesarius, (d c. 373), brother of Gregory of Nazianzus, who served the Byzantine court, and culminating in the early seventh century in the work of Alexander of Tralles (d 605) and Paul of Aegina (fl. Alexandria, c. 640). Both of the latter wrote medical treatises which were influential among the Arabs and the school of Salerno.[238]

The third line of transmission of medical knowledge was in the Latin West. Under the guidance of Gregory's friend and pupil, Jerome, wealthy Roman Christians contributed to the founding of public hospitals in the late fourth century. Augustine also encouraged the practice in northern Africa during the early fifth century.[239] A long succession of popes, prelates, and Christian princes could be added to this list.

Cassiodorus's efforts to found a medical facility at Cassiacum (southern Italy) in the sixth century are of special interest in that they were inspired by the example of the Syrian school at Nisibis. Cassiodorus advised his monks to study herbs and medicine and specified that they read the classical Greek medical works (in Latin translation). He thereby encouraged the development of medical practice within the monastic movement of western Europe that was just beginning its greatest expansion with the work of St Benedict.

The spread and development of medicine in western Europe was closely associated with Benedictine monasticism through the early Middle Ages. Constantine the African spent his last and most fruitful years (c. 1077-87) translating Greek and Arabic medical works from Arabic into Latin at Monte Cassino. At this point the Eastern (Greek, Syrian, Arabic) and Western (Latin) medical traditions came together and contributed to the medical science of the eleventh and twelfth centuries.[240]

invasion; Anne F. Dawtry, 'The *Modus medendi* and the Benedictine Order in Anglo-Norman England', in W. J. Shields, ed., *The Church and Healing* (Oxford: Basil Blackwell, 1982), pp. 28, 34.

On the Jewish contribution to Arabic medicine and its transmission to the Latin West, see I. Simon, 'Medieval Jewish Science', in René Taton, ed., *Ancient and Medieval Science* (London: Thames and Hudson, 1963), pp. 458-66.

[238]John P. Dolan and William N. Adams-Smith, *Health and Society* (New York: Seabury Press, 1978), pp. 56f.; Miller, *Birth*, pp. 168-71. For more detail on Alexander of Tralles and Paul of Aegina, see the articles in *DSB*. On Caesarius, see Gregory of Nazianzus, *Oration* VII, and, for the date of his death, see Fedwick, *Basil*, p. 13n. Gregory of Nazianzus also practised medicine (*Oration* XLIII.61) and must have influenced his brother's thinking, particularly with regard to the theological aspects of the discipline (cf. *Oration* VII.9f., 15).

[239]For references, see Gask and Todd, 'Origin of Hospitals', pp. 129f.

[240]See, e.g., Loren C. MacKinney, *Early Medieval Medicine* (Baltimore: Johns Hopkins Press, 1937), chaps. 1ff.; Dolan and Adams-Smith, *Health and Society*, chap. 4. The *Rule of St Benedict* (XXXI, XXXVI) made special provisions for the care of the sick: 'The medical achievements of the medieval clergy were destined to be a combination of the Benedictine ideal of brotherly service with Cassiodorus' plan for an intelligent comprehension of

Like any broad historical survey, this sketch of three lines of transmission might take on a certain air of inevitability, especially since it leads up to the background of our own cultural milieu. However, there was nothing inevitable about any of the three lines of development we have just traced. Each of them depended on the peculiarities of local conditions and the motives and efforts of numerous individuals. Theological convictions would be just one of the factors involved at any given point. So a thorough investigation of the interactions between theology, social conditions, and the progress of medical science from the fourth to the twelfth century would require considerable additional research. Indeed, this is a fruitful area for those who may wish to pursue the subject and make a contribution to the field. Here we may note a few of the more salient points we know something about at present.

8 Early Medieval Medicine: The Role of Creation Theology

In the case of Basil and Gregory of Nazianzus, belief in the value of medical care was closely related to belief in creation and the ethic of selfless service, as we have seen. This close relation between faith and practise seems to have continued into the fifth and sixth centuries, but it is more difficult to trace after that, at least with regard to the Christian healing ministry associated with medicine.

Ephraim of Syria (d 373) was an admirer, perhaps an acquaintance, of Basil. His best-known work in theology was his hymn-writing campaign against the Gnostic disciples of Bardaisan in which he argued for the creation and future resurrection of the body. He is also reported to have performed healing miracles in the name of Christ, and his theological understanding of that ministry can be inferred from his defense of the inclusion of Third Corinthians (Acts of Paul viii.1-3) in the Syriac canon. If Ephraim was familiar with the account of Paul's raising Frontina, he may well have made the same association between the doctrine of creation and the ministry of healing found in the Acts of Paul.[241] In any case, the basic constellation of ideas with which he worked was the same as that from which the medical philosophy of Basil and Gregory had emerged.

Ephraim's influence on Syriac Christianity was great, but it is difficult to tell whether the significance of the healing ministry was still understood in theological terms after his time. In the fifth century, the exegetical and theological works of Theodore of Mopsuestia (d 428) introduced an emphasis on Aristotelian logic into the Nestorian schools of Syria and

---

medicine based on the study of classical texts'; MacKinney, *loc. cit.*, p. 52.

[241] John Gwynn, 'Introductory Dissertation', NPNF II, vol. XIII, pp. 129ff. On the use of the Acts of Paul at Edessa to combat the Bardaisanites, see Walter Bauer, *Orthodoxy and Heresy in Earliest Christianity* (Philadelphia: Fortress Press, 1971), pp. 39-43.

Mesopotamia, and the influence of Ephraim apparently diminished.242 The extant autobiographical fragments of Hunayn ibn Ishaq (d *c.* 877), the Nestorian physician who organised the first systematic translation of Greek philosophical treatises into Arabic, give no indication of any awareness of the theological motifs we have considered.243 However, it could also be that those motifs were taken for granted as common ground with Muslim antagonists and so did not call for special discussion as much as those beliefs peculiar to Christians which Hunayn so ably defended.244

In the West, the development was slightly different. Jerome had consulted with Gregory of Nazianzus in 381 while the lattter was bishop of Constantinople. The matters they discussed largely concerned the exegesis of difficult texts, but the issue of medical care as an expression of Christian service may also have come up.245 In any case, Jerome was familiar with the cenobite monasticism of Pachomius—he translated the Pachomian rules into Latin—and he would have learned of the healing ministry, if not of its theological legitimation, from that source. In his letters, he exhorted wealthy Roman Christians who were providing care for the poor to do so in an attitude of humility as Christ had done.246

The three principal founders of Western monasticism were John Cassian, Cassiodorus, and Benedict of Nursia. After making an extensive tour of the Egyptian monasteries (385-99) and returning to Marseilles, Cassian compiled his findings in Latin for the Western brethren (420-30). His treatment of the spiritual gift of healing recorded the fundamental Christian belief that people could be healed and even raised from the dead as a demonstration of the power and mercy of God (especially in contrast to the philosophical dialectics used by heretics) but emphasised the primary value of love, humility, and personal holiness.247 Cassian, however, did not

---

[242]O'Leary, *How Greek Science*, p. 52; Segal, *Edessa*, p. 150.

[243]Max Meyerhof, 'New Light on Hunain ibn Ishaq and His Period', *Isis*, vol. viii (1926), pp. 685-90; Rachid Haddad, 'Hunayn ibn Ishaq, apologiste chrétien', in Gerard Troupeau, ed., *Hunayn Ibn Ishaq* (Leiden: Brill, 1975), pp. 292-302.

[244]The Qur'an, written in the seventh century, reaffirmed the biblical faith in the creation and future resurrection of the human body (Surahs XIII.2-6; XXXII.4-10; XXXVI.76-82; XXXVII.11-17). According to the Hadith, Muhammed taught that God had provided a medicine or treatment for every possible human ailment; Fazlur Rahman, *Health and Medicine in the Islamic Tradition* (New York: Crossroad, 1989), p. 34. The Arabic version of the Hippocratic Oath likewise affirmed that God was the 'giver of health and creator of healing and every cure'; Franz Rosenthal, *The Classical Heritage in Islam* (Berkeley: Univ. of California Press, 1975), p. 183. So there was no conscious difference between Christians and Muslims at this point.

[245]On Jerome's visit to Gregory, see J.N.D. Kelley, *Jerome* (New York: Harper and Row, 1975), pp. 69ff.

[246]Jerome, *Letters* 66.11ff.; 77.6, 10. The classical ideal of charity was viewed as a means of gaining public recognition: Martin, 'The Concept of *Philanthropia*'; Hands, *Charities*, chaps. 3, 4.

[247]Cassian, *Conferences* XV.

emphasise medical care or ministry to the poor.

Cassiodorus (d c. 575) and St Benedict (d c. 550) both made special provisions for the care of the sick in their instructions to monks. Cassiodorus was influenced by Syriac Christianity and possibly also by the monastic rules of Basil. At any rate, his affinity with Basil is particularly striking in the moving paraphrase of Ecclesiasticus 38 in his *Introduction to Divine and Human Readings*:[248]

> Learn, therefore, the properties of herbs and perform the compounding of drugs punctiliously; but do not place your hope in herbs and do not trust health to human counsels. For although the art of medicine be found to be established by the Lord, he who without doubt grants life to men makes them sound [Ecclus. 38:4-8].

The essential connection between the trustworthiness of the creator ('he who without doubt grants life to men') and the rationale for medical care was appreciated by Cassiodorus in the same sense that it had been by Basil and Gregory of Nazianzus.

Gregory the Great (pope 590-604) was second to none in his advocacy of justice for the poor and his efforts to relieve their distress. However, Gregory's advice to clerics on admonishing the sick was to encourage resignation rather than to hold out any real hope of healing. Gregory regarded health as a gift from God that could be withdrawn but said little about its restoration.[249]

The study of medicine was touched on in the monastic-school curriculum of Isidore of Seville (d 636) and given more prominence in the Carolingian curriculum designed by Alcuin of York (d 804). Cassiodorus's understanding of its theological significance continued to have influence at least through the tenth century.[250] Alcuin, for example, described the trained physician as a servant of God and God as the creator of medicinal herbs, in the tradition of Ecclesiasticus 38.[251] But there was little development until the rise of the

---

[248] Cassiodorus, *Introduction to Divine and Human Readings* I.31 (trans. L.W. Jones, p. 135). Note also the emphasis on the idea of medicine as an expression of compassion and a form of Christian ministry ('I salute you, distinguished brothers, who...perform the functions of blessed piety...you who are sad at the sufferings of others, sorrowful for those who are in danger, grieved at the pain of those who are received so that...you help the sick with genuine zeal'). Cf. Basil, *Longer Rules* LV. 2, 5, and cf. both with Ecclus. 38:1-14. Cassiodorus spent some time in Constantinople and would probably have had access to Basil's rules in the original Greek. On the Syriac connection, see O'Donnell, *Cassiodorus*, pp. 133f., 182.

[249] Gregory the Great, *Pastoral Rule* III.12.

[250] MacKinney, *Early Medieval Medicine*, notes on pp. 187, 203; 'Medical Ethics and Etiquette in the Early Middle Ages', *BHM* vol. xxvi (Jan. 1952), pp. 5-11.

[251] Alcuin stated this in a letter of around 801; E. Dummler, trans., *Monumenta Germaniae Historica: Epistolae Karolini Aevi*, vol. ii (Berlin, 1985), pp. 356-57, cited in Valerie J. Flint, 'The Early Medieval "Medicus", The Saint—and the Enchanter', *Social History of Medicine*, vol. ii (1989), p. 133.

cathedral school of Chartres in the late tenth century.

9 The Scholastic Reinterpretation: Earthly versus Heavenly Medicine

Fulbert (d 1028) became principal of the school of Chartres around 1005 and established its general direction. As one of the early champions of Aristotelian dialectic, Fulbert differentiated two distinct kinds of medicine: earthly, natural medicine, which was based on the use of herbs (accompanied by prayer to Christ for healing); and heavenly or supernatural medicine based on the miraculous powers of Christ and the saints:[252]

> As Christians, we know that there are two kinds of medicine: one of earthly things; the other of heavenly things. They differ in both origin and efficacy. Through long experience, earthly doctors learn the powers of herbs and the like, which alter the condition of human bodies....The author of heavenly medicine, however, is Christ, who could heal the sick with a command and raise the dead from the grave.

Fulbert's dichotomy of two types of medicine was closely related to the scholastic differentiation of two powers of God, *potentia ordinata* and *potentia absoluta*, which was also coming into general use in the eleventh century (sec. D). Whereas Basil, Gregory of Nazianzus, and Cassiodorus had seen the creative power of God realised in the properties of herbs and the skills of the physician, the medieval West began to view the two as antithetical.[253]

As in the case of the two powers of God, the dichotomisation of the idea of medicine had an institutional parallel: the development of professional secular medicine and spiritual care as separate disciplines. Practising physicians relied increasingly on the processes of nature as something quite distinct from the immediate creative activity of God[254] (though the older,

---

[252]Fulbert of Chartres, *Hymn in Honour of St Pantaleon* (*MPL* 151:341); trans. Katherine Park, 'Medicine and Society in Medieval Europe, 500-1500', in Andrew Wear, ed., *Medicine in Society: Historical Essays* (Cambridge: Cambridge Univ. Press, 1992), p. 64; cf. MacKinney, *Early Medieval Medicine*, p. 134. Since Fulbert advocated that the use of ointments be accompanied by prayer to Christ for healing (*Letter* 9), the dichotomy was not as clear for him as it was for William of Conches in the following century (see the previous section, esp. note 185).

[253]Katherine Park ('Medicine and Society', p. 64) claims that the early Christian apologists had contrasted earthly and heavenly medicine, but she offers no examples to support her claim. Our study has indicated that ante-Nicene theologians like Pseudo-Clement and Irenaeus emphasised miraculous healing and ignored secular medicine, not that they explicitly opposed the two. An informal differentiation between the two types of medicine was made by Arnobius (*Against the Heathen* I.48), Basil (*Longer Rules* LV), and Augustine (*On Christian Doctrine* IV.xvi.33), among others. In each case, however, it is clear that God is equally the author of both types. See D.W. Amundsen, 'Medicine and Faith in Early Christianity' *BHM*, vol. lvi (1982), pp. 335, 338f., 341f., 348.

[254]Rashi (1040-1105) commented on the tendency of (Jewish?) physicians to disavow reliance on God already in the eleventh century; Byron L. Sherwin, 'In Partnership with God:

less differentiated tradition also persisted through the seventeenth century). At the same time, as part of the Gregorian reform of the church (*c.* 1075-1122), clergy ordained to the major orders were officially discouraged from practising surgery or attending the sick except as spiritual directors.[255]

The rapid influx of new medical ideas from the Arab world during the eleventh and twelfth centuries accelerated this trend toward a dichotomisation of health care. Although there may have been a strong faith motivation behind Islamic medicine,[256] that faith was perceived as foreign by Europeans and was left behind as scientific and philosophical texts were translated and adapted to Western needs.

10 The Twelfth-Century Assessment of Technology: Hugh of St Victor

At the same time that clergy were defining themselves in opposition to professional physicians, however, the monasteries were taking an increasing interest in the development of water-powered technology, a mechanical art no less 'earthly' than medicine itself. As an example: when Arnold of Bonneval described the rebuilding of Clairvaux (1136), he was so enthusiastic about the new water-powered machinery that he neglected to mention the church! Another monastic writer adopted a more theological perspective when he thanked God that the new machines could alleviate the

---

Health, Healing and Jewish Tradition', unpublished ms., 1986. Similarly, Bernard of Clairvaux protested against the naturalistic medicine of the 'school of Galen' and argued that there were better cures in the 'school of Christ'; Dawtry, *Modus medendi*, p. 35.

Oswei Temkin argues that Hippocratic medicine was inherently naturalistic and that, as a science, it was never completely assimilated by Christianity. In other words, Hippocratic physicians of the Middle Ages had always kept their science and their religion quite separate; *Hippocrates in a World of Pagans and Christians* (Baltimore: Johns Hopkins Univ. Press, 1991), pp. 177, 250, 253-56. On this reading, the dichotomisation of the twelfth century was partly the resurfacing of an ancient Greek tension. But the Hippocratic tradition, with its emphasis on the theoretical understanding of the root causes of disease, was just one of the various streams in Western medicine. The empirical and methodist traditions were also important; Danielle Jacquart, 'The Introduction of Arabic Medicine into the West: The Question of Etiology', in Sheila Campbell et al., eds., *Health, Disease and Healing in Medieval Culture* (New York: St Martin's Press, 1992), pp. 187f. The empirical was to reassert itself in the fourteenth century with the development of medical alchemy and to culminate in the work of the Paracelsians.

[255]The most important conciliar decrees regarding the practise of medicine were those of the Second Lateran Council (1139) and the Fourth Lateran Council (1215). See Darrell W. Amundsen, 'Medieval Canon Law on Medical and Surgical Practice by the Clergy', *BHM*, vol. lii (Spring 1978), pp. 22-44; D.W. Amundsen and G.B. Ferngren, 'Medicine and Religion: Early Christianity through the Middle Ages', in M. E. Marty and K. L. Vaux, eds., *Health/Medicine and the Faith Traditions* (Philadelphia: Fortress Press, 1982), pp. 117f.; R. L. Numbers and R. C. Sawyer, 'Medicine and Christianity in the Modern World', *ibid.*, p. 140; Stanley Rubin, *Medieval English Medicine* (London: David & Charles, 1974), pp. 183-95.

The Fourth Lateran Council specifically prohibited higher clergy from performing surgery or the shedding of blood. It should be noted that the Hippocratic oath itself forbade the practise of surgery by physicians; L. Edelstein, *Ancient Medicine* (Baltimore: Johns Hopkins Press, 1967), pp. 26-33.

[256]So Rahman, *Health*, pp. 45-49.

oppressive labours of both humans and animals (a classical motif).257

The idea that various technologies or mechanical arts were God-given means of alleviating human suffering was popularised by Hugh of St Victor's overview of the arts and sciences (the *Didascalicon*) in the 1120s. Each branch of philosophy, Hugh argued, has one of two ends: 'either the restoring of our nature's integrity or the relieving of those weaknesses to which our present life lies subject'.258 The restoration of the divine likeness in human nature was of eternal consequence. It was accomplished by the theoretical sciences (theology, mathematics, and physics), which remedied human ignorance by promoting the contemplation of the truth, and also by the practical disciplines (ethics, economics, and politics), which remedied human vice by promoting the practise of virtue. The relief of human weakness, on the other hand, was accomplished by mechanical arts like fabric making, armament, commerce, agriculture, hunting, medicine, and theatre; the mechanical arts ministered to the necessities of this life rather than the next.259

So by the twelfth century technology had come to be viewed in the West as a positive, potentially liberating force in society, much as it had been in the early days of the Israelite kings, over 2,000 years earlier. At the same time, it began to be understood as a purely human endeavour, based on an understanding of the workings of nature, rather than a sacred one, related to the creative and recreative work of God.260 We have already traced the

---

[257]Lynn White, Jr., *Medieval Religion and Technology* (Berkeley: Univ. of California Press, 1978), p. 245. Cf. Palladius, *De re rustica* I.41 for the classical motif of alleviating the toil of both humans and beasts.

[258]Hugh of St Victor, *Didascalicon* I.5, trans. Jerome Taylor (New York: Columbia Univ. Press, 1961, pp. 51f.).

[259]*Ibid.*, I.8; II.1; cf. the scheme in Hugh's *Epitome of Philosophy* (see J. Taylor, 'Introduction', *Didascalicon*, pp. 10ff.; appendix A, pp. 152ff.). The list of the mechanical arts is given in *Didascalion* II.20-27.

Cassiodorus added mechanics, medicine, and architecture to the quadrivium. Isidore of Seville included mechanics, medicine, and astrology, along with the traditional quadrivium, among the seven branches of philosophy; Whitney, *Paradise Restored*, pp. 59-70. The Latin term, *artes mechanicae*, was first used to portray the manual arts in parallel to the prestigious liberal arts by John Scotus Erigena in the ninth century. It became the standard term for the crafts in general (including medicine, agriculture, and architecture, as well as mechanics) in the twelfth and thirteenth centuries; Peter Sternagel, *Die Artes mechanicae im Mittelalter* (Kallmung über Regensburg: Lassleben, 1966), pp. 13-17, 30-36, 77f.; Whitney, *Paradise Restored*, pp. 18, 27 note 17, 70ff.

[260]George Ovitt, Jr., speaks of a 'secularisation of labour'; *The Restoration of Perfection* (New Brunswick, N.J.: Rutgers Univ. Press, 1987), pp. 138, 163, 201f. Elspeth Whitney points out that Ovitt's emphasis on the devaluation of manual labour focuses on monastic writers. In contrast, Whitney's own study, which focuses on academic writers like Hugh of St Victor, traces the emergence of the mechanical arts as a legitimate form of philosophy (*Paradise Restored*, pp. 148f., but cf. pp. 79, 124) On balance, then, the mechanical arts were elevated at the expense of unskilled manual labour. Whitney agrees, however, that the rationale for the mechanical arts became more secular, particularly under the impact of Aristotelian-Arabic science, in the thirteenth century (*Paradise Restored*, pp. 115f., 120, 134,

development of this characteristically Western way of thought in the special case of medicine. A few notes will suffice to bring us up to date with regard to the other mechanical arts.

11 Background of the Twelfth-Century Assessment

We recall that technology was viewed mostly negatively by the Jews in the late Old Testament and intertestamental periods, largely because it was associated with idolatry and the oppressive power of foreign nations like the Babylonians, Greeks, and Romans. This critical assessment continued for some time, particularly in the writings of Western culture critics like Arnobius and Augustine.[261] At the same time, writers like Methodius and Lactantius began to speak of human fabrication as analagous to, or even participating in, the all-creative power of God.[262]

Basil employed several popular Platonic and Stoic themes in developing the theological meaning of the mechanical arts in his sermons on the Hexaemeron. Addressing an audience of artisans on their way to and from work, he described architecture, working in wood and brass, and weaving, and compared the products of these 'creative arts' to the world that had been fashioned by the 'supreme Artisan'.[263] The arts, Basil explained, were given to humans by God after the fall of Adam to alleviate the harshness of nature.[264] In providing a theological meaning to the mechanical arts, he anticipated Hugh of St Victor's interpretation by more than 700 years.

The two principal founders of Western monasticism, Cassiodorus and Benedict of Nursia (in the sixth century), both placed a high value on the machines and tools used by their monks.[265] Like many of the monastic fathers before him, St Benedict required his followers to spend at least six

144).

[261] Arnobius, *Against the Heathen* II.17ff.; Augustine, *City of God* XXII.24.

[262] Methodius, *On Free Will* (*ANF*, vol. VI, pp. 359f.); Lactantius, *On the Anger of God* X (*ANF*, vol. VII, p. 267b).

[263] Basil, *Hexaemeron* I.7; cf. Plato, *Sophist* 265b-e. According to Elspeth Whitney, Augustine 'more rigorously distinguishes between the productive powers of man and God, consistently stressing the inferiority of human art and the inadequacy of the analogy'; *Paradise Restored*, p. 39. I have not been able to verify Whitney's point from her citations (nn. 68f.). In fact, Augustine speaks of medicine, agriculture, and navigation as arts that 'assist God in his operations' in *On Christian Doctrine* II.xxx.47. As he explains in *On Genesis Word for Word* (VIII.ix.17f.), the deeds of husbandmen and physicians manifest the external, 'voluntary working' of providence and assist the internal, 'natural working' of providence in the natural growth of trees and the natural forces of the human body. The list of arts that occasion Augustine's more ambivalent attitude in *City of God* XXII.24, as Whitney points out, includes the production of animal traps, poisons, weapons, and other machines of defense as well as agriculture, navigation, and medicine; *Paradise Restored*, p. 54.

[264] *Longer Rules* LV.1. For the Platonic and Stoic background of this idea, cf. Plato *Protagoras* 320d-322a; Origen, *Against Celsus* IV.76.

[265] Cassiodorus, *Introduction to Divine and Human Readings* I.30; *Rule of St Benedict* XXXI, XXXII.

hours a day in manual labour.[266]

Many historians today regard this monastic emphasis on work as having been a major stimulus to the development of Western technology, but there is some disagreement as to the exact reason why. Lynn White, Jr., has argued that the Benedictines placed a high value on manual labour itself, even making it the practical equivalent of prayer.[267] Jacques Le Goff, on the other hand, argues that work was viewed as a form of penance and that the monks welcomed any opportunity to reduce its burden so as to devote themselves more to prayer and to study.[268] From the ninth century on, at least, the recurring tendency was to turn the more menial forms of labour over to the lay brothers (*conversi*) and serfs and to reserve the choir monks for spiritual exercises,[269] a dichotomisation similar to that we have already seen in the case of medicine (earthly and heavenly, secular and spiritual). In any case, the end result of the monastic enterprise was the development of labour-saving machinery and the positive assessment of the potential of technology we find in the twelfth century. The theological views of Basil, Cassiodorus, and Benedict thus provided much of the basis for the Western appreciation of the value of technology as well as medicine.[270]

Like the idea of the relative autonomy of nature, the positive valuation of technology was promoted by the creationist tradition going back to the Old Testament and intertestamental periods. It originated in the belief that God had created all things and that out of compassion for his fallen creatures

---

[266]*Rule of St Benedict* XLVIII.11; cf. *Rule of the Four Fathers* X; *Second Rule of the Fathers* V; *Rule of Marcarius* XI in C.F. Franklin et al., trans., *Early Monastic Rules* (Collegeville, Minn.: Liturgical Press, 1982).

[267]L. White, Jr., *Medieval Religion*, pp. 182ff., 241ff., 319f. In his earliest essay on medieval technology, White also cited 'the implicit theological assumptions of the infinite worth of even the most degraded human personality' and 'an instinctive repugnance towards subjecting any man to a monotonous drudgery' as contributions towards the development of labour-saving machinery in the later Middle Ages; 'Technology and Invention in the Middle Ages' (1940), cited without further documentation in Whitney, *Paradise Restored*, p. 14 n. 53, but cf. White, 'Cultural Climates and Technological Advance in the Middle Ages', *Viator* vol. ii (1971), pp. 171-201.

[268]Jacques Le Goff, *Time, Work, and Culture in the Middle Ages* (Chicago: Univ. of Chicago Press, 1980), pp. 58-62, 80-84, 110f.; cf. Marc Bloch, *Land and Work in Medieval Europe* (London: Routledge and Kegan Paul, 1967), p. 152. According to Le Goff, a more positive view of labour developed in the twelfth and thirteenth centuries in response to the rising power of artisans, merchants, and labourers; *Time, Work, and Culture*, 62-70, 110ff., 115-21.

[269]Joan Evans, *Monastic Life at Cluny, 910-1157* (Oxford: Univ. Press, 1931; Hamden, Conn.: Archon Books, 1968), pp. 87f.; David Knowles, *Christian Monasticism* (New York: McGraw-Hill, 1969), pp. 39, 52, 69-74. In the twelfth and thirteenth centuries, even the Cistercians established a distinct order of *conversi*, and the notion was developed of the three orders of society, one of which was the *laborantes*; Ovitt, *Restoration*, pp. 144, 163, 199, 200.

[270]As Cassiodorus saw the practise of medicine as an expression of Christian charity, so Bernard of Clairvaux viewed manual labor (e.g., farming) in general; Ovitt, *Restoration*, p. 147, based on Brian Stock, 'Experience, Praxis, Work, and Planning in Bernard of Clairvaux', in J.E. Murdoch and E.D. Sylla, eds., *Cultural Context*, pp. 219-68.

he had poured out his creative Spirit on the followers of Jesus. It was also constrained by the social criterion that power should be used for the benefit of others rather than in personal self-interest. By assessing technology in this light and giving it a positive value, Christians provided it with a theological legitimation that it had not enjoyed in the ancient world, among either Greeks or Jews. Since the Middle Ages was a time when the laity was emerging as a self-conscious class in Western society,[271] technology was thus provided with a strong social base without which neither it nor modern science could have developed as they have in the last seven centuries.

Still the close association between Western Christianity and technology has to be qualified in at least two ways. First, the dichotomy of earthly and heavenly callings that evolved in the eleventh and twelfth centuries was foreign to the outlook of early Christians like Basil, who viewed medicine and the other arts as the expression of the creative power of God in Christian service. And, second, the underlying Christian commitment was to the healing and restoration of the human race in accordance with God's creative intention. Technology was to be assessed—and, if need be, reassessed—in terms of its overall contribution to that public ministry. In the eleventh and twelfth centuries, technology appeared to be a force for liberation both for an artisan class emerging from a feudal society and also for Western Europe as a whole emerging as a new power in a world dominated by other, more highly developed cultures. All of this was to change in succeeding centuries with the institutionalisation of technological power and the beginnings of Western imperialism.[272] The problem of identifying a distinctively Christian attitude toward the mechanical arts was, accordingly, to become more complex.

---

[271] Chenu, *Nature, Man, and Society*, pp. 219-30.

[272] The monopolisation of technology by landlords (including monks) began as early as the tenth century and was intensified in the thirteenth and fourteenth centuries; Crombie, *Augustine to Galileo*, vol. I, p. 206; Bloch, *Land and Work*, pp. 156ff.

CHAPTER TWO

THE MEDIEVAL CHURCH AND ARISTOTELIAN SCIENCE
(Thirteenth Century to the Fifteenth Century)

A. The Reception of Aristotelian Science

The history of medieval Western European science is largely the story of the translation, assimilation, and criticism of newly discovered Greek and Arabic texts, particularly the philosophical works of Aristotle and his commentators. We have already discussed the extensive influence of Greek science on intertestamental Jewish, early Christian, and early medieval thought. So a moment of reflection may be required to see why so much of Greek science[1] should have been perceived as something radically new in thirteenth-century Europe.

It is true, as we have seen, that the early church reflected a wide spectrum of Greek philosophic thought: Pythagorean, Platonic, Aristotelian, Epicurean, Stoic, Cynic, and Hippocratic ideas were all considered and

---

[1] The terms 'science' and 'philosophy' are used interchangeably in this chapter, as in the previous one, in keeping with the usage of the Greek natural philosophers and medieval scholastics generally. There were three main branches of theoretical science (*scientia*) in Aristotle: natural philosophy or physics, mathematics (arithmetic and geometry), and first philosophy or metaphysics. In addition, there were 'subalternate' sciences like optics that were more abstract than physics, yet were dependent on the theorems of mathematics; Sir David Ross, *Aristotle* (London: Methuen, 1923; 5th ed., 1949), pp. 20, 47, 156. In the Middle Ages, mathematics was already part of the liberal arts curriculum designed by Boethius. As a result, the theoretical sciences that received the greatest attention in the thirteenth century were physics and metaphysics. The incorporation of these new sciences into the arts curriculum gave rise to the university faculty of arts and sciences; James A. Weisheipl, 'The Nature, Scope, and Classification of the Sciences', in Lindberg, ed., *Science in the Middle Ages*, p. 476.

Andrew Cunningham argues that the term, 'science', should not be used to describe the 'history of science' prior to the late eighteenth century; 'How the *Principia* Got Its Name; or, Taking Natural Philosophy Seriously', *History of Science* vol. xxix (1991), pp. 381-89. He argues instead for use of the term, 'natural philosophy', as describing a premodern discipline that focused on God and his creation. Medieval philosophers did in fact use the term *scientia*, as Cunningham points out (p. 387), but for them *scientia* was a broader study (including metaphysics) than natural philosophy, not a narrower one as Cunningham supposes. Moreover, Cunningham neglects the medieval tradition (e.g., Adelard of Bath, William of Conches, Albertus Magnus, Jean Buridan) that stressed the autonomy of nature and made a clear distinction between matters of science and matters of faith. My point against Cunningham is that the modern separation of science and theology has roots in the Middle Ages even though the separation was not complete until the late nineteenth century. On the other hand, I agree with Cunningham's argument that premodern natural philosophy presupposed the work of God in creation; cf. his 'Getting the Game Right: Some Plain Words on the Identity and Invention of Science', *Studies in the History and Philosophy of Science*, vol. xix (1988), pp. 384f.

selectively adapted. But, for the most part, this interaction took place at the level of ethical and theological considerations. There was not a great deal of interest in the specifics of Greek scientific theories mainly because the pursuit of such esoteric matters was perceived to detract from the spiritual and social objectives that Jews and Christians held to be primary. We argued in chapter 1 section E that this temporary check on the pursuit of scientific speculation was salutary for scientific and technological development in the long run. Nonetheless, it resulted in the postponement of serious consideration of scientific theories for a thousand years in the Latin West.

It was a very different story in the Greek and Syriac East. There, partly due to the greater accessibility of Greek texts and partly due to the prestige associated with philosophic pursuits in Greek culture, a strong scientific tradition survived and progressed well into the Middle Ages.[2] While 'prescientific' by modern standards, this tradition was sufficiently authoritative to impress the leaders of Arab civilisation, particularly the Abbasids of Baghdad, who became the heirs of much Greek and Syriac culture after the seventh century AD.

In order to assess the situation in Western Europe, it is helpful compare the Abbasid Muslims with the Carolingian Franks of the late eighth and early ninth centuries. The Arabs and the Franks were both tribal peoples, initially dominated by warrior classes, who had conquered and settled in the territories of earlier civilisations. Both were fiercely proud and sensitive to indications of their cultural and technological inferiority to older civilisations. Any attempt to understand the progress of natural science (whether in the Middle Ages or today) must take into account the feelings and energies of peripheral peoples like these who suddenly move into the mainstream of history. On balance, however, the Abbasid Arabs were culturally superior to the Franks, and, whether or not that superiority was really due to the assimilation of Greek science in any degree, a connection between the two indicators was clearly made in the mind of the Latin West.[3] Any significant development in Western civilisation beyond the Carolingian level of the ninth century would have to include a mastery of the science of the Greeks which had been neglected for so long. One major effect of Arab civilisation, as far as the West was concerned, was to make that point abundantly clear.

---

[2] Sambursky, *Physical World of Late Antiquity*; O'Leary, *How Greek Science Passed to the Arabs*.

[3] Christopher Dawson, *Medieval Religion and Other Essays* (New York: Sheed and Ward, 1934), pp. 66f., cites Plato of Tivoli (fl. 1132-46) on the inferiority of the Latins in the sciences. Cf. Roger Bacon's appeal for improved translations and textbooks in his treatise, *On the Errors of the Doctors* 11ff., 19ff. (Mary Catherine Welborn, trans., *Isis*, vol. xviii [1932], pp. 30-36).

From this review of the broad sweep of world history, we must now shift our focus to the actual conditions and attitudes that prevailed in the Latin West as Greek and Arabic texts began to be translated. The work of translation was a massive project carried on by numerous individuals of various nationalities and with varied motivations. Most of the work was done in Spain, South Italy, and Sicily (all former Arab colonies newly reconquered by Europeans) during the last three quarters of the twelfth and the first three quarters of the thirteenth century. A complex interplay of intellectual curiosity, personal ambition, and perceived market conditions (translations were in demand) helped produce the impressive results of this project.[4]

What was the role of the church in the initial stages of translation and assimilation? It must be said that the church neither initiated nor encouraged the process directly. In fact, a series of ecclesiastical bans from 1210 to 1263 attempted to prohibit the public reading of Aristotle's books on philosophy and their commentaries in the liberal arts curriculum at Paris.[5] The modern reader, conditioned by hundreds of years of disestablishmentarian rhetoric, must be careful, however, not to draw too simplistic a picture of the situation. For one thing, the church had been instrumental in the founding and staffing of the major Western universities, particularly those in France, Germany, and England. It was to be expected that its judicatories would make rulings concerning the arts curriculum, which was standard preparation for higher studies in theology as well as in law and medicine.[6] Indeed, most of the determinations were made with the advice and consent of representative arts masters, so it was not a violation of 'academic freedom' in the modern sense.[7]

Secondly, it has been pointed out by a number of recent historians that

---

[4] For a review, see Lindberg, 'The Transmission of Greek and Arabic Learning to the West', *Science in the Middle Ages*, chap. 2. A convenient table of translations is given by Crombie, *Augustine to Galileo*, vol. I, pp. 57-63.

[5] Copleston, *A History of Philosophy,* 9 vols. (Garden City: Doubleday, 1962-4), vol. II/1, pp. 236ff.; David Knowles, *The Evolution of Medieval Thought* (London: Longman, 1962), pp. 226ff.; Edward Grant, *Physical Science in the Middle Ages* (New York: Cambridge: Cambridge Univ. Press, 1977), pp. 24ff. See Lynn Thorndike, *University Records and Life in the Middle Ages* (New York: Columbia Univ. Press, 1944), documents 14 and 15, for the earliest of these prohibitions.

[6] Nicholas H. Steneck, *Science and Creation in the Middle Ages* (Notre Dame: Univ. of Notre Dame Press, 1976), pp. 11-15, gives a useful summary. Gordon Leff, *Paris and Oxford Universities in the Thirteenth and Fourteenth Centuries* (New York: John Wiley & Sons, 1968), chap. 3, gives detail.

[7] Edward Grant, 'Comment' on the Church and Academic Freedom in the Middle Ages, a session of the American Society of Church History, Holland, Mich., 21 April 1983. Brian Lawn points out that the 1210 and 1215 prohibitions of public lectures on Aristotle's natural philosophy at Paris did not prevent academic disputations on the subject; *The Rise and Decline of the Scholastic 'Quaestio disputata' with Special Emphasis on Its Use in the Teaching of Medicine and Science* (Leiden: Brill, 1993), pp. 32f.

the more rationalistic aspects of Aristotelian science, particularly those stressed by the Averroists, were actually detrimental to open scientific investigation. The latest trend in scholastic matters could easily become dogmatic in its own right, and, in the circumstances of the Middle Ages, it could only be checked by authoritarian measures on the part of the church.[8]

Moreover, the influence of the church should not be judged merely on the basis of official pronouncements and regulations which tend, by virtue of their function, to be restrictive. Gregory IX renewed the ban of 1210, but also in 1231 appointed a commission to expurgate the prohibited books where necessary so that they could be used for pedagogical purposes.[9] Urban IV, who again renewed the ban in 1263, was the very pope whose court (at Viterbo and Orvieto) brought together William of Moerbeke, the greatest translator of the time, and Thomas Aquinas, the greatest theologian, with the result that William was encouraged to carry out a systematic translation of the Aristotelian corpus in the early 1260s. Since William was made chaplain and confessor to the pope in 1272 and archbishop of Corinth in 1278,[10] his well known activity as a translator can hardly be said to have been done in opposition to the church!

Finally, it should be noted that only a century later, in 1366, the legates of Urban V were requiring a knowledge of all known works of Aristotle for the license to teach the liberal arts at the University of Paris.[11] Initial opposition to the translation and assimilation of Aristotle's philosophical works had thus been reversed and replaced by official approval. Indeed, present-day historians are more likely to fault the medieval church for its near canonisation of Aristotle in the fourteenth century (leading up, of course, to the condemnation of Galileo in the seventeenth century) than for its earlier opposition.[12] From a twentieth-century viewpoint, what needs to be explained is not so much the church's earlier attempt to ban the philosophical works of Aristotle as its capacity to assimilate the Aristotelian corpus and to harmonise it with its own theology.

The ability of the church to synthesise the new science with its traditional ideas is, therefore, the most remarkable feature of the thirteenth century. It is best exhibited in the masterful 'summae' (comprehensive treatises) of Thomas Aquinas, written in the third quarter of the century.

---

[8] Pierre Duhem, 'Physics, History of', *Catholic Encyclopedia*, vol. IX (New York: Robert Appleton Co., 1911), p. 50; Crombie, *Augustine to Galileo*, vol. I, pp. 75f.; Fernand Van Steenberghen, *Thomas Aquinas and Radical Aristotelianism* (Washington, D.C.: Catholic Univ. of America Press, 1980), pp. 86ff.

[9] Thorndike, *University Records*, selections 19f.

[10] Lindberg, 'Transmission', pp. 73f.

[11] Thorndike, *University Records*, selection 87.

[12] E.g., Dijksterhuis, *Mechanization*, pp. 129f.; Robert K. DeKosky, *Knowledge and Cosmos* (Washington, D.C.: Univ. Press of America, 1979), p. 59.

But it appears quite clearly among the very earliest theologians of the period. William of Auvergne (d 1249), for example, accepted the mechanistic (Aristotelian) astronomy of Alpetragius (al-Bitruji) and the formal distinction between essence and existence made by Avicenna (Ibn Sina) while, at the same time, rejecting some of the metaphysical aspects of Aristotle and Avicenna that appeared to conflict with teachings of the church.[13] Since William is generally regarded as one of the more traditional, 'Augustinian', theologians of the period, his attitude is indeed remarkable. It represents quite an advance in sophistication beyond the more recalcitrant stance toward natural philosophy exhibited by twelfth century theologians like Bernard of Clairvaux and William of St Thierry.

Indeed, William of Auvergne combined much of the naturalism of twelfth-century philosophers like Adelard of Bath and William of Conches with the religious conservatism of their critics. It is as if the centrifugal tendencies of the twelfth century and the initial breakdown of the creationist tradition had evoked an effort toward greater equilibrium and comprehension.[14]

Nonetheless, the lines were clearly drawn between church and state, revelation and reason, God and nature. The fact that the church could sanction efforts towards synthesis and make room for Aristotelian philosophy within subordinate spheres of thought could not reproduce the harmony of all things under the sovereignty of God that was portrayed in the Psalms and was taught by many of the early church fathers. The underlying bifurcation was temporarily transcended, but the strain was to prove too great, and hostilities, reminiscent of those of the twelfth century, were to break out again in the last third of the thirteenth century. We shall return to this matter in the following sections of this chapter as we discuss the effect the newly discovered science had on the development of medieval theology and the subsequent influence of theology on medieval science.

B. The Impact of Aristotelian Science on Scholastic Theology

We have just seen that Aristotelian philosophy was perceived as something radically new in thirteenth-century Europe. Of course, it was understood that this material had been known to the Greek East for more than 1500 years, but aside from a few logical treatises (the 'old logic' of Aristotle,

---

[13] Duhem, 'Physics', p. 49; Copleston, *History*, vol. II/1, pp. 247ff.; Knowles, *Evolution*, pp. 229f.

[14] For a more broadly based discussion of the issues, see Norman F. Cantor, *Medieval History* (New York: Macmillan, 1963; 2nd ed., 1969), chaps. 19, 20; esp. pp. 448, 461, 464.

Porphyry, and Boethius)[15] it had not been available in the Latin West. The seriousness with which the new ideas were treated must be understood against this background.

What was the impact this new wave of thought had on Christian theology? The most important fact was that, for the first time since the days of the early church, theologians had to defend their faith as being true in a thought-world in which their right to specify the criteria for truth was no longer uncontested. In spite of growing pluralism in both church and society, however, Europeans of the Middle Ages would not have appreciated the compartmentalisation and relativism we experience in the modern world. Truth was believed to be one: any suggestion that there could be two or more separate truths was repugnant.[16] So the challenge of the new alternative could not be ignored.

For more than a millennium the church had championed a tradition in which the world was held to have been created by an all-wise God in such a way that its operations were lawful, natural, and open to human comprehension, at least, in principle. Here now was a massive body of texts that provided an actual account of the cosmos as a rational, naturally ordered system. On the other hand, it also contained ideas like the eternity of the world that appeared to conflict with the creationist tradition. Was Aristotelianism then the fulfillment of the Christian worldview or its nemesis? From our perspective in the modern world, the science of Aristotle is outdated, but the challenge it raised is perennial: how can one reconcile a science which seemingly owes nothing to Christian faith, and may conflict with it at any point, with a faith which encourages belief in the possibility of science and values its benefits, yet cannot sanction its teachings or its applications without further scrutiny? The various ways in which medieval theology adapted to meet this challenge were, of course, closely related to the specifics of Aristotelian science, but, as we shall see, they set important precedents for similar modes of adaptation in later periods of our history.

There are basically two areas in which medieval theology adapted to meet the challenge of Aristotelian science. One area was the concept of revelation and its study, which includes the discipline of theology. The other was the concept of God, who is the primary object of theology. It was principally issues in the medieval doctrine of God that, in turn, impacted on the development of natural science, as we shall see in section C.

---

[15] Leff, *Medieval Thought*, pp. 47, 125, 175.
[16] The myth of the so-called 'double-truth' theory has been interpreted as a device of conservative polemics by Van Steenberghen, *Thomas Aquinas*, pp. 93-109. It is unfortunate that Martin Pine's fine article on the emergence of philosophy as an autonomous discipline in DHI (vol. II, pp. 31-37) is listed under the title of 'Double Truth'.

1 Revelation and Reason: Theology and Science

The principal problem confronting medievals was the fact that there now appeared to be two distinct bodies of knowledge, both of which had to be taken seriously. The relatively limited body of Aristotelian logic and science available before the thirteenth century (the 'old logic') had not presented such a problem for theologians because the principal source of human science in this case was pure reason. Augustine had effected a successful synthesis of pure reason and revelation, already in the fourth to fifth century, by portraying both as forms of illumination from God: there was a direct analogy, even a partial correlation, between the two.[17] Sense perception, particularly vision with the aid of physical light, could also be given a role at a subordinate level in this scheme with little or no threat to the hegemony of revelation. The early thirteenth-century scientist Robert Grosseteste (d 1253) worked out a system based on the light-metaphysics of Augustine in which some account could be given even of nonvisual phenomena like heat, astrological influence, and mechanical action.[18]

Significant problems like the motion of projectiles and the transmutation of the elements, however, had been treated in detail by Aristotle and his commentators, and there was no possibility here of visual analogues or an epistemology of illumination. In fact, Aristotle's scientific epistemology, in sharp contrast to that of Augustine (and Plato before him), was based on the abstraction of concepts and principles from sense data.[19]

The creationist tradition of the twelfth century had already shown strains between the concepts of God and nature, revelation and reason, even within the overall harmony of Augustinian metaphysics. The idea of abstraction, set along side of that of revelation, now threatened to introduce a complete bifurcation in both subject material and method.

The shift is significant. Rather than having to maintain an overall unity, one had to effect a totally new synthesis. Augustine had been able to choose his own ground, so to speak. Of the various schools of Hellenistic thought available, he decided to choose Neoplatonism because it came closest to meeting his spiritual needs (as it did for most other Christians of his age) and because it seemed to agree with Christian teaching at several points. The thirteenth-century scholastics did not exactly choose Aristotle, however. Through the selective interests of the Syrian Christians and Arab Muslims, Aristotle had emerged from the pack of Greek philosophers as 'the Philosopher', and it was the Arabs, as we have seen, who set the standard

---

[17] Robert E. Cushman, 'Faith and Reason', in Roy W. Battenhouse, ed., *A Companion to the Study of St Augustine* (New York: Oxford Univ. Press, 1955), pp. 291ff.

[18] A.C. Crombie, *Robert Grosseteste and the Origins of Experimental Science* (Oxford: Clarendon Press, 1953), chaps. 5, 6; *Augustine to Galileo*, vol. I, pp. 112-15; vol. II, pp. 35f.

[19] Ross, *Aristotle*, pp. 47, 54f.; Randall, *Aristotle*, pp. 40-44.

for Europeans of the high Middle Ages.

1.1 Thirteenth Century Syntheses

How, then, could one formulate a synthesis between the two bodies of knowledge, one derived from sense perception by way of abstraction and the other derived from revelation by way of illumination and faith? There was no unanimously favoured solution to this problem, but one can define four characteristic positions on a spectrum depending on the degree to which reason and revelation were integrated. At one extreme was Siger of Brabant, the Latin Averroist, who, as an arts master in the 1260s, expounded various points of Aristotelian philosophy which had implications, like the eternity of all species and the substantial unity of all human minds, which appeared directly to contradict the teachings of the church. The problem was that Siger made no effort to harmonise or synthesise these conclusions with official theology. His views were opposed by theologians like Bonaventure and Thomas Aquinas in the late 1260s and early 1270s and were condemned by two important sets of articles promulgated by the bishop of Paris, Stephen Tempier, in 1270 and 1277.[20]

A position clearly to the right of Siger, yet still making a very sharp differentiation between reason and revelation, philosophy and theology, was that of Albert the Great (d 1280). Philosophy, he held, was based on human reason unaided by God. There was a definite sphere of truth that it could establish without recourse to revelation. For instance, reason could demonstrate the existence of God as First Being, but neither his attributes nor the fact that he had created the world in time (i.e., that the world was not eternal). It could treat the intelligences Aristotle described as movers of the celestial bodies, but not the angels who were described in Scripture as intermediaries between God and humans. It could investigate those things that happen 'on the ground of causes inherent in nature', but not those miracles by which God freely chooses to manifest his power.[21] As a general rule, according to Albert (here following William of Conches), one should follow the apostles and fathers of the church in matters of faith and morals, but follow the Greek natural philosophers in matters of medicine and physics.[22]

Thomas Aquinas (d 1274) was a student and associate of Albert's around

---

[20] Van Steenberghen, *Thomas Aquinas*, pp. 1-9.
[21] *De caelo et mundo* I.iv.10 (Stiefel, *Intellectual Revolution*, pp. 101f.).
[22] Copleston *History*, vol. II/2, pp. 13f., 16; J.A. Weisheipl, 'The Celestial Movers in Medieval Physics', in Weisheipl, ed., *The Dignity of Science* (Thomist Press, 1961), pp. 162, 176; Ralph M. McInerny, *A History of Western Philosophy*, vol. II (Notre Dame: Univ. of Notre Dame Press, 1970), p. 242; William A. Wallace, 'The Philosophical Setting of Medieval Science', in Lindberg, ed., *Science in the Middle Ages*, p. 96.

the middle of the century (1245-52). Though it is difficult to tell to what extent his differences from Albert were deliberate, the fact that Thomas intentionally worked towards a greater correlation between philosophy and theology suggests that he was concerned to avoid the dichotomy that might be inferred from the writings of his mentor.

On the more positive side, Thomas sought to develop his theology along apologetic lines that could be used in missionary efforts to convert the Muslims. Aristotelian philosophy was useful here in that it provided concepts and methods that could be turned against the very people from whom the Latins had first learned them! Similar strategies were suggested by Roger Bacon and Raymond of Penafort (d 1275). In fact, it may have been the latter who suggested the idea to Thomas.[23]

For Thomas, then, a clear distinction could be drawn between philosophy and theology with respect to their methods and starting points, but there was considerable overlap in their principles and conclusions. Philosophy starts with what can be known about the world and follows reason towards divine things (such as God and the angels) as principles that must be posited in order to account for the existence and behaviour of natural things which are their proper effects. Theology, on the other hand, begins with divine things, as revealed in Scripture, and follows the rule of faith towards their effects in nature and history.[24]

But, while this overall synthesis seems harmonious enough, there are indications of underlying tension and even suggestions of dichotomisation. For one thing, the 'divine things' posited as principles by reason are not coterminus with the 'divine things' of revelation. Some theological dogmas like the Trinity, the creation of the world in time (finite age), and the future consummation were completely beyond the reach of reason and, hence, found no place in philosophy. Therefore, Thomas had to differentiate between revealed theology, which treated these particular dogmas, and natural theology, which treated those beliefs that could be established by reason without the aid of revelation. The latter category included the existence of God, divine attributes like unity, and the dependence of the world on God for its existence and consummation.[25] So Thomas allowed reason and revelation to overlap to a degree, whereas Albert had kept them quite separate.

---

[23] Roger Bacon, *Opus Maius* II.18; VII.14, Robert Belle Burke, trans., 1928 (rpt., New York: Russell and Russell, 1962), e.g., pp. 71, 632; Etienne Gilson, *History of Christian Philosophy in the Middle Ages* (New York: Random House, 1955), pp. 351f.; Anton C. Pegis, 'General Introduction', *Saint Thomas Aquinas: Summa Contra Gentiles, Book One* (Notre Dame: Univ. of Notre Dame Press, 1975), pp. 20f.

[24] Copleston, *History*, vol. II/2, pp. 30f.; R. M. McInery, *St Thomas Aquinas* (Notre Dame: Univ. of Notre Dame Press, 1982), p. 143.

[25] Kaiser, *Doctrine of God*, p. 90.

Even this limited overlap between reason and revelation has to be qualified, however, by the fact that, for Thomas, it was impossible to have both knowledge (by reason) and faith concerning the same theological truth at the same time. Once one was able to demonstrate a particular truth by reason, one ceased to hold it, in the proper sense, by faith. So truths like the existence and unity of God were not properly articles of faith, but rather a preamble to faith.[26] On the one hand, this allowed those who lacked skill in philosophy to hold them in simple faith, while, on the other, it allowed the Christian apologist to impress intellectually minded Muslims and others with the philosophical credentials of at least some of the teachings of Christianity. While these few teachings were not sufficient to establish the truth of Christianity as a whole, they did help defend the veracity of the Christian Scriptures, which were held by Muslims to have been corrupted from the original versions.

A theologian who sought a more thorough integration of faith and reason was the Franciscan, Bonaventure (d 1274, the same year as Aquinas). Using a less formal, more Augustinian notion of reason, he concluded that reason and revelation could be seen to interpenetrate almost entirely. On the one hand, theological truths like the creation of the world in time were demonstrable by means of reason alone.[27] On the other hand, due to the effects of sin, even those truths that Aquinas allowed as accessible to reason were not fully or properly understood if they were not viewed in the light of revelation. Thus, the unity of God, for instance, was not properly apprehended if it was thought to exclude the Trinity (as it was by Muslims). It followed, then, that the same truth could be held by means of both faith and reason at the same time. Indeed the cooperation of the two human faculties was essential to full understanding of any truth.[28]

Bonaventure illustrated this relationship with the image of two books. By virtue of its being created by God and its exemplifying God's ideas ('exemplarism'), nature was a book in which one could discern the divine attributes. Human ability to read this book properly had been vitiated by sin, however, so a second book of God, Holy Scripture, was needed to provide reliable knowledge of ethical and theological matters.[29]

The image of the two books of God can be traced back to the theologies

---

[26] *Summa Theologiae* I.2.2.; *On Truth* 14.9; Copleston, *History*, vol. II/2, p. 33.

[27] Copleston, *History*, vol. II/1, pp. 293ff.

[28] E. Gilson, *The Philosophy of St Bonaventure* (Patterson, N.J.: St Anthony Guild Press, 1965), pp. 81ff., 93-99, 102f.

[29] Efrem Bettoni, *Saint Bonaventure* (Notre Dame: Univ. of Notre Dame Press, 1964), pp. 32-36; Ewert H. Cousins, 'Introduction', *Bonaventure: The Soul's Journey into God* (New York: Paulist Press, 1978), pp. 23-26.

of Augustine and Francis of Assisi.[30] After Bonaventure, it was taken up by Raymond Lull (d *c*. 1316) and Raymond Sebond (or Raymond Sibiuda, d *c*. 1436).[31] It was to be used even more widely in treatments of science and theology in the Reformation and early modern periods.

For those who took the effects of human sin and the necessity of divine grace seriously, the image of the two books proved to be quite effective. Raymond Lull believed that he had been granted a vision of all truth, both scientific and theological, through divine illumination. On the basis of this vision, recorded in his *Ars magna* (*c*. 1274), believers could see the reproductive energies in plants, trees, and humans as images of the eternal generation in God through which the Father engendered the Son. The study of nature thus elevated the mind to the understanding and love of God.[32] Here there was no conflict between the relative autonomy of nature and the all-sufficiency of God.

On the other hand, the image of the two books could also be used to argue for the complete autonomy of science based on natural reason.[33] In that case, Scripture might be viewed as superfluous or even inferior to human reason. This tendency for the creationist tradition to undercut its own presuppositions became apparent already with Raymond Sebond's *Book of Creatures* of 1436 and was to contribute to the rise of rationalism and deism in the seventeenth century. Even Bonaventure's more unitive view, it appears, harboured an underlying dichotomy of reason and revelation, or of nature and grace. Nonetheless, it provided a model for the mutual questioning of reason and revelation that soon bore fruit in the natural philosophy of Franciscans like Peter John Olivi, William of Ockham, and Francis of Marchia (chap. 2, sec. C).

---

[30] According to Socrates's *Ecclesiastical History* IV.23, Antony the Great (d *c*. 350) had no written books in his possession so he looked at the 'nature of created things' around him when he wanted to read the words of God. The wording sounds like that of Eusebius (chap. 1, note 111 above), whose work Socrates was consciously continuing. For the same idea in Bernard of Clairvaux, see his Letter 107 (to Henry Murdach); cf. Roger D. Sorrell, *St Francis of Assisi and Nature* (New York: Oxford Univ. Press, 1988), pp. 29f.

[31] Gilson, *History*, pp. 352f.; J. Guy Bougerol, *Introduction to the Works of Bonaventure* (Paterson, N.J.: St Anthony Guild Press, 1964), p. 9; Clarence J. Glacken, *Traces on the Rhodian Shore* (Berkeley: Univ. of California Press, 1967), pp. 203ff., 237ff.; N. Max Wildiers, *The Theologian and His Universe* (New York: Seabury Press, 1982), p. 34.

[32] Lull, *Felix, sive libre de meravelles* (1288-89) IV.20; V.31; Anthony Bonner, ed., *Selected Works of Ramon Llull* (Princeton: Princeton Univ. Press, 1985), pp. 739f., 759ff.; cf. Frances A. Yates, *Lull & Bruno* (London: RKP, 1982), pp. 35f. Lull obviously developed the parallel to counter Muslim objections to the idea of generation in the godhead.

[33] Thus Frank and Fritzie Manuel contrast the metaphor of the two books to the more synthetic Pansophia in the seventeenth century; *Utopian Thought in the Western World* (Cambridge, Mass.: Harvard/Belknap Press, 1979), pp. 206, 211.

1.2 Concept and Method of Theology

We have reviewed four characteristic ways in which thirteenth-century scholastics (Siger of Brabant, Albert the Great, Thomas Aquinas, and Bonaventure) handled the issue of the relation between reason and revelation, or science and theology. Clearly this new concern had an effect on the concept of theology and on its method. One could no longer assume that theology in the proper sense was simply an exposition of the traditional articles of faith (e.g., those of the Apostles' Creed), for most of the first article ('I believe in God the Father Almighty, maker of heaven and earth') was now held to lie within the province of reason and even to constitute, for Thomas, at least, a preamble to faith. Thomas's system was extremely influential: he became the official doctor of the Dominican order in the early fourteenth century, and his teachings were declared normative for all Roman Catholics at the Council of Trent (1545-63). The general approach of Thomas, if not all his teachings, was also influential in Protestant scholastic circles of the late sixteenth and early seventeenth centuries.

The net effect of the synthesis with Aristotelian philosophy on the procedure of theology was threefold. First, the order in which topics were treated tended to proceed from those truths that were thought to be accessible to reason—a 'general revelation' available to all humans—to those truths that were known only on the basis of biblical revelation—a 'special revelation' unique to Christian faith. This procedure was followed by Thomas in his two great 'summas'. But the procedure was not peculiar to Thomas: it appeared in a variety of theological systems, though the location of the line between truths of natural reason and those of revelation varied widely.[34]

The second effect of the scholastic synthesis was a heightened emphasis on rigorous deduction within the treatment of revealed theology itself. In natural theology, as we have noted, one could use the abstractive method of the sciences; that is, one could begin with what is known empirically and reason inductively toward principles that must be posited in order to account for the known facts. The model here was Aristotelian natural philosophy. In revealed theology, another model became influential, however, that of Euclidean geometry. In Euclidean geometry, first principles were posited as self-evident and deductions were made in keeping with strict rules. The way had been prepared by the use of Aristotle's logic or dialectical method (the 'old logic'), which had been known in part since the tenth century and had already influenced the procedure of rational theologians like Anselm of

---

[34] After Thomas the line shifted to the point that William of Ockham (d c. 1350) held that only the existence of God was rigorously demonstrable by reason. The rest of dogma depended on pure faith; Kaiser, *Doctrine of God*, pp. 92f.

Canterbury (d 1109).[35] This trend towards rigorous deduction was reinforced, however, by various translations of Euclid's *Elements of Geometry* in the twelfth and thirteenth centuries and by the overall challenge of the newly available corpus of Aristotelian philosophy which we have been describing.[36]

The presentation of theology had always been closely related to the function of teaching, and with the rise of the universities in the thirteenth century, it had to establish a role for itself in relation to the new curricula.[37] Before undertaking the study of theology, a student had to master the arts and sciences, including both Aristotelian science and Euclidean geometry.[38] The expectations of the graduates who went on to theological studies and the questions they subsequently raised for their instructors must have had quite an impact on the way theology was taught.[39] Was revealed theology a science? If so, in what sense?[40] If it was not a science, at least not in the normative sense,[41] how could it be presented in a way that would command the respect of students trained in the new disciplines? The exact definition

---

[35] Kaiser, *Doctrine of God*, pp. 85ff.

[36] On similarities between Euclid (first LT by Adelard of Bath, *c.* 1120) and Aristotle (esp. the *Posterior Analytics*, first LT by James of Venice, *c.* 1130), see Ross, *Aristotle*, pp. 44f.; Randall, *Aristotle*, pp. 33, 40f.; Charles Burnett, 'Scientific Speculations' in P. Dronke, ed., *History of Twelfth-Century Western Philosophy*, pp. 155-59. According to Burnett, the first Western writers to apply the Euclidean method to theology were Peter of Poitiers (before 1150) and Nicolas of Amiens (*c.* 1190). Boethius had pointed the way in his work *On the Hebdomad*; *ibid.*, pp. 163ff.

[37] Wolfhart Pannenberg, *Theology and the Philosophy of Science* (London: Darton, Longman and Todd, 1976), pp. 12f.

[38] For the curriculum at Paris, see Steneck, *Science and Creation*, pp. 12f. For the somewhat different situation at Oxford, see Edith Dudley Sylla,'The Oxford Calculators', in Norman Kretzmann et al., eds., *Cambridge History of Later Medieval Philosophy* (Cambridge: Cambridge Univ. Press, 1982), pp. 542f.

[39] On Albert and his students, Pearl Kibre and Nancy Siraisi, 'The Institutional Setting: The Universities', *Science in the Middle Ages*, p. 133. For an entertaining account of the arts master Roger Bacon under student-applied pressure, see Stewart C. Easton, *Roger Bacon and His Search for a Universal Science* (Oxford: Basil Blackwell, 1952), pp. 56ff., 62f.

[40] When scholastics did allow theology the status of a science, they invariably qualified it either as an affective, practical science (one that promotes piety and strengthens faith: Alexander of Hales, Bonaventure, Albert the Great, Thomas Aquinas) or else as a 'subalternate' science (one based on a higher science, in this case God's self-knowledge as revealed in Scripture: Alexander of Hales, Bonaventure, Aquinas); M.D. Chenu, *La Théologie comme science au XIII siècle* (3rd ed., Paris: J. Vrin, 1957); Easton, *Roger Bacon*, pp. 224-31.

[41] When scholastics differentiated theology from the sciences, they generally characterised it as wisdom (*sapientia*), i.e., the study of the (eternal) cause of all (temporal) causes (Alexander of Hales, Albert the Great, Thomas Aquinas) that does not seek to reduce its first principles to ulterior causes (Aquinas). For Aquinas's views, see, e.g., Armand Maurer, 'Introduction', *St Thomas Aquinas: The Divisions and Methods of the Sciences*, 3rd ed. (Toronto: Pontifical Inst. of Medieval Studies, 1963), pp. ix-xi, W.A. Wallace, 'Science (scientia)', *New Catholic Encyclopedia*, vol. XII (New York: McGraw-Hill, 1967), pp. 1190ff.

of science was to some extent a matter of semantics, but the result for theology was the same in any case. One had to stress certain foundational principles, derived from Scripture and the rule of faith, from which one could make inferences that would decide the pertinent theological questions of the day.

Take, for instance, the procession of the Holy Spirit, an issue that divided the Christians of the West from the Eastern Orthodox. In his *Summa Against the Gentiles*, Aquinas presents a series of arguments. In the first of these, he begins with Romans 8:9 which refers to the 'Spirit of Christ' and proceeds to show (a) that the Spirit is 'of Christ' as Christ is the natural son of God and (b) that the Spirit must be 'of Christ' in the sense of having its origin from Christ because the only relations found within the godhead are relations of origin (IV.24.2). The rigour of demonstration here was not quite up to the standards of Euclid and Aristotle, but the style of presentation was similar enough, one may presume, to carry weight with students who had been trained in the arts and sciences.

The third effect of the scholastic synthesis was the subdivision of theology into distinct 'loci' as topics for consideration. To some extent, this process had already been begun with the rational procedure of Anslem and the systematic collection of texts from Scripture and the fathers by Peter Lombard (d c. 1160). Lombard's *Book of Sentences* became the principal text for the study of theology in the thirteenth century. The division of its material under the four main headings of God, creatures, the work of Christ, and the sacraments and last things provided the model not only for numerous commentaries and 'quaestiones', but also for the 'summas' of later scholastic theology.[42]

The parallel development of commentaries and 'quaestiones' on the works of Aristotle during the thirteenth century contributed to this subdivision of topics in two ways. First, the fact that both theologians and their students had to master the Aristotelian texts and be able to comment on the specialised issues involved elevated the standards of questioning and refuting alternative views in the theological faculty. Secondly, the treatment of specific issues like the potential eternity of the world and the providence of God in relation to second causes (to be discussed below), issues raised or heightened by the assimilation of Aristotelian science, called for additional sections and subdivisions in any credible presentation of theology. Separate consideration was required for what could be shown by pure reason and what was known by revelation, for what could be said with respect to God's normal activity (*de potentia ordinata*), what could be said of his special acts of power (*de potentia absoluta*), and what must be said of his

---

[42] Knowles, *Evolution*, pp. 179ff.

omnipresence irrespective of any action at all. The result, as one might expect, was a fragmentation of theological doctrine into highly specialised issues with practically no possibility for students to integrate the separately derived conclusions into a cohesive whole.[43]

In summary: scholastic theology adapted to meet the challenge of the new Aristotelian science by defining and attempting to correlate the respective roles of human reason and divine revelation within an overall synthesis. The placement of revelation in the synthesis, which corresponded to the place of theology in the university curriculum, had the effects of:

(1) establishing the order of the theological curriculum as proceeding from natural theology to revealed theology;

(2) encouraging the use of rigorous deduction in the treatment of revealed theology;

(3) increasing the isolation of distinct theological 'loci' as topics for consideration.

It is easy to see why Renaissance and Reformation theologians reacted so strongly against the scholastic synthesis, and our own attitudes will very likely be critical as well. It would be best to reserve final judgement, however, until we have reviewed the struggles of modern scholars to achieve a new synthesis and the consequences of their failures. The scholastic model survived well into the eighteenth century and even continues to provide a framework for integrative endeavours for many believers to this day.

2 God and the World

In chapter 1 we described the origins and early development of the creationist tradition and found certain paradoxes emerging that threatened to split the tradition into two camps in the late eleventh and early twelfth centuries. One party stressed the supernatural or absolute power of God (*potentia absoluta*) in both creation and providence, while the other stressed the autonomy of nature as created and ordained by God (*potentia ordinata*).

As we have seen already in this chapter, leading administrators and theologians of the church managed temporarily to transcend these differences and to effect a new synthesis in the thirteenth century. Still, the comprehensive naturalism of Aristotelian science was bound to intensify the underlying problems. During the first three quarters of the century, theology showed a marked tendency to accommodate itself to the naturalism of Aristotle. To some extent, this tendency was a continuation of the

---

[43] E. Grant, 'Cosmology', *Science in the Middle Ages*, pp. 266f.

naturalistic and secularising trends of the twelfth century.[44] A distinct shift took place during the 1270s, however when the supernaturalistic side of the creationist tradition made itself felt, and there was a massive reaction against certain aspects of Aristotelian science. This reaction, in turn, led to new directions in scientific thought in the fourteenth century, as we shall see in section C.

The creationist tradition had generally pictured God as having given being to all creatures at a particular point of time (the 'beginning' in the Greek and Latin translations of Gen. 1:1) in such a way that their subsequent behaviour was ceaselessly obedient to specific laws or commands. While continuously subject to God's ratification and amendment (where needed to fulfill moral purposes), these laws provided natural beings with a degree of autonomy that made prediction and rational investigation possible. Though it may seem paradoxical to us, the world was relatively autonomous, yet God was in direct control.

The naturalism of Aristotle was really quite different. To begin with, Aristotle was more concerned with natural change and its causes than he was with natural behaviour and its laws. Every event was seen as the realisation of a certain potency in nature under the activating effect of an agent. The agent had to be contiguous with the effect both spatially and temporally, but, in general, it was also prior in time or higher in space. If one could follow the chains of causation that lay behind terrestrial events, one would be led either backward in time (for efficient and material causes) or upward through the celestial spheres (for formal and final causes). From this vantage point, we can see three problems raised for Christian theology:

(1) the suggestion that the temporal sequence of cause and effect was eternal;

(2) the location of God at the outermost sphere of the cosmos as Prime Mover or First Cause;

(3) the problem of God's absolute power in the world of second causes.

2.1 Potential Eternity of the Universe

For Aristotle, time and change were endless, hence the chain of causation in the temporal sense (material and efficient causes) was potentially infinite,

---

[44] Historical connections are difficult to pin down here. Robert Grosseteste may well have been influenced by the school of Chartres in an arts training at Oxford or Paris. Alternatively, he may fit more directly into the native English tradition of Adelard of Bath: so R.W. Southern, *Robert Grosseteste* (Oxford: Clarendon Press, 1986), pp. 49, 62, 82, 83ff., 104, 141. Thomas Aquinas studied at the University of Naples and undoubtedly absorbed some of the secular, naturalist spirit of its founder, Frederick II (Dawson, *Medieval Religion*, p. 76).

though it could not be demonstrated to be actually infinite.[45] Hence, there was no first moment or 'beginning' of time. The world was completely stable and potentially eternal.

As one might expect, most Christian theologians in the thirteenth century rejected any notion of the actual eternity of the world. They differed, however, as to whether the contrary ideas (creation and the finite age of the universe) were demonstrable by reason or whether they could be known only by revelation. Franciscan theologians like William of Auvergne and Bonaventure gave what they saw as compelling reasons for rejecting the eternity of the world and affirming a beginning of time. Albert the Great also offered an argument for the finitude of time, but claimed that this sort of reasoning was only 'probable': philosophy alone could not demonstrate that God was the cause of the existence (as distinct from the essence) of the world. Thomas Aquinas disagreed with all of the above. Against Albert, he held that God could be shown by reason to be the efficient cause of the world's existence. But he also argued against more conservative theologians like Bonaventure that the dependence of the world on God for its existence did not entail its finite age, so that the latter was known by revelation alone after all.[46]

To some extent, the issue of the demonstrability of the finite age of the universe was simply a matter of defining the respective spheres of natural and revealed theology as discussed above. However, even if the world was believed by all parties to be temporally finite on the basis of revelation, the fact that it was allowed by some to be potentially eternal so far as rational science could determine was a significant innovation in the theological conception of nature. The paradox of potential eternity along with actual finitude was one aspect of the underlying tensions of the creationist tradition. On the one hand, God was believed to have created the world, presumably, as Genesis 1:1 seemed to indicate, a finite time ago. On the other hand, the regularity and lawfulness with which God was believed to have invested the world made it virtually impossible for reason to discover any actual 'beginning' for the processes of nature.

The creationist tradition could thus be interpreted as sanctioning the scientific quest for a natural cause behind each cause without end. Thus, William of Conches and members of the school of Chartres had attempted a naturalistic explanation of the creation account of Genesis in the twelfth

---

[45] Randall, *Aristotle*, pp. 123f., 133f., 156, 192ff.

[46] Copleston, *History*, vol. II/1, pp. 249, 293ff.; vol. II/2, pp. 16, 85ff. See further on Albert, Leo Sweeney, 'The Meaning of *Esse* in Albert the Great's Texts on Creation', in Francis J. Kovach and Robert W. Shahan, eds., *Albert the Great* (Norman: Univ. of Oklahoma Press, 1980), p. 89; Benedict M. Ashley, 'St Albert and the Nature of Natural Science', in J.A. Weisheipl, ed., *Albertus Magnus and the Sciences* (Toronto: Pontifical Inst. of Medieval Studies, 1980), pp. 86f.

century (chap. 1, sec. D).[47] Indeed, Origen and Basil had allowed for an eternal world of spiritual, if not material, creatures in order to do justice to their belief in the infinite power of God (chap. 1, sec. B). When viewed against this background, the speculations of Albert and Thomas can be seen as legitimate developments of the creationist tradition and not simply as accommodations to Aristotle. On the other hand, the potential infinity of the chain of natural causes as determinable by reason would not have been articulated so tantalisingly, as it was by Thomas, if it had not been for the impact of Artistotelian science.

What happened to the idea of the potential eternity of the world after Aquinas? The list of 219 theses condemned under Bishop Tempier of Paris in 1277 included more than twenty propositions dealing with various notions concerning the eternity of the world.[48] Four of these articles condemned the idea that the eternal existence of the world could be inferred directly from God's eternity and infinite power.[49] Two others condemned the idea that the chain of natural causes could be extrapolated back indefinitely into the past.[50] Since both of these ideas were well represented within what we have called the historic creationist tradition, it should be clear that the Condemnation of 1277 did not represent that tradition as a whole but was rather an attack, or counterattack, of one wing of the tradition on the other. In fact, the 1277 Condemnation represented that wing of the creationist tradition subsequently identified as 'conservative' or 'orthodox'.[51]

In any case, the Condemnation of 1277 did not stop speculation on the potential eternity of the world. In general, it encouraged belief in the absolute power of God and made people more aware of the incongruities that might arise when the results of pure reason were compared with the truths of revelation.[52] This left the door wide open for fourteenth-century philosophers like William of Ockham, Nicholas of Autrecourt, John of Ripa, and Nicole Oresme to argue that the most probable conclusion of

---

[47] For the thoughts of William of Conches and other twelfth-century writers on the possible eternity of the world, see R.C. Dales, 'Discussions of the Eternity of the World during the First Half of the Twelfth Century', *Speculum*, vol. lvii (1982), pp. 495-508.

[48] An earlier list of thirteen theses condemned by Tempier in 1270 included two propositions on the eternity of the world: Thorndike, *University Records*, selection 38, p. 80.

[49] Art. 24-26 and 85 in the Mandonnet numbering, Arthur Hyman and James Walsh, eds., *Philosophy in the Middle Ages* (Indianapolis: Hackett, 1973), pp. 544ff.

[50] Art. 84, 191 (*ibid.*).

[51] Pace Stanley L. Jaki, *Science and Creation* (Edinburgh: Scottish Academic Press, 1974), pp. 229f. Jaki is quite sympathetic to the speculative theology of Origen (*ibid.*, pp. 169ff.), but does not reckon with the fact that Origen's reasoning from the omnipotence of God to the eternity of the (spiritual) world was specifically mentioned in the Condemnations of 1277 (note 49 above).

[52] E. Grant, 'Late Medieval Thought, Copernicus, and the Scientific Revolution', *JHI*, vol. xxiii (April 1962), pp. 200f.

rational investigation was that the natural order is eternal, even though this conclusion was overruled by the Catholic faith.[53] Finally, in the fifteenth century, Nicholas of Cusa (d 1464) concluded that the world must be indeterminate in both beginning and end if it is to reveal the eternal God who created it.[54] Cusa served the church as bishop and cardinal, and, though highly original, he was hardly unorthodox. In this matter as in others to be discussed, his ideas give us a good indication of the tendencies in the creationist tradition at the end of the Middle Ages and the beginning of the Renaissance.

2.2 God as First Mover and Clockmaker

The feature of the new science that had the greatest influence on the doctrine of God was the Aristotelian cosmology of homocentric spheres as modified by the Arabic natural philosophers, Thebit (Ibn Qurra) and Alpetragius (al-Bitruji). As a rule there were thought to be nine or ten celestial spheres surrounding the earth: seven for the sun, the moon, and the five known planets; an eighth sphere containing the stars; an optional (ninth) sphere to allow for anomalies (either precession of the equinoxes or 'trepidation') in the motion of the stars in the eighth; and an outermost (ninth or tenth) sphere responsible for the daily rotation of the heavens.[55]

God was located, in some sense, at the boundary of the outermost (ninth or tenth) sphere. According to Aristotle's *Metaphysics*, God was the First Mover, that is, the ultimate formal and final cause, whose very presence was enough to activate the rotation of the outermost sphere of the cosmos. The latter was, therefore, the 'first movable sphere' (*primum mobile*), the only object with which God was in any kind of immediate relationship and, therefore, the one that was most active.[56] Inner spheres were moved by virtue of their proximity to outer ones, thus forming a chain of (gradually weakened) influence extending to the innermost sphere of the moon and even to the cycles of generation and corruption on earth.[57]

---

[53] Philotheus Boehner, trans., *William of Ockham: Philosophical Writings* (Edinburgh: Thomas Nelson and Sons, 1957), pp. 118-25 (pp. 132-39 in Bobbs-Merrill reprint, 1964); F. Copleston, *History,* vol. III/1, pp. 156f.; E. Grant, *Much Ado About Nothing* (Cambridge: Cambridge Univ. Press, 1981), pp. 129-34; Jaki, *Science and Creation*, p. 237.

[54] *On Learned Ignorance* II.2.101; Copleston, *History*, vol. III/2, pp. 47f.

[55] Steneck, *Science and Creation*, pp. 70f.

[56] For simplicity, we have ignored the possibility of an outermost immobile sphere introduced by Averroes; Grant, 'Cosmology', pp. 273ff.

[57] Aristotle required a set of 'counteracting spheres' for each planet in order to prevent the peculiar motion of the next higher planet being passed down to it. But the medieval West inherited the cosmology of al-Bitruji (Alpetragius, fl. 1190), in which there was only one sphere for each planet and all rotated in the same direction, from east to west, with gradually diminishing speed; cf. David C. Lindberg, *The Beginnings of Western Science* (Chicago: Univ. of Chicago Press, 1992), pp. 95f., 267.

The Aristotelian cosmology could be grafted onto the traditional Christian cosmology simply by increasing the number of heavens. The principal authority on the subject prior to the thirteenth century had been the Venerable Bede (d 735), who held that there were five visible, corporeal heavens (air, ether, Olympus, the fiery realm, and the stellar firmament). Beyond these were the 'waters above the firmament' (mentioned in Gen. 1:7; Ps. 148:4), the angelic abode (later known as the 'empyrean' or outermost created heaven), and the ultimate heaven of the Holy Trinity itself.[58] Readers of Aristotle in the thirteenth century had only to replace the four lower heavens of Bede with the four sublunar elemental zones (earth, water, air, and fire) and subdivide the stellar firmament into the nine or ten celestial spheres described above (the supracelestial 'waters', generally thought to be crystalline, were sometimes located in the ninth sphere). The result was an up-to-date cosmological model, combining the best insights of both science and theology, with a total of four sublunar elemental zones and nine or ten celestial spheres (below the empyrean).[59]

---

[58] Bede, *In Pentateuchum super Genesim* I.1 (PL 91:192); Crombie, *Augustine to Galileo*, vol. I, p. 39; Jones, ed., *Bedae Opera*, pp. 125f. For patristic antecedents of this pre-Aristotelian cosmology, see Athanasius, *Contra gentes* 36; Basil, *Hexaemeron* III.8; IV.2f.; Augustine, *De Gen. imperf.* XIV.44; *De Gen. ad litt.* III.ii.2f. For classical antecedents, see Homer, *Odyssey* VI.42-45; Lucretius, *De rerum naturae* III.18-22; Apuleius, *Liber de mundo* 33. For biblical portrayals of the height of the mountain above the clouds and the presence of fire above the mountain, see Exod. 19:18; Deut. 4:11f.; Isa. 14:13f.

Bede's cosmological scheme was followed by the *De universo* of Rabanus Maurus (d 856); cf. Aquinas, *Summa theologiae* Ia.68.4. The angelic abode was identified with the 'heaven' created on the 'first day' (Gen. 1:1) and called the 'empyrean' in the *Glossa ordinaria super Genesim*, traditionally attributed to Walafrid Strabo (d 848), but probably written by Anselm of Laon (d 1117); Grant, 'Cosmology', pp. 275f.; *idem, Planets, Stars, and Orbs: The Medieval Cosmos, 1200-1687* (Cambridge: Cambridge Univ. Press, 1994), pp. 371-72; cf. Augustine, *De Gen. ad litt.* I.i.2; iii.5; Aquinas, *Summa theologiae* Ia.lxvi.3.

[59] Ten spheres were envisioned by Thomas Aquinas, Albert of Saxony, Peter D'Ailly, and probably Roger Bacon. Robertus Anglicus, Michael Scot, and Campanus of Novarra probably envisioned nine, and Oresme allowed only eight; Grant, 'Cosmology', pp. 275-78; *Planets*, pp. 315-23. For a good illustration with ten spheres, the ninth being crystalline, see Peter Apian's *Cosmographicus liber* (1524); Toulmin and Goodfield, *Fabric of the Heavens*, Plate 9 and p. 163; S.K. Heninger, Jr., *The Cosmographical Glass* (San Marino: Huntington Library, 1977), pp. 37f. The idea that the supracelestial waters were crystalline was introduced by Josephus, Jerome, and Ambrose; O. Zöckler, *Geschichte der Beziehungen zwischen Theologie und Naturwissenschaft* (2 vols., Gütersloh, 1877-9), vol. I, pp. 63, 226; J.K. Wright, *Geographical Lore*, pp. 58f.

The Aristotelian idea of the four elements was already known in the twelfth century, but the envelopes were given an egg shape, and the sphere of fire extended beyond the moon to the edge of the universe: e.g., Peter Abelard, *Exposition of the Hexaemeron*; William of Conches, *De philosophia mundi* IV.1; J. K. Wright, *Geographical Lore*, p. 151.

Early Alexandrian Jews and Christians (before Athanasius; cf. the previous note) had attempted syntheses of Plato's homocentric cosmology with the biblical notion of heaven as the abode of God and the angels: e.g., Philo, *Preliminary Studies* 103ff.; *On Genesis* IV.110; Clement of Alexandria, *Miscellanies* II.11; V.14; Origen, *On First Principles* II.3.6; 11.6f. Late twelfth-century syntheses like those of Bernard Silvestris and Alan of Lille were also

Since God was located, symbolically at least, beyond the outermost created heaven, the effect of the Aristotelian cosmology with its nine or ten heavens was that God's action should appear to be rather more remote from terrestrial events than was traditionally thought to be the case. We speak here only of the normal mode of God's activity (*de potentia ordinata*), not of the occasional use of his absolute power (*de potentia absoluta*) in miracles which we shall discuss below. Indeed, the incorporation of the Aristotelian cosmology into the older Christian worldview would not have been possible if such a distinction had not previously been worked out in the twelfth century, as we have seen.

In Jewish and Christian cosmologies of the intertestamental and New Testament periods, there had been a number of heavens between God and humanity, ranging from three to ten in number. But these were strictly angelic heavens, which defined the spiritual chain of command from God to the physical world and the mystic path of ascent from the world to God. They were not spatial spheres within the physical world itself.[60] Even Bede's more naturalistic cosmology of the eighth century contained only two distinct levels (the stellar firmament and the waters above the firmament) between the elemental zones around the earth and the spiritual heavens of God and the angels.

So with the influx of Aristotelian thought a spatial gap threatened to open up between the regular activity of God and events on earth. Grosseteste, for instance, held that the diurnal rotation of the first moved sphere was communicated to it by God in such a way that its motion was transmitted to the lower spheres and finally to terrestrial phenomena.[61] The

based on Plato's writings. The emphasis in these cases, however, was on mystic ascent rather than on the mechanics of divine providence; cf. John E. Murdoch, *Album of Science* (New York: Scribner's, 1984), pp. 332-37, for illustrations of the Platonic and Aristotelian cosmologies, respectively.

[60] Benedikt Otzen, 'Heavenly Visions in Early Judaism: Origin and Function', in W.B. Barrick and J.R. Spencer, eds., *In the Shelter of Elyon*, *Journal for the Study of the Old Testament*, Supplement Series 31, 1984, p. 206.

Contrast the spatial arrangement of the spheres in the 'Dream of Scipio' (Cicero, *De re publica* VI.17) and *Corpus Hermeticum* I.10-26. The clear dualism of heaven and earth (parallelling that of soul and body, humans and animals) in these texts contradicts the second tenet of the creationist tradition. The importance of this distinction between spiritual and spatial hierarchies is missed by advocates of pan-Hellenism in the late Second Temple and New Testament period; e.g. James D. Tabor, *Things Unutterable* (Lanham, Md.: University Press of America, 1986), pp. 64-68.

[61] De luce (c. 1225-30); Clare C. Riedl, *Robert Grosseteste: On Light* (Milwaukee, Wis.: Marquette Univ. Press, 1942, 1978), pp. 6f., 15f. Note that God is the efficient cause of motion here, not just the formal and final cause as for Aristotle.

In contrast to this 'macrocosmic' explanation of terrestrial motion originating in the remote *primum mobile*, William of Conches and others prior to the thirteenth century had assumed more proximate 'microcosmic' causes, e.g., the upward expansion of terrestrial fire for the rotation of the celestial spheres and the flow of water from a fountain in the torrid zone for the tides and winds. On the other hand, William located the sources of terrestrial heat in the sun and stars; William J. Brandt, *The Shape of Medieval History* (New Haven: Yale Univ. Press, 1966), pp. 14f., 17, 18, 24.

same idea appears in Roger Bacon, William of Auvergne, Bonaventure, Albert the Great, and Thomas Aquinas in the thirteenth century.[62]

The remoteness of God's providence suggested by the new cosmology, of course, had to be counterbalanced, even with respect to the normal mode of God's activity, if Aristotelian science was to be acceptable to Christian faith. This was done in several ways. Bonaventure, Albert, and Thomas all attempted to restore the balance by limiting the influence transmitted through the celestial spheres to the physical, secular aspects of life. For example, the configuration of the heavens was responsible for the creation of worms and insects from putrefaction, and the radiation of the sun could influence the birth and death of higher animals, all, of course, under God's ultimate control.[63] There were two channels open, however, for the more immediate influence of God in human life under normal conditions: God could enlighten the soul or affect the will directly, and he could, and regularly did, infuse grace through the seven sacraments, particularly through the Eucharist.[64]

In effect, the normal, everyday life of medieval humans was viewed as taking place on two levels: one of nature, in which God's providence was mediated through the hierarchy of celestial spheres; and one of grace, in which God's power was mediated, for the most part, by the hierarchy of the church.[65] Thus, one of the most commonly cited features of high medieval

---

[62] Easton, *Roger Bacon*, pp. 51ff.; Weisheipl, 'Celestial Movers', pp. 162, 175f; Wildiers, *Theologian and His Universe*, pp. 46f., 53ff. Aquinas, of course, stressed that the normal application of God's power to the first moved sphere in no way compromised his omnipresence: *Summa contra Gentiles* III.68.

The independent movers or intelligences that Aristotle assigned to each sphere (*Metaphysics* XII) were in fact quite mechanical in their operation and were made entirely dependent on the First Cause by Albert, Aquinas, and other scholastics; Brandt, *Shape of Medieval History*, pp. 22f., esp. n. 74.

[63] Wildiers, *Theologian and His Universe*, pp. 53ff., 64, on Aquinas.

[64] *Ibid.*, pp. 47, 72; cf. Peter Brown, *Society and the Holy in Late Antiquity* (Berkeley: Univ. of California Press, 1982), pp. 326-31, on the eleventh and twelfth-century background. The sphere of God's immediate influence could only be an enclave within the natural order. For the individual this meant an inward sense of spirituality within the context of outwardly secular life. For the church it meant an experience of God's presence in great cathedrals and liturgies that were subsidised by the merchant and artisan classes. On the interior-exterior bifurcation in Gothic architecture and its relation to liturgy, see Arnold Pacey, *The Maze of Ingenuity* (Harmondsworth: Allen Lane, 1974), pp. 45f., 50f., 72ff.

[65] The medieval and Renaissance popes continued to regard themselves as unbound by any law. For example, Innocent VIII declared that he was not bound by his own oath to name no more than one member of his family to high office once he was elected pope (1484); Justo González, *The Story of Christianity*, 2 vols. (San Francisco: Harper, 1984), 1:372. Yet after the Gregorian reform of the eleventh century, popes increasingly legitimised their power in secular affairs by analogy to the ordinances of creation. Innocent III (pope, 1198-1216) compared the authority of the pope to the rule of the sun (in Gen. 1:16ff.) and the rule of the soul over the body (Gonzalez, *Story*, 301, 308). Giles of Rome compared papal power to the rule of the heavens over terrestrial elements as well as the rule of the soul over the body; *On Ecclesiastical Power* (1301), in William Placher, ed., *Readings*

thought, the dichotomy of nature and grace, can be understood partially as an indirect result of the impact of Aristotelian science. The cosmology of Bede had allowed an interpenetration of the two, as God and the angels were within two or three heavens of the tallest mountain (Olympus). The extensive spatialisation of the normal God-world relation brought about by the assimilation of Aristotle's cosmology, however, forced an intensification of the God-soul relation in order to legitimise piety within the new worldview. And at a time when the gap between secular life and ecclesiastical tradition was widening, the church had to consolidate its control over the channels of grace in order to counteract the apparent stranglehold of the new Aristotelian science over the realm of nature.

What happened to the idea of the restricted immediacy of God's normal action after Aquinas? The Condemnation of 1277 did not address this issue directly.[66] In fact, Bonaventure, who was the principal theological influence behind the 1277 Condemnation, was a strong proponent of the idea. As a true Augustinian, Bonaventure was content to view the natural order as exemplifying or reflecting the attributes of God from a distance and seeking direct access through mystic ascent and sacramental grace.[67]

The Aristotelian cosmology maintained its hold on Western European thought well into the Renaissance. The development of mechanical clocks in the late thirteenth and early fourteenth centuries provided the new image of clockmaker for God to take the place, or, at least, to supplement, the image of sphere-mover, but the consequences for the doctrine of God were much the same. Two figures who illustrate the transition are Nicole Oresme (d 1382) and Henry of Langenstein (d 1397).

In his commentary on Aristotle's treatise *On the Heavens* (1377), Oresme discussed at some length the question of how God could be said to be in heaven while at the same time being omnipresent. The reason for the apparent discrepancy, he concluded, was that a cause is properly said to be present where its action is most evident, and, since the heavens are the most evident effect of divine providence, God is properly said to be located in (or just beyond) them although, strictly speaking, he is everywhere.[68]

How do we know that the heavens are the most evident effect of God's providence? Oresme listed a number of indications including their great size, their permanence, their influence on terrestrial events, their orderly

in the *History of Christian Theology*, 2 vols. (Philadelphia: Westminster Press, 1988), pp. 165ff.

[66] Art. 16 (Mandonnet numbering) did warn against the thesis, 'That the first cause is the most remote cause of all things', but only in the instance that it were taken to exclude the belief that the first cause is also the most proximate of all things (Hyman and Walsh, eds., *Philosophy*, p. 543).

[67] 'A metaphysic of Christian mysticism—that is the final term towards which his thought tended': Gilson, *Philosophy of St Bonaventure*, p. 77; cf. Knowles, *Evolution*, pp. 238-48.

[68] *Book of Heaven and Earth* II.2, trans. Albert D. Menut, *Nicole Oresme: Le Livre du ciel et du monde* (Madison: Univ. of Wisconsin Press, 1968), pp. 277-85.

arrangement, and their ceaseless, regular movement. In connection with the latter two features—arrangement and movement—he introduced Cicero's analogy between the cosmos and a clock, saying that the regular movement of the heavens must depend on the power of some higher intellect just as that of a clock does, even though a clock does not have spiritual beings like angels as part of its machinery the way the real cosmos does.[69]

Oresme went on to describe how God imparted motion to the heavens at creation as follows:[70]

> ...the situation is much like that of a man making a clock and letting it run and continue its own motion by itself. In this manner did God allow the heavens to be moved continually according to the proportions of the motive powers to the resistances and according to the established order.

One can almost visualise a master craftsman balancing the wheels and weights of a fourteenth-century town clock as one reads this passage.

Oresme's idea of a quantity of motion imparted to the heavens once and for all was a relatively new one in fourteenth-century Europe and was at variance with the idea of continuously caused motion in Aristotelian physics. We shall return to this idea in section C. The idea of the remoteness of God's action from human affairs suggested by the clockwork image, however, was a direct inheritance from the thirteenth-century hierarchy of celestial spheres which combined elements of the creationist tradition with the newly discovered cosmology of Aristotle.[71]

Henry of Langenstein's *Lectures on Genesis* provide us with a representative synthesis of science and theology at the end of the fourteenth century. Though preferring Aristotle's account of the motion of the spheres to the newer idea of impetus, Henry, like Oresme, used the image of clockmaker to describe the normal action of God in relation to the world. Just as the craftsman assembles all the parts of a clock and then sets it all in motion by moving just one part, so God set the entire world in motion just by energising the angels, who in turn moved the heavens. The whole system thus formed a 'golden chain' (*catena aurea*) of efficient causes extending all the way from God to the natural phenomena that occur on earth.[72] Again the historical connection is quite clear between the

---

[69] *Ibid.* (p. 283).

[70] *Ibid* (p. 289). The French text reads 'aucunement semblable', which led A.C. Crombie to think that Oresme was denying the analogy: *Augustine to Galileo*, vol. II, p. 89.

[71] The continuity between the ideas of God as First Mover and as clockmaker is supported by evidence for the continuity in technological development between the astrolabe and equatorium inherited from the Arabs and the earliest mechanical clocks of Western Europe: Pacey, *Maze of Ingenuity*, pp. 66-71.

[72] Steneck, *Science and Creation*, pp. 90ff.

cosmology of Aristotle and the clockwork image that was gradually taking its place.

Thus the modern world was to inherit from the Middle Ages one of its basic images for describing the relationship between God and the world (the image of the two books being another). The revolution that occurred in the Renaissance did not introduce the idea of mechanism:[73] what it did was to destroy the medieval idea of the cosmos as a spatial hierarchy between humans and God.

Already in 1440, in his treatise *On Learned Ignorance*, Nicholas of Cusa had maintained that the cosmos had neither centre nor circumference (though it still had celestial spheres), for all creatures were equally proximate to the deity who was their true centre and circumference.[74] In so doing he restored the sense of the unity of heaven and earth found in the creationist tradition of the early church (section 1.3) and signalled the beginning of the end of the hold Aristotelian cosmology had had on the Western mind for over two centuries.

Nonetheless, the restricted immediacy of God's normal activity that characterised thirteenth-century thought continued to be a factor in the post-Aristotelian world. By dissolving the matrix of thought out of which the image of God as clockmaker was born, Cusa allowed the clockwork metaphor to live on in a world in which it could be applied directly to the terrestrial sphere, and even to life on earth, as well as to the heavens.[75] So, while Cusa's intent was to portray every creature as a 'created god',[76] the eventual result of his revolution was the reduction of every creature to a clocklike mechanism.

As we pause on the threshold of the mechanistic philosophy of the modern Western world, we are in a position to review the key steps that led up to this point. In chapter 1 we traced the movement of the creationist tradition through four historical stages: (1) the ancient Near Eastern view (shared by the Old Testament) of the cosmos as subject to divinely ordained laws; (2) the beginnings of the idea of the relative autonomy of nature under the impact of Hellenistic thought in the intertestamental and early Christian periods; (3) the greater emphasis on the transcendence of God and the deterministic course of nature that began with Augustine and the cultural shift from the Hellenistic East to the Latin West; and (4) the dichotomy

---

[73] Pace R.G. Collingwood, *The Idea of Nature* (London: Oxford Univ. Press, 1945), pp. 5-8.

[74] *On Learned Ignorance* II.2.11ff.; Ernst Cassirer, *The Individual and the Cosmos in Renaissance Philosophy* (Philadelphia: Univ. of Pennsylvania Press, 1972), pp. 24-28.

[75] Cusa himself used the images of an equatorium (*On Learned Ignorance* II.12.178), a spinning top (*On Actualised Possibility* 23), and a spinning globe or rolling ball (*On Games with Spheres* I).

[76] *On Learned Ignorance* II.2.104.

between God's normal role (*potentia ordinata*) and his occasional displays of supernatural power (*potentia absoluta*) that arose in the late eleventh and early twelfth centuries. In this chapter we have traced two further stages in the development: (5) the quasispatial distancing of God's normal activity under the impact of Aristotelian cosmology in the thirteenth century; and (6) the emergence of the idea of the clockwork mechanism along with the first extensive production of weight-driven clocks in the fourteenth century.

On the whole, there was continuity in the portrayal of the ceaseless regularity of the cycles of nature. What changed was the understanding of God's relationship to natural causation. From the ancient Near Eastern ideal of divine kingship to the Neoplatonic and Augustinian concept of transcendent Being, to the Aristotelian First Mover, to the late medieval Clockmaker, the idea of God's normal activity became gradually less immediate to the events of the world, leaving the relatively autonomous cycles of nature to take on the appearance of a completely autonomous mechanism.

2.3 The Problem of God's Absolute Power

The problem of God's absolute power (*potentia absoluta*) was not new with the thirteenth century. Augustine had struggled with it intermittently, and it had played a significant role in the conflict between the two wings of the creationist tradition in the twelfth century (chap. 1 sec. D).[77] The impact of Aristotelian naturalism in the thirteenth century was bound to intensify the problem, however, particularly as it was incorporated within an overall synthesis rather than championed exclusively by one party as twelfth-century naturalism was.

In our discussion of 'God as First Mover and Clockmaker', we saw that the assimilation of Aristotelian cosmology led to a greater sense of the remoteness of God's normal activity. This assimilation would not have been possible if there had not been a clear distinction in the Western European mind between the normal activity of God and his occasional miraculous displays of power, a distinction that was forged in the eleventh and twelfth centuries. Christians of the thirteenth century could assimilate the naturalism of Aristotle precisely because there was always the possibility of reverting to God's absolute power when the ideas of the potential eternity of the world and the hierarchy of natural causes threatened to compromise the sovereignty and freedom of God.

Among the first to experience the tension was Roger Bacon (d *c.* 1292), an early advocate of scientific research. While God normally worked through the medium of second causes for the sake of order, Bacon argued,

---

[77] Chapter 1, text above notes 174-90.

one must allow for God's absolute power, for which no ulterior cause or reason can be sought, both in his creation of the world at the beginning of time and in his acting without the medium of second causes at various subsequent times in history.[78]

Thomas Aquinas stressed the normal activity of God through the agency of celestial spheres and their angelic movers 'according to the order of nature'. Yet, at the same time, he tried to dispel the notion that such regularity in any way compromised God's ability to act 'apart from the order of nature' at any time or place.[79] In fact, the order of nature was, for Thomas, doubly contingent. In the first place, God could have ordained an entirely different (normal) order by virtue of his being its creator. But, secondly, God is not bound, even now, by the order he has established and which we normally observe. He can either produce ordinary effects without the precedence of their normal causes, or he can produce unprecedented effects within the normal order of things.[80]

However, for many churchmen of the late thirteenth century, the carefully balanced statements of Thomas were insufficient protection against the unrestrained speculation of arts masters like Siger of Brabant. As mentioned earlier, there was a massive reaction in the years after 1267, culminating in the Condemnation of 219 Theses by the bishop of Paris, Stephen Tempier, in the year 1277. The articles of condemnation that stressed God's absolute power as creator held that the order of the natural world was freely chosen by God and that a plurality of worlds (presumably with a plurality of 'natural' orders) could just as well have been created.[81] Articles that stressed God's absolute power even within the established order held that God can do things that are impossible according to the order of nature, e.g., that he can move the heavens in a straight line (as well as circularly) thus leaving a vacuum, and that he can act directly in nature without the mediation of second causes.[82]

The articles affirming the possibility of a vacuum and the possibility of a plurality of worlds were of particular importance. Both possibilities were ruled out by Aristotelian natural philosophy.[83] Hence, the Condemnation encouraged the exploration of new, non-Aristotelian hypotheses in fourteenth-century natural philosophy as we shall see in section C. For the moment, we need only note that the Condemnation of 1277 was a

---

[78] Easton, *Roger Bacon*, pp. 51f., 57.
[79] *Summa contra Gentiles* III.68.8.
[80] Summa Theologiae I.105.6; E. Gilson, *The Spirit of Mediaeval Philosophy* (New York: Scribner's, 1936), pp. 376f.
[81] Arts. 20, 27A (Mandonnet numbering; Hyman and Walsh, eds. *Philosophy*, p. 544).
[82] Arts. 17, 22f., 66-69 (*ibid.*, pp. 544f.).
[83] E. Grant, 'The Condemnation of 1277, God's Absolute Power, and Physical Thought in the Late Middle Ages', *Viator*, vol. x (1979), pp. 217ff.

conservative reaction to the naturalism of Aristotle as it affected the concept of God in the thirteenth century. As such, it stood in a long line of such reactions among the people of God. We have already discussed the opposition to Hellenistic culture within conservative Jewish circles in the second century BC, the critique of Greek philosophy by Irenaeus and Tertullian (second to third century AD), the reaction of Boniface to Virgil of Salzburg in the eighth century, and that of William of St Thierry to William of Conches in the twelfth. As in these earlier cases, the Condemnation of 1277 was not an attack of theology against science, but rather an attack of one wing of the creationist tradition—with its own ideal of science as well as religion—against another.

The Condemnation of 1277 had an immense influence on subsequent theological thought. In some quarters, particularly among Franciscan theologians like Duns Scotus and William of Ockham, it contributed to an increasing emphasis on the absolute power and will of God, partially at the expense of the predictability of natural processes and the reliability of human reason. Among Dominicans, on the other hand, it seems to have precipitated a reverse reaction in the adoption of Aquinas's theological synthesis as official doctrine. The net result was the well known fragmentation of scholastic theology into opposing schools (Thomists, Scotists and Ockhamists), that characterised late medieval thought.[84]

Still, we find a scientist-theologian like Henry of Langenstein in the fourteenth century trying to sort out the alternatives as best he could within an Aristotelian framework. God established the normal order of cause and effect at creation, Henry reasoned, but God could change that order at any time as he did, for instance, in the great deluge of Genesis 7.[85] For the most part, scientific reason could determine the causes of natural phenomena, but, when a problem proved recalcitrant to reason, one could always appeal to the absolute power and inscrutable will of God. For example: how could the elements have passed through the lower celestial spheres (understood as physical entities, after Aristotle) on their way to the eighth heaven in order to form the stars? By a supernatural act of God! Why do the stars twinkle while the planets do not? Because God made them that way![86]

In other words, even though Henry's universe was mechanical like a clock, there were gaps in the natural order which could only be filled by an appeal to the direct action of God. Fortunately, there were enough gaps to allow a certain credibility to the occasional exercise of God's absolute

---

[84] Gilson, *History*, pp. 408ff.; Knowles, *Evolution*, pp. 299f.
[85] Steneck, *Science and Creation*, pp. 27f., 34f., 87, 148f.
[86] *Ibid.*, pp. 61ff.

power. Four centuries would have to pass before mathematics and physics developed to the point where there would no longer appear to be room for God's direct action in nature. But the framework of thought in which God's working through second causes (*de potentia ordinata*) and his acting directly (*de potentia absoluta*) were viewed as antithetical was already well established in the Middle Ages. The almighty God of Scripture was well on his way to becoming a 'God of the gaps'.

C. THE INFLUENCE OF MEDIEVAL THEOLOGY ON NATURAL SCIENCE

The most important effect medieval theology had on the development of natural science was its legitimation of efforts to study the newly discovered texts and to assimilate their contents within the Christian worldview (chap. 2 sec. A). In spite of its shortcomings from a modern perspective, Aristotelian science was the only available body of knowledge that treated nature comprehensively and systematically. If Western science was to develop at all, it had to start somewhere, and there was no better place to start than with Aristotle. Consequently, had the theological tradition of Western Europe been such that the assimilation of the new ideas were impossible or counterproductive, it is difficult to see how any progress could have been made at all.

On the other hand, had the theological tradition of Western Europe been such that there was no adequate basis for an effective critique of Aristotelian thought, then the scientific standards of the Greeks and Arabs might have been equalled, but they could never have been surpassed. Our task in this section will be to identify those theological determinants in medieval thought that may have assisted in the development of a post-Aristotelian science.

There are basically two ways in which medieval theology affected the development of science. One was its influence on the concept and method of natural science. The other was its influence on the late medieval concept of the cosmos as God's creation, an influence which pointed beyond the cosmology of Aristotle.

1 The Concept and Method of Natural Science

The degree to which the modern understanding of the concept and method of natural science was already developed in the Middle Ages has been one of the most debated subjects of recent historical studies.[87] The present

---

[87] For a brief survey of the debate, see Grant, *Physical Science*, pp. 114f.; Dales, *Scientific Achievement*, pp. 172ff.

consensus is that several preliminary contributions were made during the medieval period. These include:
  (1) the relative autonomy allowed to the sciences in relation to theology;
  (2) the value assigned to mathematical method and quantification;
  (3) the importance of observation and experiment.
On all three of these points, the theological input had a mixed effect.

1.1 Relative Autonomy of Natural Science

The principal contribution of thirteenth-century theology was its effort to construct an overall synthesis that gave a place to the sciences along side of theology itself. In general, the sciences were means to religious and social ends, just as undergraduate studies in the arts and sciences were normally followed by postgraduate studies in theology, law, or medicine.

Still, there were variations in emphasis. Since there was no universally accepted way of constructing a synthesis, there was no universally accepted understanding of the role of the sciences. Dominicans like Albert the Great and Thomas Aquinas leaned toward a greater degree of autonomy for the sciences within their own sphere,[88] though the ultimate goal assigned was the improvement of life and the strengthening of faith.[89] Franciscans of the thirteenth century like Bonaventure and Roger Bacon placed more stress on spiritual and social ends. Bacon even suggested that the pope exercise a measure of control over scientific research in order to prevent its benefits from falling into the wrong hands and being used for anti-social (or anti-Christian) purposes.[90]

The Condemnation of 1277 was a reaction to what appeared to be too great an autonomy for science as conceived by Latin Averroists like Siger of Brabant and Dominicans like Thomas Aquinas. Paradoxically, however, its effect may have been to intensify the distinction between scientific and theological considerations. The propositions condemned for the most part dealt with theology and metaphysics, or what we would term the presuppositions of science. The result was that matters of theology were reserved as the special province of trained theologians.[91]

There was no restriction on the freedom of scientists to explore the workings of nature, however. In fact, the rejection of various Aristotelian

---

[88] Wallace, 'Philosophical Setting', p. 96; Copleston, *History*, vol. II/2, p. 151.
[89] Glacken, *Traces*, pp. 227ff.; Aquinas, *Summa contra Gentiles* II.2f.
[90] Gilson, *Philosophy of St Bonaventure*, pp. 103ff.; Dawson, *Medieval Religion*, pp. 85f., 90f.
[91] From 1272 on, the arts masters at Paris were required to take an oath that they would not 'presume to determine or even to dispute any purely theological question'; Grant, *Much Ado*, p. 331, n. 6.

propositions clearly encouraged speculation in this area.[92] In the fourteenth century, consequently, we find philosophers like William of Ockham and John Buridan, who were influenced by the Condemnation, making a clear distinction between the methods of natural science and those of theology.[93] Gradually, the two wings of the creationist tradition were turning into two separate professions or disciplines, each with its own subject matter and methodology, though the actual separation of science and theology was still a distant prospect in the late Middle Ages.

If it is difficult to achieve a clear definition of the autonomy allowed the sciences in the Middle Ages, it is even more difficult to reach an assessment of the result. Is a high degree of autonomy for the sciences a good thing? Or should spiritual and social goals be made primary and science treated as a means? Different assessments of the attitudes of medieval theologians often reflect the values of the contemporary historian as much as those of the historical subject.

Perhaps the fairest thing that can be said is that medieval theology allowed the vigorous pursuit of scientific questions while, at the same time, holding out the hope that progress on these matters would be beneficial both to the individual scientist as a person and to Western civilisation as a whole. In a society where the most ambitious youths were motivated partly by a desire for truth and partly by a desire for glory—glory for themselves and for their homelands—nothing could have provided a greater impetus for scientific development than this.

1.2 Mathematical Method and Quantification

Historians like A.C. Crombie hold that one of the most significant contributions of the Middle Ages to the development of the natural sciences was its belief in the power of mathematics for the understanding of natural processes.[94] The emphasis on mathematics came in a series of developments culminating in the fourteenth century. In the background was the creationist belief in the comprehensibility of the world epitomised in the frequently cited text of Wisdom 11:20: 'You have arranged all things by measure and number and weight.' The Latin West inherited its belief in the mathematical structure of nature through the Christian Neoplatonism of

---

[92] Grant, 'Condemnation of 1277', pp. 239ff.; *Much Ado*, pp. 116f.
[93] Copleston, *History*, vol. III/1, pp. 82-85, 166; Jaki, *Science and Creation*, p. 232.
[94] Crombie, *Augustine to Galileo*, vol. II, pp. 100, 118; 'The Significance of Medieval Discussions of Scientific Method for the Scientific Revolution', in Marshall Clagett, ed., *Critical Problems in the History of Science* (Madison: Univ. of Wisconsin Press, 1959), pp. 87f.; for a critique from the Aristotelian viewpoint, see J.A. Weisheipl, *The Development of Physical Theory in the Middle Ages* (Ann Arbor: Univ. of Michigan Press, 1959, 1971), pp. 48ff., 60ff., 87f. Weisheipl does not dispute the fact or the importance of the medieval emphasis on mathematics, however.

Augustine and Boethius[95] which was revived in the eleventh and twelfth centuries, particularly at the cathedral school of Chartres, and passed on through Robert Grosseteste and others in the early thirteenth century.[96]

The principal contributions of the thirteenth century to the mathematisation of natural science came from Robert Grosseteste (d 1253) and the so-called 'perspectivists', including Roger Bacon, John Pecham, Witelo of Silesia, and Theodoric of Freiburg. These writers excelled in their investigations of the laws of the reflection and refraction of light and their applications to the optics of lenses and to the rainbow.[97] In contrast to Aristotle and his commentators, the perspectivists were concerned more with structures and laws in nature than with efficient, formal, and final causes.[98]

Contemporary with the perspectivists, yet working more within the Aristotelian methodology, were Albert the Great and his student, Thomas Aquinas. Albert seems to have been influenced directly by Grosseteste, particularly in his treatment of the rainbow,[99] even though he rejected the subordination of natural science which he sensed in the writings of certain 'friends of Plato', who may have included Roger Bacon.[100] The influence of Grosseteste, or, at least, of Neoplatonism, was partly responsible for Albert's allowance for an 'incipient actuality' (*incohatio formae*) in prime matter, although the latter, according to Aristotelian principles, was supposed to be pure potentiality.[101]

Aquinas steered even further away from Platonism than Albert and insisted on the pure potentiality of matter. Yet, in order to avoid making nature appear to be recalcitrant to formative influence from above, he ascribed to it a 'capacity for obedience' to God's command (*potentia obedientialis*), a capacity instilled at creation by God himself. The implication for Aristotelian philosophy was that matter was susceptible to quantitative determination, that very susceptibility being educed from the pure potentiality of prime matter by an efficient cause.[102] The historian Max Jammer has argued that Aquinas's significant modification of the

---

[95] Ernan McMullin, 'Augustine of Hippo', *DSB*, vol. I, p. 337; Dales, *Scientific Achievement*, pp. 30f.

[96] Crombie, *Robert Grosseteste*, chap. 2; James McEvoy, *The Philosophy of Robert Grosseteste* (Oxford: Clarendon Press, 1982), pp. 168-80, 212ff.

[97] *Ibid.*, chaps. 5, 7ff.

[98] *Ibid.*, pp. 4f., 10f., *passim*; *Augustine to Galileo*, vol. I, pp. 83, 112ff.; vol. II, pp. 38f., 100.

[99] *Ibid.*, pp. 189-200.

[100] Weisheipl, *Development*, pp. 48ff.

[101] J.A. Weisheipl, 'The Concept of Matter in Fourteenth Century Science', in Ernan Mccmullin, ed., *The Concept of Matter in Greek and Medieval Philosophy* (Notre Dame: Univ. of Notre Dame Press, 1963), pp. 151f.

[102] Gilson, *Spirit of Mediaeval Philosophy*, pp. 377-81; Copleston, *History*, vol. II/2, pp. 46f.

Aristotelian notion of matter thus opened the way for the modern, quantitative concept of mass. Jammer has also described how Aquinas's attempts to analyse the transubstantiation of the elements in the Eucharist and the nature of the resurrection body played a role in this development.[103]

The principal contributions to the mathematisation of natural science in the fourteenth century came from Thomas Bradwardine (d 1349) and his associates and successors, known as the 'Oxford calculators', at Merton College. In his treatise *On the Proportions of Velocities in Motions*, which was written in 1328 as he began his postgraduate studies in theology, Bradwardine used the latest algebraic methods to define the concept of instantaneous velocity and to work out a new law for its dependence on motive power and resistance. 'Bradwardine's law', as it came to be known, was widely used until the seventeenth century, when it was replaced by the more accurate law of Galileo.[104]

It is not clear whether Bradwardine was influenced by Grosseteste and the thirteenth-century perspectivists to any extent.[105] Two direct theological influences that may have contributed to his appreciation for mathematics, however, were the writings of Augustine and a twelfth-century hermetic treatise called *The Book of the Twenty Four Philosophers*. The latter reintroduced the idea of God as limiting and containing all things that was so prevalent in intertestamental and early Christian thought and was to be popularised again in the fifteenth century by Nicholas of Cusa.[106]

It has been suggested that Bradwardine's faith in the power of mathematics to determine the laws of nature influenced John Buridan (d c. 1360) in his attempt to give quantitative measure to the idea of impetus.[107] If so, then both of the major trends toward the mathematisation of natural science in the fourteenth century could be seen as parts of a coherent process rooted in the creationist tradition. The work of other physicist-theologians of the later Middle Ages like Nicole Oresme (d 1382), Henry of Langenstein

---

[103] M. Jammer, *Concepts of Mass in Classical and Modern Physics* (Cambridge, Mass.: Harvard Univ. Press, 1961), chap. 4, esp. pp. 43ff. For the historical sequel, see Weisheipl, 'The Concept of Matter', pp. 163-69.

[104] Crombie, *Augustine to Galileo*, vol. II, pp. 70f.; *idem, Robert Grosseteste*, pp. 178-81; Weisheipl, *Development*, pp. 72-77; Anneliese Maier, *On the Threshold of Exact Science* (Philadelphia: Univ. of Pennsylvania Press, 1982), pp. 154-57; cf. Heiko A. Oberman, *Archbishop Thomas Bradwardine* (Utrecht: Kemink & Zoon, 1957), pp. 14ff., for the dating of Bradwardine's theological interests and studies.

[105] According to Crombie (*Robert Grosseteste*, pp. 178f.) and G. Leff (*Bradwardine and the Pelagians* [Cambridge: Cambridge Univ. Press, 1957], p. 113), he was influenced by Grosseteste. According to Weisheipl (*Development*, p. 72), he wasn't.

[106] E. Grant, 'Medieval and Seventeenth-Century Conceptions of an Infinite Void Space beyond the Cosmos', *Isis*, vol. lx (Spring 1969), pp. 42ff., 46f.; Karsten Harries, 'The Infinite Sphere: Comments on the History of a Metaphor', *JHP*, vol. xiii (Jan. 1975), pp. 7f. On the early history of the idea, see chapter 1 section B above.

[107] Crombie, *Augustine to Galileo*, vol. II, pp. 80f.

(d 1397), and Nicholas of Cusa (d 1464) gives us further evidence of the continuing vitality of the creationist belief in the power of mathematics in the period leading up to the Renaissance and the scientific revolution of the sixteenth and seventeenth centuries.[108]

1.3 Observation and Experiment

The medieval attitude towards observation and experiment was less developed than that towards the value of mathematics. The thirteenth and fourteenth centuries were a period of great technological progress: some of the instruments that were later to play a role in the scientific revolution (mechanical clock, pendulum, scales, magnifying lens, and magnetic compass) received their basic forms during this period. However, on the whole, these developments were not sufficiently advanced as yet to allow a precision of experimental testing comparable to the precision of the mathematical laws that were being worked out theoretically.[109] The thirteenth and fourteenth centuries were also a period of great advance in military technology, mining, medicine, alchemy, and navigation. For the most part, however, the new facts and ideas being discovered in these productive 'arts' were not yet incorporated into the theoretical sciences.[110]

It should not be concluded, on the other hand, that medieval writers did not attribute great value to the practical and productive disciplines in principle.[111] If they were falling behind the times in failing to make use of the results of new technological developments, they were often ahead of their times in suggesting new projects of invention and reform. The problem, in short, was that the theoretical and experimental wings of science were not well coordinated.

Traditional practical concerns of the creationist tradition like the need for calendar reform and the benefits of medicine were exhibited by Robert

---

[108]Jaki, *Science and Creation*, pp. 234-40; Steneck, *Science and Creation*, pp. 45f., 84; Dijksterhuis, *Mechanization*, pp. 230f.; Cassirer, *Individual and the Cosmos*, pp. 14, 52ff.

[109]Crombie, *Augustine to Galileo*, vol. II, pp. 109f., 127f. Notable exceptions were experiments with optical lenses and mirrors by Grosseteste and with magnets and mirrors by Peter (Peregrinus) of Maricourt: *ibid.*, vol. I, pp. 114ff.; Pacey, *Maze of Ingenuity*, pp. 69-75. Peter of Maricourt was able to determine that the Pole Star was not at the exact North Celestial Pole, an observation requiring a resolution of one degree (*Epistola de magnete, c.* 1269).

[110]A.C. Crombie, 'Quantification in Medieval Physics', *Isis*, vol. lii (1961), pp. 155ff.; Steneck, *Science and Creation*, p. 83.

[111]Even in the early church, exegetes like Augustine readily used simple experimental reasoning in their commentaries. No mathematics was involved, however. See, for example, the description of an empty jar placed upside down in water and then tilted sideways to let the air out; *De Gen. ad litt.* II.ii.5, and note the lack of mathematical training exhibited in *ibid.*, I.v.9.

Grosseteste and Roger Bacon in the thirteenth century.[112] Of particular interest from our perspective is Bacon's attempt to redeem the practical magic arts as a source of potential benefit to society. The early church opposed these arts, he argued, only because they had been used by the pagans to deceive and defraud people. Once the faith of Christ had been accepted, however, the fraudulence of magic could be purged and the arts could be put to good use.[113] Here is a typical example of the belief in the possibility of restoration tempered by the criterion of social benefit which we have found to be an integral part of the creationist tradition (chap. 1, sec. E). It should be noted, however, that the possibilities of human art were, for Bacon, limited by the effects of the fall of Adam. Thus medicine, for instance, could prolong life, but not beyond the years of our first parents. The reversal of the effects of the Fall and the attainment of immortality must await the general resurrection.[114]

In contrast to the rather limited notion of scientific experiment, the medieval aptitude for the observation of nature was impressive. In fact, some of the greatest scientific advances were made with respect to atmospheric and celestial phenomena that were subject to direct observation.

This aptitude for observation can be seen as the culmination of a tradition reaching back to the work of Bede in the eighth century and mediated by the naturalism of the school of Chartres in the twelfth century. The thirteenth century provided two new stimuli, however, that must be given due credit.

The first stimulus was the Christianised Aristotelianism of Albert the Great and Thomas Aquinas. Here the emphasis on the normalcy of God's providential action through second causes promoted the idea of the relative autonomy of natural processes and encouraged the study of nature for its own sake.[115] This was particularly true of Albert, whose personal observations in biology, botany, and geology were among the first

---

[112]Crombie, *Robert Grosseteste*, p.136; 'Grosseteste's Position in the History of Science', in D.A. Callus, ed., *Robert Grosseteste, Scholar and Bishop* (Oxford: Clarendon Press, 1955), pp. 113ff.; Roger Bacon, *Opus Maius* IV; VI, trans. Burke, pp. 290-306, 617-26.

[113]Opus Maius I.14, trans. Burke, pp. 31ff. Cf. Theophilus Presbyter's argument that, with the coming of Christ, the skills in various crafts previously used only for personal profit were to be returned to the service of God; *On Divers Arts*, trans. Cyril Stanley Smith and John G. Hawthorne (Chicago, 1963); John Van Engen, 'Theophilus Presbyter and Rupert of Deutz: The Manual Arts and Benedictine Theology in the Early Twelfth Century', *Viator* xi (1980), p. 151.

[114]On the Marvelous Power of Art and of Nature 7, trans. Tenney L. Davis (Easton, Penn.: Chemical Publishing Co., 1923), pp. 36f. So also the Franciscan, John of Rupescissa, in his *De consideratione quintae essentiae* or *Liber de famulatu philosophiae* (third quarter of the fourteenth century); L. Thorndike, *History of Magic and Experimental Science*, vol. III (New York: Columbia Univ. Press, 1934), pp. 357f., 362.

[115]James J. Walsh, *The Popes and Science* (New York: Fordham Univ. Press, 1908), pp. 295ff.; Dijksterhuis, *Mechanization*, pp. 130-33.

independent achievements of Western European science.[116] Albert studied the embryological development of insects, fish, chickens, and mammals; he dissected crickets and crabs. He observed the effects of local floods and decided (against Aristotle!) that the Milky Way was a configuration of stars rather than a sublunar vapour.[117] His characteristic manner of certifying his conclusions was to say 'I was there and I saw it for myself'.[118]

The second major stimulus to the observation of nature in the thirteenth century came from the mysticism of Francis of Assisi (d 1226). In part a reaction to the mercantilism of his native city, in part a revitalisation of the cosmic scope of the language of the Psalms, Francis's love of nature showed Western humanity how to celebrate creation as the garment of God.[119] Though Francis himself was suspicious of science, his sense of the presence of God in nature was given philosophic credentials by Bonaventure and Raymond Lull and may have inspired subsequent generations of Franciscan scholars to take the scientific study of nature more seriously.[120]

It is difficult to determine whether these stimuli had any direct influence on the scientist-theologians of the fourteenth and fifteenth centuries. The Oxford calculators were purely theoreticians. John Buridan (d c. 1360) based many of his ideas on observations involving arrows, bellows, pendulums, water wheels, and grindstones, but whether these were actual observations or merely thought experiments is difficult to say.[121] Henry of Langenstein has been credited with pursuing the study of nature for its inspirational value, but his actual observations were limited to ones such as the fact that the stars twinkle while the planets do not.[122]

In the mid-fifteenth century, Nicholas of Cusa described experiments (or games) with spinning tops and globes and also with weights on a balance scale.[123] His account of one experiment designed to demonstrate the

---

[116]Dawson, *Medieval Religion*, p. 75.

[117]Crombie, *Augustine to Galileo*, vol. I, pp. 163f.; Glacken, *Traces*, pp. 227f.; Steneck, *Science and Creation*, p. 86.

[118]Dijksterhuis, *Mechanization*, p. 133; W.A. Wallace, 'Experimental Science and Mechanics in the Middle Ages', DHI, vol. II, p. 197a. Cf. Augustine, *De Gen. ad litt.* III.viii.12: 'I myself have observed this fact, and anyone who has the occasion and the desire may do likewise'.

[119]Glacken, *Traces*, pp. 214ff.; Cousins, *Bonaventure*, pp. 23f.

[120]The case is stated effectively, if not overstated, by George Boas, 'Introduction', *Saint Bonaventura: The Mind's Road to God* (Indianapolis, Ind.: Bobbs-Merrill Co., 1953), pp. xiv-xx.

[121]Crombie, *Augustine to Galileo*, vol. II, pp. 107-10, 127f. Buridan used mirrors and tubs of water for optical experiments, according to Mary Martin McLaughlin, *Intellectual Freedom and Its Limitations in the University of Paris in the Thirteenth and Fourteenth Centuries* (New York: Arno Press, 1977), pp. 122f. Compare Maier's rather more positive assessment of fourteenth-century natural philosophers in *On the Threshold*, pp. 146f.

[122]Steneck, *Science and Creation*, p. 141; cf. pp. 55, 62f.

[123]See note 75 above. Crombie, *Augustine to Galileo*, vol. II, pp. 111f.; Dijksterhuis, *Mechanization*, p. 231.

conservation of matter by weighing a quantity of earth before planting and after harvesting was actually carried out by Helmont in the seventeenth century. But Cusa's own description of the experiment closely follows a literary convention going back to the Pseudo-Clementine *Recognitions* of the fourth century (chap. 1 sec. D).[124]

The most that can be said is that the ideals of observation and experiment were well developed in the late Middle Ages and that this was partly the result of the creationist tradition. But the realisation of these ideals had to await the sixteenth century (chap. 3 sec. A). Only in areas like alchemy and medicine do we find medieval followers worthy of the traditions of Roger Bacon and Albert the Great.

Arnold of Villanova (d 1311), one of the first physicians in the Latin West to recognise the importance of alchemy for medicine, had studied theology under the Dominicans and was later associated with the Spiritual Franciscans.[125] In the tradition of Jesus ben Sirach (Ecclus. 38:1-8), Arnold taught that medicine was a gift from God and that the true physician needed divine illumination, as well as human reason, in order to diagnose and treat diseases even when the causes were natural.[126] Arnold's writings were foundational to Paracelsus and his chemical philosophy in the sixteenth century.[127]

Guy de Chauliac (d *c*. 1370) also studied under the Dominicans and under the Franciscan, Raymond Lull, and served as private physician to three popes. He was noted for his firsthand description of symptoms of the bubonic plague and for his advice that all surgeons study anatomy based on postmortem dissections.[128]

Guy's contemporary, John of Rupescissa, was, like Arnold of Villanova, a Spiritual Franciscan who applied the experimental techniques of alchemy to medicine. He specialised in the art of extracting the spirit or essence (technically, the fifth essence) of minerals and herbs in order to render them more effective as pharmaceuticals. John's alchemical writings show clear indications of experimental procedure and careful observation. He explicitly described the steps required to extract the essences of antimony and mercury.

---

[124]Henry M. Leicester, *The Historical Background of Chemistry* (New York: Dover Books, 1971), p. 88; H.M. Howe, 'A Root of Van Helmont's Tree', *Isis*, vol. lvi (Winter 1965), p. 411; cf. above, chap. 1, note 119.

[125]Dolan and Adams-Smith, *Health and Society*, pp. 75f.; Michael McVaugh, 'Arnold of Villanova', *DSB*, vol. I, pp. 289f. The judgement of earlier historians like John Maxson Stillman (*The Story of Alchemy and Early Chemistry* [New York: Dover Books, 1960], pp. 289f.) was rather more negative.

[126]Walter Pagel, *Paracelsus* (Basel: Karger, 1958), pp. 253f.; McVaugh, 'Arnold of Villanova', pp. 290f.

[127]Leicester, *Historical Background*, p. 96.

[128]Dolan and Adams-Smith, *Health and Society*, pp. 75ff.; Vern L. Bullough, 'Guy de Chauliac', *DSB*, vol. III, p. 218.

And, in recommending one of his prescriptions, he called on his readers to 'believe one who has tried it because I have tested it'.[129]

In general, the fourteenth and fifteenth centuries were less productive than one might have expected in view of the striking advances of the twelfth and thirteenth centuries. Not until the sixteenth century were experimental techniques sufficiently advanced to realise the possibilities inherent in the programs of observation and experiment suggested by Roger Bacon and Albert the Great.[130] The works of alchemist-physicians like Arnold of Villanova and John of Rupescissa were studied as classics for centuries, however, and helped pave the way for the rise of early modern chemistry.

2 Beyond the Natural Philosophy of Aristotle

We turn now to examine various ways in which medieval Christian theology may have challenged the physics and cosmology of Aristotle in such a way as to lead science beyond it. It should be realised at the outset that Aristotelian science was a comprehensive account of natural phenomena that was rather well suited to reality as it was experienced at the level of technology characteristic of the ancient and medieval world. Consequently, the actual replacement of Aristotle's cosmology could only take place in the event that (a) developments in instrumentation opened the way to a new 'perception' of the natural world and (b) a new account of natural phenomena could be devised that was as suited to the new perception of reality as Aristotle's was to the old.

In other words, there was no question of creationist theology causing a scientific revolution by itself. The most it could do was to suggest alternative hypotheses and to sanction the efforts of some of the more creative scientists of the day. As we saw in our discussion of the relative autonomy of natural science within the medieval synthesis, theology in the thirteenth and fourteenth centuries was in a position to do just that. In fact, it was in a better position than it had been before the thirteenth century or would be after the Renaissance. Before the thirteenth century, there had not been a sufficiently coherent and autonomous body of scientific knowledge in the Latin West with which theology could interact. After the Renaissance, the autonomy of natural science would be much greater and the credibility of theology much less.

A second point must be kept in mind. Despite an ordinance of the University of Paris in 1272 aimed at preventing arts masters from disputing theological questions,[131] theology was not yet the special preserve of trained

---

[129]John of Rupescissa, *De consideratione*; Thorndike, *History*, vol. III, pp. 358-63.
[130]Leicester, *Historical Background*, p. 84; Crombie, *Augustine to Galileo*, vol. II, p. 114.
[131]See note 91 above.

theologians any more than natural philosophy was the special preserve of a class of trained specialists in science. Even arts masters like John Buridan, who were not trained in the higher faculty of theology, were believing Christians, and their understanding of the faith was a determinant in their handling of scientific questions in spite of the fact that they were not allowed to address the theological questions in their own right. This working arrangement in the universities was the functional equivalent of the medieval synthesis, in which science and theology overlapped to a degree, while each maintained autonomy within its own sphere.

There were three basic developments in late medieval science that show the direct influence of theological doctrines such as creation and the omnipotence of God. They are:

(1) the possibility of a void;

(2) the gradual articulation of the idea of impetus;

(3) the suggestion of alternatives to the geocentric cosmology of Aristotle with its dualism of heaven and earth.

It is important to note the logical connection between these three developments as viewed against the background of Aristotelian physics: in the order we have stated them, they are directed against progressively more foundational tenets of the latter. A brief review of the logic of Aristotle's natural philosophy, begining with the most foundational tenet, will show this.

The key to the Aristotelian system of natural philosophy lay in the concept of natural place. Each of the four terrestrial elements had a natural place, with earth at the centre since it was the heaviest. The celestial bodies had no weight but were kept in their orbits by an attraction towards higher spheres culminating in God, the Unmoved Mover. The celestial bodies were thus different from sublunar ones: they were composed of a fifth element, and they showed signs of intelligence and desire, evidence of the role of spiritual beings in their motions. This foundational picture had become deeply embedded in the scholastic synthesis, as we saw in section B, and it would be the hardest aspect of Aristotelian physics to dislodge.

From the Aristotelian notion of natural place we can infer the idea of natural (intrinsically caused) motion and its opposite, forced (extrinsically caused) motion. Celestial bodies naturally moved in circles (epicycles could be added to account for anomalies). But sublunar bodies, at least simple ones composed of just one element, remained in their natural place and naturally moved back to that place if displaced. Any motion away from a sublunar body's natural place must, therefore, be continuously forced. Hence, the late medieval idea of impetus, which meant that unnatural motion could be sustained in the absense of an external force, undermined the distinction between natural and forced motion, and even called into question the concept of natural place that lay behind it.

On the basis of the ideas of natural place and forced motion, Aristotle denied the idea of a void space which had been posited by the Greek atomists. Among the many paradoxes that the existence of a void would lead to was the thought that any body in a void, once set in motion, would move to infinity. Moreover, it would have to move at infinite velocity since there would be no medium to resist it. Not only were these results counterintuitive, but they contradicted the rules of natural and forced motion mentioned above.[132] So, if one wanted to attack the Aristotelian system as a whole, one way to start would be to insist on the possible existence of a void.

2.1 The Possibility of a Void

One of the earliest contributions to the discussion was made by Thomas Aquinas. Thomas did not allow the existence of a void, but he did contend that the natural motion of a body in a hypothetical void would not be instantaneous. Even in the absense of a resisting medium, the quantitative magnitude (*corpus quantum*) which was educed (along with its natural form) from the pure potentiality of matter in the body would be enough to constitute a resistance.[133] As we have seen, the theological idea that lay behind Aquinas's interest in the quantification of matter was the absolute obedience of all things to the determination of God as exemplified in the transubstantiation of the eucharistic elements and the resurrection of the body.[134]

A more direct challenge to the Aristotelian denial of the void came from the Condemnation of 1277. As noted earlier, the Condemnation was a reaction of the conservative wing of the creationist tradition to the assimilation of Aristotle's cosmology in the scholastic synthesis. One of the articles explicitly stated that God could move the world in a straight line, thus leaving a vacuum or void. Another stated that God could create a plurality of worlds; this also suggested the possibility of a void space between the separate worlds.[135] The theological motive behind these bold suggestions was the desire to acknowledge the absolute power of God and

---

[132]E. Grant, 'Motion in the Void and the Principle of Inertia in the Middle Ages', *Isis*, vol. lv (Sept. 1964), pp. 265f.; *Much Ado*, pp. 6f.; J.A. Weisheipl, 'Motion in a Void: Aquinas and Averroes', in A.A. Maurer, ed., *St Thomas Aquinas, 1274-1974* (Toronto: Pontifical Inst. of Medieval Studies, 1974), pp. 473ff.

[133]Grant, 'Motion in the Void', pp. 270f.; *Much Ado*, pp. 38f.; Weisheipl, 'Motion in a Void', pp. 480f.

[134]See text above notes 101-2.

[135]Arts. 66 and 27A, respectively, in the Mandonnet numbering (Hyman and Walsh, eds., *Philosophy*, pp. 544f.); Art. 49 and 34 in the original numbering (Grant, *Physical Science*, p. 28).

the contingency of the natural order: God could have created a different order than the one he did, and he was not bound by the present order but could alter or even annihilate and recreate any portion of it.[136]

It is a well-documented fact that the Condemnation of 1277 influenced a number of scientists, as well as theologians, to speculate on the possible existence of a void. Just to give some idea of the variety of speculation, one should mention the following: Henry of Ghent (d 1293) argued that if God created a body outside our world (i.e., beyond the outermost celestial sphere), there would be an intervening three-dimensional vacuum.[137] Thomas Bradwardine (d 1349) inferred from the fact that God could move the world that the divine nature must exist beyond space and time in an uncreated, dimensionless infinity which is itself the 'place' of the world.[138] John of Ripa also started from the fact that the world could be moved, but he concluded instead that it must exist in a created void space of infinite proportions: even though infinite and distinct from God, this void space would be infinitely exceeded by the immensity of God![139]

All three of these examples concern the possibility of an extracosmic void, whether spatially extended or dimensionless. Albert of Saxony (d 1390), university rector and bishop, speculated also on the possibility of a void space *within* the cosmic order based on the power of God to annihilate any or all of the matter within it.[140]

It should be noted that emphasis on the omnipotence of God could also lead to the opposite conclusion: Duns Scotus (d 1308) argued (against Thomas Aquinas) that God could produce an effect at a distance and, therefore, that his omnipresence was not prerequisite to the exercise of his absolute power. Consequently, one could not assume the existence of an infinite void simply on the basis of God's ability to create (or move) the world wherever he might please![141] For these, or similar, reasons, the majority of scholastics affirmed the absolute power of God, and hence the possibility of an infinite void, yet concluded either that such a void was

---

[136] Text above notes 81-83.

[137] E. Grant, 'Place and Space in Medieval Physical Thought', in Peter K. Machamer and Robert G. Turnbull, eds., *Motion and Time, Space and Matter* (Columbus: Ohio State Univ. Press, 1976), p. 151.

[138] Grant, 'Medieval and Seventeenth-Century Conceptions', pp. 44ff.; 'Place and Space', pp. 144f.; *Much Ado*, pp. 135ff., 142f. On Augustine's view of God's transcendence, which undoubtedly influenced Bradwardine, see Leff, *Bradwardine*, pp. 28-31, and compare Kaiser, *Doctrine of God*, pp. 79ff.

[139] Grant, 'Place and Space', pp. 146-50; *Much Ado*, pp. 129-34. If Bradwardine's position was similar to Augustine's, Ripa's is reminiscent Origen's; see chapter 1 section B, text above notes 53-55.

[140] Grant, 'Motion in the Void', pp. 280f.; *Much Ado*, pp. 47f.

[141] Grant, *Much Ado*, p. 146; cf. Copleston, *History*, vol. II/2, p. 251.

unnecessary or else, like John Buridan, that there was no evidence to support it.[142]

It was not until Nicholas of Cusa (d 1464) that anyone suggested that the universe itself might be potentially infinite (from the point of view of God's unlimited power), since this more radical step required the rejection of the fundamental Aristotelian notions of natural place and a geocentric cosmos.[143] In the meantime, medieval speculations on the possibility of a void weakened the hold of Aristotle and made possible the discussion of the concept of a permanent impetus of motion.

2.2 The Idea of Impetus

Terrestrial motion that was 'natural' in the Aristotelian view required no explanation beyond the object's tendency to return to its natural place. The cause was purely internal to the body, or, perhaps, relational between the body and its natural place. There was no significant change on this issue in the late Middle Ages.

Motion that was either circular or away from a natural place did require an explanation, however. There had to be some means by which a force could be continuously applied as long as the circular or 'unnatural' motion lasted. The cause was, therefore, purely external to the body in the Aristotelian view. In the case of a projectile, for example, the initial thrust lasted for only a very short time. Hence, for Aristotle, the continuing upward motion of the projectile was due to an external agency like the surrounding air which was itself set in motion along with the projectile by the initial thrust.

Behind this Aristotelian physics lay the metaphysics of matter and form. Matter was purely passive in itself. The characteristics of a body came exclusively from its natural form. Hence, a body could contribute nothing to the determination or sustenance of an unnatural motion.

Thomas Aquinas exemplifies the Christian Aristotelian of the thirteenth century. He accepted the matter-form metaphysic of Aristotle and the consequent distinction between natural and forced motion. He introduced two slight deviations from Aristotle, however, that paved the way for future developments. First, for Aquinas, the heavens were continuously moved by God (as efficient cause): they did not simply move by desire for him (as final cause).[144] Thus he opened the way to direct comparison between the

---

[142]Grant, 'Place and Space', pp. 140f., 150.

[143]Copleston, *History*, vol. III/2, pp. 47f.; Grant, *Much Ado*, pp. 139f.; Tyrone Lai, 'Nicholas of Cusa and the Finite Universe,' *JHP*, vol. xi (April 1973), pp. 161-67.

[144]Gilson, *Spirit of Mediaeval Philosophy*, pp. 75f.; Weisheipl, 'Celestial Movers', pp. 186f; 'The Commentary of St Thomas on the *De caelo* of Aristotle', in William E. Carroll, ed., *Nature and Motion in the Middle Ages* (Washington, D.C.: Catholic Univ. of America Press, 1985), pp. 194f.

motion of the heavens and cases of forced motion on earth (e.g., spinning wheels).

Second, Aquinas attributed to matter a susceptibility for quantitative determination and magnitude (*corpus quantum*) based on his belief in the sovereignty of God and the properties of the Eucharist and the resurrection body.[145] While adhering to the letter of Aristotelian metaphysics, this deviation violated the spirit of Aristotle enough to provide new insight into the possible contribution of a body to its own (unnatural) motion: the conception of an inherent quantity of matter was fundamental to the later idea of an inherent quantity of impetus.

An alternative way of working around the Aristotelian categories was offered by Robert Kilwardby (d 1279), a Dominican like Aquinas who was, however, an Augustinian in the theological tradition of Grosseteste and Bacon. Kilwardby developed the idea of God as the efficient cause of the motion of the heavens to the point of denying that he was the immediate, present mover of the heavens (as formal cause). God had set the heavens in motion in the beginning in such a way that they now continued their respective motions 'by their own inclinations and tendencies'. The motion of the heavens was, in fact, as spontaneous and natural as the motion of sublunar bodies moving to their natural places.[146] Kilwardby thus attributed an even more active role to bodies (celestial bodies, at least) than did the more strictly Aristotelian Aquinas.

One of the first medieval scientist-theologians to challenge the Aristotelian view of projectile motion was the Spiritual Franciscan, Peter John Olivi (d 1298). Olivi's own position is difficult to determine with precision, but his speculations give some insight into the intellectual and spiritual dynamics that were changing ideas about the natural world in the late thirteenth century. For one thing, he reported the opinions of others (other Spirituals?) that projectile motion was due to an 'inclination' or 'species' (in the Augustinian sense of a secondary form) that was impressed on the moving object directly by the initial thrust of the mover. This impulse along the direction of initial thrust apparently continued until it was overcome by the tendency of the object to return to its natural place. In response to this position, Olivi suggested an even more radical view, arguing that motion required no cause at all! Motion was simply a 'mode of being situated', just like being at rest. This was very close to the germinal

---

[145] See text above notes 101-2, 132-33. Historically, there lay behind Aquinas's speculations the mechanics of Avempace (Ibn Bajjah) which, in turn, derived from the natural philosophy of Basil and John Philoponus; see chapter 1 section D, text above notes 121-29. Cf. Olaf Pedersen, 'The Development of Natural Philosophy, 1250-1350', *Classica et Mediaevalia*, vol. xiv (1953), pp. 113f.

[146] Weisheipl, 'Celestial Movers', pp. 160, 177-83.

idea behind the law of inertia as formulated by Galileo and Newton in the seventeenth century.[147]

Olivi exemplified a characteristic of Franciscans in the tradition of Bonaventure in that he attacked the Aristotelian framework, motivated in part by his faith, but he also gave reasons for his rejection and did not appeal exclusively to faith. William of Ockham (d c. 1350), perhaps influenced by Olivi, also denied that an object required a continuously impressed force to keep it in motion: God could, by his absolute power, produce an effect without any mediating cause at all.[148] Ockham also gave scientific arguments to refute the idea of an mediating force, however. For example, if projectiles were kept in motion by the surrounding air currents, then two arrows moving in opposite directions and passing in mid-air would interfere with each other's motion.[149] The falsity of the conclusion implied the falsity of the premise.

Another Franciscan contribution was made by Francis of Marchia around 1320. Drawing on an analogy between God's setting the heavens in motion and the infusion of grace through the sacraments (from Bonaventure),[150] Marchia reasoned that an impressed force should remain (temporarily) in a projectile even after its initial thrust, just as a residual power to confer grace remained in the Eucharist even after the moment of consecration. Hence, like Olivi and Ockham, though for different reasons, he rejected the Aristotelian idea that air currents kept the projectile in motion.[151]

The Dominican and Franciscan speculations we have mentioned might have led nowhere if they had not been drawn together and reformulated by an arts master like John Buridan. Like Kilwardby, Buridan returned to the historic creationist idea of God setting the heavens in motion once and for all by imparting an impetus that would keep them moving indefinitely.[152] Like Olivi, he also treated the forced motion of sublunar objects in terms of impressed force or impetus. Impetus was defined as the product of the quantity of matter in motion times its velocity and was viewed as lasting

---

[147]Crombie, *Augustine to Galileo*, vol. II, pp. 72f.; Maier, *On the Threshold*, pp. 82ff.

[148]Crombie, *Augustine to Galileo*, vol. II, pp. 76ff.; Gilson, *History*, p. 497; Wallace, 'Philosophical Setting', pp. 108f.

[149]Copleston, *History*, vol. III/1, p. 170.

[150]Wildiers, *Theologian and His Universe*, pp. 46f.

[151]Crombie, *Augustine to Galileo*, vol. II, p. 73; Weisheipl, *Development*, pp. 70f.

[152]*Quaestiones super octo libros Physicorum Aristotelis* VIII.xii.7; Marshall Clagett, *The Science of Mechanics in the Middle Ages* (Madison: Univ. of Wisconsin Press, 1959), p. 561. In support of this notion, Buridan cited the traditional exegesis of Genesis 2:2, according to which God rested after the six days of creation, having imbued creatures with the energies needed to accomplish their tasks on their own; E.A. Moody, 'Jean Buridan', *Studies in Medieval Philosophy, Science, and Logic* (Berkeley: Univ. of California Press, 1975), pp. 448f.; cf. chap. 1, note 107 above for Aristobulus's view of God's rest on the seventh day.

until it was overcome by a contrary tendency to return to a natural place.[153] Like Olivi and Ockham, Buridan cited the power of God to maintain unnatural motion even in the absense of a mediating cause.[154] And, like Ockham, he gave reasons for rejecting the need for air currents to keep a sublunar object in motion: a millwheel, for instance, once set in motion, would continue to spin even if it were surrounded by a closely fitted cover to keep the surrounding air away.[155]

Aside from these insights from earlier workers, Buridan's own distinctive contribution was twofold. First, by he developing the analogy between the rotating heavens and a spinning millwheel on earth, he elevated the idea of impetus to a general principle that cut across the Aristotelian dichotomy between heaven and earth.[156] Second, he realised that the idea of a hypothetical vacuum, discussed above, implied that an impetus once imparted would last forever if there were no counteractive tendency to return to a natural place.[157] For the first time in Western European history, it was realised that a body could theoretically maintain its state of motion by itself, without any external force and without any supernatural act of God to keep it moving.

As we saw in chapter 1 section D, the basic idea of impetus had deep roots in the creationist tradition of the intertestamental and early Christian periods. But it only began to come to scientific fruition with the work of Buridan in the fourteenth century. Buridan's concept of a permanent impetus based on the idea of God's setting the heavens in motion like a spinning millwheel was reformulated by Albert of Saxony (d 1390). Albert's works were later published in the early sixteenth century and had a formative influence on Galileo, who developed the idea of inertia that lies at the basis of modern Newtonian mechanics.[158]

While the late medieval concept of impressed force or impetus had its

---

[153]Crombie, *Augustine to Galileo*, vol. II, pp. 80-85; Moody, 'Galileo and Avempace', pp. 408f. (= *Studies*, p. 271).

[154]Maier, *On the Threshold*, p. 159; Grant, 'Condemnation of 1277', pp. 234f.

[155]Crombie, *Augustine to Galileo*, vol. II, p. 80.

[156]*Ibid.*, pp. 85f.; Dales, 'De-Animation', p. 548. The key passage is *Questiones de caelo et mundo* II.xii.6f. (see chap. 1, n. 130 for translation). Even here Buridan's insight was partly anticipated by Marchia's analogy between the rotating heavens and a potter's wheel (compare the texts in Clagett, *Science of Mechanics*, pp. 530 and 561). The millwheel (*mola*) analogy was common in late medieval literature (e.g. Dante, *Paradise* XII.3) and may have been derived from the hexaemeral tradition; cf. chap. 1 sec. D above, esp. notes 122, 128 and 131.

[157]Grant, 'Motion in the Void', pp. 275f., esp. notes 32-33, corrects the misinterpretation of Maier, *On the Threshold*, pp. 90ff., 161f.

[158]Duhem, 'Physics', pp. 51f.; Crombie, *Augustine to Galileo*, vol. II, p. 137; Dales, *Scientific Achievement*, p. 111. Already in the late fourteenth century Buridan's concept of a permanent impetus was regarded as part of 'the common philosophy' by Henry of Langenstein: Steneck, *Science and Creation*, p. 142.

roots in the creationist tradition, there is at least one significant difference between the two that should be noted at this juncture: in the biblical and patristic literature, the seemingly perpetual motions of nature had their origin in the word or command of God rather than in a mechanical thrust. Basil had used the illustration of a spinning object later taken up by Buridan, but he used it to illustrate the power of God's word as exhibited in all the cycles of nature.[159] But with the establishment of Aristotelian cosmology in thirteenth-century Europe, the idea of God as the efficient cause of the motion of the celestial spheres gained prominence, as we saw in section B. As a result, the late medieval idea of impetus was a good deal more mechanical, and the corresponding notion of God a good deal more deistic, than in the earlier period.

2.3 Moving Heaven and Earth

The basic cosmology of Aristotle was largely untouched by speculations on the void and on impetus. The earth was still assumed to be at rest at the centre of the cosmos; the heavens were still assumed to be rotating concentric spheres made of a substance unlike anything found in the sublunar realm. The initial challenge to this outlook came in the form of a revival of ancient Greek speculation on the possibility of many worlds, which, in Aristotelian terms, meant many centres of gravity or many earths.

Already in the early thirteenth century, at the time of the first major influx of Greco-Arab ideas, Michael Scot reported that some philosophers (presumably contemporaries) held that God could make other worlds, even an infinite number of them. Michael Scot, William of Auvergne, Roger Bacon, and Thomas Aquinas all argued against this possibility as implying the existence of a void space between the worlds.[160] In fact, the assimilation of the Aristotelian cosmology into the more traditional Christian worldview led to the entrenchment of the geocentric, geostatic outlook during this period.

The Condemnation of 1277 not only affirmed the possibility of a void, based on the idea that God could move the cosmos, but it also insisted on the possibility of a plurality of worlds based on the absolute power of God in creation.[161] Fourteenth-century natural philosophers like Bradwardine, Buridan, and Oresme were all directly influenced by this decision in their

---

[159] Chapter 1 section D, text above note 122.

[160] Grant, 'Condemnation of 1277', pp. 217ff. Among Jewish texts on creation, *Seder Rabbah de Bereshit* and *Midrash Aleph Beth* 3:5 (7th-8th cent. AD) interpret Ezekiel 48:35 ('around [the city] are 18,000') as stating that there are 18,000 worlds surrounding our own; N. Sed, *La cosmologie juive* (Paris, 1980), pp. 180ff.; Deborah F. Sawyer, *Midrash Aleph Beth* (Atlanta: Scholars Press, 1993), pp. 108, 115.

[161] See note 135 above.

speculations.

Among later medieval philosophers who argued for the possibility (not the actuality) of a plurality of worlds on the basis of God's omnipotence were Henry of Ghent (d 1293), Richard of Middleton (fl. c. 1294), William of Ockham (d c. 1350), John Buridan (d c. 1360), Nicole Oresme (d 1382), and Henry of Langenstein (d 1397).[162] This line of reasoning culminated in the speculations of Nicholas of Cusa (d 1464), in which the cosmos was unbounded and had no unique centre or circumference.[163]

A second line of reasoning was developed by Nicole Oresme based on the idea that God could move the entire cosmos in a straight line and the associated idea of an extracosmic void. If the centre of the cosmos moved, Oresme asked, what would be the reference point for defining motion? It would have to be the infinite, immovable void space beyond the cosmos, which is none other than the infinite immensity of God. Consequently, the definition of local motion did not require a fixed earth at the centre of the cosmos as Aristotle had supposed.[164] Although Oresme still concluded that the earth was fixed at the centre of the cosmos, his speculations opened up new possibilities, and, less than a century later, Nicholas of Cusa could dispense with the idea of a fixed reference point for motion almost entirely.[165]

But perhaps the most powerful corrosive to the Aristotelian cosmology was the development of mathematical laws and physical models that cut across the dichotomy of heaven and earth. The unity of heaven and earth had been one of the basic tenets of the creationist tradition, as we saw in chapter 1 section C. But opinion was divided with regard to the possible role of angels in directing the motions of the heavens. With the assimilation of the Aristotelian cosmology in the thirteenth century, the case for intracosmic angels was virtually secured: Aquinas, for one, identified the immediate movers of Aristotle's celestial spheres with one of the hierarchies of angels described by Pseudo-Dionysius.[166] This, together with the Aristotelian idea of a fifth element unique to the heavens, made a unified physics of heaven and earth virtually impossible.

The first effort towards a more unified view came from the mathematical

---

[162]Grant, 'Place and Space', p. 151; 'Condemnation of 1277', pp. 220-23; Steneck, *Science and Creation*, p. 28.

[163]Copleston, *History*, vol. III/2, pp. 47f. On the possible influence of Cusa's decentralised ecclesiology, see Dolan, 'Introduction', *Unity and Reform*, pp. 23ff.

[164]*Book of Heaven and Earth* I.24; II.8, trans. Menut, pp. 177, 365-73; Crombie, *Augustine to Galileo*, vol. II, pp. 78f., 90f.

[165]Dijksterhuis, *Mechanization*, pp. 232f.; Toulmin and Goodfield, *Fabric of the Heavens*, pp. 168f. Cf. text above notes 74-76.

[166]DeKosky, *Knowledge and Cosmos*, pp. 58f.

physics of Thomas Bradwardine, which we discussed earlier.[167] In his treatise *On the Proportions of Velocities in Motion*, composed in 1328, Bradwardine attempted to develop an abstract algebraic formula that would be applicable to all types of motion, whether linear (as on earth) or circular (as in heaven).[168] It was probably not just a coincidence that Bradwardine became a theological student about that time, for his theological writings of later date stress the ubiquity of God's providence as the coefficient of all natural events.[169] This sense of the immediacy of God's normal activity had been lost under the impact of Aristotelian cosmology in the thirteenth century, as we saw in section B. Thomas Bradwardine signalled a process of recovery that was to continue with fifteenth-century theologians like Cusa and culminate in reformers of science and theology in the sixteenth century.

John Buridan developed his concept of impetus based on the analogy we have discussed between the rotation of the heavens and the spinning of a millwheel on earth.[170] In doing so, he developed the beginnings of a unified dynamics of heaven and earth to correspond to the more mathematical kinematics of Bradwardine and his successors at Oxford.[171] One consequence of this treatment was that there was no longer any need for angels to direct the movements of the celestial spheres.[172]

Of Buridan's successors, Albert of Saxony followed faithfully in dismissing the angels, while Nicole Oresme retained them as a means of providing a kind of inertia to prevent the celestial bodies from moving too fast![173] So the angels made it into the early Renaissance, but not by much. Their role was severely restricted by Henry of Langenstein (d 1397) and was completely ignored by Nicholas of Cusa (d 1464).[174] Langenstein and Cusa also affirmed the unity of heaven and earth in another way by rejecting the idea of a fifth element and positing the universality of the four primary elements for both heaven and earth.[175]

This concludes our review of ways in which theological considerations influenced the development of natural science in the late Middle Ages and prepared the way for the rise of modern science in the Renaissance. There have been enough points of contact to indicate that theological ideas

---

[167]Text above notes 104-6. The ground for Bradwardine's unification may have been prepared by Grosseteste's modifications of Aristotle in the thirteenth century; McEvoy, *Philosophy*, pp. 182-88, 216.
[168]Weisheipl, *Development*, pp. 73-77.
[169]Leff, *Bradwardine*, pp. 50ff.; Oberman, *Archbishop Thomas Bradwardine*, pp. 57f., 77f.
[170]Text above notes 155-56.
[171]Crombie, *Augustine to Galileo*, vol. II, p. 86; Dales, 'De-Animation', p. 548.
[172]Moody, 'Galileo and Avempace', pp. 408f. (= *Studies*, p. 271).
[173]Duhem, 'Physics', p. 52; Maier, *On the Threshold*, pp. 92f., 96ff.
[174]Steneck, *Science and Creation*, pp. 92ff.
[175]*Ibid.*, pp. 60f.; Crombie, *Augustine to Galileo*, vol. II, pp. 60f.

associated with the ancient creationist tradition were a real factor both in the assimilation and in the revision of Aristotelian science in Western Europe.

2.4 Summary and Analysis

We are in a position to offer the following summary of the interaction between theology and science through the Middle Ages. In the biblical and patristic periods, nature was conceived as governed by laws that were authored and administered by God and, hence, were subject to divine ratification and amendment as a matter of course. This picture was transmitted to the Latin West, though with significant changes, as we saw in chapter 1, section D. A sharp differentiation between the absolute creative power of God and the normal course of nature arose in the twelfth century, partly as a response to the rise of an increasingly naturalistic science. That differentiation, in turn, facilitated the assimilation of Aristotle's cosmology in the thirteenth century by a church which might otherwise have been so resistant as to impede further progress.

The result of the medieval synthesis was twofold. On the one hand, considerations of God's normal exercise of providence through second causes led to a replacement of the biblical image of God as cosmic legislator by the idea of God as First Mover (the sense of the latter shifting meanwhile from formal to efficient cause of motion). This modified creationist image of God as setting the heavens in motion once and for all was, in turn, instrumental in the development of the late medieval ideas of impressed force and impetus that provided the background for early modern classical mechanics.

On the other hand, a conservative reaction to the naturalism inherent in the new synthesis led to a renewed emphasis on God's absolute power both in establishing the normal course of nature and in superseding it at any time. Specific assertions like the possibility of the cosmos being moved, the possibility of a vacuum, and the possibility of other worlds challenged the authority of Aristotle in natural philosophy and encouraged efforts to develop a unified mechanics applicable to both heaven and earth.

Even a brief summary such as this one will show that the historical relationship between theology and science was not one of direct causation. Theology neither directly impeded nor caused the rise of modern science. Rather, the two interacted, with changes in each making changes in the other more feasible. We may speak of certain scientific ideas like impetus or the unity of heaven and earth having their roots in the creationist tradition, but we must not think of that tradition as being unchanging over the centuries. It provided a continuity of ideas, but it was not constant. It adapted to meet the challenges of science in the twelfth and again in the thirteenth and fourteenth centuries. Adaptations in the creationist tradition, in turn, provided insight and inspiration to natural philosophers for whom theological belief was still an important part of life.

Even allowing for variability and adaptation, it is difficult to say to what degree the creationist tradition was responsible for the progress of natural science through the Middle Ages. We have argued that the Aristotelian paradigm was the most comprehensive system of natural philosophy available, hence, that the creationist tradition served the progress of science well both in making the assimilation of Aristotle possible and in stimulating constructive criticism that could lead beyond Aristotle.

We do not know whether modern science could have been built on any other historical base or even whether a different kind of modern science could have been developed on the same base.[176] Any real understanding of the role of the creationist tradition in the development of science must await the outcome of efforts in nonwestern cultures to graft modern science on their traditions and the subsequent contributions of those traditions to the further progress of science.

History is the best exegesis of theology. So the full meaning of our beliefs will not be determined completely until history reaches its own conclusion. In the meantime, those beliefs will continue to be needed in the interpretation of history and the anticipation of future developments.

---

[176]Joseph Needham has compared the history of science in China through the Middle Ages with that in the Latin West and concluded that a noncreationist tradition like the traditional Chinese was not so conducive to the development of modern science; *The Grand Titration* (London: Allen and Unwin, 1969), pp. 35ff., 46, 322-27. Other factors, like geographic and political conditions, were determinative as well (*ibid.*, pp. 184-89, 196-202, 210ff.). Two recent studies that take all these factors into account both reach the same conclusion: Kenneth Stunkel, 'Technology and Values in Traditional China and the West, Part I', *Comparative Civilizations Review*, No. xxiii (Fall 1990), pp. 77, 85; Derk Bodde, *Chinese Thought, Society, and Science* (Honolulu: Univ. of Hawaii Press, 1991), p. 344.

To place these conclusions in perspective, however, one should note that the concept of science that is assumed by all three writers is that of mechanistic, classical physics. Chinese thought may yet be found to be more conducive to post-Newtonian modes of scientific research.

CHAPTER THREE

RENAISSANCE, REFORMATION, AND EARLY MODERN SCIENCE
(Fifteenth Century Through the Seventeenth Century)

A. Renaissance Science Through Copernicus and Paracelsus

The sixteenth century stands out in the history of the Western European church as a time in which Christians tried to get back to the basic truths of their faith. Its importance for our study stems from the degree to which basic creationist themes were rediscovered and reaffirmed.

The Protestant Reformation was one aspect of this broader movement—we shall turn to it in the following section (B). In this section, we shall look briefly at the influence of fundamental theology on the development of natural science up to the middle of the sixteenth century, a period during which the outlook of the principal figures was clearly still independent of the teachings of the Protestant reformers. Several of the modern sciences took their basic form during this period or, at least, were tending in the direction we recognise today as 'modern'. The fully modern features of natural science were completed in the seventeenth century—based on a synthesis of influences from the Renaissance and Reformation of the fifteenth and sixteenth centuries. We turn to that synthesis in section C.

The fifteenth and sixteenth centuries are generally called the 'Renaissance' because they witnessed a rebirth and a new growth in the arts and sciences that raised Western civilisation from a cultural backwater to a dominating force in world events. It was a period of world exploration and of scientific experimentation. It was also the period during which the invention of printing allowed the wide dissemination of scientific works, both old and new. Exploration, experimentation, and printing together opened up an unprecedented wealth of natural history to the reading public and provided the framework for significant advances.

The history of natural science becomes far more complex during the Renaissance than it was during the Middle Ages. In place of a relatively stable data base derived from common experience and shaped by literary tradition, we find a growing body of new data based largely on individual observation made possible by the development of new technologies. In place of a single dominant philosophical tradition like Aristotelianism, we have an interweaving of several classical traditions together with new currents of thought that disavowed speculative philosophy altogether. Even in the relatively conservative field of theology there were new developments in theosophy and mysticism as well as a renewed interest in traditional doctrines.

In spite of all this novelty, the reader who is familiar with the basic themes in the creationist tradition and the medieval mutations of those themes will find little that is completely new in the Renaissance, other than the sheer variety of ideas brought to focus at one time. Ancient traditions were celebrated without any sense that they were out of date, because in many cases they were being rediscovered or refined in ways that provoked new insight. Here we shall discuss three major Renaissance traditions—the scholastic, the Neoplatonic, and the experimental. All three of these had roots in the Middle Ages and have been described in chapter 2. We shall look here for the theological factors that may have contributed to their revitalisation and fruitfulness during the Renaissance.

1 The Scholastic Tradition to Vesalius

The Renaissance has so often been portrayed as a reaction *against* the scholasticism of the Middle Ages that we are liable to overlook the important contributions made by the scholastics, particularly in Italy, France, and Switzerland. Since most of the ideas involved have been treated in chapter 2, they need only be mentioned briefly here. But due recognition of their existence is needed in order to establish the importance of continuity with the Middle Ages, particularly in issues for which Copernicus is often regarded as being revolutionary.

1.1 A Variety of Scholastic Traditions

There was an ongoing discussion of questions concerning scientific method in the tradition of Latin Averroism and the Parisian school of John Buridan, particularly at the University of Padua and the College of Rome.[1] Buridan's idea of impetus was also developed by Leonardo at Milan and by Calcagnini at Ferrara. Independently of Copernicus, Calcagnini argued (*c.* 1530) that it would be perfectly natural for the earth to rotate on its axis.[2]

Marliani (d 1483) revised the law of motion originally formulated by Thomas Bradwardine and also spoke of experiments that refuted Aristotle's earlier version of the laws of motion.[3] Bradwardine's speculations on the possibility of a void and the 'mean-speed theorem' of the Oxford calculators

---

[1] John Herman Randall, *The Career of Philosophy*, vol. I (New York: Columbia Univ. Press, 1962), chaps. 1, 2, and 11; William Wallace, *Prelude to Galileo* (Dordrecht: Reidel, 1981), pp. 129-59, 192-252.
[2] Crombie, *Augustine to Galileo*, vol. II, pp. 135ff.; Dreyer, *History*, pp. 292ff.
[3] Crombie, *Augustine to Galileo*, vol. II, p. 113

were thoroughly discussed in the school of John Major (1469-1550) at Paris. A Spanish participant in Major's circle of students, Domingo de Soto, concluded correctly (1551) that acceleration should be uniform for a freely falling object and that the speed of fall should be proportional to the time elapsed after the onset of free fall.[4]

In the field of astronomy, Italian scholars like Fracastoro (1535) tried to carry out Aristotle's original programme for representing the motion of the heavens in terms of rotating spherical shells. This modelling programme ruled out the use of either the eccentrics or the epicycles introduced by Ptolemy in the second century. It indicates an imperative for simple geometry and realistic physics—an imperative similar to the one that led Copernicus to his revolutionary hypothesis—yet it comes directly out of the scholastic tradition.[5]

In the fields of anatomy and surgery, the traditional ideas of Galen continued to be studied by Vesalius and his successors at Padua and by Felix Platter and Theodore Zwinger at Basel. Vesalius's treatise, *On the Fabric of the Human Body*, published the same year as Copernicus's major work on astronomy (1543), was the first major correction of Galenic ideas and is generally regarded as the foundation of modern anatomy.[6]

1.2 Creationist Motifs in Renaissance Scholasticism

Even from this brief survey, one can see that the technical expertise of late medieval scholasticism continued to be scientifically fruitful in at least half a dozen different ways. But what of the theological motifs which, as we we have found, played such a large role in medieval scholasticism?

In some cases traditional creationist motifs were clearly in evidence. John Major, for instance, based his speculations on the existence of a void space on the power of God to create or destroy matter anywhere, even beyond the known stars: God could even create an infinite number of worlds beyond our own, he argued.[7] Another example is Domingo de Soto, who combined his scientific research with theological studies as Bradwardine had done two centuries earlier.

Those like Leonardo and Calcagnini, who were attracted to the idea of impetus, were influenced, as were their late medieval predecessors, by the traditional idea of God having imposed laws on the world at the very moment of creation. Calcagnini's references to the source of these laws as 'Nature' or 'Providence' and Leonardo's references to 'Reason' as the

---

[4] Wallace, 'Experimental Science', p. 204.
[5] Dreyer, *History*, pp. 296f., 301ff.; cf. Toulmin and Goodfield, *Fabric*, pp. 169f.
[6] W.P.D. Wightman, *Science and the Renaissance*, vol. I (Edinburgh: Oliver and Boyd, 1962), chap. 12.
[7] Grant, *Much Ado*, pp. 149-52.

helmsman of nature suggest that this creationist idea had been taken over with very little sense of its biblical origins.[8] On the other hand, it should be kept in mind that the convenience of defining nature as distinct from God and reason as distinct from revelation was by this time a venerable theological tradition dating back to the eleventh and twelfth centuries (chapter 1 section D). So Calcagnini and Leonardo could still be taken to represent the creationist tradition, at least in its more radical, secularised wing.

Of the scientists listed, Fracastoro and Vesalius seem to exhibit the least theological motivation in their work. Fracastoro served Pope Paul III and was appointed official physician (1546) to the Council of Trent. Yet it has been said that there is little or no trace of religious motivation in his scientific work.[9] Vesalius described the human body as the product of divine craftsmanship, and it has been claimed that this belief was an important factor in the passion with which he pursued his dissections.[10] But his references to the 'Great Artificer of all things' may simply have been a formality, particularly in his dedicatory letters addressed to monarchs.[11] On the other hand, Fracastoro's concept of a unifying principle of 'sympathy' in nature[12] and Vesalius's idea of design in the fabric of the human body clearly indicate their creationist roots. It is not until we reach the eighteenth century that we find anything like a genuinely secular, nontheological tradition of science.

2 The Neoplatonic Tradition to Copernicus

The Neoplatonic tradition is distinguished from the scholastic in that it was not based on the works of Aristotle and his commentators. Although it received new attention in the Renaissance, its roots go back to the days of the early church—long before the major impact of Aristotelian thought. In fact, up to the thirteenth century, it was the dominant force in Western European thought. Even during the two centuries (the thirteenth and fourteenth) of Aristotelian ascendancy, there was a strong Neoplationist movement among the Franciscans, particularly among the followers of Robert Grosseteste, and during the fifteenth century there were several new schools of thought that brought it back into the forefront. We shall look

---

[8] Dreyer, *History*, pp. 293f.; Randall, *Career*, vol. I, pp. 305f.
[9] In contrast to Paracelsus: Walter Pagel, 'Religious Motives in the Medical Biology of the XVIIth Century', *BHM*, vol. iii (Feb. 1935), p. 108.
[10] Crombie, *Augustine to Galileo*, vol. II, p. 278.
[11] E.g., Vesalius's Dedicatory Letter to Phillip II of Spain, attached to his *Epitome* of 1543, trans. L.R. Lind, *The Epitome of Andreas Vesalius*, Yale Medical Library Publication No. 21 (New York: Macmillan, 1949), p. xxxiv.
[12] Bruno Zanobio, 'Fracastoro', *DSB*, vol. V, p. 106b.

briefly at two of these, the Florentine perspectivists and the hermetic-cabalist tradition, as background for a study of Nicholas Copernicus.

2.1 The Florentine Perspectivists

Paolo Toscanelli (1397-1482) was the founder of this movement; Brunelleschi (1377-1446) and Alberti (1404-1472) were his most important students and collaborators. Continuing the work of the medieval perspectivists (e.g., Bacon, Pecham, Witelo) and inspired by the practical mathematical and experimental emphases of Cusa, these mathematician-artists pioneered a geometrical mapping of space that affected all areas of science and technology from crystallography to geography and astronomy, and from painting to architecture and city planning.[13]

The Florentine perspectivists provide background for the work of Copernicus in at least two ways. First, Toscanelli's astronomical studies led him to express dissatisfaction with the equants and eccentrics of Ptolemy. The departure from perfectly circular motion did not accord with his Platonic ideal of simplicity.[14] One of the chief representatives of this Neoplatonic approach to astronomy was Domenico Novara (1454-1504), Copernicus's professor at Bologna during the last years of the fifteenth century.[15]

The second way in which Florentine perspectivism may have influenced Copernicus was in its uniform, nonhieratic conception of space. Rather than representing things from a transcendental viewpoint in which spatial location and magnitude were determined by intrinsic value, artists began systematically to portray objects as they appear from a particular point of view on a level with the landscape itself. It has been suggested that this practise of perspectival representation prepared the way for Copernicus's perspectival shift from a geocentric to a heliocentric representation of the heavens.[16]

Even aside from the question of its impact on Copernicus, the work of Toscanelli, Brunelleschi, and Alberti was clearly revolutionary in its effect.

---

[13] Eugenio Garin, *Science and the Civic Life in the Italian Renaissance* (Garden City, N.Y.: Doubleday, 1969), chap. 2; Joan Gadol, *Leon Battista Alberti* (Chicago: Univ. of Chicago Press, 1969), chap. 4; Cecil J. Schneer, 'The Renaissance Background to Crystallography', *American Scientist*, vol. lxxi (May 1983), pp. 257f.

[14] Pacey, *Maze*, p. 99.

[15] The classic statements of this influence are Edwin Arthur Burtt, *The Metaphysical Foundations of Modern Physical Science* (Garden City, N.Y.: Doubleday, 1954), p. 54; Thomas S. Kuhn, *The Copernican Revolution* (Cambridge, Mass.: Harvard Univ. Press, 1959), p. 129; Alexandre Koryé, *The Astronomical Revolution* (London: Methuen, 1973) pp. 21f. For a cautionary note, see Paolo Rossi, 'Hermeticism, Rationality and the Scientific Revolution', in M.L. Righini and W.R. Shea, eds., *Reason, Experiment, and Mysticism in the Scientific Revolution* (New York: Science History Publications, 1975), pp. 267-73.

[16] E.g., Gadol, *Alberti*, pp. 156f., 198f.

Although it was an effort conceived in largely secular terms,[17] it would be useful to have a detailed examination of its theological aspects. The shift from a hieratic to a uniform conception of space undercut the traditional concepts of sacred location and preferred direction which lay at the root of traditional religion.[18] So the work of the Florentine perspectivists can hardly be regarded as lacking in theological significance.

One way to try to assess the perspectivist revolution theologically would be to look at its background; another would be to look at its effects. As to the former, even the brief historical sketch given above shows that there was significant theological input. The medieval perspectivist tradition was firmly rooted in the idea that God created all things in accordance with measure, number, and weight (Wisd. 11:20), and this aspect of the creationist tradition clearly lived on in the aesthetics of Florentine perspectivists like Alberti.[19] While this notion of mathematical order is usually referred to as Neoplatonic or Neopythagorean, it should not be forgotten that it was championed and cultivated by Christians who also saw it as profoundly biblical.[20]

Also in the background was the influence of Nicholas of Cusa, a contemporary of Toscanelli—in fact, the two knew each other as students at Padua (c. 1420). Cusa's vision of nature as a book of God written in the language of mathematics and his nonhieratic view of the cosmos were also rooted in the idea of creation, as we saw in chapter 2.[21] Even Alberti's apparent disavowal of metaphysics in favour of the mechanical arts was very much in the spirit of Cusa's pitting the wisdom of the uneducated artisan against the university-trained scholastic—an ideal (or anti-ideal) derived from the Franciscan tradition of the thirteenth century.[22] If Francis of Assisi, Roger Bacon, Raymond Lull, and Nicholas of Cusa may be said to have

---

[17] Richard Olson describes the tradition of Alberti as secular and antiphilosophical; *Science Deified and Science Defied*, vol. I (Berkeley: Univ. of California Press, 1982), pp. 221, 230. More balanced studies of Alberti indicate that he was a mystical humanist, not a secular one; Mark Jarzombek, *On Leon Baptista Alberti, His Literary and Aesthetic Theories* (Cambridge Mass.: MIT press, 1989), pp. xiv, 61, *passim*.

[18] Gadol, *Alberti*, pp. 150f.

[19] Dorothy Koenigsberger stresses the twofold effect of creation in Alberti's work: divinely created harmony in the world, and the divine image in humanity allowing it to recognise the former; *Renaissance Man*, pp. 10, 12, 13, 21f., 25f., 26f., 50.

[20] See chapter 2 section C1.2, 'Mathematical Method and Quantification'; cf. chap. 1, notes 33 and 37.

[21] Chap. 2, notes 74-76, 108, 123. Dorothy Koenigsberger stresses the influence of Cusa's homogeneous, nonhieratic view of space on the Florentines, *Renaissance Man*, pp. 105-8. More likely, the two are parallel developments.

[22] Gadol, *Alberti*, pp. 202f.; cf. Cassirer, *Individual*, pp. 48ff., and Koenigsberger, *Renaissance Man*, p. 103, for a similar influence of Cusa on Leonardo; and J.N. Hillgarth, *Ramon Lull and Lullism in Fourteenth-Century France* (Oxford: Clarendon Press, 1971), pp. 52-55, 270-73, on the Franciscan ideal of untrained simplicity.

laicised Christian spirituality, then Alberti may be said simply to have carried the process a step further and secularised it.

What then were the theological effects of the perspectivist revolution? Did the process of secularisation destroy the very spirituality it worked with, as critics have often said? Not exactly. It would be more accurate to say that the desacralisation of space required the interiorisation of spirituality rather than its destruction. The locus of ethical and spiritual value shifted from the object to the subject. However, interiorisation was a tendency present in Christian Neoplatonism from as early as the fourth century, and the pressure towards secularisation was clearly in evidence as early as the twelfth century.[23] So whatever epistemological difficulties there may be in the subject-object dichotomy of modern Western thought are not simply the result of the perspectival revolution of the fifteenth century itself. And the positive results that came from that revolution—the more realistic representation of nature, the more accurate mapping of the continents, and the vision of well-planned cities—can hardly be said to be lacking in theological value as far as the creationist tradition is concerned.

2.2 The Hermetic-Cabalist Tradition

In contrast to the naturalism of the Renaissance perspectivists, the numerology of the Renaissance cabalists may appear to be mystical and entirely unscientific. We could clearly state the scientific import of the work of Toscanelli and Alberti; the question we had to ask was whether it had any theological content. In contrast, the ideas of Ficino, Pico, Reuchlin, and Agrippa were clearly theological. The question is whether they had any scientific impact.

The founders of Renaissance Cabalism were Marsilio Ficino (1433-99) and Pico of Mirandola (1463-94), whose careers, like those of the perspectivists a generation earlier, centred in the Republic of Florence. They and their associates are often referred to as the 'Florentine Neoplatonists'. Whereas the Florentine perspectivists were influenced primarily by Western Neoplatonism in the traditions of Grosseteste and Cusa, the Florentine Neoplatonists drew much of their inspiration from newly discovered hermetic and cabalistic texts purporting to contain wisdom so ancient that it even antedated that of the Greeks.

Ficino's major contribution was the translation of the Greek hermetic writings (the *Corpus Hermeticum*, the first part of which is known as 'Poimandres' or 'Pymander') that had been presented to the Florentine financier, Cosimo de' Medici, around 1460. Pico enriched this

---

[23] Kaiser, *Doctrine of God*, section 4.1; and this work, chapter 1, section D. Cf. our comment on the idea of the 'two books of God', chapter 2, section B, text above notes 30-33.

Neoplatonic-hermetic mixture by introducing the ideas of the Jewish Cabala (or Kabbalah, meaning 'Tradition') and correlating its numerological ideas with Pythagorean mathematics.[24] The result was a heady concoction of mathematics and magic, mysticism and machinery, that has challenged the analytical skills of historians of our generation for explanation and assessment.

As far as the creationist tradition is concerned, the rediscovery of hermetic texts did not really introduce anything new. In fact, hermetic ideas were known in the Middle Ages and were particularly influential with Bradwardine and Cusa, as we have seen.[25] What Ficino's work did was revitalise non-Aristotelian ideas that had been peripheral and give them respectability in some Christian circles. We shall mention a few of these.

The principal feature in Ficino's hermeticism was a hierarchical model of the cosmos that was not a strictly spatial hierarchy and that made the influence of spiritual entities much more immediate to human concerns than Aristotle's did.[26] In contrast to later cabalist models (Pico et al.), Ficino's cosmology was rather simple. From top to bottom it consisted of:

(1) the divine Mind with its eternal ideas;

(2) the world soul (*anima mundi*), which contained the seminal principles or causes;

(3) the world itself (*mundus animatus*), which was itself a (spatial) hierarchy extending from the material heavens to the earth.[27]

---

[24] The classic and still indispensable reference on this subject is Frances A. Yates, *Giordano Bruno and the Hermetic Tradition* (Chicago: Univ. of Chicago Press, 1964); see pp. 12f., 84f., 146.

[25] Chapter 2, text above note 106. On the relationship between Hermeticism and Judaism, see C.H. Dodd, *The Bible and the Greeks* (London: Hodder & Stoughton, 1935), chaps. 6-7; Birger A. Pearson, 'Jewish Elements in *Corpus Hermeticum* 1 (Poimandres)', in R. van de Broeck and M.J. Vermaseren, eds., *Studies in Gnosticism and Hellenistic Religions* (Leiden: E.J. Brill, 1981), pp. 336-48; Moshe Idel, 'Hermeticism and Judaism,' in I. Merkel and A.G. Debus, eds., *Hermeticism and the Renaissance* (Washington D.C.: Folger Books, 1988), pp. 59-76. Even though Pearson concludes that 'Poimandres' was a gnosticising reinterpretation of Jewish traditions ('Jewish Elements', pp. 346ff.), the parallels with 2 Enoch that he points out place it well within the creationist traditon (with the exception of the heaven-earth dualism mentioned in chap. 2, note 60). For historical purposes, it is better to view Judaism, Christianity, and Gnosticism as three sibling developments of the Jewish tradition than as incompatible worldviews.

[26] Chapter 2, section B2.2, 'God as First Mover and Clockmaker'.

[27] Yates, *Bruno*, pp. 64, 119; Wayne Shumaker, *The Occult Sciences in the Renaissance* (Berkeley: Univ. of California Press, 1972), pp. 121f. In his *Three Books on Life* (1489), Ficino also envisioned a *spiritus mundanus* or 'world spirit' that mediated between the world soul and the material world itself (III.1, 3 *passim*). This world spirit radiated from the heavens and could be attracted and focussed on earth by the use of suitable images like talismans; Carol V. Kaske and John R. Clark, *Marsilio Ficino: Three Books on Life* (Binghamton: Center for Medieval and Early Renaissance Studies, 1989), pp. 43ff., 49; John S. Mebane, *Renaissance Magic and the Return of the Golden Age* (Lincoln: Univ. of Nebraska Press, 1989), pp. 28ff.

The entire Aristotelian system of sublunar zones and celestial spheres fit within this third category, so Ficino's cosmology was definitely weighted toward the invisible and spiritual rather than toward the material and spatial, as Aristotle's had been.

Within this hermetic cosmology the positions of the sun and of humanity took on particular significance. In the standard medieval cosmology, the earth was at the centre of the cosmos, but the sun also had a strategic location in that its sphere was the fourth, midway between the moon (the nearest 'planet') and Saturn (the farthest), or, equivalently, midway between the earth and the stars. Moreover, the sun's light illuminated all the other planetary bodies. Indeed it was frequently viewed as a key symbol of God's reign over the created world.[28]

Ficino's hermetic cosmology gave the sun an even greater role in terms of spiritual influence. From the standpoint of the world soul, the location of the sun was central; it mediated the influence of the world soul, and through it the influence of God and the angels, to all the visible world.[29] It is true that the earth was still at the centre in the spatial sense, but the sun could now be viewed as central in dynamical terms. In fact, the hermetic idea of the world soul was not so far away from the later, Newtonian notion of universal gravitation.[30] On the other hand, it should also be noted that Ficino's 'natural magic', with its use of solar talismans and hymns addressed to the sun as a lesser god, could only arouse hostility in orthodox Christian circles and prejudice church officials against the notion of heliocentrism even when it was offered as a purely scientific hypothesis.[31] Like many theological ideas, Ficino's hermeticism could work for and against science at the same time.

The central role of humanity in the cosmos was also heightened in Ficino's cosmology. For one thing, human beings were viewed as participating in all three levels: (1) human intellect corresponding to the divine Mind; (2) soul (associated with the subtle, astral body) corresponding to the world soul; and (3) material body at the mundane level. In other words, there was a direct parallel between the macrocosm (the cosmos) and

---

[28] Dante, *Paradise* X.53f.; XIV.73-78, 96; XV.76; XX.1; XXIII.28ff.; cf. Keith Hutchison, 'Towards a Political Iconology of the Copernican Revolution', in Patrick Curry, ed., *Astrology, Science and Society* (Woodbridge, Suffolk: Boydell Press, 1987), pp. 104f.

[29] Yates, *Bruno*, p. 120; Shumaker, *Occult Sciences*, p. 124; Robert S. Westman and J. E. McGuire, *Hermeticism and the Scientific Revolution* (Los Angeles: William Andrews Clark Memorial Library, 1977), p. 16.

[30] The Stoic-hermetic idea is sometimes referred to as 'general gravitation' to distinguish it from the later Newtonian principle: Koryé, *Astronomical Revolution*, p. 110 note 4; cf. note 12 above on Fracastoro's idea of cosmic 'sympathy'.

[31] Yates, *Bruno*, pp. 63, 75-78, 82, 152f.; Hugh Kearney, *Science and Change, 1500-1700* (New York: McGraw-Hill, 1971), pp. 101-4.

the microcosm (humanity).³² Such a parallel had been discussed by theologians as early as Philo and Clement of Alexandria and had been a regular feature of Christian Platonism in the early Middle Ages.³³ The idea of humanity as a microcosm implied that the human intellect was capable of comprehending God's creation, particularly the more rational (celestial) parts of it, by direct intuition. With the ascendancy of Aristotelian inductivism in the later Middle Ages, this idea had subsided, but Ficino's hermeticism helped bring it back back into vogue in the Renaissance and revitalise the creationist notion of the comprehensibility of the world that was to inspire early modern scientists like Kepler (section B below).

Another way in which Ficino heightened the role of humanity was with respect to its creative role in the cosmos. As one who participated in all levels of existence, even the divine (as intellect), humanity could, in principle, know all things and create all things. The dynamical hegemony of the sun in the order of nature was paralleled in the technological potential of human arts. Humanity was the true vicar of God upon whom the freedom and responsibility of restoring and completing the work of creation fell.[34]

While such utopian ideas are sometimes regarded as 'secular-humanist' today, it should be clear from our discussion in chapter 1, section E that they also had deep roots in biblical and patristic thought. What we find missing in Ficino's hermetic humanism from the perspective of the creationist tradition is an *equal* emphasis on the harnessing of human creativity for altruistic social ends. Ficino's drive toward individual enlightment, even if interpreted in corporate terms as human domination over nature, could be a dangerous and destructive force in history if not

---

[32] Shumaker, *Occult Sciences*, pp. 121f. In Neoplatonic thought, the 'astral body' is the vehicle that the soul acquires during its descent through the planetary spheres into its material body on earth. It is intimately related to the *idolum*, the lowest of the three levels of the human soul (*mens, ratio,* and *idolum*), which is responsible for phanatasy and sense perception. It is also somehow related to the medical spirit, which mediates between the life of the soul and the body as the *spiritus mundanus* mediates between the world soul and the material world; Kaske and Clark, *Marsilio Ficino*, pp. 42f.; Gary Tomlinson, *Music in Renaissance Magic: Toward a Historiography of Others* (Chicago: University of Chicago Press, 1993), pp. 107-110.

[33] The macrocosm-microcosm correspondence was primarily associated with the Platonist tradition: Plato, *Timaeus* 44d, 47bc; Philo, *On Creation* 69, 82, 137, 146; *Who Is the Heir?* 281ff.; *Q. on Exodus* I.23; 2 Enoch 30:8; Clement of Alexandria, *Protrepticus* 1 (*ANF* 2:172b); *Stromateis* II.11 (*ANF* 2:359); cf. Hutchison, 'Towards a Political Iconology', p. 101, note 11; J.K. Wright, *Geographical Lore*, pp. 148ff., 185, note 17. It survived in the later Middle Ages also in the encyclopedist and Lullian traditions; Paolo Rossi, 'The Legacy of Ramon Lull in Sixteenth-Century Thought', *Mediaeval and Renaissance Studies*, vol. v (1961), pp. 189f.; Heninger, *Cosmographical Glass*, pp. 107f.

[34] Ficino, *Platonic Theology* XIII.3, Josephine L. Burroughs, trans., *JHI*, vol. v (1944), pp. 233f.

checked by the biblical injunctions to self-sacrifice in the service of others.[35]

Pico of Mirandola amplified the ideas of Ficino by drawing on the mystical literature of the Jewish Cabala (Kabbalah). Like Hermeticism, Cabala was an offshoot of the ancient creationist tradition that developed independently of Latin Western thought and then helped revitalise the latter during the late Middle Ages and the early Renaissance. In Cabala, however, the idea of creation is stressed more than it is in Hermeticism. The basis of all creation is said to be a set of Hebrew letters or divine names that provide the basic laws for all natural phenomena. This theme can be traced from the Old-Testament and intertestamental idea of the law or word of God in creation, through Jewish apocalyptic and rabbinic literature, to the *Sefer Yezirah* ('Book of Creation') of the third to sixth century AD, which became a foundational text for medieval Jewish Cabala.[36]

Two further developments in cabalism that influenced Pico and other Renaissance Neoplatonists occurred in the late thirteenth century. First, Moses of León compiled a lengthy treatise called the *Sefer ha-Zohar* ('Book of Splendor', 1275-86), which developed a multi-level model of the cosmos similar to that of Hermeticism. The cabalist model was more complex, however, in that each level in turn manifested a tenfold structure patterned after the ten ciphers (*sefirôth*) which were based on the names or attributes of God.[37] And, second, Abraham Abulafia (*c*. 1240-92) developed what he regarded as an even higher form of Cabala, which allowed immediate communion with the angelic world and the exploration and control of the cosmos by the mental permutation of the letters of the divine names.[38]

---

[35] The biblical criterion of seeking the benefit of others is cited in Ficino's *Apology*; Walter Pagel, *Paracelsus* (Basel: Karger, 1958), pp. 222f. Cf. above, chap. 1, sec. E, for the biblical and early Christian background.

[36] See chap. 1, sec. D, text above notes 29-33, 104-9; Christian D. Ginsburg, *The Essenes...The Kabbalah* (London: RKP, 1955), pp. 147-59; Gershom Scholem, *Kabbalah* (Jerusalem: Keter, 1974), pp. 23-28; *idem*, *Origins of the Kabbalah* (Philadelphia: Jewish Publication Society, 1987), pp. 24-35; Gruenwald, *Apocalyptic and Merkavah Mysticism*, pp. 10f., 104ff. For a translation of the first sixteen chapters of the *Sefer Yezirah*, dealing with the ten primordial ciphers (*sefirot belima*), see *idem*, 'Some Critical Notes on the First Part of Sefer Yezira', *Révue des études Juives*, vol. cxxxii (1973), pp. 475-512.

[37] Ginsburg, *Essenes*, pp. 104-7; Leo Schaya, *The Universal Meaning of the Kabbalah* (London: Allen & Unwin, 1971), pp. 68ff. The idea of the ten *sefirôth* was first used in the *Sefer Yezirah*. The early thirteenth-century *Sefer ha-Bahir* ('Book of Brilliance') and Rabbi Isaac the Blind were the first to use the idea of the ten *sefirôth* in the later, cabalistic sense of dynamic emanations of God—possibly under the influence of ancient Gnostic sources; Joseph Dan, ed., *The Early Kabbalah* (New York: Paulist Press, 1986), pp. 5ff., 7f., 13, 28, 32.

[38] Scholem, *Major Trends in Jewish Mysticism* (New York: Schocken Books, 1946), pp. 136f., 144f.; Perle Epstein, *Kabbalah* (Garden City, N.Y.: Doubleday, 1978), pp. 76-100; Moshe Idel, *Kabbalah: New Perspectives* (New Haven: Yale Univ. Press, 1988), pp. 147f.; *idem*, *The Mystical Experience in Abraham Abulafia* (Albany, N.Y.: SUNY Press, 1988), pp. 19ff., 31-41, 109ff.; *Golem: Jewish Magical and Mystical Traditions on the Artificial*

Pico's own version of the Cabala was still rather rudimentary compared to that of some later Renaissance cabalists we shall consider shortly. He adopted a simplified three-level model of the cosmos (corresponding to the three levels described in Genesis 1:6ff.) and identified it as the framework of Ficino's 'natural magic' based on the power of the world soul. The detailed correspondences between the higher and lower levels (as in the *Zohar*) then allowed him to understand this natural magic as a 'marrying' of things in heaven with things on earth. But Pico also developed what he called a 'spiritual magic' (some called it 'demonic') based on the direct invocation of the angels, much as Abraham Abulafia had added his higher Cabala to the 'way of the Sefirôth' of the *Zohar*.[39] The resulting possibilities for access to the divine being, and hence for human knowledge and creative power, were even greater than those conceived by Ficino.

Whatever one's sense of the value of these ideas may be, they cannot be ignored as part of the background of the scientific and theological developments of the sixteenth century. In fact, some of the key figures in the immediate background of both the Protestant Reformation and the scientific revolution were keenly interested in them, as we shall see.

Lefèvre d'Étaples (or Faber Stapulensis, 1455-1536), the leading pre-Reformation thinker in France, published two editions of Ficino's translation of the hermetic corpus, one in 1494 and another in 1505.[40] Though Lefèvre avoided any association with the kind of natural magic espoused by Ficino and Pico,[41] hermetic influence can clearly be seen in his *Introduction to Astronomy* (1503). For instance, he compared the human effort to reconstruct (in the imagination) the heavens and their motions with the original creation of the heavens by God.[42] Human science was thus dignified as part of the image of the divine in humanity.

It would be helpful to know whether the more immediate sense of the spiritual world allowed by the hermetic cosmology (in comparison to the Aristotelian cosmology of the Middle Ages) was a factor in Lefèvre's downplaying the significance of works and the sacraments as channels of

---

Anthropoid (Albany, N.Y.: SUNY Press, 1990), pp. 96-104. According to Idel, the 'way of the *sefirôth*' (as in the *Zohar*) and the 'higher Cabala' of Abulafia were two competing trends in late thirteenth-century Jewish Cabala: the first was theosophical, theurgical, theocentric, nomian, and committed to a *visio rerum omnium in Deo*; the other was ecstatic, unitive, anthropocentric, 'anomian', and committed to a *visio Dei in omnibus rebus*; *Kabbalah*, pp. xi-xvi, 154.

[39] Yates, *Bruno*, pp. 84-96, 100f., 121ff.

[40] Yates, *Bruno*, p. 170; Shumaker, *Occult Sciences*, pp. 202, 207.

[41] D.P. Walker, *The Ancient Theology* (London: Duckworth, 1972), pp. 105f.; Eugene F. Rice, Jr., 'The *De Magia Naturali* of Jacques Lefèvre d'Étaples', in Edward P. Mahoney, ed., *Philosophy and Humanism* (New York: Columbia Univ. Press, 1976), pp. 21, 25.

[42] Pierre Duhem, *To Save the Phenomena* (Chicago: Univ. of Chicago Press, 1969), pp. 56f. Cassirer notes also the influence of Cusa here; *Individual*, pp. 67ff. Lefèvre published an edition of Cusa's works in 1514; Albert Hyma, *The Christian Renaissance* (New York: Century, 1925), pp. 261f.

grace in salvation.⁴³ Lefèvre's regard for theologians like Augustine and Cusa would have to be considered here as well, but such a major shift in the understanding of salvation and of the sacraments suggests more than just a matter of theological preference. It may well reflect a corresponding shift in cosmology, and Hermeticism could have have provided this for Lefèvre and others who were rethinking their theology in the late fifteenth and early sixteenth centuries.⁴⁴

A second pre-Reformation thinker with such interests was Johann Reuchlin (1455-1522). As Lefèvre introduced the hermeticism of Ficino into France, Reuchlin introduced the cabalistic ideas championed by Pico into southern Germany. His treatise *On the Wonder-Working Word*, published in the same year as Lefèvre's first edition of the hermetic corpus (1494), was the first thoroughly cabalistic book ever written by a non-Jew.⁴⁵ Reuchlin's Christianity showed primarily through his identification of that wonder-working word with the name of Jesus which, as in the New Testament, had taken the place of Yahweh, the Old Testament name for God.⁴⁶

---

⁴³ Hyma, *Christian Renaissance*, pp. 277ff.; H.A. Enno Van Gelder, *The Two Reformations in the 16th Century* (The Hague: Martinus Nijhoff, 1961), pp. 190f. Cf. chap. 2, text above notes 63-65.

⁴⁴ In the early years of the sixteenth century, Lefèvre repudiated his interest in 'magic of any sort'; Prefatory Epistles (dated February 1504) to Lefèvre's edition of the Pseudo-Clementine Recognitions; Rice, *'De Magia'*, pp. 28f. If his subsequent evangelicalism (as of 1509) was already nascent in 1504 and if, as argued here, it fulfilled a spiritual need similar to that of his earlier hermetic interests, it is understandable that it may have helped gradually to suppress the latter. Lefèvre's *Fivefold Psalter* of 1509 compared those who dabbled in magic to those who had been weaned from the grace of God; Philip Edgcumbe Hughes, *Lefèvre* (Grand Rapids, Mich.: Eerdmans, 1984), pp. 23f. This statement may reflect the fact that Lefèvre himself had been weaned from magic as he turned to the grace of God.

However, the *Fivefold Psalter* still contained unqualified praise for Reuchlin's *On the Wonder-Working Word* (1494); Hughes, *Lefèvre*, p. 22. Could his emphatic repudiation of magic in 1504 have had anything to do with the incipient shift in patronage from the esoterically inclined Germain de Ganay to the more conventionally pious Briconnet family?; cf. Rice, *'De Magia'*, p. 21. In the Prefatory Epistles of 1504, Lefèvre specifically rebutted those who legitimate natural magic by differentiating it from demonic magic. This looks very much like a repudiation of the position Lefèvre assumed in his dialogue with Ganay in *De magia naturali*; Rice, *'De Magia'*, pp. 21f. and note 8. At about the same time (1503 or 1504) Lefèvre's pupil, Charles de Bovelles, reported Trithemius to Ganay for the angelic invocations he found in his *Steganographia*; Noel L. Brann, 'The Shift from Mystical to Magical Theology in the Abbot Trithemius', *Studies in Medieval Culture*, vol. xi (1977), p. 147. Lefèvre, on the other hand, in the Prefatory Epistle to his 1505 edition of the *Corpus Hermeticum*, addressed to Denis Briconnet, Bishop of Saint Malo, felt it necessary to plead that he was 'doing something not unwelcome to the piety of [Briconnet's] spirit'; Hughes, *Lefèvre*, p. 25.

⁴⁵ Scholem, *Kabbalah*, p. 198.

⁴⁶ *On the Wonder-Working Word* II.9; cf. *On the Art of the Kabbalah* (1517), trans. Martin and Sarah Goodman (New York: Abaris Books, 1983), pp. 73, 77, 113ff. In Hebrew, Jesus or Joshua is YHSWH. The consonants are the same as those of YHWH except for the middle one, S (*shin*). The wonder-working power of the name Jesus is due to the fact that it contains the ineffable name of the Father (YHWH) and makes it audible by adding the letter S in the middle: so Jerome, Nicholas of Cusa, Pico, and LèFèvre before Reuchlin; Hughes, *LèFèvre*,

Reuchlin also saw the four Hebrew letters in the name of God (YHWH) as the basis for Pythagoras's numerology based on the number four (the tetractys or quaternion). Pico had already suggested such an association,[47] but Reuchlin carried it much further and even referred to his own work as 'Pythagoras reborn'.[48] Apparently, the Pythagorean connection became quite prominent in the minds of Reuchlin's readers: an anonymous work entitled *Letters of Obscure Men* (1515-17) stated that Reuchlin's treatise on cabalism contained the sayings of the ancient sage Pythagoras, who was known to have practised the unlawful art of necromancy.[49] Significantly, Reuchlin approved of the esoteric science of Pythagorean numerology and championed the Pythagorean counsel of secrecy because such ideas could not be understood by the common people.[50] This was enough to guarantee the disapproval of the leading Protestant Reformers for hermetic and cabalistic ideas in general, as we shall see in section B.

Similar ideas appeared in the works of Trithemius of Sponheim (1462-1516) and his student Agrippa of Nettesheim (1486-1535). The latter was particularly important for his emphasis on the basic points of the creationist tradition in his famous treatise, *On Occult Philosophy* (first draft written by 1510; published 1531-33): the creation of all things by number, weight, and measure (Wisd. 11:20); the creation of the human soul with a faculty of reason by which it could comprehend the structure of the cosmos; the unity of all nature, both heaven and earth (here due to the universal presence of the world-soul); the infusion of all creation with law-abiding energies; the application of mathematics to the mechanical arts; the mandate of humanity to cooperate in the work of divine creation; and even the directive that this mandate be carried out for the benefit of one's fellow humans.[51] The criterion of social benefit was, admittedly, secondary: the point was made in a preliminary address to the reader in which Agrippa attempted to stave off the charge of heresy.[52] It is a useful reminder,

pp. 19-22.

[47] 'But in those parts [of the Cabala] which concern philosophy you really seem to hear Pythagoras or Plato,' Pico, *On the Dignity of Man* (written in 1486 as a preface to his 900 Theses); Ernst Cassirer et al., *The Renaissance Philosophy of Man* (Chicago: Univ. of Chicago Press, 1948), p. 252.

[48] *On the Art of the Kabbalah* (trans. Goodman), p. 39; Lewis W. Spitz, *The Religious Renaissance of the German Humanists* (Cambridge, Mass.: Harvard Univ. Press, 1963), pp. 67f., 75f.

[49] *Epistolae obscurorum vivorum* II.69, trans. Francis Griffin Stokes (New York: Harper & Row, 1964), pp. 247f.

[50] Spitz, *Religious Renaissance*, pp. 77. Contrast Abulafia's claim to have made the Kabbalah accessible to ordinary people; Epstein, *Kabbalah*, pp. 76-80, 85.

[51] Shumaker, *Occult Sciences*, pp. 134-37, 153; Randall, *Career*, vol. I, pp. 192f.; Yates, *Bruno*, pp. 134ff., 139, 147.

[52] Willis F. Whitehead, ed., *Three Books of Occult Philosophy*, rev. ed. of 1651 English trans. (New York: Samuel Weiser, 1971), pp. 25ff. Trithemius had similarly appealed to the exercise of love as a guarantee that he was not opening himself to the influence of demons in his pursuit of knowledge; Letter to Johannes Capellarius, 1505; Brann, 'Shift from Mystical to

however, of the kind of ethical stance any science would have to assume in order to win public approval in a Christian society.

The Florentine perspectivists and the hermetic-cabalists represent two very different forms of the creationist tradition, one bordering on secular naturalism and the other on the verge of uncontrolled superstition. Together they indicate the richness and potential for productivity in the creationist tradition of the Renaissance period. But they also indicate the need for a more balanced approach that could check the extremes of secularism, on the one hand, and superstition, on the other. The attempts of the early Reformers of the sixteenth century to achieve such a balance is best understood against this background; so are the attempts of leading scientists of the sixteenth century like Copernicus and Paracelsus.

2.3 Nicholas Copernicus (1473-1543)

Copernicus's treatise *On the Revolutions of the Celestial Spheres*, written in stages between 1512 and 1542 and published in 1543, is one of the most important works in the entire history of science. For the most part it is a technical work with only occasional references to the philosophical and theological ideas we are interested in here. Taken on their own, these comments would be almost impossible to interpret, and scholars have differed widely in their conclusions. Following the approach of Alexandre Koryé, we shall interpret Copernicus as 'a man deeply imbued with the entire, rich culture of his period...a *humanist* in the best sense of the word'. He was trained in canon law, medicine, and philosophy as well as in mathematics and astronomy. The latter two subjects he studied at Bologna under Domenico Maria Novara, himself a humanist influenced deeply by the Neoplatonic ideas we have discussed in the paragraphs above.[53]

For these reasons we judge that Copernicus's occasional remarks about philosophy and theology are to be taken seriously. They are not just concessions to the eclectic tastes of the time but represent the true convictions of the author. And where Copernicus indicates that theological ideals helped motivate him to initiate and pursue his programme of research, we may take these confessions at face value. (Motives involved in his final decision to publish his manuscript will be considered later.)

It is helpful to begin with a brief look at the *Commentariolus*, a short prospectus which Copernicus circulated privately among his friends while the larger work was still in the early stages of preparation.[54] Several features stand out here which help us understand Copernicus's motivation. The most prominent is Copernicus's insistence that all the planetary orbits (both

Magical Theology', p. 155.

[53] Koryé, *Astronomical Revolution*, pp. 20ff.

[54] Edward Rosen, trans., *Three Copernican Treatises*, 3rd ed. (New York: Octagon Books, 1971), p. 59. On p. 345 Rosen suggests 1511-12 as a date for the *Commentariolus*.

deferents and epicycles) be perfectly circular and that all motions around these circles be perfectly regular and uniform with respect to their proper centres.[55]

Copernicus's persistence in adhering to strict principles and his clear-headed application of them to the 'very difficult and almost insoluble problem' of planetary motion is truly awesome. In general, it bespeaks a profound conviction in the rationality ('a more reasonable arrangement') of the world, and hence in the reality one must ascribe to the mathematical entities one is dealing with theoretically. The specific kind of rationality and reality Copernicus subscribed to was derived from the Aristotelian-scholastic tradition. The planets moved as a result of being embedded in an interconnected set of perfect spheres: hence the circular orbits. Copernicus realised, however, as a result of the failure of previous attempts of this kind, like that of Fracastoro,[56] that these spheres could not all have the same centre and that epicyclic spheres would have to be allowed.

Still, it was not the mere insistence that orbits be circular that differentiated Copernicus's procedure from Ptolemy's: it was the added stipulation that the motions of the planets in their circular orbits be perfectly regular and uniform about their own proper centres.[57] This followed from the aforementioned Aristotelian notion of celestial spheres on the supposition that the motion of the spheres is naturally and invariably a uniform rotation. Here we have the long-term effect of the idea of impetus developed by John Buridan and his school in the fourteenth century, an idea that was still cultivated in the Renaissance scholasticism Copernicus was exposed to at Cracow and elsewhere.[58]

Several other features of the *Commentariolus* reflected the Renaissance Neoplatonism that influenced Copernicus, particularly during his studies at Bologna. First there was his dissatisfaction with Ptolemy's system on the grounds of its lack of simplicity: it was 'neither sufficiently absolute [Aristotelian emphasis] nor sufficiently pleasing to the mind [Neoplatonic]'. This dissatisfaction led him to attempt a solution 'with fewer and much simpler constructions than were formerly used'.[59] In the final version published in 1543, there was actually little or no gain in simplicity over Ptolemy's system as far as geometrical constructions were concerned: at least forty-six different circles were required to account for all the planetary

---

[55] *Ibid.*, pp. 57f., 71, 76.
[56] Above, note 5.
[57] Rosen, trans., 'Introduction', p. 29.
[58] Koryé, *Astronomical Revolution*, pp. 55-59; Paul W. Knoll, 'The Arts Faculty at the University of Cracow', in R.S. Westman, ed., *The Copernican Achievement* (Berkeley: Univ. of California Press, 1975), pp. 149f.
[59] Rosen, trans., pp. 57f.

motions. But in the *Commentariolus*, Copernicus had hoped to make his system work with only thirty-four circles.[60] In any case, Copernicus eliminated the equants that were required by Ptolemy's system and provided a semblance of uniform circular motion.

A second 'Neoplatonic' feature reflected the discipline of mathematical perspective developed by Toscanelli and his students in early fifteenth-century Florence. Copernicus repeatedly expressed his satisfaction in explaining a variety of phenomena as they appear from the earth as a result of the earth's motion around the sun.[61] There is a preference for simplicity of explanation here, but also an aptitude for perspective-shift which would not have been likely prior to the work of the Florentine perspectivists.[62]

A third possible 'Neoplatonic' feature of Copernicus's proposed system was the location of the sun at the centre of the universe. It has been suggested by several historians that this innovation was based on the dynamical centrality of the sun in Hermeticism and in late medieval Neoplatonism generally.[63] Such a connection is by no means proven historically, however, and other historians have rejected it.[64]

The motivating ideas of the *Commentariolus*, then, are clearly related to the creationist tradition of the late Middle Ages and the early Renaissance. Still, there are no explicit references to God or to the idea of creation itself. Such comments appear for the first time in the preface and first part he wrote for his treatise, *On the Revolutions*, when preparing it for its publication in 1543. It is difficult to know precisely how to interpret these late references to the idea of creation. On the one hand, they appear to be only concessions to Christian piety: Copernicus was keenly aware of his writing for a larger audience, both mathematicians and theologians, learned and unlearned alike.[65] On the other hand, no one would doubt that Copernicus really believed in the theological points he made.

One of the references to the idea of creation appears in the preface addressed to Pope Paul III. After describing the inadequacies of the Aristotelian and Ptolemaic systems already discussed in the

---

[60] *Ibid.*, p. 90; Koryé, *Astronomical Revolution*, pp. 26f., 43-51; Toulmin and Goodfield, *Fabric*, p. 175.

[61] Rosen, trans., pp. 58f., 61, 64f., 77f.

[62] Koryé, *Astronomical Revolution*, p. 49; cf. above, note 16.

[63] Kuhn, *Copernican Revolution*, pp. 129-41; Koryé, *Astronomical Revolution*, p. 65; cf. above, notes 28-30; Harold P. Nebelsick, *Circles of God* (Edinburgh: Scottish Academic Press, 1984), pp. 211-18, 246f.

[64] Jaki, *Science and Creation*, pp. 259f.

[65] A convenient English translation by John F. Dobson and Selig Brodetsky can be found in Milton K. Munitz, ed., *Theories of the Universe* (New York: Free Press, 1957), pp. 149ff. Cf. A.M. Duncan, trans., *Copernicus: On the Revolutions of the Heavenly Spheres* (Newton Abbot: David & Charles, 1976), pp. 23ff.

Commentariolus, Copernicus says that he went through a prolonged period of uncertainty as to which system was the best. Then, at last, he says, he 'began to chafe that the philosophers could by no means agree on any one certain theory of the mechanism of the universe, wrought for us by a supremely good and orderly Creator....'[66] Although the wording here is rather stereotyped, it may accurately reflect Copernicus's mental struggle at a turning point in his personal development. The allusion to the Creator is not gratuitous: it comes precisely at the point of Copernicus's frustration with the inconsistency of the Hellenistic natural philosophers. The sequel, in this case, was not the repudiation of natural philosophy as such but a resolve to read more widely in the works of philosophers, particularly the Pythagoreans. The sense is that somewhere among the ideas of the ancients there must be a clue to a solution that would uphold the regularity, uniformity, and symmetry that befitted the work of God. Belief in the ultimate compatibility of biblical faith and Greek science and criticism of the diversity and unverifiability of the views of the secular philosophers—both features of the creationist tradition since the second century AD (chap. 1 sec. A)—thus sustained Copernicus in his quest for deeper understanding of the cosmos.

Two further references to the idea of creation occur in the first part of *On the Revolutions*. The first of these is made in an attempt to counter the Aristotelian notion of natural place. The earth must be at the centre of the cosmos, according to Aristotle, because it is natural for the heaviest element to gravitate towards the geometric centre. In contrast, Copernicus affirms that there can be many centres of gravitation because 'gravity is but a natural inclination, bestowed on the parts of bodies by the Creator....'[67] Here again we have one of the basic ideas of the creationist tradition: the laws of nature are not intrinsic and cannot be deduced a priori: rather they are imposed or infused by God in such a way that they appear to operate automatically. This idea had been diluted to some extent during the thirteenth century but was reaffirmed in connection with the Condemnation of 1277 and was influential in the fourteenth-century antecedents of Copernicus.[68] The relatively conscious adherence of Copernicus to that tradition may be seen by a comparison with the more secular interpretation of Fracastoro described earlier.[69]

The other reference to the idea of creation appears in relation to the idea of the immensity of the cosmos. Copernicus concluded that the dimensions of the universe must be at least 2,000 times larger than previously

---

[66] Munitz, ed., p. 151.
[67] *On the Revolutions* I.9 (p. 164); cf. I.7 (pp. 160f.).
[68] Koryé, *Astronomical Revolution*, pp. 55ff.
[69] Above, notes 9 and 12.

thought.[70] He even suggested the possibility of an infinite universe or, at least, of an infinite void space beyond the stars. The issue could not be resolved on the basis of mathematical astronomy alone, he concluded; it was a matter for the natural philosophers (i.e., physicists) to settle. Yet the immeasurable distance of the fixed stars was consistent with the greatness of 'this divine work of the great and noble Creator'.[71] These speculations are clearly rooted in the ideas of the infinity and omnipotence of God as developed in the late medieval period and continued in various places during the Renaissance.[72]

We assumed at the outset that Copernicus was conversant with the philosophical and theological ideas of his time. The results of our brief study have vindicated this assumption in finding Copernicus to be a faithful son of the creationist tradition. In fact, a review of the different aspects of that tradition as they were reaffirmed in the late medieval period (chap. 2, sec. C) will show that practically all of the contributions of that period had a part to play in this Renaissance man's thought. A similar review of our treatment of the early Renaissance will show that he drew on both the Aristotelian-scholastic and the Neoplatonic-perspectivist traditions. The hermetic-cabalist tradition was probably not a significant factor, however. Copernicus did refer to the Pythagoreans, but only to their views on the motion of the earth, not to their numerology.[73] He also cited the hermetic writings, as Ficino had done, to support the idea of the dynamical centrality of the sun.[74] But in the next breath he also cited Sophocles and Aristotle, so we may infer that all these citations were merely a way of literary illustration. Copernicus was neither as secular in outlook as Alberti and Fracastoro, nor as mystical as Ficino and Pico. He should be located somewhere near the centre of the creationist tradition as it was represented in the context of Renaissance thought.

3 The Experimental Tradition to Paracelsus

Of all the varied aspects of the medieval creationist tradition, the one that was least developed was the appreciation of the relevance of industrial and technological innovation for the progress of basic science.[75] As a result, nature was conceived as existing in an ideal state of equilibrium rather than as the malleable substance we think of today. Aristotle's mechanics

---

[70] Koryé, *Astronomical Revolution*, p. 106 note 12.
[71] *On the Revolutions* I.6, 8, 10 (pp. 159-62, 170).
[72] Above, chap. 2, 'Possibility of a Void'; chap. 3, notes 4, 7.
[73] *Commentariolus* (trans. Rosen, p. 59); *On the Revolutions*, preface; I.5 (Munitz, ed., pp. 151, 158), cf. Yates, *Bruno*, pp. 153ff.
[74] *On the Revolutions* I.10 (p. 169); Yates, *Bruno*, p. 154.
[75] Chapter 2, section C1.3, 'Observation and Experiment'.

described things as having a natural context outside of which they did not behave normally and towards which they would return if displaced. Similarly, Galen's medicine described the functioning of the body as a balance of humours which had to be maintained in health and restored in the event of disease.[76]

In contrast, the new ideal of scientific knowledge introduced in the Renaissance was that we only know things as artifacts. That is, we only truly know things that we have made ourselves and natural things that we have succeeded in measuring or analysing into parts and reconstructing.[77] The new ideal of analysis did not entirely displace the old—indeed, many modern epistemologies have attempted to synthesise the two ideals—but there was a major shift in the new direction in the fifteenth and sixteenth centuries: first in the mechanical arts and astronomy, and then in metallurgy and medicine. And along with this shift came a renewed sense of the ethical problems raised by technology and a restatement of the biblical criteria for technology assessment.

3.1 Artisans and Astronomers

The interest in measuring instruments itself was not new. It was encouraged by the belief that God had created all things by measure, number, and weight (Wisd. 11:20), a belief that was revitalised by the Neoplatonic revival of the fifteenth century. Somehow this motivation had to be combined, however, with the actual use of instruments like mechanical clocks and navigational tools which were being produced with increasing skill and accuracy during the Renaissance.

One of the principal centres of production was Nuremberg in Bavaria. Nicholas of Cusa is said to have purchased three scientific instruments there in 1444.[78] Even if Cusa was not much of an experimentalist himself, his influential vision of the possible benefits of experimentation to science lived on in George Peurbach (1423-61) and his associate, John Regiomontanus (1436-76), the pioneers of modern observational astronomy, in Vienna. In 1457 they planned to observe a lunar eclipse and found that the time predicted by the astronomical tables then in use (the Alfonsine Tables of 1252) was in error by eight minutes.[79] Measurement of changes in the apparent size of the moon as it moved around the earth also led the

---

[76] Ernan McMullin, 'Medieval and Modern Science: Continuity or Discontinuity?', *International Philosophical Quarterly*, vol. v (1965), pp. 118f.; Kearney, *Science and Change*, p. 116.

[77] Paolo Rossi, *Philosophy, Technology, and the Arts in the Early Modern Era* (New York: Harper & Row, 1970), Appendix I; cf. Glacken, *Traces*, pp. 462-71.

[78] Wightman, *Science and the Renaissance*, vol. I, p. 22n.

[79] C.D. Hellman and N.M. Swerdlow, 'Peurbach', *DSB*, Suppl. I, pp. 473f.

two to criticise Ptolemy's system of planetary motions, and this observation later came to the attention of Copernicus at a time when he was beginning to develop his own system of the heavens.[80]

The production of new instruments and the composition of new tables of data were to be a major factor in the rise of modern science. In 1471 Regiomontanus chose Nuremberg as the site of his own observatory due to the accessibility of instruments there. After his death in 1476, his work was carried on by his associate, Bernard Walther, who performed the first uninterrupted series of astronomical observations. The data produced were used by Tycho and Kepler a century later.[81]

If there was anything lacking in the experimental philosophy of the Nuremberg school, it was the ideal of investigating and publishing for the benefit of others rather than merely as a self-aggrandising response to market conditions. There was great demand for Regiomontanus's improved astronomical tables (*Ephemerides*, 1474), primarily due to their usefulness in navigation and astrology. However, after Regiomontanus's death, Walther refused to circulate most of the more recent astronomical observations that had been made. Copernicus owned a copy of the Regiomontanus-Walther solar tables (1490), but he did not have access to the other data until much later in his work.[82]

In 1945 the historian Edgar Zilsel published an article in which he argued that the understanding of science and technology as a real benefit to humankind was essential to scientific progress in the early modern period, and that this perception grew out of the artisan tradition of the late Middle Ages and early Renaissance.[83] Among the examples of the latter he gave are the following. Around 1400 the Florentine painter Cennini composed a handbook on pigments and painting in the reverence of God, the Virgin, and the saints and 'for the use and profit of any one who wants to enter the craft'. Similar motivations were expressed in a more secular vein by Roriczer (1486), Dürer (1528), Apian (1532-3), and Tartaglia, among others.[84]

---

[80] E. Rosen, 'Regiomontanus', *DSB*, vol. XI, p. 349a. Novara, Copernicus's teacher at Bologna, was a student of Regiomontanus; *ibid.*, p. 352a.

[81] Wightman, *Science and the Renaissance*, vol. I, p. 22; A. Pannekoek, *A History of Astronomy* (London: Allen & Unwin, 1961), pp. 180f.

[82] Dreyer, *History*, p. 289; Wightman, *Science and the Renaissance*, vol. I, pp. 108, 111f. According to Pannekoek, John Schoner published some of Walther's data in time for Copernicus to use them in his final version; *History*, pp. 183, 196.

[83] E. Zilsel, 'The Genesis of the Concept of Scientific Progress', *JHI*, vol. vi (June 1945), pp. 325-49; cf. above, chap. 1, sec. E, text above notes 257-70. For similar motives among fourteenth-century physicians, see Darrel W. Amundsen, 'Medical Deontology and Pestilential Disease in the Late Middle Ages', *JHMAS*, vol. xxxii (Oct. 1977), pp. 418f.

[84] *Ibid.*, pp. 333-38, 343f.; cf. Rossi, *Philosophy, Technology, and the Arts*, pp. 70ff., which is partly based on Zilsel's article.

Tartaglia's two pioneering books on ballistics (1537, 1546) were prefaced by discussions of the possible dangers and benefits to society resulting from the improved use of artillery. Under normal circumstances, he reasoned, the publication of such information would be 'cruel and deserving of no small punishment by God', but under the threat of Turkish invasion, he felt compelled to make his findings public.[85] As it happened, the benefits accruing to both science and society from Tartaglia's insights into the dynamics of falling objects far exceeded the advantage of short-term improvements in aiming cannon derived from his tables of gun elevations. But the fact that early modern scientists saw themselves, and came to be seen by the general public, as contributing to the common good was of utmost importance for the development of strong social support for the relatively new and untried enterprise.

The most impressive instance of the influence of Christian concern for the public good on scientific development took place in conversations between Copernicus and Tiedemann Giese, Bishop of Kulm at the bishop's castle in 1539. According to his young associate, George Joachim Rheticus, Copernicus wanted to publish astronomical tables and rules of calculation based on his new system of the heavens for 'common mathematicians' and to keep the underlying hypotheses and proofs to himself and his personal associates in accordance with the 'Pythagorean principle' of education.

Bishop Giese, on the other hand, had 'mastered with complete devotion the set of virtues and doctrines required of a bishop by Paul' (1 Tim. 3; cf. 1 Cor. 7-16), and he realised 'that it would be of no small importance to the glory of Christ if there existed a proper calendar of events in the Church and a correct theory and explanation of the motions'. He therefore argued with Copernicus that the astronomical tables would be 'an incomplete gift to the world' without the supporting theory. Something of an amateur astronomer himself, Giese complained that the existence of tables without explanations had caused great inconvenience and many errors in the past and that the Pythagorean counsel of secrecy had absolutely no place in mathematical science as practised by Christians.[86]

Such was the pressure under which Copernicus allowed Rheticus to publish a 'first report' on his theory (1540) and finally to copy the entire manuscript of his treatise *On the Revolutions of the Celestial Spheres* for printing (1542-43). The place of publication, significantly, was Nuremberg, where Cusa had bought his instruments nearly a hundred years earlier.

---

[85] Stillman Drake and I.E. Drabkin, trans., *Mechanics in Sixteenth-Century Italy* (Madison: Univ. of Wisconsin Press, 1969), pp. 68f.

[86] On Rheticus, see Rosen, trans., *Three Copernican Treatises*, pp. 192-95. Cf. Augustine's citation of 1 Tim. 3 in relation to the duties of a bishop, *City of God* XIX.19. In his biography of Copernicus, Rosen seems to ignore Giese and give Rheticus all the credit; *ibid.*, pp. 393f.

156 CHAPTER THREE

We have treated the theological motives behind the publication of Copernicus's work under an entirely different heading from that of the motives that led him to do his research in the first place. Copernicus's motives for research were derived primarily from the scholastic and Neoplatonic traditions of the late Middle Ages and Renaissance, traditions which were rooted in the creationist ideas of the comprehensibility, unity, and relative autonomy of nature (chap. 1, secs. B-D). Giese's motives in urging him to publish were derived more directly from the biblical ideal of mutual service in the church of Christ (chap. 1, sec. E), reinforced by the late medieval and early Renaissance artisan tradition of art and technology in the service of humanity. At this critical juncture of history, at least, these different aspects of the creationist tradition worked in marvellous harmony.[87]

3.2 Metallurgy and Medicine

Scientific instruments and tables of data were one way in which the Renaissance contributed to the new view of nature. Advances in the technology of mining and in the medical use of chemicals were another. Here the Neoplatonic emphasis on mathematics was of little use at the time, but the importance of the Christian idea of creativity through the human arts was as great as that of the artisan tradition itself. There were two prominent figures in this sector, Agricola and Paracelsus, who represented opposite ends of the scale socially and financially.

George Bauer Agricola (1494-1555) worked as a town physician in Bohemia and Saxony and studied the extensive mining operations in his area. On the basis of his medical and mining expertise, Agricola corrected the natural history of the classics at several points. He also achieved considerable wealth and status in society. Yet he took a genuine interest in the welfare of the miners, argued for safer working conditions, and exposed himself and his family to considerable danger while tending the sick during the outbreak of plague in 1552-53.[88]

Agricola's importance for our study lies primarily in his discussion of the propriety of disturbing the earth in order to extract its resources. It may come as a surprise to the modern reader to learn that the industry of mining was under heavy criticism already in the sixteenth century for its destruction

---

[87] Harold Dorn has recently made the dialectic of free inquiry (for Dorn, the Hellenic Greek tradition) and social utility as determined by public institutions (Marx's "Oriental tradition") the framework for interpreting the entire history of science; *The Geography of Science* (Baltimore, Md.: Johns Hopkins Press, 1991), pp. 94f., 119, 176. On neither side does Dorn recognise the contribution of the creationist tradition, but this is understandable in view of his primary interest in geographic and ecological factors.

[88] *De re metallica* I, trans. H.C. and L.H. Hoover (New York: Dover Books, 1950, pp. 3-6); Helmut M. Wilsdorf, 'Agricola', *DSB*, vol. I, pp. 77f.

of farm land, consumption of wood, extinction of wildlife, pollution of water supplies, and disruption of country life, to say nothing of the use of its products for warfare.[89] In view of the perennial nature of these problems, Agricola's response and its bearing on the creationist tradition should be of particular interest to those engaged in 'technology assessment' today.

There are three points in Agricola's response. First, he argued that the wisdom of God in creating the earth was to be seen in his making a variety of things available for human use. So the location of metals underground was no more a reason for leaving them there than the location of fish in the water or birds in the air was sufficient reason for leaving them there. According to Scripture, the ground was given to humanity to cultivate (Gen. 2:15), and this could be done through mining as well as through agriculture.[90]

To state Agricola's thought in modern terms: what is natural is a matter of social and cultural definition. Hunting, fishing, and farming were forms of technology that had been accepted as necessary to society and sustained a social structure favouring the rural land owners. Mining, on the other hand, was a growth industry under the control of the new merchant-bankers like the Fuggers of Augsburg. So the economic and social change of the Renaissance was as much a factor in dislodging the Aristotelian idealisation of the 'natural' place of things as was the more theoretical challenge of academic theology and philosophy. Agricola was an heir of the creationist tradition insofar as he saw the work of creation as an ongoing process partly entrusted into the hands of humans and realised that human technology was neither good nor evil in itself but had to be assessed in terms of particular criteria.

Agricola's second point was to specify those criteria and to show how, in his experience, the practise of mining satisfied them. First criterion: any human enterprise should benefit those who are directly engaged in it—here Agricola had to deal with the problems of risk: financial risk for the owners and occupational hazards for the workers.[91] Secondly, the enterprise should be beneficial to society as a whole—here Agricola argued that to say the Creator had placed metals in the earth to no social purpose would be to accuse him of wickedness. On the contrary, he pointed out, the use of metals was essential to all the other arts, especially to those like agriculture that were held in such esteem by the critics of mining, and hence metals could be used in the protection of life and the preservation of health, as God had willed.[92]

---

[89] *Ibid.*, pp. 8-11.
[90] *Ibid.*, p. 12. I have reorganised Agricola's own presentation.
[91] *Ibid.*, pp. 4ff.
[92] *Ibid.*, pp. 12ff., 18ff.

Agricola's third and concluding point was that all God's gifts were good and that it was the responsibility of humans to use them for the good of others as well as for themselves. To reject good things simply on the basis of their possible misuse would be blasphemy against their Creator.[93]

To modern eyes haunted by the memory of technological abuse, Agricola may appear to have been merely a 'company man'. He never really dealt with the problems of the ecological damage and social disruption caused by the growth of mining even in his own time, and he implied that companies should never be held responsible for the harm caused by their products when misused. But these were matters beyond his personal influence, probably even beyond his personal understanding. In the arena that Agricola did influence and understand, the health care and working conditions of the miners, his ideas were quite progressive. For instance, he opposed the kind of exploitation of miners practised in Greco-Roman times.[94] Moreover, he must be credited with having attempted a comprehensive Christian ethic that did justice to the needs of employer, employee, and society as a whole. Faulty as modern industrial practises were to be, they could be held accountable, at least, to an ethic whose credibility had been established over a period of two millennia and which is still with us today, more than four centuries after the seminal word of Agricola.

Very much more clearly on the side of the poor was Philip Theophrastus Bombast of Hohenheim, better known as Paracelsus (*c*. 1493-1541). Like Agricola, Paracelsus was trained as a physician and worked among miners and peasants, chiefly in Austria and Switzerland. Unlike Agricola, he was a hopeless eccentric and was never able to hold a job for long. He spent most of his life on the brink of poverty as a wandering lay preacher repeatedly attacking the academic medical authorities of his time. His epitaph simply stated that he had devoted his life to the art of healing and wished his goods to be distributed to the poor upon his death.[95] Though he was very nearly heretical from the viewpoints of later scientific and theological orthodoxy, Paracelsus understood at least some of the basic truths of both science and theology as well as anyone in European history.

Paracelsus's two principal contributions to science itself were his doctrine of the 'three basic principles' (*tria prima*) of matter and his view of physiology as governed by relatively autonomous organ systems. Both of these ideas had roots in the alchemical-medical tradition of the Middle Ages and were nourished by the hermetic-cabalist tradition of the Renaissance.

---

[93] *Ibid.*, pp. 18f.
[94] On the working conditions in Greek and Roman mines, see K.D. White, *Greek and Roman Technology* (Ithaca: Cornell Univ. Press, 1984), pp. 35f.
[95] W. Pagel, 'Paracelsus', *DSB*, vol. X, pp. 304ff.; George Sarton, *Six Wings: Men of Science in the Renaissance* (Bloomington: Indiana Univ. Press, 1957), pp. 108ff.

Their importance lay in the alternatives they offered to the Aristotelian doctrine of the four elements and the Galenic physiology based on four humours. In other words, both contributed to the breakdown of the medieval view that conceived of nature in terms of equilibrium states.

The 'three principles' of matter, according to Paracelsus, were mercury, sulphur, and salt. These were not quite the same as the chemicals we know by those names today, but rather three kinds of product resulting from a typical chemical reaction: a vapour (smoke), a flame (light), and a solid residue (ash). For Paracelsus these three categories also corresponded to three levels of human existence: spirit, soul, and physical body. Since spirit was the highest and most powerful level of being, Paracelsus and his followers stressed the importance of mercury (or mercuric oxide) in the treatment of disease.[96] An unorthodox worldview resulted in unorthodox but powerful medicine. Some people were actually healed by it!

Even though Paracelsus did not completely do away with the Aristotelian conception of the elements (earth, water, air, and fire), his 'three principles' did offer an alternative way of looking at matter. It was really quite revolutionary to think of the world in terms of components that are not normally visible but can only be produced through chemical processes. Previously, the only alternative to Aristotle's system had been the atomic hypothesis of the Epicureans. Atoms, of course, were completely invisible, and their production was far beyond the technology of the sixteenth century. Paracelsus put science on the trail of discovering the basic constituents of matter, however, that eventually did lead to the discovery of atoms and beyond.

Paracelsus's physiology was based on the idea that a human being is a microcosm or miniature version of the cosmos, an idea that we have already mentioned in our discussion of Ficino. One of the implications of the microcosm theory was that there could be detailed correspondences between the organs of the human body and the planetary spheres of the cosmos: for instance, the sun corresponded to the heart, Mercury to the lungs, and Venus to the kidneys. The planets did not affect the human organs directly, but celestial events and human diseases had definite correspondences based on their common dependence on forces at work at the level of the world soul.

As a result, Paracelsus saw diseases like the plague as coming from outside the body by infection and regarded them as affecting particular organs

---

[96] Pagel, *Paracelsus*, pp. 100-104, 267ff.; Allen G. Debus, *The English Paracelsians* (London: Oldbourne, 1965), pp. 27ff.; cf. above, note 32, on Ficino's three-level hermetic anthropology.

In traditional alchemy, there were just two principles: sulphur was the active, spiritual principle, and mercury was the passive, material principle (Lindberg, *Beginnings*, p. 288 note 12). So Paracelsus added salt to provide the material principle and elevated mercury to a higher level than sulphur.

rather than as disturbances in the humoural balance of the body, as Galenic medicine had taught. The result was a more analytical approach to the structure and function of the body and a more interventive kind of medicine using chemicals targeted at specific organs.[97]

These contributions encouraged a new, more experimental, kind of science in which humans understood nature in terms of the ways they could influence it, rather than in terms of what it was in its undisturbed state. Like Ficino, Pico, Reuchlin, and Agrippa, Paracelsus was exhilarated by the new sense of human freedom and power over nature. But more clearly than any of his cabalist predecessors, he also had a deeply Christian sense of his work as a healing ministry ordained by God. It was God who had created herbs and minerals, he said, recalling the words of Jesus ben Sirach (Ecclus. 38:1-4), and it was God who gave humans the skill to create medicines out of these raw materials and the wisdom to know how to use them at the appropriate time.[98]

To some extent, according to Paracelsus, this skill could be mastered by any physician who was called by God, regardless of religious persuasion. This follows from the fact that God implanted seminal causes in nature at creation and that these were generally accessible to human understanding and control. All that was requried was an attitude of humility combined with a persistence in seeking that would allow nature to reveal itself so that the physician could read it like an open book. Compared to the inspired medicine of the Hebrew prophets and Christian saints, the gentile arts based on the 'light of nature' were just crumbs from the table of the Lord (Mark 7:28), but even the children of God depended on the light of nature and needed to learn from the gentile arts, especially when they had lost their natural ability to use that light.[99]

As a devout Christian, however, Paracelsus believed that deeper levels of wisdom were available to humans than those implanted by nature. Moreover, since spiritual and even divine influences were involved in many forms of illness, deeper levels of wisdom were mandatory. For cases such as these, the invitation of Christ was open: 'Come to me...and learn from me, for I am gentle and lowly in heart' (Matt. 11:28f.). From Christ flowed the very foundation of truth.[100]

---

[97] Pagel, 'Religious Motives', pp. 99-106; *Paracelsus*, pp. 38, 67-72, 76ff., 107ff., 127ff., 137-44, 215ff.

[98] Kurt F. Leidecker, trans., *Volumen Medicinae Paramirum of...Paracelsus* (Baltimore: Johns Hopkins Press, 1949), pp. 56-63; Jolande Jacobi, ed., *Paracelsus: Selected Writings* (New York: Pantheon Books, 1951, 1958), pp. 55f., 69.

[99] H.E. Sigerist, ed., *Four Treatises of...Paracelsus* (New York: Arno Press, 1979), pp. 24-29; Pagel, *Paracelsus*, pp. 54-61, 109f.

[100] *Ibid.*, p. 25; Pagel, *Paracelsus*, pp. 65f.

In what exactly did this divine wisdom consist? Unless the sickness was intended by God to lead to death, there would be a predetermined time at which the issue could be resolved one way or the other (cf. Ecclus. 38:13f.). As this time of 'purgatory' passed, the patient might be healed miraculously if he had faith in God. If faith is lacking, however, 'the physician works nevertheless the miracle which God would have accomplished wonderfully if the patient had had faith'. Apparently, the faith and compassion of the physician made it possible for him to discern the right time and prescribe the correct treatment.[101] At this point the boundary between the natural and the supernatural became a bit fuzzy, but that seems to have been the way Paracelsus wanted it. According to our historical analysis, a strict division between nature and supernature only entered into Western theology around the twelfth century.[102] Paracelsus's inability (or unwillingness) clearly to differentiate the two may simply derive from his desire to circumvent scholasticism and return to a more biblical theology.

From his reading of the Bible, Paracelsus derived two things that he felt distinguished the Christian physician from the unbeliever: an article of faith and a sense of calling. The article of faith was that there was no illness that could not eventually be cured. For us today in an age of the promise of 'miracle drugs', this may sound commonplace, but Paracelsus lived in a time when diseases like epilepsy seemed to be beyond cure, and new diseases like the plague were at their height. Along with new diseases, however, would come new techniques and new chemicals, he believed. Why? Because Christ himself had said that 'the sick have need of a physician' (Mark 2:17), and he would not have said this if the requisite skills and medicines could not be developed. In fact, the very curse of death had been nullified through the work of Christ in vanquishing the principalities and powers that inflicted death. Thus there was no limit to the possibilities of the healing art when practised by a believer.[103]

Paracelsus described his personal sense of calling as a physician as coming from the commandment of Jesus that we love our neighbours as much as we love ourselves (Mark 12:31). This meant not only that every effort should be expended in seeking a cure for our neighbours' illnesses, but that the physician should serve his fellow humans out of true love and not just for money.[104] The work of the physician was, therefore, a ministry of

---

[101]Leidecker, trans., *Volumen Paramirum*, pp. 57-62. On the possibility of inspired medicine, cf. Pagel, *Paracelsus*, pp. 253f., on Arnald of Villanova.

[102]Chapter 1, section D11 and 12, and chapter 1, section E9.

[103]Sigerist, ed., *Four Treatises*, pp. 12-15; Jacobi, ed., *Paracelsus*, pp. 4f.; Leidecker, trans., *Volumen Paramirum*, pp. 59f. Cf. the less optimistic view of Roger Bacon, chap. 2, note 114, above.

[104]Sigerist, ed., *Four Treatises*, pp. 14, 29-33.

God's grace as much as the preaching of the gospel was: after all, the welfare of the soul was bound up with that of the body. So the doctor acted not for himself but for God, and he must never take advantage of his position or use his skill to harm his patients.[105]

Like Copernicus, Paracelsus represented something close to the centre of the creationist tradition. But his emphasis was on the ministry of healing and restoration ordained by Christ rather than on the logic and impetus ordained by God at creation. In Copernicus, the autonomy of nature and the independence of science were nearly absolute and needed to be balanced by the sympathetic pastoral direction of Bishop Tiedemann Giese. In Paracelsus, on the other hand, nature and God, science and theology, were all mixed together. What was required in this case was a reworking of his insights with due respect for the rationality and autonomy of nature. Johann Andreae, Francis Bacon, and Robert Boyle were to supply this need in the seventeenth century. All three were deeply influenced by the teachings of the Protestant Reformation and their implications, both positive and critical, for natural science.

B. Renaissance Science and Reformation Theology Through Kepler and Bacon

The Protestant Reformation was a cluster of movements during the first half of the sixteenth century that attempted to restore the teachings and practise of the early church. Principal continental European reformers were Martin Luther and Philip Melanchthon in Wittenberg; Martin Bucer, Wolfgang Capito, and Johann Sturm in Strassburg; Ulrich Zwingli and Heinrich Bullinger in Zurich; John Oecolampadius in Basel; and Guillaume Farel in Neuchâtel and John Calvin and Theodore Beza in Geneva. This is just a short list of names and places, but it will serve to remind us of the diversity of personalities and contexts.

All of these figures had a broad experience of Renaissance culture, and each one must have reflected on the issues raised by the science of the time. On this subject, however, only the thoughts of Luther (1483-1546), Melanchthon (1497-1560), Zwingli (1484-1531), and Calvin (1509-64) are now sufficiently well known for us to make any meaningful observations, and even for these four there is much that we do not yet fully understand. The reason is that most of the Reformers' efforts were directed towards

---

[105]Jacobi, ed., pp. 55f., 66ff., 111f. As Pagel points out, Paracelsus was influenced by Ficino's ideal of the priest-physician here; *Paracelsus*, p. 222f. On the tension between the two callings, healing and preaching, see Owen Hannaway, *The Chemists and the Word* (Baltimore: Johns Hopkins Univ. Press, 1975), pp. 6ff., 44ff.

contemporary ecclesiastical and social issues. Moreover, most of the research of modern historians of the Reformation has been focused in the areas of theology and ethics as distinct from issues of science and technology, thus imposing a present-day compartmentalisation on the writings of people who may have responded more holistically to the culture of their time, even if they did not often address issues of science and technology directly. There is room for much more original research to be done.

Three general questions will help us focus attention on the interaction of science and theology in the Reformation period: the question of the continuity of Reformation theology with the historic creationist tradition; the question of the possible influence of the special issues of medieval and Renaissance science on Reformation theology; and the question of the possible influence of the distinctive emphases of Reformation theology, in turn, on the development of the physical sciences down to the time of Kepler and Bacon (the early seventeenth century).

In general, the Reformers reiterated the four basic themes of the creationist tradition as we defined them in chapter 1. Such continuity with the historic tradition is just what one would expect in view of the Reformation emphasis on the teachings of church fathers like Basil and Augustine, who were regarded as the foremost exegetes of Scripture.

A complementary source of creationist ideas was the renewed interest in the classics which the Reformers shared with their humanist contemporaries. As in the intertestamental, patristic, and medieval periods, there was a complex interplay of biblical and Greco-Roman ideas that often defies historical analysis. Also, as in the earlier periods, there were varying emphases and even differing views on some of the particular problems that the idea of divine creation raised and left unresolved. We shall point out some of these nuances as we review the four themes of the historic creationist tradition.

1 Comprehensibility of the World

In the creationist tradition it was believed that the world could be understood by humans, at least to the extent that the limits of human exploration allowed. Underlying this belief were the doctrines that God had created the world in accordance with his wisdom or rationality and that he had created humanity in his image. As humans participated in God's rationality, they could in principle understand the plan of creation (chap. 1, sec. B).

The basic ideas of the creation and comprehensibility of the world were reiterated by the Reformers in continuity with the teaching of the church of all ages. Luther, for example, regarded reason as 'something divine' which permitted humans to understand matters as recondite as the motions of the stars.106 Melanchthon taught that humans, created in the image of God, have the ability to observe, calculate, and control how one thing follows from another in nature.107 And Calvin pointed to the arts and sciences as evidence of traces of the image of God even in fallen humanity.108

In opposition to supporters of papal authority, the Reformers emphasised the divine ordination of Christian laity in secular matters like civil government and the mechanical arts—this was one of the motivations that lay behind their stress on the secular implications of the doctrine of creation. Various scholars have pointed to the influence of William of Ockham and the late medieval nominalist tradition, which was often associated with a healthy respect for the independence of the secular powers in an overall covenantal framework. In this respect, the Reformers were in line with the left wing of the medieval creationist tradition even though they were 'conservative' in many other ways.

The net result was that an important reversal of orientation took place in communities influenced by Reformation thought: a high value on the work of artisans and scientists now became associated with the new orthodoxy, whereas in the Middle Ages it had been championed primarily by theological radicals. The Reformation stress on the divine calling of the laity was not a new one, but it received a new, more powerful legitimation.

The claim that the Reformation brought about a re-evaluation of secular activities like scientific research naturally raises the question of whether the sciences progressed more readily in Protestant than in Roman Catholic cultures as a result. Scholars have debated this question rather heatedly with no clear resolution of the issue. Certainly, the re-evaluation of the role of the laity in mainstream Protestantism was not necessarily a greater stimulus to innovation than the preservation of that same valuation in the more radical wing of Roman Catholicism. The researcher must look at specific cases to see how theologically formulated values actually functioned.

When one considers the consistency with which the idea of the comprehensibility of the world recurred among Protestant scientists, one is inclined to believe that the Reformation emphasis on the idea of creation was indeed a factor in the history of science. The idea was stated variously by a series of Lutheran astronomers and alchemists leading down to Johannes Kepler (1571-1630) and Johann Andreae (1586-1654), all of whom were influenced by the ideas of Luther and Melanchthon. It was also

---

[106] *Disputation Concerning Man*, Thesis 4 (1536; LW 34:137); *Lectures on Genesis* 1:26f. (1535; LW 1:66ff.).

[107] *Loci Communes* 2, 3 (1555; Clyde L. Manschrek, trans., *Melanchthon on Christian Doctrine*, New York: Oxford Univ. Press, 1965, pp. 13, 39f.); cf. Charlotte Methuen, 'The Role of the Heavens in the Thought of Philip Melanchthon', *JHI*, vol. lvii (July 1996), pp. 400ff.

[108] *Institutes* I.xv.4; II.ii.12-17.

articulated in various ways by a number of scientists in the English Reformation tradition leading down to Francis Bacon (1561-1626), many of whom were influenced by the ideas of Calvin and other Swiss reformers.[109] But the idea occurred as well in the writings of reforming Catholic naturalists like Giordano Bruno (*c.* 1550-1600), Tomasso Campanella (1568-1639), and Galileo Galilei (1564-1642). It should also be noted that most of these scientists, both Catholic and Protestant, were influenced to a degree by the ideas of Plato and the Neoplatonic tradition, though such philosophic interests were not necessarily in conflict with the creationist tradition, as the two had interacted positively through most of their history.

On the basis of the historical evidence, then, we must affirm that the Reformation emphasis on the creation and comprehensibility of the world was a factor, but not a determinative factor, in the history of science. It was not determinative in that it did not by itself make Protestantism a more suitable environment than Roman Catholicism for innovative science. From an historian's perspective, it is enough to demonstrate that Reformation theologians were cognisant of scientific issues and that late sixteenth-century Protestant scientists approached their work with a revitalised theological perspective that provided inspiration and meaning to their efforts.

The influence of the Reformers' doctrine of creation was not determinative for the history of science in another sense: it did not produce a consensus on some of the major issues in the philosophy of science of the later sixteenth and early seventeenth centuries. Foremost among these was the question of the value of mathematics and of mathematical models, in particular, as a means of understanding the world God had created. Astronomers in the Copernican tradition like Rheticus (1514-74) and Kepler

---

[109] A striking exception is Lambert Daneau, a French Reformed divine who taught theology and physics at the Geneva Academy (along with Beza). Daneau argued that general physics (dealing with the heavens and the terrestrial elements) must be derived from Scripture, not from the Greek philosophers. (However, he managed to interpret the Bible as teaching that the earth was round.) In fact, most of the secrets of nature have been placed by God beyond human reach, e.g., high in the heavens, or in uninhabited parts of the earth, or in the depths of the sea; *Physica Christiana* (1576); ET by Thomas Twyne, *The Wonderfull Workmanship of the World* (1578); Francis R. Johnson, *Astronomical Thought in Renaissance England* (Baltimore: Johns Hopkins Press, 1937), p. 186, where Daneau is incorrectly said to have based all science on Scripture alone; John S. Mebane, *Renaissance Magic and the Return of the Golden Age* (Lincoln: Univ. of Nebraska Press, 1989), pp. 103f.; Donald Sinnema, 'Aristotle and Early Reformed Orthodoxy: Moments of Accommodation and Antithesis', in Wendy E. Helleman, ed., *Christianity and the Classics* (Lanham, Md.: Univ. Press of America, 1990), pp. 137-40.

In Daneau's defence, it should be pointed out that the telescope had not yet been thought of and that many parts of the earth remained inaccessible in spite of the advances in circumnavigation. Daneau assumed that scholars who speculated on such matters were invoking demonic assistance to get around the limits imposed on humanity by God; *A Dialogue of Witches* (ET, 1575); Mebane, *Renaissance Magic*, p. 104.

tended to value mathematics more highly than did some physicians like Peter Severinus (*c.* 1540-1602) in the tradition of Paracelsus. Those like John Dee (1527-1608) and Thomas Digges (1546-95), who were influenced by Platonic thought, naturally valued it more highly than those like Francis Bacon, who reacted against Platonic and Neoplatonic influences.

In other words, there were both speculative and empirical emphases in post-Reformation thought just as there had been in the Middle Ages. On the other hand, the differences in emphasis ought not to be overstressed: even 'empiricists' like Severinus and Bacon valued mathematics as a tool of experimental quantification. Their reaction to speculative mathematics was occasioned by its association with seemingly reactionary forces like Aristotelian logic and Neoplatonic hermeticism.

Not only was there significant disagreement as to the *means* of comprehending the world; there were also differing opinions about the implications of the doctrine of creation for the nature of the cosmos and its interconnections. Foremost among these was the question of the possible infinity of the cosmos, an idea that had already been discussed by late medieval philosophers like John of Ripa and Nicholas of Cusa (chap. 2, sec. C).

The Reformers themselves held that the cosmos was finite and bounded by an outermost sphere (the *primum mobile*) in the tradition of medieval Aristotelian cosmology. Melanchthon rejected the very possibility of an infinite universe.[110] Calvin at one point ridiculed the idea of an extra-cosmic void.[111] So comprehensibility entailed finitude for the most influential teachers of the Reformation.

In the post-Reformation period, however, Thomas Digges, a radical Protestant and the first true Copernican in England, took the radical step (1576) of treating the realm of the stars as extending endlessly far outwards from the planetary realm. Such an infinite extension made sense theologically, Digges argued, in as much as the realm of the stars was the court of God 'to whose infinite power and majesty such an infinite place...only is convenient'.[112] The reason the attribute of infinity did not make the world incomprehensible to Digges was apparently that there was still a fixed point of reference in the sun, which was located at the centre.

An even more radical Copernican position was taken by the Italian philosopher, Giordano Bruno, about a decade later (1584). Originally a Dominican priest, Bruno experimented with Reformation theology at

---

[110]Preface to 1537 edition of Euclid's *Elements of Geometry* in Marian A. Moore, trans., 'A Letter of Philip Melanchthon to the Reader', *Isis*, vol. 1 (June 1959), p. 146.

[111]*Institutes* I.xiv.1 (1559 edition).

[112]*Perfit Description of the Caelestiall Orbes* (1576; Alexandre Koryé, *From the Closed World to the Infinite Universe*, Baltimore: Johns Hopkins Press, 1957, p. 38).

Geneva and Tübingen and was influenced by Protestant thinkers like Thomas Digges in Elizabethan England.[113] He developed the idea that the universe reflects the attributes of God to the point that it bordered on pantheism: not only was the structure of the universe a reflection of God's wisdom, but the universe itself was an expression of his immensity. Since the world was not God himself, but only a multiple image of God's simple immensity, Bruno concluded, there must be an infinite number of worlds, as the Epicureans had taught, and there was no fixed centre to things, not even the sun. Evidently, the prospect of an infinite universe suggested new possibilities of human comprehension to Bruno: it was not necessarily the threat to intelligibility that it had seemed to the ancients and even to the Reformers.[114]

Bruno is a peripheral figure for the historian of the Reformation as well as for the historian of science. The importance of his case stems from two factors: the seeming modernity of Bruno's views—he was the first Western European to picture the universe as an immense plurality of worlds the way we do today—and the fact that he was tried and executed for unorthodox beliefs by the Roman church (Rome, 1600). Consequently, some historians have portrayed him as a martyr for modern ideas at the hands of a reactionary church.

It is true that one of the charges made against Bruno was that he taught an infinity of different worlds, but this teaching was not disallowed by the Roman church as a matter of (hypothetical) science so much as on the basis of established doctrine. From a scientific point of view, the idea of a plurality of worlds was pure speculation at the time, while theologically it seemed to call into question the uniqueness of the Incarnation. Along the same line, Bruno was condemned for holding that Moses and Jesus performed miracles by magic arts and that the Holy Spirit was the world soul as portrayed in Neoplatonic Hermeticism.[115] This is hardly the kind of philosophy most modern critics of the church would want their martyrs to espouse!

---

[113]Paul Oskar Kristeller, *Eight Philosophers of the Italian Renaissance* (Stanford: Stanford Univ. Press, 1964), p. 128; the influence of Digges's notion of the heavens as the court of God is clearly seen in *On the Infinite Universe and Worlds* (1584); Dorothea Waley Singer, *Giordano Bruno: His Life and Thought* (New York: Henry Schuman, 1950), p. 245.

[114]Koryé, *Closed World*, pp. 39-53.

[115]Yates, *Giordano Bruno*, pp. 354f. Roger Bacon had already attributed miracles of the saints (Moses included) to magic: *Opus Majus*; cf. Theodore Otto Wedel, *The Medieval Attitude Toward Astrology* (New Haven: Yale Univ. Press, 1920), p. 72. Edward Gosselin argues that Bruno was also under suspicion of acting as an agent of Henry III in seeking compromise with English Protestants on the interpretation of the sacrament. In fact, Bruno appparently tried to shift the issue to his avowed Copernicanism because he felt that it was a safer topic! 'Bruno's "French Connection"', in Merkel and Debus, eds., *Hermeticism*, 175ff.

From our perspective Bruno and his accusers represent differing emphases within the creationist tradition—Bruno drew his inspiration from the idea of the immensity of God as Origen had done in the third century and Nicholas of Cusa had done in the fifteenth; his accusers were understandably concerned about the implications of Hermeticism and about the danger that the idea of creation would be emptied of any meaning if the universe had no recognisable limits or structure. The idea that a world created by God was intelligible to humans created in his image raised important questions about the possible infinity (and eternity) of that world, but it did not settle them as far as the historian can see. Nor can the intolerance of Bruno's accusers be invoked to discriminate between Protestant and Catholic versions of the creationist tradition. Michael Servetus (*c.* 1511-1553), a Catholic influenced by radical Protestant ideas who speculated on the role of the world soul in human physiology, was condemned and executed (1553) for his unorthodox theological views at the recommendation of John Calvin in Protestant Geneva.

We have cited the examples of Digges and Bruno to illustrate the flexibility of the idea of the comprehensibility of the world even within the sphere of Protestant influence. The issue of the possible infinity of a created, comprehensible world was still unresolved in the early seventeenth century, as is shown by the response of Kepler to the speculations of Bruno and the new discoveries of Galileo.

Like Bruno, Kepler understood the cosmos, or, at least, its geometrical structure, as an image of God.[116] But, in the tradition of Martin Luther and in keeping with Protestant orthodoxy, which was battling various groups of antitrinitarians at the time, he drew his inspiration from the idea of the Trinity rather than the divine attributes of immensity and simplicity. Kepler championed the ideas of Copernicus, which he imbibed from a line of Lutheran astronomers through his teacher, Michael Maestlin (1550-*c.* 1633).

---

[116] '...Geometry, coeternal with God and shining in the divine Mind, gave God the pattern...by which he laid out the world so that it might be best and most beautiful and finally most like the Creator'; *Harmonices mundi* III.1 (1619); W. Von Dyck, Max Caspar, F. Hammer, and M. List, eds., *Gesammelte Werke* (Munich: C.H. Beck, 1937-) 6:104f.; J.V. Field, *Kepler's Geometrical Cosmology* (Chicago: Univ. of Chicago Press, 1988), p. 123.

'Since geometry is co-eternal with the divine mind before the birth of things, God himself served as his own model in creating the world (for what is there in God which is not God?), and he with his own image reached down to humanity'; *Harmonices mundi* IV.1; *Gesammelte Werke* 6:223; Gerard Simon, 'Kepler's Astrology'in A. and P. Beer, eds., *Kepler* (Oxford: Pergamon, 1975), p. 447; Bernhard Sticker, *Erfahrung und Erkenntnis* (Hildesheim: Verlag Gerstenberg, 1976), p. 122n.

Cf. Kepler's *Dissertatio cum Nuncio Sidereo* (an open letter addressed to Galileo, 1610): 'Geometry is one and eternal, shining in the mind of God. That share in it accorded to humans is one of the reasons [*causae*] that humanity is the image of God'; *Gesammelte Werke* 4:308; J. V. Field, 'Astrology in Kepler's Cosmology' in P. Curry, ed., *Astrology*, p. 154; *idem*, *Kepler's Geometrical Cosmology*, p. 78.

In fact, he was, after Rheticus, the first true Copernican among Lutheran astronomers.

Unlike Bruno, Kepler regarded the sun as the geometric and dynamical centre of the cosmos: it was thus a suitable image of God the Father, who was the origin of the godhead in trinitarian theology. The outer sphere of the stars, which like the sun was stationary in the Copernican system, was the image of God the Son, and the space enclosed by the stellar sphere (the realm of the earth and planets) was the image of the Holy Spirit. Hence the basic geometry of the universe—a sphere with fixed center, periphery, and space between—could be comprehended as a finite, created image of the infinite, uncreated Trinity.[117]

So Kepler developed the idea of the comprehensibility of the world in a way very different from that of Bruno, who focused on the immensity and simplicity of God. Even Digges's idea of a sphere of stars extending outward to infinity would not have fit Kepler's notion of the Trinity. The coequality of Father, Son, and Spirit required the cofinitude of the centre, periphery, and space of the cosmos. The stellar realm could not be infinite while the sun and planetary space were not.

Kepler's finitist cosmology was challenged by two astronomical discoveries and the speculation they engendered in the early seventeenth century. The appearance of novae (previously unknown stars) in 1572 and 1604 led some to suggest that space had immeasurable depths from which stars could be transported towards the earth periodically on immense cosmic wheels. Then, in 1610 came the first reports of Galileo's new discoveries made with a primitive telescope: a large variety of stars that could not be seen with the naked eye were revealed for the first time. Again it was speculated that the universe had infinite depths and perhaps even an infinite number of planetary systems. Galileo himself entertained such infinitist notions in private correspondence, influenced no doubt by the speculations of Cusa and Bruno.[118]

Kepler took these suggestions seriously but consistently rejected them. His reasons were partly scientific. An infinite universe with no boundary, Kepler argued, could not have a unique centre. But the sun appeared to be located at the centre of an enormous cavity in which there were only planets and beyond which there were only stars. So there was insufficient evidence to convince the astronomer that the universe was indeed infinite; appearances

---

[117]E.g., Kepler's letter to Maestlin, 3 Oct. 1595, *Gesammelte Werke* 13:23; Crombie, *Augustine to Galileo*, vol. II, p 195; Arthur Koestler, *The Sleepwalkers* (New York: Macmillan, 1959), pp. 261f.; Koryé, *Astronomical Revolution*, pp. 138, 143f., 154, 284ff.; Sticker, *Erfahrung, p. 124;* Job Kozhamthadam, *The Discovery of Kepler's Laws* (Notre Dame: Univ. of Notre Dame Press, 1994), pp. 29, 140.

[118]Koryé, *Closed World*, pp. 95-99.

suggested just the contrary. Science aside, the obvious discomfort the suggestion of an infinite universe occasioned Kepler and the ingenuity with which he constructed arguments to refute it betray an underlying assumption about the comprehensibility of the world rooted in the creationist tradition.[119]

In spite of the diversity of interpretation of the comprehensibility of the world, there was continuity and consistency on the basic point. Kepler illustrates the underlying idea magnificently. As he wrote in 1599, near the beginning of his career, he believed that the laws of the universe could be discovered by human science:[120]

> Those laws are within the grasp of the human mind; God wanted us to recognize them by creating us after his own image so that we could share in his own thoughts....Only fools fear that we make humanity godlike in doing so; for God's counsels are impenetrable [cf. Rom. 11:33],[121] but not his material creation.

This creationist faith motivated Kepler to seek a simple way of describing the planetary orbits in mathematical terms. It sustained him through seemingly endless trials and errors[122] and enabled him at length to discover an harmonious system of the planets which included, among other things, the three laws (1604-1619) that bear his name to this day:

(1) the orbit of each planet is an ellipse with one focus located at the sun;

(2) the angular velocity of any planet at any given time is inversely proportional to the square of its distance from the sun (equivalent to the law of equal areas in equal times);

(3) the period of revolution of each planet is also related to the size of its orbit by a simple power law.[123]

In accordance with his creationist faith, Kepler joyously gave all the credit for his discoveries to God:[124]

---

[119]Koryé, *Closed World*, chap. 3, esp. pp. 58-76.

[120]Letter to Herwart von Hohenburg, 9 April 1599 (*Gesammelte Werke* 13:117; the Latin is also given in Koryé, *Astronomical Revolution*, p. 379 note 15; ET in Crombie, *Augustine to Galileo*, vol. II, p. 195; cf. Max Caspar, *Kepler* (London: Adelard-Schuman, 1959), p. 93).

[121]'God's counsels' here evidently refer to the decrees of election (and reprobation); cf. Augustine, *On the Gift of Perseverence* 8(18).

[122]*Epitome of Copernican Astronomy*, Introduction to Book IV (Jaki, *Science and Creation*, p. 268); cf. *Harmonices Mundi*, Introduction to Book V (Koestler, *Sleepwalkers*, pp. 393f.).

[123]The role of Kepler's theology in his discovery of these laws is detailed by Job Kozhamthadam, *Discovery of Kepler's Laws*, pp. 179-80, 185-86, 191-93, 198, 213-15, 240-45.

[124]*Harmonices mundi* V; ET modified from Rudolf Haase, 'Kepler's Harmonies,' in A. and P. Beer, eds., *Kepler*, p. 526.

> I give you thanks, Creator and God, that you have given me this joy in thy creation, and I rejoice in the works of your hands. See I have now completed the work to which I was called. In it I have used all the talents you have lent to my spirit. I have revealed the majesty of your works to those who will read my words, insofar as my narrow understanding can comprehend their infinite richness.

Kepler's 'three laws' became the basis of modern astronomy and were the starting point from which Isaac Newton derived the inverse-square law of universal gravitation later in the seventeenth century. If there was ever a point in Western history at which the progress of science depended on belief in the doctrine of creation, it was in the work of the Lutheran, Johannes Kepler. The reinforcement of basic Christian doctrine brought about by the Protestant Reformation, particularly in the circle of Philip Melanchthon, must be credited with this much at least.

2 Unity of Heaven and Earth

The second theme in the creationist tradition as we have presented it was the belief that God created all things with the same basic material (itself created out of nothing) and imposed on them the same basic laws of motion and change. From the beginning of the patristic period (second century AD), this belief was directed primarily against the Aristotelian idea that the heavens were fundamentally different from the earth. Specifically, Aristotle taught that the heavens were more directly subject to the providence of God than the earth was, that they moved circularly rather than up or down like the sublunar elements, that they were immutable in substance unlike the four elements of the sublunar realm, and, consequently, that they must be made of a fifth element (the ether) different from the four elements found on earth (chap. 1, sec. C).

These ideas were challenged in various ways by late medieval natural philosophers like Bradwardine, Buridan, and Cusa who developed their own ideas from their belief in the absolute (as well as the ordained) power of God over all things (chap. 2, sec. C). But the teachings of Aristotle still provided the only comprehensive framework for understanding the varied phenomena of nature, and the alternatives posed by these philosophers were generally regarded as mere speculation or hypothesis. The situation was to change dramatically with the birth of a new physics in the seventeenth century. We wish to know what contribution, if any, Reformation theology may have made in this direction.

The fact of the matter is that the leading Reformers accepted Aristotelian cosmology as part of the established science of the time. Luther himself seems not to have paid much attention to it. He ignored fundamental ideas like that of solid celestial spheres made out of a fifth element unique to the heavens.125 Luther's view of nature was based on the biblical notion of the all-pervading energies of God and was not troubled by the current problems of cosmology.126 But Melanchthon and Calvin, both of whom were trained in the humanist tradition, particularly valued Aristotle and other classics as the basis for college and university curricula they were responsible for initiating. Their immediate successors, Caspar Peucer (1525-1602) and Theodore Beza (1519-1605), followed in their footsteps, and converts from Roman Catholicism like Peter Martyr Vermigli (1499-1562) and Girolamo Zanchi (1531-90), who were trained in Thomistic thought, reinforced the establishment of Aristotelian ideas. As a result, the cosmology of Aristotle became deeply ingrained in the more educated and influential strata of Protestant culture.127

In view of this Protestant conservatism in matters of science education, we would not expect much by way of a stress on the unity of heaven and earth or a direct challenge to the Aristotelian dichotomy. It is certainly true that Melanchthon and Calvin both accepted Aristotelian science as a valid description of the normal course of nature. On the other hand, they both argued, as did their medieval forebears, that this course was neither necessary nor inviolable from God's point of view (*de potentia absoluta*). It was a manifestation of God's ordained laws for nature (*de potentia ordinata*).128

Calvin, moreover, softened the edges of the heaven-earth duality in several ways. For one thing, he did not follow Aristotle in teaching that the heavenly bodies were made of a unique fifth element but adopted instead the Stoic notion that they were made of fire.129 For another, he stressed the biblical idea that the earth was as immediately controlled by the providence of God as was the outermost heaven (the *primum mobile*).

In fact, Calvin's doctrine of universal and particular providence had the effect of bringing out the parity between heaven and earth in this respect. God's government of terrestrial affairs by the rotation of the celestial spheres and the influence of their luminaries on the earth was viewed by Calvin as an instance of universal providence.130 On the other hand, his stabilisation

---

[125]*Lectures on Genesis* 1:6, 14 (LW 1:23-32, 42f.). The pedagogical remarks in this passage (at 1:27f., 28f., 31f.) were probably interpolated by Melanchthon or one of his pupils.

[126]Heinrich Bornkamm, *Luther's World of Thought* (Saint Louis, Mo.: Concordia, 1958), pp. 176-94.

[127]John Dillenberger, *Protestant Thought and Natural Science* (Garden City, N.Y.: Doubleday, 1960; London: Collins, 1961), pp. 51-64.

[128]Melanchthon, *Loci Communes* 3 (1555; trans. Manschrek, p. 42f.); Calvin, *Institutes* I.xvi.1ff.

[129]*Commentary on Genesis* 1:15.

[130]*Institutes* I.xvi.1; *Commentary on Jeremiah* 10:12f.; *Warning Against Judiciary Astrology*, trans. Mary Potter in *Calvin Theological Journal*, vol. xviii (Nov. 1983), pp. 166-70.

of the earth against rotational forces[131] and his restraint of the seas and rains lest they inundate the earth were seen (within an Aristotelian framework) as instances of immediate or particular providence.[132]

Thus Calvin directly challenged the idea that God's immediate control was limited to the outermost heavens, and he did this not by resorting to the idea of God's absolute power (*potentia absoluta*) or by appealing to the possibility of miracles. Rather, he found gaps in the fabric of Aristotelian naturalism that seemed to indicate the supernatural activity of God in the ordinary course of nature (*potentia ordinata*). This was as radical a challenge as could be made without rejecting the basic Aristotelian framework, in which the earth was at the centre of the cosmos in a realm of falling bodies and changing substances while the planets and stars moved on celestial spheres with circular motion and unchangeable substance. As a reformer and one who wished to distance himself from every suggestion of being a revolutionary, Calvin was unwilling to take such a bold step.

But even these partial challenges by Calvin had no effect, as far as we know, on the history of astronomy in the latter part of the sixteenth century. Instead, the gradual undermining of belief in a heaven-earth duality was occasioned by purely scientific developments like the discovery of stars (novas) that changed in magnitude, the determination that comets were celestial rather than atmospheric phenomena, and the growing realisation that Copernicus had been right in placing the sun at (or near) the centre of the planetary system.

Many of the astronomers who contributed to this revolution were Protestants: e.g., Thomas Digges and William Gilbert (1544-1603) in Elizabethan England and Tycho Brahe (1546-1601), Christoph Rothmann (1550-*c*. 1650), and Johannes Kepler in the continental Lutheran tradition. Apparently only in the case of Kepler was theology a significant factor:[133] Kepler's repeated attempts to find a universal law that would account for the motion of both earth and the planets was inspired in part by his analogy between the created cosmos and the uncreated Trinity, as we have seen. One result of this model was that the source of cosmic motion was shifted from the periphery of the universe (the *primum mobile*) to the centre (the sun). But the precise meaning and origin of Kepler's trinitarian speculations are

---

[131] *Twelfth Sermon on Psalm 119* (*CO* 32:620); *Commentary on Psalms* 93:1; 119:90.

[132] *Commentary on Genesis* 16-9; 7:11f.; *Commentary on Psalm* 104:5-9; *Second Sermon on Job 26* (*CO* 34:434f.).

[133] William Gilbert rejected Aristotle's heaven-earth duality on the basis of the hermetic idea that all the cosmos was alive—the earth as well as the heavens; *De magnete*, V.12; Kocher, *Science and Religion*, 181; Kearney, *Scientific Change*, 110. As far as I know, there is no evidence that Gilbert was influenced by Calvin in this respect. As we shall see below, Calvin took a dim view of hermetic ideas generally. However, Francis Bacon synthesised hermetic-Neoplatonist and the Calvinist-Puritan motifs at about the same time as Gilbert.

not clear: they may have had as much to do with Neoplatonic thought as with Protestant theology.[134]

At most, we could argue that one of the reasons for Kepler's attraction to Neoplatonic and hermetic ideas was a dissatisfaction with the apparent distance of God's normal activity implied by Aristotle's system. If so, he had much in common with fellow Lutherans who were interested in the analogy between macrocosm and microcosm found in mystical alchemy: for example, Valentin Weigel (1533-88), Heinrich Khunrath (1560-1605), and Johann Andreae (with whom Kepler exchanged correspondence).[135] A common source of inspiration in the writings of Luther is a possibility worth considering.

However, radical Catholic natural philosophers like Bernardino Telesio (1509-88), Francesco Patrizi (c. 1530-97), and Giordano Bruno also challenged the Aristotelian dichotomy of heaven and earth on purely philosophical grounds, inspired by late medieval critics like Cusa and by the hermetic thought of Ficino.[136] So belief in the unity of the cosmos as the object of God's creation and immediate providence can be accounted for simply in terms of the creationist tradition (including sympathies with Neoplatonism in opposition to the hegemony of Aristotelian thought) common to a variety of thinkers in the sixteenth century, both Protestant and Catholic. It is difficult to trace any direct influence of the magisterial Reformers here, even in the case of Kepler's undoubtedly religious sense of the unity of the cosmos.

If the sixteenth- and seventeenth-century developments that led to the scientific idea of the unity of heaven and earth were relatively free of theological determinants, that does not mean that such developments were without theological meaning. Though medieval Aristotelianism made the providence of God seem remote, at least it affirmed immediate providence for the outermost heavens. If the Reformers had had their way, such immediate providence would have been affirmed for the earth as well as for the heavens. Rheticus and Kepler, however, both justified the unification brought about

---

[134]On Kepler's Platonism, as distinct from Pythagoreanism or Hermeticism, see Field, *Kepler's Geometrical Cosmology*, pp. 99, 188ff. A remnant of Aristotelian dualism appeared in his notes added to the second edition of *Mysterium cosmographicum* (1621): 'the heavens, the first of God's works, were laid out much more beautifully than the remaining small and common things [on earth]'; J.V. Field, 'A Lutheran Astrologer: Johannes Kepler', *AFHES*, vol. xxxi (20 Dec 1984), p. 219 = 'Astrology', p. 166. I take this to be merely a poetic way of explaining Kepler's failure to complete the application of his astrological science to terrestrial events.

[135]Frances Yates, *The Rosicrucian Enlightenment* (London: RKP, 1972), chaps. 3, 4, 11; John Warwick Montgomery, *Cross and Crucible* (The Hague: Martinus Nijhoff, 1973), vol. I, pp. 12-20.

[136]Randall, *Career*, vol. I, pp. 203f., 208ff., 331ff.; Kristeller, *Eight Philosophers*, pp. 98-103, 112-25, 131-35; Yates, *Bruno*, pp. 181f.

by the Copernican system in terms of the late medieval image of God as a clockmaker.[137] As we argued in the case of Nicholas of Cusa (chap. 2, sec. B), an equalisation of heaven and earth in mechanistic terms could have the effect of making God seem equally remote from all things, rather than equally near. This, in fact, was the theological tendency that prevailed in the late seventeenth and eighteenth centuries.

3 Relative Autonomy of Nature

The third feature of the creationist tradition was its profound sense of the relative autonomy of nature. The operations and effects of inanimate bodies were viewed as the result of God's creative word or decree which established a specific law of behaviour for each object created. They were, therefore, regular and yet contingent on the governing word or decree of God. Their autonomy was relative, not absolute.

In the Middle Ages, this subtle interplay of biblical motifs was formalised in the scholastic distinction of *potentia Dei ordinata* and *potentia Dei absoluta*. *Potentia ordinata* was the power of God as expressed in the ordinary course of nature and history. *Potentia absoluta* was the freedom God exercised in establishing that course and which he retained to alter it on occasion (e.g., in miracles or at the eschaton). Under the successive influences of Neoplatonism and Aristotelian natural philosophy, particularly in the left wing of the creationist tradition, the normal course of nature came to be viewed as a virtually autonomous entity referred to as 'Nature'. The regular operations of Nature were clearly distinguished from the occasional interventions of God. In response to this apparent limitation of the sovereignty of God in his creation, more conservative theologians tended to stress the absolute power of God (*potentia absoluta*) and to limit the prerogatives of reason in natural science (chap.1, sec. D, and chap. 2, sec. C).

In general, leading Reformation theologians tended to shift the emphasis back to God's *potentia ordinata* while limiting the efficacy of second causes and the scope of human reason within that sphere. The shift of emphasis had to do primarily with two theological doctrines: those of revelation and salvation. Both doctrines were clearly aimed at the teachings of the Roman Catholic Church, particularly as defined by the medieval scholastics. We must review them briefly before passing on to their counterparts in natural philosophy in order to keep a reminder of the historic Reformation emphasis.

---

[137]Rheticus, *Narratio Prima* (trans. Rosen, pp. 137f.); Kepler, Letter to Herwart von Hohenburg, 10 Feb 1605 (Crombie, *Augustine to Galileo*, vol. II, pp. 202f.; Koestler, *Watershed*, p. 340; the Latin is given by Koryé, *Astronomical Revolution*, p. 378 note 8).

The Roman church viewed new revelations as a recurring phenomenon of history. God continued to reveal himself in miracles, visions, and occasional refinements of doctrine (based on an original, apostolic deposit) through the auspices of the church. In contrast, Luther and other Reformers emphasised the canonical Scriptures as the locus of all divine revelation (*sola scriptura*). To seek to know God and God's will by any other means—whether by speculative reason, private revelations, or the formulation of apostolic tradition—even under the auspices of the church—was to spurn God's sufficient self-disclosure in the Bible. The Catholic idea of revelation as a recurring act of God through the church was replaced by the Protestant one of a fixed body of teaching to which the church itself was subject. The idea of revelation thus shifted from the sphere of *potentia absoluta* to that of *potentia ordinata*.[138]

The Roman Catholic idea of salvation, on the other hand, was already located within the sphere of *potentia ordinata*. Salvation was effected by God's grace channelled through and conditioned by various second causes: some under the control of the individual, like natural reason and good works; and some belonging to the priestly function of the church, like baptism, absolution, and the eucharist. The Reformers had to recognise the validity of such exercises since they were prescribed in Scripture; in fact, they went to great lengths to affirm the value of good works and the sacraments against the teachings of radical Protestants like the Antinomians, Libertines, and Anabaptists. But the Reformers denied any causal efficacy to human efforts, whether individual or ecclesiastical. Salvation was a gift of God by grace alone (*sola gratia*), and it was available through faith in Christ alone (*sola fide, sola Christo*).

The result as in the issue of revelation was an affirmation of God's revealed will and a rejection of biblically unauthorised ways of seeking salvation, such as the invocation of saints or angels. But, whereas the Roman church based its claims on the notion that the administration of salvific means had been committed to the priesthood, the Reformers defended the idea that God could act in ordained ways that were not reducible to the operation of second causes.

While the principal concerns of the Reformers had to do with theological doctrines like revelation and salvation, they also had implications for scientific issues of the time. In general, the results were rather paradoxical. The operations of nature and the investigations of science were affirmed as

---

[138] According to Heiko A. Oberman: 'Together with the humanist quest for authentic sources (*fontes*), the insistence on nothing but God's commitment [in revelation], the *sola potentia ordinata*, may evolve into a *sola scriptura*, the Reformation principle, "Scripture alone"'. 'Reformation and Revolution', in Owen Gingerich, ed., *The Nature of Scientific Discovery* (Washington, D.C.: Smithsonian Institution Press, 1975), p. 149.

part of the ordinances of creation and reaffirmed as part of God's will for a renewed humanity. On the other hand, they were challenged whenever they appeared to conflict with the express teachings of Scripture, particularly those concerning God's care for and salvation of humans.

In other words, the medieval dialectic of *potentia ordinata* and *potentia absoluta* was largely replaced by a Reformation dialectic of creation- and salvation-ordinances within the sphere of *potentia ordinata*.[139] Still, it amounted to a differentiation of nature and supernature, or, to be more exact, a clear distinction between God's indirect operation through second causes and his direct operation with or without the cooperation of second causes. The result was a series of conflicts in which scientists and theologians struggled to establish their respective authorities. That struggle was temporarily resolved in the work of Francis Bacon, work that was to guide the progress of science in seventeenth-century England. We shall review these conflicts briefly to assess the effects of Reformation theology.

There were three fairly distinct streams of scientific research in the sixteenth century: scholastic Aristotelian science, a burgeoning Copernican astronomy, and hermetic-cabalistic investigations in mathematics and alchemy (chap. 3, sec. A). Of the three, Aristotelian science was best known and was received most favourably by the Reformers. We shall consider their treatment of Aristotelian naturalism here and postpone discussion of their reactions to Copernicanism and hermetic-cabalistic thought until our consideration of the fourth theme of the creationist tradition.

Aristotle's stress on the order of the universe suited the Reformers' interest in God's *potentia ordinata* as it had suited similar interests in the twelfth and thirteenth centuries (chap. 2, sec. A). On the other hand, Luther, Zwingli, and Calvin (not Melanchthon so much)[140] agreed with conservatives of the Middle Ages in rejecting the completeness of natural causation. But they did not appeal to God's *potentia absoluta*, as medieval scholastics had done and as Pope Urban VIII was to do in his instructions to Galileo (1624, 1631).[141] Instead, they cited apparent gaps in the web of second causes, gaps which were evidence of the direct action of God even within the sphere of *potentia ordinata*.

There are numerous examples of this attempt to find gaps within the domain of second causes. Luther (or Melanchthon), for example, concluded

---

[139] E.g., Calvin, *Institutes* II.ii.12f.

[140] Preface to 1531 edition of Sacrobosco's *Sphere* (Bruce Thomas Moran, 'The Universe of Philip Melanchthon', *Comitatus*, vol. iv, 1973, p. 10).

[141] Duhem, *To Save the Phenomena*, pp. 110ff. (where the warning is dated before the accession of Urban in 1623); Crombie, *Augustine to Galileo*, vol. II, pp. 220f.; Jerome J. Langford, *Galileo, Science and the Church*, rev. ed. (Ann Arbor: Univ. of Michigan Press, 1971), p. 114; Winifred Wisan, 'Galileo and God's Creation', *Isis*, vol. lxxvii (Sept. 1986), pp. 480ff.

(1535) from the etymology of the Hebrew word for sky (*shamayim*) that the heavens were composed of a watery substance (*mayim*). This inference from Scripture was confirmed observationally by the blue colour of the sky and the humidity of the atmosphere. Since the remarkable stability of the heavens and the constancy of their motions could not be accounted for on the basis of the natural properties of water, they could then be taken as a sign of the power of God's sustaining word.[142]

Another example given by Luther and later developed by Calvin (1554) was the gathering of the terrestrial waters into seas (Gen. 1:9). According to Aristotelian principles, the Reformers argued, earth should gravitate as much as possible to the centre of the cosmos and be completely covered by water. The fact that the waters were confined to seas and there was dry ground suitable for habitation was, therefore, a perpetual miracle: it was part of the established order of things, but it could not be accounted for on purely naturalistic principles.[143]

In reaction to the naturalism of Aristotelian philosophers of their time, Luther, Zwingli, and Calvin all came to view efforts to achieve complete causal explanations as a threat to the sense of God's providence appropriate to Christian piety.[144] Only Zwingli came close to rejecting the operation of second causes as such,[145] but all three disallowed the power of second causes to effect all that happens in nature and history.

In fact, it was in order to secure the immediate, regular operation of God in nature and history that Calvin articulated his doctrine of particular providence, which we discussed earlier. Particular providence was God's continuous, direct intervention within the sphere of *potentia ordinata*, an

---

[142]*Lectures on Genesis* 1:6 (*LW* 1:23-26). The Melanchtonian character of the redaction of Luther's commentary (1544) is noted by Pelikan in his introduction to vol. 1 of *Luthers Works* (*LW* 1:x-xii). For the traditional Jewish derivation (partially) of *shamayim* from *mayim*, see Num. Rab. XII.4, 2 vols. (London: Socino Press, 1939), 1:457.

[143]*Lectures on Genesis* 1:9; 7:11f. (*LW* 1:34f.; 2:93f.); *Commentary on 2 Peter* 3:5f. (*LW* 30:194); Calvin, *Commentary on Genesis* 1:9; *Commentary on Psalm* 33:7; 104:5-9; *Commentary on Jeremiah* 5:22. Susan E. Schreiner has a fine discussion of Calvin's thinking on this theme; *The Theater of His Glory* (Durham: Labyrinth Press, 1991), pp. 24ff.

The idea that the containment of the waters was a perpetual miracle appeared in several medieval scholastics: e.g., William of Auvergne (Pierre Duhem, *Le système du monde*, 10 vols. [Paris: Hermann, 1913-59], vol. IX, pp. 109ff.) and Thomas Aquinas's *Literal Exposition of Job* (Schreiner, *Theater*, p. 24 note 117).

[144]Paul Althaus, *The Theology of Martin Luther* (Philadelphia: Fortress Press, 1966), pp. 105-15; Thomas F. Torrance, *Calvin's Doctrine of Man* (London: Lutterworth, 1949), pp. 27ff., 61ff.; William J. Bouwsma, 'Calvin and the Renaissance Crisis of Knowing', *Calvin Theological Journal*, vol. xvii (Nov. 1982), pp. 202f.; Dillenberger, *Protestant Thought*, pp. 34ff.; Keith Hutchison, 'Supernaturalism and the Mechanical Philosophy', *History of Science*, vol. xxi (1983), pp. 314-18.

[145]*On Providence* 3 (1530); S. M. Jackson, trans., *The Latin Works of Huldreich Zwingli*, vol. II (Durham, N.C.: Labyrinth Press, 1983), pp. 138-59. But Zwingli allowed a force and power in nature as distinct from immediate divine causation in *On Providence* 7 (*ibid.*, pp. 208ff.).

intervention which assured his continuous control of the strategic forces affecting human life.¹⁴⁶ Instances of particular providence included regular phenomena like the containment of the waters and the stability of the earth against rotational forces. They also included variations in normal phenomena like the seasons and human heredity. Since 'accidents' like these could not be accounted for entirely by the principles of Aristotelian natural philosophy, or even by those of astrology which were based on Aristotelian naturalism, they too could be taken as indications of God's immediate and continuous providence.¹⁴⁷

On the other hand, Calvin also taught a doctrine of universal providence, and this provided some justification for the theoretical investigations of natural philosophers, as well as for the more practical skills of physicians and astrologers. As long as scientific endeavour did not contradict the express teachings of Scripture (not just the popular idiom sometimes used by Moses) or infringe on the territory of particular providence, it could serve the glory of God and the benefit of humanity. The art of medicine, for example, was based on knowledge of the natural properties of herbs and the influences of the stars that determined the most propitious times for their use.¹⁴⁸ All such arts and sciences relied on the universal providence of God operating through second causes and were valued as part of the ordinances of creation.

The teaching of the Reformers was, therefore, both restrictive and affirmative with respect to scientific endeavour. Within the Lutheran tradition, these two emphases were embodied in the respective teachings of Luther and Melanchthon. The important school of natural philosophers who looked to Melanchthon as their theological mentor carried on the Aristotelian tradition of seeking the underlying causes of natural phenomena, and this tradition bore much scientific fruit, culminating in the work of Johannes Kepler.¹⁴⁹ Even though the geocentric cosmology of Aristotle and Ptolemy was gradually replaced by the heliocentric astronomy of Copernicus, Melanchthon's stress on the relative autonomy of nature based on the ordinances of creation was continued to good effect. Luther's

---

¹⁴⁶François Wendel, *Calvin* (London: Collins, 1963), pp. 179ff.

¹⁴⁷*Institutes* I.16.2f.; *Warning against Judicial Astrology* (trans. Potter), pp. 166, 170, 176ff., 183.

¹⁴⁸*Against the Libertines* 14, 24 (1545; Benjamin Wirt Farley, trans., *John Calvin: Treatises Against the Anabaptists and Against the Libertines* [Grand Rapids: Baker, 1982], pp. 242f., 322f.); *Warning Against Judicial Astrology* (trans. Potter), pp. 166ff.

¹⁴⁹On Melanchthon, see William Hammer, 'Melanchthon, Inspirer of the Study of Astronomy', *Popular Astronomy*, vol. lix (June 1951), pp. 308, 313-18; Francis Oakley, 'Christian Theology and the Newtonian Science' *Church History*, vol. xxx (Dec. 1961), p. 445; Moran, 'Universe', p. 10. On Rheticus, Reinhold, and Peucer, see Duhem, *To Save the Phenomena*, pp. 64, 70, 76; Koryé, *Astronomical Revolution*, pp.31f., 41. On Kepler, see Crombie, *Augustine to Galileo*, vol. II, pp. 188-203.

more critical attitude towards Aristotelian naturalism was apparently not a significant obstacle.

Within the Calvinist tradition, on the other hand, both the affirmation and the restriction of science continued side by side and often gave rise to conflicting emphases. This diffraction of the Calvinist synthesis is best seen in Elizabethan England, where differing attitudes towards the established church tended to force the issue. In the mid-sixteenth century, prior to the Elizabethan settlement, Robert Recorde (c. 1510-58) exemplified a Protestant approach to science that balanced faith with reason and valued astronomical research as a demonstration of the providence of God in creation. Like the Reformers themselves, Recorde rejected a literalistic interpretation of the Bible in matters of astronomy. Yet he saw no need to reinterpret Scripture to accommodate new ideas like those of Copernicus.[150] Such an unproblematic approach to astronomy could still be followed by Thomas Hood as late as 1588.[151]

The general situation became more polarised, however, after the accession of Mary Tudor (1553), who tried to re-establish Roman Catholicism as the official religion of England. Hundreds of Protestants emigrated to southern Germany and Switzerland. Some developed stronger sympathies with the Calvinist theology and church polity of Geneva, while others followed the more moderate line of Heinrich Bullinger at Zurich. When Elizabeth succeeded Mary in 1558, she established the episcopal form of church government as a 'middle way' and gave ecclesiastical positions to moderate clergy who were willing to support her policies. These moderates tended to emphasise the side of Reformed thought that showed regard for universal providence and the working of God through second causes. A few examples will suffice.

Thomas Cooper (c. 1517-94) published a volume of sermons in 1580 while serving as bishop of Lincoln. He followed Calvin in affirming that all seeming chance or fortune is really the result of divine providence. But providence for Cooper was just the operation of God through second causes. Conversely, what we call nature 'is nothing but the very finger of God working in his creatures'. The idea of God working outside the regular channels established at creation must have been as repugnant to Cooper as the practise of Puritan radicals who conducted 'prophesyings' outside the auspices of the established church.[152]

---

[150]Paul H. Kocher, *Science and Religion in Elizabethan England* (San Marino, Calif.: Huntington Library, 1953), pp. 155f., 191f.

[151]Francis R. Johnson, 'Thomas Hood's Inaugural Address as Mathematical Lecturer of the City of London (1588)', *JHI*, vol. iii (Jan. 1942), pp. 103-6.

[152]Keith Thomas, *Religion and the Decline of Magic* (New York: Scribners, 1971), pp. 79f.; *DNB*, vol. IV, pp. 1074ff.

A better known example is Richard Hooker (*c*. 1554-1600), the protégé of John Jewel and John Whitgift, two of the pillars of the Elizabethan church. After his appointment as Master of the London Temple (1585), Hooker debated the merits of the Anglican church order with the eloquent Puritan, Walter Travers, and out of this altercation came his major defense, *Of the Laws of Ecclesiastical Polity* (1593-7). Like Bishop Cooper, Hooker stressed that all God's operations are limited by the laws established at creation and so are not violent or casual, but regular. In fact, there was a direct analogy in his mind between the laws of nature and a 'kingdom rightly ordered'.[153]

Similar views were held by establishment-supported laymen like Walter Bailey, physician to the Queen. In a treatise he wrote on the medicinal properties of the bathing waters of the realm (1587), Bailey stressed that God had ceased from new acts of creation after the sixth day of creation (Gen. 2:3), having given to his creatures 'a nature and power by which they stand and fall'. At present, then, God works mostly by natural means: to say that he acts supernaturally would be to subvert both nature and philosophy.[154]

These Elizabethan moderates were following Calvin in stressing the *potentia Dei ordinata*. They wished to restrict the possibility of new revelations and miracles that might subvert the new order they were trying to establish: in this, too, they were taking a leaf out of Calvin's book.[155] But they tended to reduce Calvin's teaching about *potentia ordinata* to the doctrine of universal providence through second causes, and they gave less attention to his teaching about particular providence, which would have made nature more directly dependent on the continuous activity of God and left slightly less scope for the autonomy of nature.

It was the more radical Protestants, particularly the Puritans, who stressed the doctrine of particular providence. The best known example is William Perkins (1558-1602), the leading Puritan theologian of the late sixteenth century. While he adopted a scholastic method of presentation to meet the standards of contemporary scholarship, Perkins differed from Aristotle in stressing the contingency of all natural processes. Even though various qualities and virtues had been implanted in things at creation, he argued, they could not function or even sustain themselves against decay without the quickening power of God. Normally, it was true, God did act

---

[153]*Laws of Ecclesiastical Polity* I.2f.; *DNB*, vol. IX, pp. 1183f.
[154]Kocher, *Science and Religion*, pp. 108f.
[155]*Institutes* IV.i.5; xix.18; *Against the Libertines* 9 (trans. Farley), p. 222; *Warning Against Judicial Astrology* (trans. Potter), pp. 181f.; *Concerning Scandals*, trans. John W. Fraser (Grand Rapids: Eerdmans, 1978), p. 44; *Second Sermon on the Ten Commandments* in B.W. Farley, trans., *John Calvin's Sermons on the Ten Commandments* (Grand Rapids: Baker, 1980), p. 55.

through natural means, but he also worked without means, either above or even against nature; in miracles, for example.[156]

Perkins was not an extremist by any means. He was loyal to the Church of England, and as early as 1588 he denounced the presbyterian and congregationalist schismatics for undermining the authority of the state.[157] It was probably from the schismatic radicals that the most vocal challenge to the idea of natural causation came. Most of our evidence for Puritan strictures on the scientific quest for second causes comes from the early seventeenth century. However, we may surmise that such pious concerns were voiced already in the late sixteenth century because Francis Bacon had to defend his programme for scientific research against such 'a religion that is jealous' as early as 1603.[158]

Bacon saw both the established Aristotelianism of the universities and the hostility to science of the radical Puritans as threats to the progress of science. His instinctive efforts to forge a middle way were thus parallel to Calvin's similar efforts half a century earlier, though Bacon was more emphatic than Calvin both in his rejection of Aristotle and in his affirmation of natural causation.

In his criticism of the stranglehold he felt Aristotelian thought had on the universities, Bacon may have been influenced by the strictures of Puritan preachers and lecturers, several of whom had been fellow students at Trinity College, Cambridge. Aristotelian philosophy had brought about a second fall, Bacon contended. Instead of listening to and interpreting nature to discover her laws, Aristotle and his followers had committed the sin of prescribing laws for her. But, rather than seek the causes of phenomena in scholastic manuals of philosophy, scientists should go to the book of nature itself.[159]

---

[156] Perkins, *An Exposition of the Creede* (1595); Kocher, *Science and Religion*, pp. 103, 108; Ian Breward, ed., *The Works of William Perkins* (Appleford: Sutton Courtenay Press, 1970), pp. 53, 592f.; *DNB*, vol. XV, p. 894a.

[157] Leonard J. Trinterud, *Elizabethan Puritanism* (New York: Oxford Univ. Press, 1971), pp. 11ff.; Breward, ed., *Works*, p. 9.

[158] *Of the Interpretation of Nature* 1, 25 (1603; James Spedding et al., eds., *The Works of Francis Bacon* [London: Longman et al., 1857-74], vol. VI, pp. 29f., 75); cf. Kocher, *Science and Religion*, pp. 14-17; Geoffrey Bullough, 'Bacon and the Defence of Learning', *Seventeenth Century Studies* (Oxford: Clarendon Press, 1938), pp. 2ff.; reprinted in B. Vickers, ed., *Essential Articles for the Study of Francis Bacon* (Hamden, Conn.: Archon Books, 1968), pp. 94ff. Note also the reference to 'the blind and immoderate zeal of religion', 'the simpleness of certain divines', and 'the simpleness and incautious zeal of certain persons' who viewed the scientific search for second causes as a threat to religion; *Novum Organum* I.89 (1620; ET, Fulton H. Anderson, ed., *Francis Bacon: The New Organon* [Indianapolis, Ind.: Bobbs-Merrill, 1960], pp. 87ff.).

[159] *The History of the Winds* (1623; Benjamin Farrington, *Francis Bacon* [New York: Henry Schuman, 1949], pp. 148ff.); *Thoughts and Conclusions* (1607; B. Farrington, *The Philosophy of Francis Bacon* [Chicago: Univ. of Chicago Press, 1966], pp. 77f.).

Bacon's criticism of the radical Puritan extreme was no less severe. The doctrine of providence should not be invoked, as was done by the more zealous divines, in such a way as to discourage the scientific quest for natural causes, for 'certain it is that God worketh nothing in nature but by second causes'.[160] Still, the contemplation of nature need not detract from our sense of dependence on God. Though the study of physics may at first incline some people to atheism, further progress will bring them back to religion as they find the wisdom and power of God reflected in his creation.[161]

Though not a major contributor to the sciences himself, Francis Bacon provided his age with a vision of the viability of science in a Christian culture. Rather than trying to carve out a place for piety within the framework of Aristotelian natural philosophy, as Calvin and the other Reformers had done, Bacon projected a new view of nature which allowed both the full operation of second causes and the full dependence of all things on God. In this new order, there would be no gaps in natural explanation, so God would have to be seen to function immediately in the whole of nature or else not at all. Bacon's hope was that the new science would not displace God's providential activity the way Aristotelian science had done. Unfortunately, this was not to be the case, at least, not in the long run.

Bacon's solution to the sixteenth-century conflict of science and religion was effectively to separate the two antagonists. Nature and grace were two separate kingdoms or departments of the *potentia Dei ordinata*: the kingdom of nature was accessible through the arts and sciences based on human reason and observation; the kingdom of God was accessible through the forgiveness of sins based on the teachings of Scripture. Ultimately the two were united in God: one was based on God's works and revealed his power; the other on God's word and revealed his will.[162] But the clear separation of the two in this life implied a far greater autonomy of nature and far more individualism in religion than either Scripture or the Reformers had ever intended.

---

[160]*Interpretation of Nature* (1603); *Advancement of Learning* (1605); *Works*, vol. VI, pp. 29f., 75, 91f., 96; *Thoughts and Conclusions* (1607); Farrington, *Philosophy*, pp. 77f.

[161]*Sacred Meditation* 10; *Student's Prayer, Works*, vol. XIV, pp. 93, 101; *Interpretation of Nature, Works*, vol. VI, p. 33; *Thoughts and Conclusions* in Farrington, *Philosophy*, pp. 78f., 95.

[162]*Confession of Faith, Works*, vol. XIV, pp. 49ff.; *Novum Organum* I.89; II.52 (1620); *Works*, vol. IV, pp. 247f.; Anderson, ed., *Novum Organum*, pp. xvif., 88f., 267f.; cf. Gary Bruce Deason, 'The Philosophy of a Lord Chancellor', Ph.D. diss. (Princeton Theological Seminary, 1977), pp. 312ff.; Jeffrey Barnouw, 'The Separation of Reason and Faith in Bacon and Hobbes, and Leibnitz's *Theodicy*', *JHI*, vol. xlii (Dec. 1981), pp. 610ff.

4 Ministry of Healing and Restoration

The fourth theme in the creationist tradition also related to the power of God, but not the power of God in creation and providence so much as the power God had placed in human hands in order to renew the fallen members of that creation. In the first few centuries after Christ, this power was believed to manifest itself primarily in miraculous healings that demonstrated the outpouring of God's Spirit on the Church. The monastic movement beginning in the fourth century developed the skills of ancient medicine and technology as part of the Christian ministry of renewal, though a secularising process set in as early as the twelfth century. Still, important practitioners of medicine like Arnold of Villanova and Paracelsus viewed their work as a ministry of healing in the medieval and Renaissance periods (chap. 1, sec. E; chap. 2, sec. C). We wish to know what impact the Reformation had through its revitalisation of biblical and patristic teachings, particularly on the idea of the social value of the sciences later popularised by Johann Andreae and Francis Bacon. We shall trace first the Lutheran tradition down to Andreae, then the Calvinist tradition down to Bacon.

4.1 The Lutheran Tradition to Andreae

Luther (and/or Melanchthon) followed patristic and medieval commentators in viewing the human arts as restoring at least a semblance of Adam's original dominion over nature.[163] In fact, Luther foresaw a new era of scientific and technical progess consequent upon the reformation of religion and the renewal of the image of God in humanity.[164] Physicians and alchemists were viewed as contributors to this restoration due to the wonders they could perform based on their knowledge of the secret powers of nature.[165] Such secular vocations were ordained by God as part of the work of creation, and their practitioners were, therefore, coworkers with God.[166] But, though Luther saw an analogy between human arts and the works of God, he regarded the arts as functions of human nature, not functions of the Spirit working through human skills.[167] This was a result of his clear distinction between creation- and salvation-ordinances, which we discussed earlier.

Also in keeping with the creationist tradition, Luther and Melanchthon applied the Christian ideal of service to the arts and sciences as well as to

---

[163] *Lectures on Genesis* 1:26 (*LW* 1:67).
[164] *Ibid.* (pp. 64f.); *Table Talk* 1:1160 (Bornkamm, *Luther's World*, p. 184).
[165] *Kirchenpostille* (1522; WA 10/1:560 [Bornkamm, *Luther's World*, pp. 181f.]).
[166] Gustav Wingren, *Luther on Vocation* (Philadelphia: Muhlenberg Press, 1957), pp. 129f.
[167] *Commentary on the Fifteen Psalms* (Kocher, *Science and Religion*, p. 8); S.F. Mason, 'The Scientific Revolution and the Protestant Reformation, II', *Annals of Science*, vol. ix (30 June 1953), pp. 155, 162.

other secular activities like politics. Self-exalting ambition like that attributed to various popular charismatic leaders was strongly discouraged. Instead, the arts and sciences were sanctioned in so far as they contributed to social and moral development of the community as a whole.[168]

The danger of this criterion of social benefit was that it could lead to a conservative, or even reactionary, attitude towards new scientific ideas. This is most clearly seen in the case of Luther's and Melanchthon's attitudes towards the new astronomy of Copernicus.

In 1539 the first reports of Copernicus's work filtered back to Wittenberg through the correspondence between Melanchthon and Copernicus's assistant, Rheticus (chap. 3, sec. A). The only recorded reaction that was attributed to Luther himself is found in his *Table Talk* for that year. As recorded by Anthony Lauterbach, it was rather negative:[169]

> So it goes now. Whoever wants to be clever must agree with nothing that others esteem. He must do something of his own. This is what that fellow does who wishes to turn the whole of astronomy upside down. Even in these things that are thrown into disorder I believe the Holy Scriptures, for Joshua commanded the sun to stand still and not the earth [Josh. 10:12].

Clearly the issue for Luther was not a technical question of the merits of heliocentric theory, but the seeming ambition of the astronomer and the possibly disruptive effect his teachings might have on a Christian society. The same was true of Melanchthon's initially hostile reaction.[170]

We must remember that Luther and Melanchthon were deeply alarmed by the social upheavals of the 1520s and were sensitive to any criticism of the educational programs they had instituted, particularly at the university level. Melanchthon's insistence on the proper order and subdivision of topics for instruction was particularly vulnerable to ideas as novel as those of Copernicus.[171]

So the criterion of social benefit which was part of the creationist tradition could have a restraining effect on the development of science. But, in the long run, it could also be beneficial. If, as we have argued, the main

---

[168]Wingren, *Luther*, pp. 127f.; Melanchthon, Preface to 1537 ed. of Euclid's *Elements* (trans. Moore, p. 147); Preface to 1552 ed. of Regiomontanus's *Tabulae Directionum* (Moran, 'Universe', p. 16); Hannaway, *Chemists*, p. 117.

[169]*Table Talk* 4:4638 (recorded by Anthony Lauterbach, 4 June 1539; *LW* 54:359). Lauterbach's version of the *Table Talk* is generally thought to be more reliable than Aurifaber's. Both versions show indications of literary dependence on Cicero's *Academica*; C.B. Kaiser, 'Calvin, Copernicus, and Castellio', *Calvin Theological Journal*, vol. xxi (April 1986), pp. 8, 20f. with notes. These classical touches may indicate Melanchthon's influence.

[170]Letter to Mithobius, 1541 (CR 4:679); *Initia doctrinae physicae* (1549 ed.; CR 13:216ff.); Moran, 'Universe', pp. 13f.

[171]Robert Stupperich, *Melanchthon* (London: Lutterworth, 1966), pp. 55ff., 69, 72; C.L. Manschrek, *Melanchthon the Quiet Reformer* (Nashville: Abingdon, 1958), pp. 76, 83f., 122, 130ff.; Moran, 'Universe', p. 4.

reason for the failure of Greco-Roman science to sustain long-term growth was its lack of strong social support, a major advance would require the construction of a new image of the sciences as contributing to recognised goals like the amelioration of the human condition. So the strictures of the Reformers (based on those of the church fathers in relation to Greek science) could serve the long-term interests of science by forcing it to adopt a new public-mindedness that would help gain the kind of social support it needed.

Melanchthon himself illustrates this tendency to look more favourably on new astronomical ideas once they were seen to serve the interests of the community. As early as 1549, he spoke more positively of Copernicus's work.[172] As he put it a few years later (1552), the inquisitiveness and zeal of astronomers like Copernicus can widen the scope of human knowledge:[173]

> Therefore we must not refrain from investigating the wisdom in the work of God....We cannot overlook the fact that the sciences are a gift of God in order to recognise him and thereby to maintain life in a wiser order.

Following the leadership of Melanchthon, a series of Lutheran astronomers studied the work of Copernicus. At first they were only pragmatically interested in the possibility of improved calculations, but gradually they came to accept the underlying hypothesis that the sun was really stationary at (or near) the centre of the planetary system. Kepler's early acceptance of heliocentric cosmology (*c.* 1590) led to his pioneering efforts to discover the mathematical laws that describe the motions of the planets around the sun. In accordance with Christian teaching, Kepler felt himself under obligation to publish his results for the glory of God, even though he realised that others might then use them to make advances he might otherwise have reserved for himself.[174]

While Copernican astronomy finally passed the test of social utility, the hermetic and alchemical traditions came increasingly under fire. In fact, the reason that Copernicanism found more favour in Protestant communities

---

[172]*Oration on Caspar Cruciger* (1549; *Corpus Reformatorum* 11:839); 1550 ed. of the *Initia doctrinae physicae*; Dillenberger, *Protestant Thought*, p. 41; Moran, 'Universe', pp. 14ff.

[173]Letter to the Fuggers, 24 Feb. 1552 (*Corpus Reformatorum* 7:951; Moran, 'Universe', p. 16). Methuen warns against overestimating Melanchthon's flexibility with respect to Aristotelian cosmology; 'Role of the Heavens', pp. 402f.

[174]'If this [*Mysterium cosmographicum*] is published, others will perhaps make discoveries I might have reserved for myself. But...I strive for the glory of God, who wants to be recognised from the book of Nature, that these things may be published as quickly as possible. The more others build on my work, the happier I shall be', Letter to Maestlin, 3 Oct. 1595 (*Gesammelte Werke* 13:39f.). J.V. Field uses this point to differentiate Kepler's attitude from that of the Pythagoreans and characterises it as 'an entirely "modern" attitude to publishing one's results'; *Kepler's Geometrical Cosmology* (Chicago: Univ. of Chicago Press, 1988), p. 189.

than in Catholic circles of the early seventeenth century was that it had been championed by a school of Lutherans who were not associated with the extremes of Renaissance Hermeticism and Cabalism (chap. 3, sec. A).[175] In contrast, the case of Giordano Bruno imprinted an association between Copernican ideas and Pythagorean exaltation on the minds of prominent Catholic officials like Cardinal Bellarmine. Bellarmine's experience with Bruno made him suspicious of Galileo's work as early as 1611,[176] even though Galileo had even less sympathy with the occult sciences than Kepler had![177]

Luther had rejected the idea of the Renaissance magus popularised by Ficino, Reuchlin, and Agrippa.[178] The ideas of individual self-exaltation and Pythagorean secrecy associated with these arts were clearly incompatible with his sense of the solidarity of the church. Nonetheless, a series of Lutheran scholars and physicians did cultivate occult sciences like alchemy: for example, Andreas Osiander (1498-1552), Valentin Weigel (1533-88), Heinrich Khunrath (1560-1605), Michael Maier (1568-1622), and Johann Andreae (1586-1654). Their inspiration may well have come from Luther's sense of the energies of God diffused through all creation, as well as from the direct influence of Reuchlin and Paracelsus (chap. 3, sec. A).[179]

The principal critic of the occult sciences in the Lutheran tradition was Andreas Libau (or Libavius; c. 1560-1616). In the tradition of Melanchthon, Libau stressed the importance of corporate accountability for the sciences in contrast to the individual enthusiasm of alchemists like Paracelsus and his followers. Libau studied at Basel (1588) where, through the earlier teaching

---

[175] On Kepler's early interest in and later distancing from hermetic and cabalistic ideas, see Yates, *Rosicrucian Enlightenment*, pp. 222f.; W.P.D. Wightman, *Science in a Renaissance Society* (London: Hutchinson, 1972), pp. 149ff., 162f.; Robert S. Westman, 'Nature, Art, and Psyche: Jung, Pauli, and the Kepler-Fludd Polemic', in Brian Vickers, ed., *Occult and Scientific Mentalities in the Renaissance* (Cambridge: Cambridge Univ. Press, 1984), 204ff.

[176] Giorgio de Santillana, *The Crime of Galileo* (Chicago: Univ. of Chicago Press, 1955), 28f.; DeKosky, *Knowledge and Cosmos*, pp. 186f. Kepler himself wrote Galileo praising him for 'following Plato and Pythagoras, our true teachers' (13 Oct. 1597; Koestler, *Sleepwalkers*, p. 359).

[177] Randall, *Career*, vol. I, p. 356; Stillman Drake, *Galileo Studies* (Ann Arbor: Univ. of Michigan Press, 1979), p. 277.

[178] Louis Israel Newman, *Jewish Influence on Christian Reform Movements* (New York: Columbia Univ. Press, 1925), p. 623; John G. Burke, 'Hermeticism as a Renaissance Worldview', in Robert S. Kinsman, ed., *The Darker Vision of the Renaissance* (Berkeley: Univ. of California Press, 1974), p. 102; F.A. Yates, *Lull & Bruno*, collected papers, vol. I, (London: RKP, 1982), p. 213.

[179] Manschrek, *Melanchthon*, pp. 22, 25; Bruce Wrightsman, 'Andreas Osiander's Contribution to the Copernican Achievement', *Copernican Achievement* (Westman, ed.), pp. 217, 223ff.; Montgomery, *Cross and Crucible*, pp. 17-20; Steven E. Ozment, *Mysticism and Dissent* (New Haven: Yale Univ. Press, 1973), pp. 209, 214; Yates, *Rosicrucian Enlightenment*, pp. 38f., 145-54.

of Reformed physician-theologians like Theodore Zwinger[180] and Thomas Erastus,[181] the name of Paracelsus had become synonymous with disrespect for the ancients and an exaltation of human powers over nature. In the early seventeenth century, Libau wrote vehemently against the Paracelsians (particularly Oswald Croll, whom we shall consider later) for being socially irresponsible and for using his supposed knowledge for personal gain.[182]

However unfair these criticisms may have been to Paracelsus, who was an eccentric at worst, they were salutory for the image of the alchemical tradition as a whole. The danger was that they might also encourage an attitude of resignation in the face of the recalcitrant structures and awesome powers of nature.[183] In this respect, the optimism of the hermeticists and alchemists was a helpful counterbalance. The two sides of the creationist tradition, the charismatic and the socially responsible, thus complemented each other, as was later pointed out by Andreae. Responding to the criticisms of Libau, Andreae wrote a description of the ideal Christian society (*Christianopolis*, 1619) as one in which alchemical research was done for the benefit of medicine and technology. Thus the strictures of Lutheran orthodoxy helped to purge alchemy of the pride and arrogance associated with the image of the Renaissance magus and paved the way for a more fruitful science of chemistry in the seventeenth century.[184]

4.2 The Calvinist Tradition to Bacon

A similar development occurred in the Reformed tradition of southern Germany and Switzerland. Calvin followed Luther quite closely in his estimate of the arts and sciences. Like Luther, he saw in secular disciplines like medicine a God-given skill based on knowledge of the hidden

---

[180] On Zwinger's critique of Paracelsianism and his influence on Erastus, see Peter G. Bietenholz, *Basle and France in the Sixteenth Century* (Geneva: Librairie Droz, 1971), p. 70.

[181] On Erastus's critique of Paracelsianism, see Pagel, *Paracelsus*, pp. 313f., 331; 'Erastus, Thomas', *DSB* 4:387b; Debus, *Chemical Philosophy*, vol. I, p. 132. According to Jole Shackelford, Erastus focussed his attack on the theory of the Paracelsian, Petrus Severinus, that the seeds of disease were planted in creation by God; 'Early Reception of Paracelsian Theory: Severinus and Erastus', *SCJ*, vol. xxvi (Spring 1995), pp. 128ff., esp. note 27. According to an article by Chad Gunnoe (which I have not yet seen for myself), Erastus was trying to assert his own religious orthodoxy as much as to criticise that of Severinus, with respect to the doctrine of creation. Erastus's real concern was that the Paracelsians' insistence on the spiritual illumination of physicians was desecularising, i.e., it failed to separate medicine from religion; 'Thomas Erastus and His Circle of Anti-Paracelsians', in Joachim Telle and Wolf-Dieter Mueller-Jahncke, eds., *Analecta Paracelsica* (Stuttgart: Steiner Verlag, forthcoming), cited in Shackelford, 'Early Reception of Paracelsian Theory', p. 133n. In any case, it was Erastus's charges of innovation and 'disgraceful ambition', that most influenced Libavius.

[182] Hannaway, *Chemists*, pp. 83f., 87f., 95f., 99f., 104, 112, 117.

[183] Hannaway, *Chemists*, pp. 84f., 87f., 101f., 115.

[184] Betty Jo Teeter Dobbs, *The Foundations of Newton's Alchemy* (Cambridge: Cambridge Univ. Press, 1975), p. 58; Olson, *Science Deified*, pp. 271-77.

correspondences of nature.[185] Like Luther, Calvin viewed such skills as based on the gifts of God in nature and clearly differentiated them from the more important work of God in salvation, but unlike Luther he regarded them as part of the work of the Spirit.[186] In fact, Calvin's distinctive insistence that miracles of healing had ceased after apostolic times led him to view the arts as God's ordained means of ministering to the body in the present age.[187]

So Calvin had a fair sense of the biblical ministry of healing, if not a millenarian anticipation of the restoration of all things through science and technology. He advocated the idea of social utility as a check to unbridled enthusiasm. In this respect, Calvin was influenced by the humanist sociopolitical ideal of the common good (as was his elder colleague Martin Bucer, the leading reformer at Strassburg).[188] But Calvin's sense of Christian vocation immeasurably deepened the social aspect of his teaching. Since the arts and sciences were gifts of God's Spirit, they were to be used for the common good of humanity. They were good in themselves, but, if used for personal ambition or greed, they became evil. Taking the gifts of God in vain was just as much a violation of the Third Commandment as taking the name of God in vain (Exod. 20:7).[189]

Calvin combined the dialectic of individual human skill and social responsibility with the differentiation of creation- and salvation-ordinances we discussed earlier. While the arts and sciences were based on nature as created by God, the love of neighbour needed to direct their use was a supernatural gift that had been lost at the Fall and was restored through the grace of regeneration, which alone can 'erase from our minds the yearning to possess, the desire for power, and the favour of men'.[190]

It will come as no surprise, then, that Calvin spoke harshly of the hermetic and cabalistic strains of Renaissance philosophy. Agrippa of

---

[185]*Warning Against Jucicial Astrology* (trans. Potter, pp. 166ff.); Marcel, *Calvin et Copernic, Revue réformée*, vol. xxxi (1980), p. 56.

[186]*Institutes* II.ii.14ff.

[187]*Institutes* IV.xix.18; Marcel, *Calvin et Copernic*, pp. 122, 126.

[188]Hans Baron, 'Calvinist Republicanism and Its Historical Roots', *CH*, vol. viii (1939), pp. 30-42; E. Harris Harbison, *The Christian Scholar in the Age of the Reformation* (New York: Scribners, 1956), pp. 160-64; Heiko Oberman, *Masters of the Reformation* (Cambridge: Cambridge Univ. Press, 1981), pp. 276-81, 285, 292ff.

[189]*Institutes* II.ii.16; viii.22; *Commentary on 1 Corinthians* 8:1 (*CNTC* 9:171f.). Michael Monheit argues that Calvin's anxiety about personal ambition was the result of his struggle with his own earlier motivations as a humanist scholar; '"The ambition for an illustrious name": Humanism, Patronage, and Calvin's Doctrine of the Calling', *SCJ* vol. xxiii (Summer 1992), pp. 267-87.

[190]*Institutes* II.ii.12; III.vii.2-7 (quote from III.vii.2 [*LCC*, vol. XX, John T. McNeill, ed., London: SCM Press, 1960, p. 691]).

Nettesheim was singled out for particular censure,[191] but Calvin undoubtedly had the entire tradition of Ficino, Pico, and Reuchlin in mind. Their ideas were criticised on both counts mentioned above: they used the name and mysteries of God for personal ambition (thus transgressing the commandment of love),[192] and they sought access to God and knowledge of his decrees through the hierarchy of angels, a practise that was forbidden in Scripture on penalty of death (Exod. 22:18; Lev. 20:6, 27; Deut. 18:10).[193] Calvin specifically rejected the hermetic notion that Moses and Jesus performed miracles by magic arts, one of the teachings for which Bruno was condemned and executed in 1600.[194]

The continental Calvinist tradition did not produce a school of occult natural philosophy comparable to that of the Lutheran tradition. But there were isolated figures like Oswald Croll (*c.* 1560-1609), whose medical training was at Heidelberg, Strasbourg, and Geneva. One recent historian, Owen Hannaway, has claimed that Calvin's ideas stimulated Croll's adaptation of Paracelsian teachings to the mainstream of Reformation theology.[195] Actually, Croll was far closer to Ficino and Paracelsus than to Calvin in his spiritualistic interpretation of the physician's craft. Whereas Calvin made regeneration the prerequisite for the attitude of humility and self-sacrifice needed by arts like medicine, Croll made it prerequisite to the skill of healing itself.[196] Croll thus combined the spheres of nature and grace in the work of the physician in a way that Calvin would have rejected as a 'mingling of heaven and earth'.[197]

On the other hand, Croll could claim to pass the basic double criterion posed by Calvin. He challenged the aspiring physician to seek new medicines for the benefit of humanity,[198] and he made it clear that the practise of alchemy was a form of Christian discipleship, not a means to salvation.[199] It may be that the strictures of Calvin did exert some influence on Croll in spite of significant differences between the two. If so, the

---

[191]*Concerning Scandals* 201 (1550; trans. Fraser, p. 61); *Commentary on 1 Peter* 1:25 (1551; *CNTC* 12:254).

[192]*Institutes* II.viii.22 (1536-9).

[193]*Institutes* I.xiv.4, 10, 12; *Warning Against Judicial Astrology* (trans. Potter), pp. 187f.; *Commentary on the Pentateuch*, Exod. 22:18, trans. Charles William Bingham, (Edinburgh: Calvin Tract Society, 1853) vol. II, pp. 90ff.

[194]*Concerning Scandals* 201: 'execrable blasphemies against the Son of God' (1550; trans. Fraser), p. 61); *Commentary on Acts* 7:22 (1552); *Institutes* I.8.6 (1559).

[195]Hannaway, *Chemists*, pp. 1, 5, 10, 47-54.

[196]Preface to *Basilica chymica* (1609; Hannaway, *Chemists*, pp. 8f., 11, 18-21, 48ff., 56, 106f.).

[197]*Institutes* II.2.13; IV.xvii.14f., 30; *Commentary on Genesis* 1:6.

[198]Hannaway, *Chemists*, p. 4; Allen G. Debus, *The Chemical Philosophy* (New York: Science History Publications, 1977), vol. I, pp. 22f.

[199]Hannaway, *Chemists*, pp. 49, 52f.

long-term effects of these strictures can only be judged to have been beneficial for science: Croll's reformulation of the ideas of Paracelsus have been credited with gaining academic recognition for the medical value of many chemicals that otherwise would not have been used.[200]

Finally, we must trace the lines of the debate in Elizabethan and early Stuart England which culminated in the reform programme of Francis Bacon. Both the idea of the creative powers of humanity and the criterion of social benefit were already present on the English scene, at least in humanist circles, as parts of the general Renaissance worldview.[201] Differing emphases developed, however, between Neoplatonists and the stricter Protestant clergy in the later sixteenth century.[202] Bacon had to straddle the two positions, just as he had to straddle the Aristotelian and Puritan positions with respect to the relative autonomy of nature, in order to commend the scientific enterprise to James I and to English society as a whole.

The major representatives of the Neoplationist tradition in English science were John Dee (1527-1608),[203] William Gilbert (1544-1603), Sir Walter Raleigh (1552-1618), Thomas Harriot (1560-1621), and Robert Fludd (1574-1637). All were influenced by the hermetic-cabalist ideas of Ficino, Pico, and Agrippa. Accordingly, they espoused the Renaissance magus image of the potential of humanity, according to which initiates could perform miracles by exploiting the hidden energies of nature.[204]

The ideal of social service was commonplace in Renaissance England, but it was the Protestant clergy (both Anglican and Puritan) who popularised the ideal among the middle classes (the merchants and the

---

[200] Gerard Schroder, 'Crollius', *DSB* 3:472a.

[201] Margo Todd, *Christian Humanism and the Puritan Social Order* (Cambridge: Cambridge Univ. Press, 1987), pp. 16-21, 32-43, *passim*.

[202] Thomas, *Religion*, p. 268.

[203] According to Nicholas H. Clulee, Dee's early work (*Propaedeumata aphoristica*, 1558) was in the Neoplatonic-Aristotelian tradition of Robert Grosseteste and Roger Bacon; his Neoplatonist interest in the ontological status of mathematics came later (following Proclus, 'Mathematicall Praeface', 1570), as well as his Neoplatonist-hermetic-cabalist interest in spiritual magic (following Trithemius, *Monas hieroglyphica*, 1564; 'Libri mysteriorum", 1583-89); Clulee, *John Dee's Natural Philosophy* (London: Routledge, 1988), pp. 16, 71f., 232-36.

Dee shared the vision of social reform with the Renaissance humanists and Protestant Reformers to be discussed below, but he identified such improvements closely with the extension of the British Empire under Elizabeth and with his own personal preferment; John S. Mebane, *Renaissance Magic and the Return of the Golden Age* (Lincoln: Univ. of Nebraska Press, 1989), pp. 84-87.

[204] Peter J. French, *John Dee* (London: RKP, 1972), pp. 90-93, 106; Pyarali M. Rattansi, 'Alchemy in Raleigh', *Ambix*, vol. xiii (1965), p. 127; Robert Hugh Kargon, *Atomism in England from Hariot to Newton* (Oxford: Clarendon Press, 1966), pp. 27f.; Yates, *Bruno*, pp. 144f. (Plate 10).

gentry)[205] and who took the Neoplatonists to task on this score. A few examples will suffice.

There was already a strong emphasis on social reform in the preaching of early English Reformers like Thomas Starkey (d. 1538), Hugh Latimer (d. 1555), Nicholas Ridley (d. 1555), and John Hooper (d. 1555), who were influential during the reigns of Henry VIII (d. 1547) and Edward VI (d. 1553).[206] At this stage the principal continental influences were Luther, Melanchthon, Bucer, Zwingli, Oecolampadius, and Bullinger, all of whom had stressed the obligation to fulfill the Christian commandment of love.

Latimer and his pupil, Thomas Becon (d. 1567), were particularly vociferous in calling attention to the needs of the poor: they championed the practise of true religion as care for orphans and widows (Jas. 1:27) in contrast to the hypocrisy they associated with Catholic monasticism.[207] Hooper interpreted the third commandment as forbidding the use of God's name for personal glory or profit, especially in the occult arts and sciences. He specifically criticised the printing of the hermetic writings of Trithemius and Agrippa, due to their advocacy of conjuring spirits and their Pythagorean secrecy. In so far as the arts and sciences could be dissociated from magic and astrology, however, Hooper valued them as gifts from God for the maintenance of life and the restoration of health (Jas. 1:17).[208] His views thus followed those of Calvin almost exactly.

The impact of early Reformation social teaching on science can be seen in the case of the leading English naturalist of the mid-sixteenth century, William Turner (d. 1568). Turner combined the two callings of medicine and the priesthood throughout most of his career. A close friend of Latimer and Ridley during his student days at Cambridge, he was ordained by Ridley in 1552 and was an ardent advocate of religious reform all his life. He was one of those who went into exile for his beliefs during the reign of the Catholic, Queen Mary. Turner wrote a number of controversial theological works and published an English translation of the Heidelberg Catechism in 1572.[209]

---

[205] W.K. Jordan, *Philanthropy in England, 1480-1660* (London: Allen & Unwin, 1959), pp. 151-55.

[206] Felicity Heal and Rosemary O'Day, eds., *Church and Society in England* (London: Macmillan, 1977), p. 9; Jordan, *Philanthropy*, pp. 156-63. On Starkey, see also Todd, *Christian Humanism*, pp. 32f., 36, 40-43, 45, 124ff.

[207] Latimer, 'Fifth Sermon on the Lord's Prayer' (1552; G.E. Corrie, ed., *Sermons*, Parker Society Works, vol. 28 [Cambridge: Cambridge Univ. Press, 1844], p. 392; Philip Edgcumbe Hughes, *Theology of the English Reformers*, new ed. [Grand Rapids: Baker, 1980], p. 98); Becon, 'Preface' (1564; John Ayre, ed., *Early Works*, Parker Society Works, vol. 2 [Cambridge: Cambridge Univ. Press, 1843], p. 10; Jordan, *Philanthropy*, p. 163).

[208] *Declaration of the Ten Holy Commandments*, chap. 6 (Zurich, 1548; Samuel Carr, ed., *Early Writings*, Parker Society Works, vol. 20 [Cambridge: Cambridge Univ. Press, 1843], pp. 325-33; Thomas, *Religion*, p. 268 note 3).

[209] *DNB* 19:1290ff.; *DSB* 13:501f.

During his travels on the Continent, Turner became interested in the work of Paracelsus on the healing properties of mineral waters. In fact, the earliest reference to Paracelsus we have in the English language is in a tract on mineral waters which Turner wrote in Basel in 1557.[210] In his travels in Switzerland and Germany, Turner probably also heard criticisms of Paracelsus for his supposed arrogance and greed.[211] Later, back in England, he stressed the moral obligation of scientists to work for the benefit of humanity in his treatise on the medicinal properties of wines (1568).[212]

Another example of Protestant social conscience in the practise of medicine is the work of William Bullein (d. 1576) and his brother Richard (d. 1563). Having taken Holy Orders about the same time as Turner (1550), William resigned them in order to practise medicine at Durham soon after the accession of Queen Mary (1553).[213]

Bullein castigated the irreligion and unscrupulous practise of some contemporary physicians in his portrayal of Dr. Tocrub the Nullafidian ('unbeliever') in his *Dialogue Against the Fever Pestilence* (1565). By way of contrast, he described his brother Richard as the ideal Christian physician, 'a zealous lover in Physicke, more for the consolacion and help of th'afflicted sicke people beyng poore, than for the lucre and gaine of the money of the welthie and riche'.[214] A Latin inscription over the two brothers' tomb in London states that William devoted his own medical services to rich and poor without distinction as well.[215]

The preaching and practise of Christian charity thus continued unabated in the Elizabethan period, even though the focus of theological controversy gradually shifted to matters of church polity.[216] One further example of interest is that of the Puritan Robert Johnson (1540-1625), who served for a while (*c.* 1570) as chaplain to Sir Nicholas Bacon, the father of Francis Bacon. Johnson continued the Reformed tradition of clerical social service by founding schools and re-endowing hospices during his tenure as rector in Rutland (1584).[217]

Late sixteenth-century Puritans were critical of the same two features of the occult arts and sciences that John Hooper had singled out: the lack of

---

[210] Allen G. Debus, 'The Paracelsian Compromise in Elizabethan England', *Ambix*, vol. viii (June 1960), p. 75.

[211] On Turner's itinerary from 1540 to 1546, his meeting with Gesner at Zurich and visit to Basel in 1543, see *DNB* 19:1290b. On Gesner's own attitude to Paracelsus, see Paul Kocher, 'Paracelsian Medicine in England: The First Thirty Years', *JHMAS*, vol. ii (1947), p. 454 note 8; Allen Debus, 'Paracelsian Compromise', p. 74.

[212] Kocher, *Science and Religion*, p. 27.

[213] *DNB* 3:244a.

[214] Kocher, *Science and Religion*, pp. 240ff.; *DNB* 3:245f.

[215] A.H. Bullen, *Elizabethans* (New York: Russell & Russell, 1962), pp. 157.

[216] Jordan, *Philanthropy*, pp. 165-79.

[217] *DNB* 10:914f.

Christian humility associated with the invocation of angels and the Pythagorean tradition of secrecy.[218] It is at this point that Calvin's influence is probably the most noticeable: in 1561 there appeared an English edition of his *Warning Against Judicial Astrology*, in which he criticised Hermeticism, Cabalism, and alchemy, as well as astrology.[219] The typically Calvinist contention that the power to alter the substance of nature belongs only to God, not to alchemists, is found, for instance, in the works of William Perkins around the turn of the century.[220]

Such were the diverging currents of the creationist tradition in the late sixteenth century when Francis Bacon formulated his programme for the advancement of the sciences. Recent historians have pointed out that Bacon derived much of his inspiration from the hermeticism of Ficino, Telesio, and Bruno and from the alchemical tradition of Paracelsus. In the mechanical arts and natural sciences, he saw the promise of a restoration of Adam's original dominion over nature—a vision common to the hermeticists, but shared only by Luther among the Reformers.[221] He also adopted the Paracelsian view of the plasticity of nature: the substance of things could be altered by humans to bring out the virtue of properties hidden since creation.[222] Thus, in addition to the restoration of lost innocence through Jesus' words of forgiveness, Bacon saw a restoration of human dominion foreshadowed in Jesus' deeds, particularly his healing diseases and subduing the forces of nature.[223]

But Bacon was a political realist, not an enthusiast. He realised that the arts and sciences could not command the social support they needed as long as they were associated with personal ambition and the quest for individual

---

[218]Kocher, *Science and Religion*, pp. 14ff.; Deason, *Philosophy*, pp. 297ff., 306f.
[219]Thomas, *Religion*, pp. 367ff., cf. 269f.
[220]*Discourse of the Damned Art of Witchcraft* (late 1590s?, Breward, ed., *Works*, pp. 581, 592f.); *Treatise of Callings in Works* (1605; Kocher, *Science and Religion*, p. 66 note 11). On the relation of Perkins to the Presbyterian Puritans, see Breward's Introduction to his edition of the *Works*, pp. 9-16.
[221]D.P. Walker, *Spiritual and Demonic Magic from Ficino to Campanella* (London: Warburg Inst., 1958), pp. 199f.; Farrington, *Philosophy of Bacon*, pp. 27ff.; Deason, *Philosophy*, pp. 112f., 116f.
[222]Paolo Rossi, *Francis Bacon: From Magic to Science* (London: RKP, 1968), pp. 15-19; Deason, *Philosophy*, pp. 251f.
[223]*Sacred Meditation* 2 ('Of the Miracles of Our Saviour', *Works* 14:82).

power.[224] Accordingly, he revised the hermetic vision in a number of ways to answer the criticisms raised by the clergy.

First of all, Bacon stressed the fact that the arts and sciences were dependent on the grace of God even though their aim was only temporal. A 'great instauration' or restoration of human dominion had been promised by God to Adam and again to Daniel (Gen. 1:28; Dan. 12:4).[225] Prayer was thus an important part of any effort toward the advancement of science. The student of science was encouraged to pray specifically for a revival of learning following Bacon's own example:[226]

> To God the Father, God the Word, God the Spirit, we pour forth most humble and hearty supplications; that he, remembering the calamities of mankind and the pilgrimage of this our life, in which we wear out days few and evil, would please to open to us new refreshments out of the fountains of his goodness, for the alleviating of our miseries.

Thus it was to God's grace, not to human ingenuity alone, that the sciences were to look for help.

Second, Bacon made a clear distinction between the wonders to be accomplished through science and the original creative acts of God by referring back to the hexaemeral tradition of the church fathers. God took six days to create things by supernatural means, and then he rested from his

---

[224] As Julius Martin has suggested, comparison should be made, for instance, with the image of the magician in Marlowe's *Doctor Faustus*: 'O what a world of profit and delight/ Of power, of honour, of omnipotence/ Is promis'd to the studious artisan!' (I.i.54ff.); *Francis Bacon, the State, and the Reform of Natural Philosophy* (Cambridge: Cambridge Univ. Press, 1991), p. 173.

Martin has reinterpreted Bacon's concerns about the ambitions of alchemists and other scientists as a straightforward projection of his anxieties about 'voluntarist' (i.e., radical Puritan) threats to the power of the Tudor/Stuart monarchy; *Francis Bacon*, pp. 173f.; *idem*, 'Natural Philosophy and Its Public Concerns', in Stephen Pumfrey, Paolo Rossi, and Maurice Slawinski, eds., *Science, Culture, and Popular Belief in Renaissance Europe*, (Manchester: Manchester Univ. Press, 1991), 105-113.

However, Martin has vastly oversimplified the Puritans on this issue (as being all radicals and zealots; 'Natural Philosophy', p. 107), thereby neglecting the positive social views of Puritans undoubtedly known to Bacon (e.g., Robert Johnson and, probably, William Perkins). *Pace* Martin, Bacon should be approached, not just from the establishmentarian ('Elizabethan settlement') side—strong as that was—but in terms of a dialectic and synthesis of establishmentarian *and* nonconformist (potentially revolutionary) elements both in science and religion. In effect, Bacon was a moderate Puritan or an evangelical Anglican.

[225] Farrington, *Francis Bacon*, p. 146; Bacon, *The Interpretation of Nature* (1603; *Works* 6:32); *The Refutation of Philosophies* (1608; Farrington, *Philosophy*, pp. 131f.).

[226] 'Student's Prayer' (*Works* 14:101; cf. Farrington, *Francis Bacon*, p. 147); cf. Bacon's prayer in the preface to the *Great Instauration* (1620):

> ...at the outset of the work I most humbly and fervently pray to God the Father, God the Son, and God the Holy Ghost that, remembering the sorrows of mankind and the pilgrimage of this our life wherein we wear out days few and evil, they will vouchsafe through my hands to endow the human family with new mercies (Anderson, ed., *The New Organon*, p. 14).

CHAPTER THREE

creative work. In spite of the partial curse placed on nature following the fall of the first human pair (Gen. 3:14-19), the laws of nature had remained unchanged from that day to the present, and they would continue unchanged until the eschaton when there would be a new creation. In the meantime God was not idle. God preserved creation by a universal providence and did many works of healing and restoration.[227] Thus God was the 'Creator, Preserver and Restorer of the universe'.[228] But whereas the works of creation and preservation (and salvation of the soul) belonged to God alone, the opportunity to serve as the agents of healing and restoration was open to all. As Bacon put in a prayer to be used by writers:[229]

> We humbly beg that this mind may be steadfastly in us, and that thou, by our hands and also by the hands of others on whom thou shalt bestow the same spirit, wilt please to convey a largesse of new alms to thy family of mankind.

In both of these points, the attribution of all constructive human activity to the grace of God and the restriction of the arts and sciences to the bounds of nature established by God, Bacon was following the line of Calvin. We do not know just how much of Calvin's writings he had read, but there can be little doubt of Calvin's influence, whether direct or indirect. Such influence is clearly suggested by the way Bacon delt with the double criterion Calvin had established for assessing the arts and sciences: the criterion of social benefit and the strict separation of creation- and salvation-ordinances.

First, Bacon stated that all human knowledge should be used, not for personal gain, but for the benefit of humanity. A passage from the essay, *Of the Interpretation of Nature* (1603), illustrates the way in which the Reformers' teaching on the 'law of love' helped to shape this social awareness:[230]

> But yet evermore it must be remembered that the least part of knowledge passed to man by this so large charter from God must be subject to that use for which God has granted it, which is the benefit and relief of the state and society of man; for otherwise all manner of knowledge becometh malign

---

[227]'Confession of Faith' (*Works* 14:49-51).
[228]Preface to *The History of the Winds* (1623; Farrington, *Francis Bacon*, p. 150).
[229]'Writer's Prayer' (*Works* 14:102).
[230]*Works* 6:33f.; cf. 6:94 (1 Cor. 13:2 is also cited); Farrington, *Francis Bacon*, pp. 148ff.; idem, *Philosophy*, pp. 28ff. Note that Bacon's 1591 letter to Lord Burghley seems to show more humanist influence ('all knowledge', *philanthropia*) than his writings a decade later. The turning point may have been the failure of his 1594 proposals for reform (*Gesta Grayorum*) because 'My zeal was taken for ambition' (*Interpretation of Nature*); cf. Rossi, *Francis Bacon*, pp. 23f. As Burghley and others at the Elizabethan court were patrons of moderate Puritans like Henry Smith and John Dodd (William Haller, *Elizabeth I and the Puritans* [Ithaca: Cornell Univ. Press, 1964], pp. 33-36), Bacon would have been forced to take their views more into account.

and serpentine...; as the Scripture saith excellently, *knowledge bloweth up, but love buildeth up* [1 Cor. 8:1].

As Bacon put it in his essay *Of the Advancement of Learning* (1605), science should not be a shop for profit or sale, but a storehouse for both the glory of the Creator and the relief of the human estate.[231] In the ideal scientific community (the 'House of Solomon' in the New Atlantis), workers daily sang hymns of praise to God and prayed to him 'for the illumination of our labours and the turning them into good and holy uses'.[232] Those, like the monks (and even some presbyterian Puritans, according to Bacon), who concentrated on the praise of God to the exclusion of caring for their neighbours were guilty of neglecting the second table of the Law and hypocritically violating the teachings of the New Testament (Jas. 1:27; 1 John 4:20).[233] Their self-discipline and communal rule could be emulated, but they would be put to better use by the new order of God's servants, the research scientists.

Following the example of earlier Protestant critiques, Bacon criticised Renaissance Hermeticism and Paracelsian alchemy as being motivated by personal ambition and for restricting the benefits of knowledge to an elite. Instead, he envisioned an international network of research facilities exchanging information with one another in a common effort to benefit all nations.[234] However, Bacon did not rule out ambition; he tried to harness and prioritise it. Ambition to increase one's own power over others (like that associated with Hermeticism and alchemy) was the most vulgar and violent form; ambition for one's country was more noble than personal ambition, yet still basically covetous and disruptive; only ambition to increase the shared power of all humanity over the universe was sufficiently altruistic and peaceful to be an appropriate sign of the kingdom of God (1 Kgs. 19:12; Luke 17:20).[235]

---

[231] *Works* 6:134.

[232] *New Atlantis* (1627; *Works* 3:166; Glacken, *Traces*, p. 475; Elliott M. Simon, 'Bacon's New Atlantis', *Christianity and Literature*, vol. xxxviii [Fall 1988], p. 56).

[233] *Sacred Meditation* 7 ('Of Hypocrites', *Works* 14:89f.); cf. *Advertisement Touching the Controversies of the Church of England* (*The Works of Lord Bacon* [London: William Ball, 1837], 1:349b), where the same two texts are cited; and *Thoughts and Conclusions* (1607; Farrington, *Philosophy*, p. 93); Deason, *Philosophy*, p. 97.

Calvin had made a similar comparison of the two tables of the Law in *Inst.* II.ii.24; viii.11; *Comm. on Matthew* 22:37, 39. But Calvin associated the precepts of the second table with the 'preservation of civil society' (*Inst.* II.ii.24), not with its amelioration.

[234] Rossi, *Francis Bacon*, pp. 18, 23, 27-33; Yates, *Rosicrucian Enlightenment*, pp. 120, 174f.; Deason, *Philosophy*, Abstract and pp. 114, 299f.

[235] *Interpretation of Nature* (1603; *Works* 6:36); *Thoughts and Conclusions* (1607; Farrington, *Philosophy*, pp. 92f.); *Novum Organum* I.129 (1620; *Works* 4:114; F. H. Anderson, ed., pp. 118f.). So the Elder of Solomon's House abandons the traditional 'Laws of Secrecy' and instructs the visiting narrator to publish the new gospel of science and technology 'for the good of other nations'; *New Atlantis* (*Works* 3:166; cf. Simon, 'Bacon's

198   CHAPTER THREE

The second indication of Calvin's influence is Bacon's differentiation of creation- and salvation-ordinances in his critique of the excesses of Renaissance Hermeticism. Bacon rejected the notion (found in hermetic texts like the *Asclepius*) that the perfection of alchemy and ceremonial magic would lead to spiritual rebirth.[236] He also eliminated the idea of a system of correspondences between the heavens and the human body based on the presence of the world soul in order to differentiate more clearly between the Spirit of God and the forces of nature.[237] With ethical and theological safeguards like these, Bacon hoped, the sciences would embark on an era of unprecedented growth and all humans would benefit as a result.

5 Conclusion

We have now reviewed Reformation theology and its impact under the four headings of the creationist traditon inherited from the early church. What, on balance, can we say the effect of the Reformation was on the development of the sciences through the early seventeenth century?

Positively, we may say the following: the teachings of the Reformers gave the secular arts and sciences unprecedented legitimacy as evidences of the image of God in humanity, as witnesses to universal providence, and as aspects of the Christian ministry of healing and restoration. They also popularised the fundamental criterion of public benefit without which a socially acceptable programme for the advancement of the sciences such as Bacon's would not have been possible.

On the more negative side, the Reformers were conservative in ways that occasioned later conflicts. Most of them adhered to Aristotelian cosmology as the best science available for pedagogical and apologetic purposes. As a result, they were not open to new ideas like the heliocentric system of

---

New Atlantis', p. 56). And 'authors of inventions' were revered as gods by the ancients because 'the benefits of discoveries may extend to the whole race of man', and 'discoveries...confer benefits without causing harm or sorrow to any'; *Novum Organum* I.129 (*Works* 4:113; F.H. Anderson, ed., p. 117). For a more critical reading of Bacon, based on the post-modern critique of science by Horkheimer and Adorno, see James Holstun, *A Rational Millennium* (New York: Oxford Univ. Press, 1987), pp. 50-54, and Martin, 'Natural Philosophy' (who completely misses the point made here; pp. 111f.). The inconsistencies of Bacon's position were already evident to Swift and Blake in the eighteenth century (chap. 4, sec. C below).

[236]Walker, *Spiritual and Demonic Magic*, p. 201; Deason, *Philosophy*, pp. 109f., 115. Robert M. Schuler has shown that there were isolated cases in which orthodox English Calvinists of the sixteenth and early seventeenth centuries associated alchemical purification with spiritual regeneration; 'Some Spiritual Alchemies of Seventeenth-Century England', *JHI*, vol. xli (April 1980), 303-8.

[237]*De Dignitate et Augmentis Scientarum* (1623; *Works* 4:367, 379f.); Stanton J. Linden, 'Francis Bacon and Alchemy', *JHI*, vol. xxxv (Oct. 1974), pp. 549, 555f.; Graham Rees, 'Francis Bacon's Semi-Paracelsian Cosmology', *Ambix*, vol. xxii (July 1975), pp. 97f. Bacon never mentioned Fludd by name, but he seems to have rejected many of Fludd's ideas; William H. Huffman, *Robert Fludd and the End of the Renaissance* (London: RKP, 1988), pp. 171ff.

Copernicus and the possibility of an infinite universe. Some tried to eliminate the Aristotelian dualism of heaven and earth and place all things under the direct sovereignty of God, but they failed to alter the late medieval tendency to view the cosmos as a unity in increasingly mechanical terms.

Moreover, the Reformers introduced a new dichotomy into the sphere of God's ordinances (*potentia Dei ordinata*): the clear differentiation between creation- and salvation-ordinances. Calvin tried to bridge the two with his doctrine of particular providence, but in order to give concrete examples he had to find gaps in the order of natural causation so as to make room for the direct action of God in everyday life. As belief in the existence of such gaps declined in the seventeenth and eighteenth centuries, the locus of God's immediate influence was gradually to become restricted to inward experience.

Finally, in their concern for communal solidarity and social justice, the Reformers made insufficient allowance for individual genius, and they took an unduly negative stance towards the occult arts and sciences that had contributed so much to the creativity of the Renaissance. The long-term effect of their strictures may have aided the shift of interest to more exact sciences like chemistry. But the increasingly mechanistic science of the seventeenth century and the Romantic revival of the late eighteenth century must also be seen as partial consequences of their strictures against hermetic and cabalistic thought.

For all that, the Reformation provided scientists of northern Europe with a worldview that encouraged the development of alternatives to both Aristotelian naturalism and Neoplatonic Hermeticism. Whereas late medieval philosophers had only Aristotle to work with, and Renaissance philosophers had to choose between Aristotle and Plato, Francis Bacon had three dialogue partners: Aristotelian naturalism, Neoplatonic Hermeticism, and the Protestant critique of both. The progress of the sciences in the seventeenth century benefitted from all three of these components in a way that it could not have done if there had been no Reformation in the sixteenth century.

C. The Seventeenth Century: Spiritualist, Mechanist, and Platonic Traditions through Leibniz, Boyle, and Newton

The seventeenth century was ushered in by the generation of Galileo, Kepler, and Bacon; it culminated in the work of Boyle, Newton, and Leibniz. By the beginning of the eighteenth century, the fundamentals of modern physics had been established, and the 'scientific revolution' was well on its way.

Several general characteristics of the seventeenth-century development should be noted. A major feature of the period was the shift of the focus of activity in the sciences to the northwestern part of Europe, particularly France, Germany, the Low Countries (Belgium and the Netherlands), and England. In religion there was a gradual shift of emphasis after the Thirty Years' War (1618-48) from a heightening of confessional differences (among Catholic, Lutheran, and Reformed) to a common quest for tolerance and political stability. This was particularly true in England, where much of our story takes place. Christians saw themselves either as loyal to the established church of their nation or as nonconformists. They no longer defined themselves along the explicitly confessional lines laid down in the sixteenth century.[238] Their chief concerns were to avoid the extremes of fanatical enthusiasm, on the one hand, and atheistic materialism, on the other. In so far as various scientific ideas were associated with either of these extremes, common theological beliefs were activated that helped articulate and legitimate alternative scientific ideas.

In terms of the major themes of the creationist tradition, the theology of the seventeenth century was continuous with that of the late medieval and Reformation periods. There was very little that was new. Consequently, in this section our study will focus on the ways in which the motifs of the creationist tradition were put to use. Many of the fundamental ideas of modern science evolved out of concepts that had formerly been largely theological or metaphysical.[239]

However, the ways in which various scientists used creationist themes differed dramatically. Historically it is no longer possible for us to treat the creationist tradition as a unified whole, or even as a spectrum with different wings (as in the Middle Ages). Instead, it must be subdivided into derivative, more specialised traditions that define themselves over against each other as much as in terms of traditional themes. For convenience, we shall group the major contributors to the period into three broad families—the spiritualist, the mechanist, and the Platonic—and treat them separately, while pointing out ways in which various members of these traditions developed their own individual syntheses.[240]

---

[238]G.R. Cragg, *The Church and the Age of Reason* (Harmondsworth: Penguin Books, 1969), pp. 9-15.

[239]Among the pioneering studies that broke the old positivist notion of the origins of modern science were E.A. Burtt, *The Metaphysical Foundations of Modern Physical Science* (London, 1925); Max Jammer, *Concepts of Space* (Cambridge, Mass., 1954), *Concepts of Force* (1957), and *Concepts of Mass* (1961); Alexandre Koryé, *From the Closed World to the Infinite Universe* (Baltimore, Md., 1957).

[240]The historiographic ideal of balancing common intellectual resources with individual appropriations is described by Michael Hunter, *Science and the Shape of Orthodoxy: Intellectual Change in Late Seventeenth-Century Britain* (Woodbridge: Boydell Press, 1995), pp. 13-18. As Hunter argues, historical families of ideas should be regarded as ideal types that were recognised by all parties, even though few individuals fit them exactly.

Aristotelian natural philosophy continued to command some respect in the seventeenth century, but it gradually fell out of favour. The place of Aristotelianism as the dominant paradigm of science was taken over by the mechanical philosophy, associated primarily with the names of Descartes, Gassendi, and Boyle. In competition with the mechanical philosophy, as with Aristotelian tradition before it, was the spiritualist approach to nature, based on the ancient tradition of Hermeticism and the work of Paracelsus and Andreae. In the seventeenth century the spiritualist tradition was associated primarily with the name of Helmont, but it merged with an ongoing quest for an encyclopedic or unified science known as 'pansophism':[241] we shall treat the two together as one continuous tradition. The roots of pansophism went back as far as the encyclopedic art of Raymond Lull in the late thirteenth and early fourteenth centuries.[242] In the early modern period, it was particularly associated with the names of Comenius, Hartlib, and Leibniz, all of whom had ties with seventeenth-century spiritualism.

Finally, steering a middle course between the mechanical and spiritualist traditions in British natural philosophy was a Platonist tradition that flourished at Cambridge University and culminated in the work of Isaac Newton. In the early eighteenth century, Newtonianism was to become the dominant paradigm of physical science, supplanting and partly assimilating the mechanist tradition as the mechanist tradition had supplanted the Aristotelian tradition before it.

The differences among these three traditions can best be seen by considering one of the major concerns of the seventeenth century—whether theology and science should be regarded as separate disciplines or whether they should be integrated to some extent into a comprehensive wisdom. Put in another way, should the immediate action of spiritual agents (human or divine) be categorised as separate from the action of matter, or should the two be integrated in some way? The three traditions we shall examine were

---

[241]The term, 'pansophism', was borrowed by Comenius from Peter Laurenberg's *Pansophia, sive paedia philosophica* (1633); Manuel and Manuel, *Utopian Thought*, p. 207. An alternative term—mostly used by critics of spiritualism, Hermeticism, and alchemy—was 'theosophy': so Kepler in his critique of Fludd; Henry More of Agrippa, Paracelsus, Fludd, and Boehme; and Diderot's *Encyclopedia* of Paracelsus, Weigel, Croll, Fludd, Helmont, and Khunrath; Brian P. Copenhaver, 'Natural Magic, Hermeticism, and Occultism in Early Modern Science', in David C. Lindberg and Robert S. Westman, eds., *Reappraisals of the Scientific Revolution* (Cambridge: Cambridge Univ. Press, 1990), pp. 284f. Copenhaver's preference for this term may reflect his own effort clearly to demarcate Hermeticism from natural magic in relation to the scientific revolution. However, the practical impossibility of making such a clear demarcation is brought out by the overlap of philosophical and popular Hermeticism Copenhaver describes on p. 277 and in note 45.

[242]Manuel and Manuel, *Utopian Thought*, pp. 207f. The Lullian art was presented as a 'pansophist art of discourse' in northern Europe by Agrippa of Nettesheim in his early *Commentary on the Ars Brevis of Raymond Lull* (not printed until 1531); Anthony Bonner, ed., *Selected Works of Ramon Lull* (Princeton, N.J.: Princeton Univ. Press, 1985), p. 81.

divided on this issue, with the mechanists generally opting for a clear separation, the spiritualist-pansophists preferring a degree of integration, and the English Platonist tradition trying to steer a middle course.

The choice made regarding the relation of matter and spirit had clear implications for one's view of matter and hence for science itself. Mechanical philosophers like Descartes and Boyle regarded matter as entirely passive, whereas spiritualists regarded matter as inherently active due to the immanence of individual spirits or energies. The English Platonists and Newtonians saw matter as passive in itself but activated by spiritual or supramechanical principles which governed groups of bodies.

As the protagonists perceived these issues, there were real dangers on either side. A clear separation of matter and spirit could lead to a view of matter as seemingly autonomous and make the spiritual redundant as far as science was concerned. On the other hand, any inclusion of spirit in natural philosophy could lead to a naturalistic explanation of the spiritual and encourage pantheism or even outright atheism. Or, in an attempt to avoid the implication of atheism, either side could produce an appeal to a 'God of the gaps', that is, a concentration of God's activity in those phenomena for which naturalistic explanations were lacking.

While there was little that was new theologically in the seventeenth century, there was ample concern for the viability of the inherited themes of the creationist tradition. In so far as pious Christian natural philosophers were unable to resolve these issues, many eighteenth-century philosophers abandoned the creationist tradition as an interpretive scheme for their life and work.

So examining the controversies of the seventeenth century is more than an academic exercise. One cannot help but be impressed with the degree to which theology was a vital factor in scientific development throughout this phase of the scientific revolution. One must also ask why the theological tradition that inspired such accomplishments had become so shaky that it rapidly lost its credibility in the following century.

1 The Spiritualist/Pansophist Tradition from Helmont to Leibniz

1.1 Helmont

The most important figure in the development of early seventeenth-century chemistry was the Fleming, Joan Baptista van Helmont (1579-1644). Helmont was a Paracelsian. Though he earned a medical degree from the University of Louvain in 1599, he rejected the standard Galenic medicine of the time and found new direction in the unorthodox writings of the sixteenth-century alchemist Paracelsus.[243] But he avoided some of the more speculative notions of Paracelsian alchemy, like the treatment of human nature as a microcosm with detailed correspondences to the structure of the stellar realm.[244] Helmont's positive contributions to chemistry include the discovery of carbon dioxide, the development of the idea of a specific 'gas' corresponding to each chemical substance, and an early formulation of the law of the conservation of matter.

Many of Helmont's major contributions were motivated by theological ideas embedded in the creationist tradition. One of these ideas was the ancient notion of seminal principles or 'ferments' implanted by God in the primordial waters of creation. Augustine had used this idea, and Paracelsus had developed it into the notion of an *archeus* or formative spirit in each organic entity (animal, vegetable, or, in alchemical thought, even mineral). The *archeus* was present in the seed of each organism, ruled its growth, and survived its death in a gaseous discharge.

Helmont developed the idea of seminal principles in empirical terms by postulating a 'spirit' or 'gas' as the smoke given off when a chemical is burned. Each chemical substance had its own distinctive 'gas' that could be produced and identified; e.g., *gas carbonum* (what we call carbon monoxide) and *gas sylvester* (carbon dioxide and nitrous oxide). The original substance could be reconstituted experimentally, according to Helmont, if its gas were recombined with water. Water vapour was not a 'gas', however, since water was the raw material used by the Spirit in forming the creation (Gen. 1:2).[245] The value of this philosophy was its recognition of an irreducibly unique character in each chemical substance and each organism. Its weakness

---

[243] *Oriatrike or physicke refined* (John Chandler, trans., London, 1662), p. 7; Walter Pagel, *Joan Baptista van Helmont* (Cambridge: Cambridge Univ. Press, 1982), pp. 1-7.

[244] Allen G. Debus, 'Mathematics and Nature in the Chemical Texts of the Renaissance', *Ambix*, vol. xv (Feb. 1968), p. 22; Pagel, *Helmont*, pp. 206f.; Brian Vickers, 'Analogy vs. Identity: The Rejection of Occult Symbolism, 1580-1680', in Vickers, ed., *Occult and Scientific Mentalities in the Renaissance* (Cambridge: Cambridge Univ. Press, 1984), pp. 144-49.

[245] Walter Pagel, 'The Debt of Science and Medicine to a Devout Belief in God', *Journal of the Transations of the Victoria Institute*, vol. lxxiv (1942), pp. 105-8; Stephen Toulmin and June Goodfield, *The Architecture of Matter* (Harmondsworth: Penguin Books, 1965), pp. 168ff.; Eugene M. Klaaren, *Religious Origins of Modern Science* (Grand Rapids, Mich.: Eerdmans, 1977), pp. 63-71.

Walter Pagel gives a detailed chart of the relationships among Helmont's various concepts in *The Religious and Philosophical Aspects of van Helmont's Science and Medicine*, Supplements to *BHM*, No. 2 (Baltimore: Johns Hopkins Press, 1944), p. 21. According to Pagel, the *archeus* is both matter (water) and form (like Aristotle's concept of substance). The *archeus* and the ferment and life given by God together constitute the semen or seed planted at creation. The 'gas' or spirit is somehow contained in the *archeus* but is distinct from the material component of water. I doubt that the exact relationships among Helmont's concepts can be reconstructed from his writings.

was its inability to show that all material beings, even living ones, have some fundamental constituents (later called chemical elements) in common.

The remarkable continuity of the creationist tradition to this point is illustrated by Helmont's famous willow-tree experiment, designed to show the primacy of water as the substratum of all living things. Helmont's experimental procedure was basically the same as that discussed by Nicholas of Cusa in the fifteenth century and described in the Pseudo-Clementine *Recognitions* as far back as the fourth century (chap. 1, sec. D).[246]

In addition to such minor motifs, the major themes of the creationist tradition also appear clearly in the work of Helmont. Underlying the idea of seminal principles, for instance, is the theme that God determined the properties and laws of all things through the same decree by which he created them, constituting what we have termed the 'relative autonomy' of nature (chap. 1, sec. D). What people call 'nature' is just the effect of that decree. As Helmont confessed in his *Oriatrike or Physicke Refined*:[247]

> I believe that Nature is the command of God, whereby a thing is that which it is, and doth that which it is commanded to do or act. This is a Christian definition [as opposed to an Aristotelian one] taken out of the Holy Scripture.

On the basis of God's creative decree, all beings have the power to operate in accordance with their own properties and laws:[248]

> For that most glorious Mover [God's Spirit] hath given powers to things, whereby they of themselves and by an absolute force may move themselves or other things.

A characteristic feature of the spiritualist tradition from Helmont to Leibniz was a correspondence between the decree of God and the relatively autonomous work of creatures.

On the more practical side of the creationist tradition, Helmont was strongly motivated by the ideal of the Christian physician who received his calling and wisdom directly from God (Ecclus. 38) and dispensed medicines without charge to the poor (Matt. 10:8)[249]—what we have termed the 'ministry of healing and restoration' (chap. 1, sec. E). The image of God, he argued, was found not in human reason (as the scholastics held), but in charity and humility. This theme of science as a form of charity will appear again in Comenius and Hartlib.

---

[246] Herbert M. Howe, 'A Root of Van Helmont's Tree', *Isis*, vol. lvi (Winter 1965), pp. 408-419.
[247] *Oriatrike* VII.2f. (Klaaren, *Origins*, p. 62).
[248] *Oriatrike* XXIV.1 (*ibid.*, p. 66).
[249] Debus, *The Chemical Philosophy* (New York: Science History Publications, 1977), 2:358; Pagel, *Van Helmont*, pp. 6f.

As an advocate of the Christian ideal of charity, Helmont clearly distinguished between Aristotelian 'reason' and empathetic understanding. Reason, according to Helmont, attempts to dominate the thing known and discerns only multiplicity. It originated with the fall of Adam and tended towards death as Helmont's experience with the guardians of scholasticism amply demonstrated. Understanding, on the other hand, attends to the unique nature of a thing and discerns its vital principle and unity. Understanding is a gift from God tending towards life and depends continually on divine grace. Characteristically, Helmont saw the physician, not the priest, as the one called by God to exercise this healing ministry.[250] Through a reform of the medical profession, the effects of the Fall could be reversed, and the original perfection of things could be restored.[251] This utopian theme is also characteristic of the spiritualist tradition in the seventeenth century.

The wide influence of Helmont is due largely to the efforts of his son. Francis Mercury van Helmont (1614-99) was responsible for the publication of his father's works after the latter's death in 1644, and he disseminated his father's ideas through his contacts and travels in England and Germany. As a close friend of two of the major figures of the pansophist tradition, Hartlib and Leibniz, Francis van Helmont encouraged various programmes aimed at the construction of a unified science and the reunion of the Protestant churches.[252]

The eventual decline of Helmontian and Paracelsian ideas in England toward the end of the century was bound up with the political associations of spiritualist ideas that were formed during the Civil War (1642-49) and the Interregnum (between Charles I and Charles II, 1649-60). The ideas of the German spiritualist, Jacob Boehme (1575-1624), were championed by various Seekers, Familists, and Quakers.[253] John Webster, an Independent who attacked the universities of Oxford and Cambridge for neglecting the ideas of Paracelsus, Helmont, and Boehme (1654), was suspected of

---

[250]Pagel, *Religious and Philosophical Aspects*, pp. 9ff.; Allen G. Debus, *Man and Nature in the Renaissance* (Cambridge: Cambridge Univ. Press, 1978), p. 126.

[251]Klaaren, *Origins*, pp. 77-80.

[252]R.W. Meyer, *Leibnitz and the Seventeenth-Century Revolution* (Cambridge: Bowes & Bowes, 1952), pp. 63f.

[253]Rufus M. Jones, *Spiritual Reformers in the 16th and 17th Centuries* (Boston: Beacon Press, 1959), chap. 12; Keith Thomas, *Religion*, pp. 375f.; Christopher Hill, *The World Turned Upside Down* (New York: Viking Press, 1972), pp. 141, 154; Nigel Smith, *Perfection Proclaimed* (Oxford: Oxford Univ. Press, 1989), chap. 5.

Paradoxically, the translator who first introduced Boehme's works to England (John Sparrow, trans., *XL Questions Concerning the Soule*, 1647) did so in order to provide an antidote to religious radicalism. And the first recorded public espousal of Boehme's ideas occurred at commencement exercises at Cambridge (Charles Hotham, *Ad philosophiam Teutonicam manuductio*, 1646, published in 1648). But sectarians immediately took up his ideas for their own purposes; J. Andrew Mendelsohn, 'Alchemy and Politics in England, 1649-1665', *Past & Present* 135 (May 1992), pp. 34ff.

Familist and Leveller sympathies.[254] Francis Mercury van Helmont, himself, was briefly asociated with the dreaded Quakers during his visit to England in 1670-78.[255]

These nonconformists valued the spiritualist tradition for its universalist and democratic implications for church and state. In the mind of the moderate Puritan and Anglican public, on the other hand, the sectarian groups were associated with anarchy and materialistic pantheism. Richard Baxter, for instance, attributed the willingness of some sectaries to shed blood to the disruptive influence of Paracelsian ideas.[256] One should not associate spiritualism exclusively with the English sects, however.[257] Prominent Helmontians like Nicholas Le Fèvre and Thomas Shirley served as personal physicians to Charles II after the Restoration of 1660.[258] Elias

---

[254]*Academiarum examen* (1654); cf. P.M. Rattansi, 'Paracelsus and the Puritan Revolution', *Ambix*, vol. xi (Feb. 1963), pp. 28f.; Hill; *World*, pp. 66ff.; Yates, *Rosicrucian Enlightenment*, pp. 185f.; Debus, *Man and Nature*, pp. 135f.

[255]Meyer, *Leibnitz*, p. 63; Carolyn Merchant, 'The Vitalism of Francis Mercury Van Helmont: His Influence on Leibniz', *Ambix*, vol. xxvi (Nov. 1979), p. 171. But Mendelsohn reminds us that F.M. Helmont's ideas were strongly opposed by George Fox; see Alison Coudert, 'A Quaker-Kabalist Controversy: George Fox's Reaction to Francis Mercury van Helmont', *Journal of the Warburg and Courtauld Institutes*, vol. xxxix (1976), pp. 171-89; Mendelsohn, 'Alchemy and Politics', p. 70.

It appears that Paracelsian medicine was also practised among the Ranters and their sympathisers. In fact, John Chandler, who translated the works of Joan Baptista van Helmont into English (1662), was a Ranter convert to Quakerism. Albert Otto Faber, a German iatrochemist who was invited to England by Charles II in 1660/1, joined the Quakers, was arrested and imprisoned for attending their meetings and propagating their teachings in 1664, and was deported in 1667; Peter Elmer, 'Medicine, Science, and the Quakers: The 'Puritanism-Science' Debate Reconsidered', *Journal of the Friends Historical Society*, vol. liv (1981), pp. 270-73, 278; idem, 'Medicine, Religion, and the Puritan Revolution', in Roger French and Andrew Wear, eds. *The Medical Revolution of the Seventeenth Century* (Cambridge: Cambridge Univ. Press, 1989), pp. 22ff., 38. Mendelsohn counters that Faber later returned to England, served as physician to aristocrats, and attended (the Laudian) Gilbert Sheldon, archbishop of Canterbury, on his deathbed in 1677; Mendelsohn, 'Alchemy and Politics', pp. 65, 76.

[256]*Christian Directory* (written, 1664-65); cf. P.M. Rattansi, 'The Social Interpretation of Science in the Seventeenth Century', in Peter Mathias, ed., *Science and Society, 1600-1900* (Cambridge: Cambridge Univ. Press, 1972), p. 25; David Kubrin, 'Newton's Inside Out!', in Harry Woolf, ed., *The Analytic Spirit* (Ithaca, N.Y.: Cornell Univ. Press, 1981), pp. 103f. Baxter lumped 'Behmenists, Paracelsians, and all Enthusiasts' together as radicals already in his *One Sheet against the Quakers*, 1657; Mendelsohn, 'Alchemy and Politics', p. 36.

Earlier associations of spiritualist alchemy and anarchist forces were forged by Ben Jonson's *The Alchemist* (1610); cf. Mebane, *Renaissance Magic*, 137-42. On the other hand, Johnson made use of alchemical imagery himself in a masque at the court of James I; cf. Mendelsohn, 'Alchemy and Politics', pp. 39 notes 45, 52.

[257]The strength of the association between alchemy and radical politics in the mid-seventeenth-century English mind has been questioned by Mendelsohn, 'Alchemy and Politics', pp. 30-78.

[258]Klaaren, *Origins*, p. 80; Toulmin and Goodfield, *Architecture*, p. 171. According to Hartlib, 'Some whisper that the K[ing] should be a Teutonicus [a follower of Boehme] and lover of Chymistry'; Hartlib to John Worthington, 4 June 1660; quoted in Mendelsohn, 'Alchemy and Politics', p. 59.

Ashmole, another alchemist sympathetic to the ideas of Helmont, was a founding member of the Royal Society of London (1660).[259]

1.2 Comenius and the Hartlib Circle

However, a more socially acceptable form of spiritualism, commonly referred to as 'pansophism', was introduced into England by Comenius and Hartlib. Johann Amos Comenius (1592-1670) was a member of the Bohemian Brethren and served in his later years as the last bishop the Brethren had as an independent church. He studied at the Reformed University at Herborn, where he was influenced by the encyclopedic, Lullist science of Johann Heinrich Alsted, and at Heidelberg, where he was impressed by the efforts of David Paraeus towards reunion between the Lutheran and Reformed churches. Through his reading, he also imbibed the utopian visions of Johann Andreae and Francis Bacon, discussed in the previous section.[260]

One of Comenius's most famous works, *The Labyrinth of the World* (written in 1623, published in Czech in 1631), looked forward to a 'reformation of the whole world' based on the restoration of human science to its pristine, prelapsarian state.[261] Though Comenius concluded this work on a pessimistic note, reflecting the reverses suffered by his people during the Thirty Years' War,[262] his proposals for a universal Christian science or 'pansophy' were well received and published by Samuel Hartlib in London in 1637-42.[263] With the approval of the Long Parliament,[264] Comenius was invited to visit to England to help Hartlib found a research institute,

---

[259]Yates, *Rosicrucian Enlightenment*, p. 198; Webster, *Paracelsus*, p. 64. Ashmole published his *Fasciculus chemicus* in 1650. Among those to whom he gave a copy of his book and generous financial support was the Familist leader, John Pordage; Mendelsohn, 'Alchemy and Politics', pp. 39f. note 46.

[260]Yates, *Rosicrucian Enlightenment*, pp. 156f., 179; Hans Aarsleff, 'Comenius', *DSB* 3:359; Bonner, ed., *Selected Works of Ramon Lull*, p. 83.

[261]*The Labyrinth of the World and the Paradise of the Heart*, trans. Matthew Spinka (Chicago: National Union of Czechoslovak Protestants in America, 1942) chap. 13, p. 56; Yates, *Rosicrucian Enlightenment*, pp. 161f.

[262]Harold J. Grimm, *The Reformation Era, 1500-1650* (New York: Macmillan, 1954, 1973), pp. 418f.

[263]Meyer, *Leibnitz*, p. 65; H.R. Trevor-Roper, *Religion, the Reformation and Social Change* (London: Macmillan, 1967), pp. 277f., 286; James R. Jacob, *Robert Boyle and the English Revolution* (New York: Burt Franklin & Co., 1977), p. 30; Aarsleff, 'Comenius', p. 360a; Manuel and Manuel, *Utopian Thought*, pp. 325f.

[264]Parliament was challenged to extend the invitation in a sermon by John Gauden, 'The Love of Truth and Peace' (1640), probably on the instructions of John Pym, a friend and fervent supporter of Hartlib. Comenius (and John Dury) were hailed, not as scientists, however, but as 'two great and public spirits who have laboured much for truth and peace'; Trevor-Roper, *Religion*, pp. 257, 261f.

modelled on Bacon's projected 'House of Solomon'.[265] During his stay in London (1641-42), Comenius wrote his classic, *The Way of Light* (not published until 1668), a pansophist manifesto in which he anticipated an era of unprecedented progress in the sciences and in their beneficial impact on society.[266]

Throughout Comenius's work there is an emphasis on programmes for public education and international cooperation that would involve the populations of all nations in the labours and fruits of the scientific enterprise.[267] One recent biographer has aptly summarised Comenius's programme as 'a prescription for salvation through knowledge raised to the level of universal wisdom, or pansophy, supported by a corresponding program of education'.[268]

Samuel Hartlib (*c.* 1600-1662) was an expatriate from Polish Prussia who settled in London (1628) after his homeland was occupied by imperial armies during the Thirty Years' War.[269] Like Comenius, he was inspired by the utopian visions of Andreae and Bacon; he had two of Andreae's works translated into English and published.[270] As we have seen, he was responsible for the publication of Comenius's pansophic manifesto in 1637 and for the latter's visit to London in 1641. He also published the utopian work of his colleague Gabriel Plattes, *A Description of the Famous Kingdome of Macaria* (1641). The latter was addressed as an appeal to the Long Parliament for the promotion of scientific research, universal education, and socialised medicine which would 'lay the corner stone of the world's happiness'.[271]

Though Hartlib received no financial support from Parliament, he used what private funding he had to subsidise the work of promising alchemical adepts like Thomas Vaughan, George Starkey, Frederick Clodius, and

---

[265] As Charles Webster has pointed out, Hartlib was even more impressed by Bacon's proposal for a college 'Dedicated to Free and Universal studies of Arts and Sciences', which was addressed to James I in his *De augmentis scientarum* (1623; Book II, Intro.); *Samuel Hartlib and the Advancement of Learning* (Cambridge: Cambridge Univ. Press, 1970), pp. 31f.

[266] Yates, *Rosicrucian Enlightenment*, pp. 177f. When Comenius published the *Via lucis* in 1668, he dedicated it to the newly organised Royal Society of London in order to support his claim to be its true, spiritual founder; Trevor-Roper, *Religion*, p. 289.

[267] Yates, *Rosicrucian Enlightenment*, p. 179; Jacob, *Boyle*, p. 35.

[268] Aarsleff, 'Comenius', p. 361.

[269] Yates, *Rosicrucian Enlightenment*, pp. 175f.; cf. Grimm, *Reformation*, p. 420.

[270] Yates, *Rosicrucian Enlightenment*, pp. 155, 180f.; Jacob, *Boyle*, p. 36. In his *Ephemerides* of 1653, Hartlib stated that Andreae's *Christianopolis* and Campanella's *City of the Sun* were more Christian than Bacon's *New Atlantis*; P.M. Rattansi, 'The Intellectual Origins of the Royal Society', *Notes and Records of the Royal Society of London*, vol. xxiii (1968), note 30. This dissonant note in Hartlib's Baconianism should be explored.

[271] Charles Webster, 'English Medical Reformers of the Puritan Revolution', *Ambix*, vol. xiv (Feb. 1967), pp. 21ff.; Yates, *Rosicrucian Enlightenment*, p. 177.

Robert Boyle. Starkey is noteworthy for his defense of the medical techniques of Paracelsus and Helmont against the criticisms of the Galenic College of Physicians and for his refusal to desert the poor of London during the plague of 1665.[272]

Like Comenius, Hartlib stressed the need for communication and cooperation in scientific research. Through his 'Office of Address' for the exchange of letters, he promoted the communication of useful discoveries among scientists of Europe and America for the benefit of the general public. An important result was the requirement that alchemical ideas be discussed in a common language that was comprehensible to all.[273] In the context of the late seventeenth-century reaction to spiritualist groups in general, this discipline imposed on alchemical work was necessary for the viability of the emerging science of chemistry.

The criterion of communicability and social utility was deeply embedded in the creationist tradition and had been applied to alchemical studies in both Lutheran and Reformed traditions in the late sixteenth century. One of Hartlib's closest associates, John Dury, updated the traditional theme in an apology for the 'Office of Address' in terms of the Christian duty to share all knowledge, both religious and scientific.[274] Robert Boyle (c. 1647) also called for the open communication of medical knowledge which God intended 'for the good of all Mankind'. Like Dury he appealed to the Christian duty of charity citing 'our Saviour's prescription...*Freely ye have received, freely give*' (Matt. 10:8 AV).[275] To a degree this humanitarian ideal lived on in the writings of Dury's son-in-law, Henry Oldenburg (c. 1618-77), who served as corresponding secretary of the Royal Society of London after 1662.[276]

---

[272] P.M. Rattansi, 'The Helmontian-Galenist Controversy in Restoration England', *Ambix*, vol. xii (Feb. 1964), pp. 19ff.

[273] Dobbs, *Foundations*, pp. 64, 69.

[274] Charles Webster, *The Great Instauration* (London: Duckworth, 1975; New York: Holmes & Meier, 1976), p. 508; cf. Dobbs, *Foundations*, p. 70, on the similar views of Plattes.

[275] 'An Epistolical Discourse of Philaretus to Empericus...inviting all the true lovers of Vertue and Mankind, to a free and generous communication of their Secrets and Receits in Physick', in Hartlib ed., *Chymical, Medicinal and Chyrugical Addresses*, 1655); cf. Webster, *Samuel Hartlib*, p. 40; Dobbs, *Foundations*, pp. 68f.; Jacob, *Robert Boyle*, pp. 32ff.

The deep impression members of the Hartlib circle made on Boyle is evidenced in his description of them as practising 'so extensive a charity, that it reaches unto everything called man' and taking 'the whole body of mankind for their care'; Letter to Francis Tallents, 20 Feb. 1646/47; *Works* 4:xxxiv-xxxv; Rattansi, 'Intellectual Origins', p. 134.

[276] 'All for the glory of God, for the honor and advantage of this kingdom and the universal good of mankind' (closing words of Oldenburg's preface to the first issue of the *Philosophical Transactions*, 1660); Rossi, *Philosophy, Technology, and the Arts*, p. 97. Cf. Webster, *Great Instauration*, pp. 501f. On Oldenburg's spiritualist and millenarian sympathies, see *ibid.* and James R. Jacob, 'Boyle's Circle in the Protectorate', *JHI*, vol. xxxviii (Jan. 1977), p. 136.

Hartlib's utopian and pansophist associations did not prevent him from valuing the developing mechanical philosophy. In fact, he was responsible for introducing Robert Boyle to Gassendi's atomism in the 1640s.[277] As we shall see later, Boyle's natural philosophy developed along more mechanist lines in the 1650s, and he did not remain in the spiritualist or pansophist tradition. For a possible successor to the ideals of Helmont and Comenius, we must look back to continental Europe, and forward in time to Leibniz.

1.3 Leibniz

Gottfried Wilhelm Leibniz (1646-1716) is one of the most profound and difficult philosophers in Western history. Like other major thinkers, he does not fit neatly into any tradition; he defines a tradition or philosophy of his own. Leibniz was influenced by the work of Boyle, whom he met in London in 1673, as well as that of other mechanical philosophers like Descartes, Gassendi, and Huyghens. As pointed out earlier, the three traditions we are considering are not separate entities, but overlapping, interacting aspects of one continuous stream of development. After a brief interest in the mechanical philosophy in the late 1660s and early '70s, however, Leibniz rejected its dualism of matter and spirit and developed his own distinctive views.[278] Several features of his thought make sense only when viewed in the context of the spiritualism of Helmont and the pansophism of Comenius. We shall consider these here.

Whereas many nonconformists and political radicals valued the spiritualist tradition for its levelling implications for society, Leibniz valued it for the sense of order and harmony it brought to a world of turmoil and fragmentation. In particular, he stressed the value of reason against the claims of enthusiasts to new revelations and repudiated the followers of Jacob Boehme as fanatics.[279] Some historians have concluded that Leibniz betrayed his spiritualist roots in trying to make the truths of faith palatable to human reason.[280] As we shall see, this is only partly true.

Leibniz was a close friend of Helmont's son, Francis Mercury, whom he met in Mainz in the year 1671: the two had common interests in alchemy, cabalistic number theory, the development of a universal science, the integration of faith and knowledge, and the reunion of the Protestants in a

---

[277] Dobbs, *Foundations*, p. 63.

[278] Leroy E. Loemker, 'Boyle and Leibniz', *JHI*, vol. xvi (Jan. 1955), pp. 23f., 31f., 35, 38.

[279] *New Essays on Human Understanding* (c. 1704) XIX.1-5 (Meyer, *Leibnitz*, pp. 69, 73f.).

[280] Meyer, *Leibnitz*, pp. 75f., 152.

universal church.281 Though precise lines of influence are difficult to trace, there are enough points of contact between the two Helmonts and Leibniz to suggest that Leibniz's peculiar notion of active principles in nature was partly rooted in the Helmontian idea of vital principles implanted in matter by God.282

In his mature philosophy, Leibniz called these active principles 'monads', and he developed a highly original metaphysic based on the orchestration of the monads through a harmony preestablished by God. Of more immediate significance for the history of physical science was the earlier version of Leibniz's active principles, his notion of *vis viva,* which was a vital (or motive) force that inheres in a moving body. In the case of a falling body, it is the product of the weight and the height through which the body has fallen.283 In modern terms it is the kinetic energy of the moving body. Leibniz's *vis viva* is similar to the late medieval notion of 'impetus' and conveys the same sense of a vitality intrinsic to the nature of an object as created and set in motion by God. Leibniz stressed the idea of vitality in contrast to the notion of mere inertia that had developed through the work of Descartes and Newton.284

One of the most important issues of natural philosophy in the seventeenth and eighteenth centuries was what happened to the quantity of motion in the event of a collision between two bodies. John Wallis (1668) showed that the sum of the products of the mass and velocity (i.e., the momenta) of the bodies, in any given direction, was conserved. Christiaan Huyghens (1669) showed that the sum of the products of the mass and the square of the velocity was conserved in elastic collisions.285 But what

---

[281] Meyer, *Leibnitz,* pp. 61-64; Merchant, 'Vitalism', p. 171. For a chronology of contacts between Leibniz and F.M. van Helmont, see Allison P. Coudert, *Leibniz and the Kabbalah* (Dordrecht: Kluwer, 1995), pp. xiii-xvii.

[282] Ludwig Stein, *Leibniz und Spinoza* (Berlin, 1890), pp. 209-213; Walter Pagel, 'The Religious and Philosophical Aspects of van Helmont's Science and Medicine', *BHM,* suppl. no. 2 (1944), pp. 1-43; Meyer, *Leibnitz,* p. 64; Mason, *History,* pp. 353f.; Toulmin, *Architecture,* pp. 172f.; Pagel, *Paracelsus,* pp. 36, 108; Anne Becco, 'Leibniz et Francois-Mercure Van Helmont: bagatelle pour des monades', *Studia Leibnitiana,* vol. vii (1975), pp. 119-41; Merchant, 'Vitalism', pp. 170ff., 180. Margaret Alic argues for the influence of Lady Anne Conway's organismic philosophy, recorded in her notebooks of 1671-75 and edited and published (without Conway's name) by F.M. van Helmont in Latin translation in 1690 (English translation published in 1692); *Hypatia's Heritage* (Boston: Beacon Press), pp. 4-8.

[283] Richard S. Westfall, *The Construction of Modern Science* (Cambridge: Cambridge Univ. Press, 1977), pp. 135ff.; C.D. Broad, *Leibniz: An Introduction* (Cambridge: Cambridge Univ. Press, 1975), p. 65.

[284] *Specimen dynamicum* (1695; Gary C. Hatfield, 'Force (God) in Descartes' Physics', *SHPS,* vol. x [1979], p. 139); *De ipsa natura* (1698; Leroy E. Loemker, *Struggle for Synthesis* [Cambridge, Mass.: Harvard Univ., Press, 1972], p. 210).

[285] Wallis, 'A Summary Account of the General Laws of Motion', *Philosophical Transactions of the Royal Society of London* (1668); Huyghens, 'Sur les regles du mouvement dans la rencontre des corps', *Journal des scavants* (1669); cf. Zilsel, 'Physical Law', p. 272.

happened to the latter sum (equivalent to Leibniz's *vis viva*) in inelastic collisions, e.g., when the bodies stuck together? Clearly it was not conserved in any outwardly visible way.

On the basis of his belief in the inherently active character of matter and without experimental evidence, Leibniz postulated that the quantity of *vis viva* was, in fact, conserved, even in inelastic collisions. Somehow it was absorbed into the dynamics of the constituent particles of the colliding bodies. Later, in the nineteenth century, it was discovered that the lost mechanical energy was converted into heat, by then understood as another form of energy. Thus Leibniz anticipated the law of the conservation of energy, one of the basic conservation principles in classical modern physics.[286]

The inherently dynamic nature of matter in Leibniz's natural philosophy was an important consideration in his dispute with the Newtonians in the early eighteenth century. Both Newton and Leibniz were concerned to see nature as the product of the activity of God, but in differing ways. Whereas Newton and his disciples saw the activity of God in his use of supramechanical principles and repeated intervention in the activity of matter, Leibniz found it in the operation of the original divine decree by which matter was invested with an energy that would continue indefinitely and undiminished in quantity. As he wrote in 1715:[287]

> ...the same force and vigour remains always in the world, and only passes from one part of matter to another, agreeably to the laws of nature, and the beautiful pre-established order....Whoever thinks otherwise, must needs have a very mean notion of the wisdom and power of God.

Implicit in the Leibniz's account was the supposition that even inanimate creatures exhibit a degree of intelligence. All creatures are the subjects of God, not just in the passive sense of receiving commands, but also in the active sense of executing them flawlessly in concert with one another. God, according to Leibniz, is like a king who not only provides laws, but also educates his subjects and endows them with the capacity to fulfill them.[288] Moreover, the coordinated fulfillment of such decrees was inherently telological and could not be accounted for in strictly mechanistic terms.

We noted earlier that participation in the spiritualist tradition was not incompatible with an interest in the mechanical philosophy. In some

---

[286]George Gale, Jr., 'Leibniz' Dynamical Metaphysics and the Origins of the Vis Viva Controversy', *Systematics*, vol. xi (Dec. 1973), pp. 186f., 201; Carolyn Iltis, 'Leibniz and the Vis Viva Controversy', *Isis*, vol. lxii (1971), pp. 22, 32ff.; J. Mittelstrass and E.J. Aiton, 'Leibniz: Physics, Logic, Metaphysics', *DSB* 8:152f.
[287]Letter to Princess Caroline, Nov. 1715; cf. H.G. Alexander, ed., *The Leibniz-Clarke Correspondence* (Manchester: Manchester Univ. Press, 1956), p. 12.
[288]Answer to Clarke's First Reply (Alexander, ed., pp. 19f.).

respects, Leibniz's concept of nature was more like the mechanical philosophy than the spiritualist one. For instance, he rejected Newton's use of supramechanical forces as an illegitimate appeal to occult qualities.[289] And, whereas the Helmontian idea of the *archeus* was a supramechanical force that governed the behavior of material substance, Leibniz's *vis viva* was conceived as entirely mechanical in the sense that it was a function of matter in motion, though neither passive like inertia nor a merely linear measure of motion like momentum.

In the terms of the creationist tradition, Leibniz thus stressed the relative autonomy of nature almost to the point of suggesting complete independence from its Creator.[290] Even miracles were not fresh interventions by God for Leibniz, but rather instances of laws or reasons of a higher order than those presently known to us but ordained by God from the beginning. Effects which seem to contradict the normal properties of substances must be viewed as resulting from their original God-given essence.[291]

It is important to note, however, that for Leibniz the 'essence' of a thing included its 'union with God himself' as its creator as well as its obedience to the laws of mechanics; it was not autonomous in the sense of being separate from God.[292] As Leibniz put it in his correspondence with the pietist leader, Philip Jacob Spener:[293]

> Thus I hold that, even if individual effects in nature can be explained mechanically, nevertheless, even mechanical principles and their effects, all order and all physical rules in general, arise not from purely material determinations, but from the contemplation [*consideratio*] of indivisible substances, and especially of God. Thus I think I am able to satisfy those prudent and pious thinkers who rightly fear that the philosophy of certain men among the moderns [i.e., the mechanical philosophers] is too material, and that it prejudices against religion.

In the final analysis, Leibniz remained faithful to the spiritualist vision of all things existing in God.

The spiritualist tradition was itself a vital force in seventeenth-century science. Based on the creationist idea of the relative autonomy of nature, it contributed in significant ways to the development of both chemistry and physics, particularly through the work of Helmont and Leibniz. It also

---

[289] Richard S. Westfall, *Never at Rest: A Biography of Isaac Newton* (Cambridge: Cambridge Univ. Press, 1975, 1980), p. 730.

[290] J.E. McGuire, 'Boyle's Conception of Nature', *JHI*, vol. xxxiii (Oct. 1972), pp. 538f.; Francis Oakley, *Omnipotence, Covenant, and Order* (Ithaca, N.Y.: Cornell Univ. Press, 1984), p. 92.

[291] *Discourse on Metaphysics* (1686) 16; cf. G.H.R. Parkinson, ed., *Leibniz: Philosophical Writings* (London: Dent, 1973), pp. 28f.

[292] *Discourse* 7 (Parkinson, ed., p. 245).

[293] Letter of July 1687 (Meyer, *Leibnitz*, pp. 145f.).

contributed through the work of Comenius and Hartlib to the formation of social and moral values of the emerging scientific community, based on the creationist ideal of the ministry of healing and restoration. The great strength of the spiritualist tradition was its ability to generate powerful organising principles like gas, matter, and force (energy) and to give them empirical meaning in quantifiable form. The major weakness of the tradition was its inability to press beyond the uniqueness and independence of natural entities to probe their common constituents and determine their laws of interaction.

2 The Mechanist Tradition from Descartes to Boyle

The principal figure in the early development of the mechanical philosophy was René Descartes (1596-1650). Descartes is so well known as the founder of modern philosophy in general and of the rationalist school in particular that it is easy to lose sight of his roots in scholastic theology. He was trained in the classical liberal arts curriculum at the Jesuit College of La Flèche. One of the standard texts used in the Jesuit schools (as well as in many Protestant universities) of the seventeenth century was the *Metaphysical Disputations* (1597) of Francesco Suarez. Since Suarez was an important formative influence on Descartes, we shall take a brief look at his thought.[294]

2.1 The Scholastic Background: Suarez

After the Council of Trent (1545-63) declared the modified Aristotelianism of Thomas Aquinas normative for all Catholics, a new movement of scholastic theology developed in the Roman Catholic church. Francesco Suarez, who taught at the University of Salamanca in Spain, was one of the most important representatives of this neoscholasticism.

We recall that one of the principal issues in scholastic theology was the relationship between *potentia absoluta* and *potentia ordinata*, the absolute and ordained powers of God. It was generally agreed that the laws of nature were ordained by God as the normal pattern of natural events. The problem was that the idea of *potentia ordinata* could lead to the suggestion that the laws of nature were somehow self-operating or automatic, once established. Late medieval theologians had countered this tendency by stressing *potentia absoluta*, the absolute power of God to choose, suspend, and even alter the laws of nature.

The contribution of Suarez was an interpretation of the laws of nature designed to avoid this dilemma. The laws of nature, he held, were not rules

---

[294]Meyer, *Leibnitz*, p. 186, note 135; Loemker, *Struggle*, p. 20; J.L. Heilbron, *Elements of Early Modern Physics* (Berkeley: Univ. of California Press, 1982), p. 100.

God imposed on nature, but rules God imposed on himself. In contrast to the spiritualist tradition, Suarez viewed inanimate creatures as incapable of either understanding or obeying God's laws. So the laws of nature had to be executed by God himself operating on a material substance that was completely passive.[295] These two themes, an all-active God and a passive matter, are the characteristic theological ideas we find (in differing contexts) in mechanical philosophers like Descartes and Boyle. It is important to see that they were developed as a way of stressing nature's *dependence* on God, even though they were sometimes later taken to imply an *autonomous* view of nature.

2.2 French Mechanical Philosophy: Descartes, Mersenne, and Gassendi

The context in which Descartes developed his ideas was entirely different from that of Suarez. The scholastics had worked within an Aristotelian frame of reference, even though many of them had been critical of various particulars of Aristotle's philosophy. In Aristotle's worldview the earth stood still, the universe was finite, and the planetary spheres were moved by their desire to match the eternity of God. With the adoption of Copernicus's model of the universe, however, the consistent structure of the scholastic framework crumbled, and in its place there emerged a paradox: all things were now in motion, even the earth, but there was no general mechanism for generating this motion comparable to the role of the *primum mobile* (the outermost celestial sphere) in Aristotle. The entire question of the relation of God to matter and motion had to be rethought.

Descartes was the first to offer a consistent system of natural philosophy to replace that of Aristotle. He developed the idea of the passivity of matter in such a way as to stress the immutability as well as the omnipotence of God. Matter was incapable not only of behaving in accord with God's laws, but even of continuing in existence without the continual creative action of God. In other words, there was no real difference for Descartes between the work of God in sustaining the world and his original act of creation; each instant of time was a new creation.[296] In philosophical terminology, Descartes's position is often described as a form of 'occasionalism'.[297]

Since matter was entirely passive for Descartes, like the unformed matter of Aristotle it had no innate qualities, no character of its own. Positively this meant that matter was entirely receptive to the mathematical laws

---

[295]*Metaphysical Disputations* XX.4; XXX.17; *On Laws* I.1; II.2 (Zilsel, 'Physical Law', p. 279; Oakley, *Omnipotence*, pp. 57, 88).

[296]*Meditations* (1641) III; *Principles of Philosophy* (1644) I.21 (E.S. Haldane and G.R.T. Ross, trans., *The Philsophical Works of Descartes*, Cambridge: Cambridge Univ. Press, 1911, 1:168, 227f.).

[297]Crombie, *Augustine*, 2:313ff., 318f.; Hatfield, 'Force', pp. 135f.

imposed on it by God. Hence, Descartes concluded—as Aquinas had done in the thirteenth century—that matter must be quantifiable. It must have geometric extension. But Descartes went further than Aquinas in making geometric extension the very essence of matter[298] and concluding that it must be made up of corpuscles (not necessarily indivisible atoms) with definite sizes, shapes, and speeds.[299] Matter was, therefore, completely different from mind, the latter being unextended and indivisible. This Cartesian dualism of mind and matter was perfectly consistent with the traditional view of the soul as a simple, immaterial substance. But it did seem to contradict the Catholic doctrine of transubstantiation, as Antoine Arnauld and others pointed out to Descartes:[300] if the substance of a body could not be distinguished from its geometric extension as Descartes held, the flesh and blood of Christ could not possibly be the true substance of the eucharistic hosts offered in churches around the world.[301]

Negatively, since matter was entirely passive, Descartes would not allow it any innate qualities (other than geometric size and shape) or capacities for influencing other bodies. There was no such thing as weight or gravity or magnetism for Descartes, and there were no causal relations among events. There was only a continuum of material bodies in relative motion sustained by God's continual recreation. In fact, the only element of contingency in Descartes's system was due to the presence of human minds: through free choice, minds could influence the direction (though not the speed) of motion of some of the corpuscles in their bodies and thereby have a role in the determination of history.[302]

Since the existence and behaviour of matter depended entirely on God, the laws of physics were a direct expression of the attributes of God. For Descartes the most important attributes of God were his eternity and immutability. Not only was God immutable in his nature, but he was also immutable in his action. In other words, God was entirely consistent in the

---

[298] *Principles* II.4, 10f. (*Philosophical Works*, 2:255f., 259).

[299] James Collins, *Descartes' Philosophy of Nature* (Oxford: Basil Blackwell, 1971), pp. 22ff.

[300] Arnauld's Objections of 1648. In his earlier correspondence with Denis Mesland (1645-46), Descartes argued that the bread and wine become the body and blood of Christ by being united with, and informed by, his soul. In 1669, Arnauld rejected this explanation (as stated by the Benedictine Cartesian, Desgabets) because it made the soul of Christ present in the eucharist rather than his human body; Steven M. Nadler, 'Arnauld, Descartes, and Transubstantiation', *JHI*, vol. xlix (April 1988), pp. 233-37.

[301] Pietro Redondi, *Galileo Heretic* (Princeton, N.J.: Princeton Univ. Press, 1987), pp. 212f. (Aquinas), 215f. (Ockham), 218 (Nicholas of Autrecourt), 223f. (Suarez), 283-86 (Descartes).

[302] Copleston, *History* 4:145.

way he acted, always producing the same effects under the same circumstances.[303]

As a result of the immutability of God's being and acts, two things could be concluded. First, the laws governing physical nature could never change nor be violated in any way. The laws of physics, therefore, were eternal and immutable like God. After all, they were simply the rules God had imposed upon himself, and, since God always acted in accordance with the same rules or laws, he always produced the same kinds of effects.[304] Thus there could be no miracles in the sense of violations of the laws of physics. Once God had decreed the laws governing events, in Descartes's universe, there were no exceptions.[305] On the other hand, all events were new creations of God: they were not the direct result of previous events as causes. Paradoxically, the role of God in nature was ubiquitous and yet highly constrained (by God's own nature) at the same time. Providence was so universal that it could not be recognised in any particular instance.[306]

Second, the immutability of God implied for Descartes the conservation of the amount of matter and the amount of motion in the universe. The amount of motion was the sum of the products of the quantity of matter (i.e., extension or bulk) and the speed of motion (in whatever direction) of all the bodies in the universe. Descartes expressed this principle of conservation in three laws:

(1) each body continues in its given state of rest or motion unless it collides with another body;

(2) each body will move in a straight line unless constrained to move otherwise;

(3) when one body collides with another there is a transfer of motion from the stronger to the less strong in accordance with seven special rules of impact.

According to Descartes, all the phenomena of nature could be explained in terms of these laws, even the seemingly 'occult' phenomena of gravitation, magnetism, and light.[307]

As Edgar Zilsel pointed out in 1942, Descartes established the modern concept of 'laws of nature' by joining the theological idea of God's ordaining laws for all creatures with the new empirical laws of

---

[303]*Principles* II.36 (Hatfield, 'Force', p. 122; E. McMullin, ed., *Evolution*, p. 25).
[304]*Le monde* 7 (written 1637; James Collins, *Descartes*, p. 50).
[305]*Le monde* (Mahoney, trans., p. 77; McMullin, ed., *Evolution*, p. 25).
[306]Wallace E. Anderson, 'Editor's Introduction' to *The Works of Jonathan Edwards, Vol. 6: Scientific and Philosophical Writings* (New Haven: Yale Univ. Press, 1980).
[307]*Le monde* (Mahoney, trans., p. 59; McMullin, ed., *Evolution*, p. 25); *Principles* II.26-64 (Zilsel, 'Physical Law', p. 268); Westfall, *Construction*, pp. 35ff.

post-Aristotelian science.[308] The idea of laws of nature was popularised later in the seventeenth and eighteenth centuries and eventually became quite secular in meaning. The eventual elimination of any reference to God led to the notion that the laws were possessed by nature rather than prescribed for nature by God.

For Descartes, then, the continued existence and behaviour of matter was predictable because it was based on the eternity and immutability of God. It was so predictable, in fact, that Descartes could speculate about possible ways in which the present universe could have evolved in accordance with the laws of nature from an original chaos of material particles. A primordial (hypothetical) assemblage of matter in motion would evolve into the universe as we know it through the mutual collisions of the particles in accordance with the laws of motion. There was no need to appeal to the miraculous or to occult qualities of any kind. Even plants and animals could be viewed as machines with what appear to be natural movements determined by the motions of hidden strings and fluids.[309]

Without denying that the world was in fact formed miraculously, i.e., by God's *potentia absoluta*, as taught by Scripture and the church, Descartes imagined a hypothetical cosmos coming to look just like the present one solely by God's *potentia ordinata*, as understood by unaided human reason.[310] Paradoxically, the assimilation of divine providence to the idea of creation (occasionalism) could thus be turned around into an assimilation of divine creation into universal providence (naturalism). The creationist tradition thus led Descartes to the idea of the possible eternity of the world, as it had led the medieval scholastics before him (chap. 2, sec. B). Conversely, just as the stress on *potentia ordinata* in the Middle Ages produced a reaction in the late thirteenth century, Cartesian philosophy produced various reactions both within the mechanist tradition and beyond it in the late seventeenth century, as we shall see.

The story of the reception of Descartes's mechanical philosophy in the seventeenth century is very similar to that of the initial resistance, gradual assimilation, and subsequent reaction to the natural philosophy of Aristotle

---

[308] Zilsel, 'Physical Law', pp. 269-72; cf. Jane E. Ruby, 'The Origins of Scientific "Law"', *JHI*, vol. xlvii (July 1986), p. 358. One can appreciate Zilsel's positive observations of the development of the idea of laws of nature in the ancient Near East and in the seventeenth century without subscribing to his notion that the idea was dormant in the Middle Ages or that its seventeenth-century revival was sociologically determined; cf. John R. Milton, 'The Origin and Development of the Concept of the "Laws of Nature"', *Archive of European Sociology*, vol. xxii (1981), pp. 178-83.

[309] *Discourse on Method* 5; Genevieve Rodis-Lewis, 'Limitations of the Mechanical Model in the Cartesian Conception of the Organism', M. Hooker, ed., *Descartes* (Baltimore, Md.: Johns Hopkins Univ. Press, 1978), pp. 152-70.

[310] *Discourse* 5 (Haldane and Ross, trans., 1:107ff.); cf. McMullin, ed., *Evolution*, pp. 21ff.).

in the thirteenth century (chap. 2). Ironically, it was primarily the continued adherence to Aristotle in the churches and universities of the seventeenth century that provided the initial resistance to Cartesian ideas. During the 1640s, the (Reformed) universities of Utrecht[311] and Leiden[312] ruled against the teaching of anti-Aristotelian ideas in Descartes. In 1662, the arts faculty of the (Catholic) University of Louvain issued a similar ruling. In 1650 the Ninth General Congregation of the Jesuits condemned fifteen propositions related to the teachings of Descartes. In 1663, at the instigation of the Jesuits, Descartes's philosophical writings were placed on the Index of Prohibited Books until such time that they were suitably corrected. In 1671, the Court of the University of Paris commanded its theologians to enforce a royal edict prohibiting any philosophy but Aristotle's. The universities of Angers and Caen became officially anti-Cartesian later in the 1670s.[313]

Much of this initial reaction was due to the the inherent conservatism of the overseers and professors of the institutions involved and should not be taken as reflecting the ecclesiastical or academic communities as a whole.[314] As early as 1628, Descartes was encouraged by Cardinal de Berulle, founder of the French Oratorians, to apply his method to the problems of medicine and mechanics—'the one would contribute to the restoration and conservation of health, and other to some diminution and relief in the labours of mankind'.[315]

---

[311]Voetius took the lead in opposing Cartesian thought at Utrecht in 1639 (*Disputationes de atheismo*; publ. in the first volume of *Disputationes selectae theologicae*, 1648). Under his leadership, the university senate decreed in 1642 that the new philosophy was insolent and an affront to those who taught the truth. In 1643, the Utrecht council forbade the production or distribution of Descartes's writings within the city's precincts. The charge of 'atheism' really meant the encouragement of skepticism through methodological doubt and rejection of the traditional proofs for the existence of God; Ernst Bizer, 'Reformed Orthodoxy and Cartesianism', in Robert W. Funk, ed., *Journal for Theology and the Church*, vol. II (Tübingen: J.C.B. Mohr, 1965), pp. 22-26, 32ff., 38. On Melchior Leydekker, Voetius's disciple and successor at Utrecht, see *ibid.*, pp. 73-81.

[312]Trigland and Revius took the lead at Leiden by raising charges in a disputation of 1647. The University forbade its professors to mention Descartes's name in lectures and disputations and commanded Descartes (who had addressed letters defending himself against the charges to the trustees) to keep silent on the issues raised (Pelagianism and blasphemy, respectively). However, further disputations were held without using the name of Descartes. In 1656 and 1676 new decrees were issued forbidding the defence of Cartesian theses; Bizer, 'Reformed Orthodoxy', pp. 39f. Cf. Edward G. Ruestow, *Physics at Seventeenth and Eighteenth-Century Leiden* (The Hague: Martinus Nijhoff, 1973), chap. 3: 'Tumult over Cartesianism' and chap. 4: 'Johannes de Raey: The Introduction of Cartesian Physics at Leiden'.

[313]Heilbron, *Elements*, pp. 27ff.; Trevor McLaughlin, 'Censorship and Defenders of the Cartesian Faith in Mid-Seventeenth Century France', *JHI*, vol xl (Oct. 1979), pp. 565-69; Nadler, 'Arnauld', pp. 238ff.

[314]On the conservatism of university professors, see Ruestow, *Physics*, 1f., 140f., 146.

[315]Adrien Baillet, *La vie de Monsieur Descartes* (Paris, 1691), 2:165 (A.C. Crombie, 'Descartes', *DSB* 4:51f.). Descartes's reiteration of de Berulle's words at the end of his *Discourse* is cited by Peter Gay as foundational to Enlightenment thought!; *The*

The primary concerns of conservative theologians seem to have been Descartes's methodological skepticism, his separation of natural philosophy from positive theology, and, for Catholics, the implications of his equation of matter with geometric extension for the doctrine of transubstantiation. Many moderate theologians believed that the essentials of Christian faith were not threatened.

As early as 1645, the city council of Utrecht forbade anyone to write either for or against Descartes.[316] At Leiden, the Cartesian methodology was ably defended by Christoph Wittich in the 1670s[317] and was generally supported by theological followers of John Cocceius (1603-69), who sought a purely scriptural basis of theology that complemented Descartes's procedure of separating natural philosophy from theology.[318] In Calvinist Geneva, Francis Turretin (1623-87) took a positive interest in Descartes's natural philosophy and only objected to the application of his methodical doubt to theological issues.[319] Catholic theologians, particularly some of the Jesuits, charged that Descartes's equation of material substance with geometric extension would undermine the dogma of transubstantiation.[320] On the other hand, there would have been no need for the widespread efforts to discourage interest in Descartes's teachings if the latter had not also

---

Enlightenment, 2 vols. (London: Wildwood House, 1973), 2:6.

[316]Bizer, 'Reformed Orthodoxy', p. 27. Subsequently Utrecht theologians focussed their attacks against Cartesians on the Copernican issue; *ibid.*, p. 53.

[317]Wittich, *Theologia pacifica* (1671); *Consensus veritatis in Scriptura divina et infallibili revelatae cum veritate philosophica a Renato Des Cartes detecta* (1682); *Theologia pacifica defensa* (published posth. 1689). The key to Wittich's defence was the Cartesian distinction between philosophy (based on reason) and theology (based on revelation); Bizer, 'Reformed Orthodoxy', pp. 52f., 60f.

Wittich had already defended Descartes's Copernicanism against opposition at Leiden and Utrecht in three disputations of 1659ff. He stood trial before the Synod of Gelder in 1660, but his orthodoxy was confirmed in 1661. Just before moving to Leiden, Wittich defended his Cartesian teaching at Nymwegen against Maresius; Bizer, 'Reformed Orthodoxy', pp. 52f., 58, 59f. Heidanus had also defended the Cartesian methodical dualism and doubt at Leiden but was removed from his chair of theology, c. 1676; *ibid.*, pp. 40f., 59.

[318]Cocceius was initially opposed to Cartesian ideas; Bizer, 'Reformed Orthodoxy', pp. 45, 74. But Dutch Cartesians like Heidanus and Wittich found Cocceius's biblical positivism useful in their cause; Bizer, 'Reformed Orthodoxy', pp. 58f., 63.

[319]Michael Heyd, 'Orthodoxy, Non-Conformity and Modern Science', in M. Yardeni, ed., *Modernité et non-conformisme en France à travers les ages* (Leiden: Brill, 1983), p. 104.

[320]The corpuscular physics of *Principles of Philosophy* was denounced by the Jesuit Thomas Compton Carleton, professor of theology at Liège, in 1649. Descartes's equation of substance with extension was repudiated by the Jesuits Theophile Raynaud (1665) and Le Valois (1681), by the Jansenist du Vaucel (1681), and by M. le Moine (by 1680). His rejection of substantial forms was criticised by M. Claude Morel, dean of the faculty of theology at the University of Paris, in 1671 (Redondi, *Galileo Heretic*, pp. 285f., 301; Nadler, 'Arnauld', pp. 240-44). According to Redondi, the attack on Descartes's physics was a continuation of an earlier campaign against Galileo's corpuscular philosophy; *Galileo Heretic*, pp. 280, 301, 307f., 314f. The main difference between the two cases was that Descartes did not allow a vacuum in his physics and so was in less conflict with Aristotelianism and Tridentine eucharistic teaching; *Galileo Heretic*, pp. 291f.

achieved a degree of popularity in the colleges and universities under ecclesiastical control.

By the mid-eighteenth century, even the strongholds of conservatism had altered their earlier opposition. In 1704, Jean-Alphonse Turretin (the son of François) praised Descartes's *Discourse on Method* as containing the best precepts for the pursuit of knowledge.[321] Through the efforts of the Oratorian Nicholas Malebranche (1674-5), and the Jansenist Antoine Arnauld (1671-81), both strong Augustinians, aspects of Descartes's philosophy came to be regarded by many French Catholics as supportive of official church dogma.[322] Some Jesuit schools also adopted Cartesian principles in the eighteenth century, partly as a means of counteracting the traditional Augustinianism of the Jansenists.[323] In 1706, the Fifteenth General Congregation of the Jesuits allowed the defence of the Cartesian system as an hypothesis and praised it for its overall completeness and consistency, if not for its correctness. Descartes's ideas were officially recognised even at the University of Paris in 1720.[324] As in the case of the reception of Aristotle in the thirteenth century, the remarkable thing about the church was not so much its initial resistance to new scientific ideas as its ability to reevaluate and assimilate them in keeping with the historic creationist tradition.

The two principal French churchmen to develop the mechanical philosophy of Descartes in the early seventeenth century were Marin Mersenne (1588-1648) and Pierre Gassendi (1592-1655). A primary objective for both men was to protect the mathematical sciences from association with the hermetic numerology of Ficino, Pico, Bruno, Fludd, and the so-called Rosicrucian Brotherhood.

Mersenne was concerned about what he saw as a confusion of matter and spirit, the natural and the supernatural, and a limitation of the power of God

---

[321]Heyd, 'Orthodoxy', pp. 107f.
[322]Malebranche, *De la recherche de la verité* (1674-75); Arnauld, *Plusiers raisons pour empêcher la censure ou la condemnation de la philosophie de Descartes* (1671?); *Examen d'un écrit* (1680); *Apologie pour les Catholiques* (1681?); Aram Vartanian, *Diderot and Descartes* (Princeton, N.J.: Princeton Univ. Press, 1953), pp. 34-40; Gillispie, *Edge*, p. 159; Lionel Rothkrug, *Opposition to Louis XIV* (Princeton, N.J.: Princeton Univ. Press, 1965), pp. 47ff., 56-60, 83ff.; Margaret Jacob, *The Radical Enlightenment* (London: Allen & Unwin, 1982), pp. 45f.; Nadler, 'Arnauld', pp. 240-45. Arnauld's defence was based on the strict separation of matters of philosophy or reason (*questions de fait*) from matters of faith (*questions de droit*), which rest on the absolute power of God; Nadler, 'Arnauld', pp. 234f., 241f., 244. Malebranche's *Traité de la nature et de la grâce* was placed on the Index in 1689 and was there as late as 1704; Redondi, *Galileo Heretic*, pp. 317f.
[323]Vartanian, *Diderot*, p. 39; Colm Kiernan, *The Enlightenment and Science in Eighteenth-Century France* (Banbury: Voltaire Foundation, 1973), pp. 146, 157.
[324]Heilbron, *Elements*, pp. 28f.; McLaughlin, 'Censorship', p. 569.

in the hermetic appeal to natural magic.[325] In particular, he opposed some of Robert Fludd's ideas about spiritual sympathies between bodies: e.g., between a weapon and the body it had wounded.[326] By way of contrast, Mersenne championed Descartes's idea of treating living systems as machines,[327] though he was not as confident as Descartes of the human mind's ability to determine the inner mechanisms of bodies—the omnipotence of God meant for him that one could never know for sure how God had designed things.[328] Curiously, Mersenne's anxiety about the dangers of Hermeticism and, indirectly, his support for the mechanical philosophy were partly inspired by the Jesuit campaign against Rosicrucianism in the early 1620s[329]—a point worth noting in order to show that a conservative religious movement can have 'progressive' as well as reactionary effects with respect to a new philosophy.

Mersenne advocated the disciplining of alchemy by the elimination of mysticism and secrecy, the use of clear terminology, and the foundation of public alchemical academies for the improvement of human health. Thus he played a role in French science comparable to that of Hartlib in England, though Mersenne was less supportive of alchemical studies and stressed the importance of mathematics more than Hartlib.[330] Both men were as concerned to organise and encourage the work of others as to develop their own scientific ideas.

Pierre Gassendi gave a distinctively new direction to the mechanical philosophy in three ways. First he treated continued existence and the power of causal influence as inherent properties of material bodies, even allowing

---

[325]Dobbs, *Foundations*, pp. 56f.; Heilbron, *Elements*, p. 20; A.C. Crombie, 'Mersenne', *DSB* 9:316ff.

[326]Debus, 'Chemical Philosophers', pp. 246f.

[327]As a matter of fact, the young Descartes accepted the validity of occult phenomena like the antipathy between two drums, one covered with a lamb's skin and the other covered with that of a wolf (*Compendium musicae*, 1619). In later years (perhaps as the result of Mersenne's influence?), he attempted to reduce all such phenomena to mechanical interactions; Shea, *Magic*, pp. 112f.

[328]Crombie, 'Mersenne', pp. 317f.; Peter Dear, *Mersenne and the Learning of the Schools* (Ithaca, N.Y.: Cornell Univ. Press, 1988), pp. 6, 53, 108, 227. Still, Mersenne held to the notion of the comprehensibility of the phenomenal world, based on the manifestation of divine wisdom in nature and the image of God in humanity; *Quaestiones celeberrimae in Genesim* (1623); Dear, *Mersenne*, pp. 108. According to Dear, Mersenne derived his 'Platonism' from Augustine; *ibid.*, pp. 99f., 109, 115f., 226.

[329]Yates, *Rosicrucian Enlightenment*, pp. 104ff., 111ff.; *The Occult Philosophy in the Elizabethan Age* (London: RKP, 1979), pp. 172f. Mersenne's critical attitude toward Paracelsus himself was partly derived from the earlier criticisms of Thomas Erastus, on whose references to Paracelsus he relied in his 1623 *Quaestiones celeberrimae in Genesim*; William L. Hine, 'Mersenne and Alchemy', in Z.R.W.M. von Martels, ed., *Alchemy Revisited* (Leiden: Brill, 1990), p. 189.

[330]Debus, 'Chemical Philosophers', p. 246; *Man and Nature*, pp. 124f.; Dobbs, *Foundations*, p. 57. For Mersenne's stress on mathematics, see Dear, *Mersenne*, pp. 71-79, 96-109, 225-28.

for weight and internal energy as well as mere bulk. Thus he shifted the course of the mechanical philosophy away from the occasionalism of Descartes toward naturalism and a clearcut division between rational science and revealed religion.[331]

Second, Gassendi incorporated the atomism of Epicurus and Lucretius into his mechanical philosophy. He claimed that all physical phenomena could be explained in terms of the motions of indivisible corpuscles or atoms of varying weight, size, and shape. He was not the first to do so: Galileo, Hariot, and Beeckman had already advocated the idea of atoms in their early versions of the mechanical philosophy.[332] But Gassendi's two major Epicurean treatises, published in 1649, made the ideas of classical atomism more widely available.[333] Moreover, by limiting the amount of matter in the universe to a finite quantity and by crediting God with the original endowment of atoms with motion, he made atomistic ideas more palatable to Christian minds.[334]

Third, Gassendi advocated the idea of an infinite void space that was later utilised by Newton. In part, this idea was a byproduct of Gassendi's interest in atomism, which postulated an infinite void as the framework for the activity of the corpuscles of matter. It also reflected the influence of the Italian naturalists, Patrizzi and Campanella, and earlier medieval speculations about the possibility of an extra-cosmic void (chap. 2, sec. C).[335]

Gassendi's Epicurean version of the mechanical philosophy was to have an immense impact on the development of the mechanical philosophy in England.

---

[331] Crombie, *Augustine*, pp. 133f.; Hutchison, 'Supernaturalism', p. 300; Bernard Rochot, 'Gassendi', *DSB* 5:286b. According to Barry Brundell, Gassendi intended to harness classical learning to the Christian faith but ended up separating the two and granting natural philosophy greater autonomy; 'Gassendi between Religion and Science', *Quadricentenaire de la naissance de Pierre Gassendi, 1592-1992*, vol. I (Digne-des-Bains: Société Scientifique et Littéraire des Alpes de Haute-Provence, 1994), pp. 121-32.

[332] Robert H. Kargon, 'Atomism in the Seventeenth Century', *DHI* 1:133; DeKosky, *Knowledge*, pp. 286ff.; R. Hooykaas, 'Beeckman', *DSB* 1:567a. Epicurean atoms differed from Aristotelian 'natural minima' in that their dynamics were strictly physical—based on strictly quantitative properties like size and shape—rather than chemical—based on qualitative properties like those of the four Aristotelian elements; N. Rattner Gelbart, 'The Intellectual Development of Walter Charleton', *Ambix*, vol. xviii (1971), p. 157.

[333] Gassendi, *Philosophiae Epicuri syntagma*; idem., *Animadversions on the Tenth Book of Diogenes Laertius*.

[334] Robert H. Kargon, 'Walter Charleton, Robert Boyle, and the Acceptance of Epicureanism in England', *Isis*, vol. lv (June 1964), p. 184; idem, 'Atomism', pp. 135f.; Rochot, 'Gassendi', pp. 285f. As Brundell points out, Gassendi's primary intention was not to compromise Christian faith in any way, but to counteract the alchemical theories of philosophers like Robert Fludd and to find an alternative to the Aristotelian natural philosophy; *Pierre Gassendi* (Dordrecht: Reidel, 1987), pp. 114, 129, 135, 137ff.

[335] Jammer, *Space*, pp. 85-94, 110; Grant, *Much Ado*, pp. 199f., 195ff.

2.3 English Mechanical Philosophers: Digby to Boyle

The progress of the mechanical philosophy after Descartes, Mersenne, and Gassendi has largely to do with its reception and subsequent development in England.[336] The main theological and social stimulus towards the reception of the mechanical philosophy was a growing tendency to associate spiritualism with pantheism and anarchy and the consequent need to find a viable alternative conducive to ecclesiastical peace and political stability. There were four principal figures involved: Kenelm Digby (1603-65), Thomas Hobbes (1588-1679), Walter Charleton (1620-1707), and Robert Boyle (1627-91). We shall touch on Digby, Hobbes, and Charleton briefly before discussing the culmination of the seventeenth-century mechanical tradition in the work of Robert Boyle.

Digby and Hobbes represent two opposing ways of interpreting the mechanical philosophy: Digby stressing the immateriality of the soul and Hobbes reducing it to a manifestation of matter in motion.

Sir Kenelm Digby was a devout Catholic and a moderate Aristotelian in philosophy. As a sincere Christian and a royalist, loyal to Charles I during the English Civil War (1642-49), he was deeply disturbed by the proliferation of radical Protestant sects during the English Civil War. In particular, he wished to refute the millenarians' doctrine of mortalism—the teaching that the soul ceases to exist at death and awaits the final resurrection for its regeneration. As an exile in Paris (1643-54), Digby met Descartes and other members of Mersenne's circle. He saw in the mechanical philosophy and its matter-spirit dualism a powerful weapon against the radical doctrines of materialism and mortalism.[337] Accordingly, he followed Descartes in viewing plants and animals as machine-like, though, like Gassendi, he held that matter consisted of atoms endowed with efficient causality.[338]

---

[336]Earlier seventeenth-century English atomistic and mechanical ideas, mixed with Aristotelian and alchemical ones, also continued to be influential, particularly in the cases of Kenelm Digby, Thomas Hobbes, and Margaret Cavendish; see Stephen Clucas, 'The Atomism of the Cavendish Circle: A Reappraisal', *Seventeenth Century*, vol. ix (Autumn 1994), pp. 247-73.

[337]*Two Treatises: Of Bodies; Of the Immortality of Man's Soul* (Paris, 1644; London, 1645); cf. William B. Hunter, 'The Seventeenth Century Doctrine of Plastic Nature', *HTR*, vol. xliii (July 1950), p. 210; Richard S. Westfall, *Science and Religion in Seventeenth-Century England* (New Haven, Conn.: Yale Univ. Press, 1958; Ann Arbor: Univ. of Michigan Press, 1973), pp. 35, 80; John Henry, 'Atomism and Eschatology: Catholicism and Natural Philosophy in the Interregnum', *BJHS*, vol. xv (1982), pp. 213f., 222ff.

[338]Keith Thomas, *Man and the Natural World* (New York: Pantheon Books, 1983), p. 35; Crombie, *Augustine*, 2:313; M.B. Hall, 'Digby', *DSB* 4:96a. Gassendi had christianised Epicureanism by reversing Epicurus's teaching on the mortality of the soul as well as in the ways noted above; Margaret J. Osler, 'Baptizing Epicurean Atomism: Pierre Gassendi on the Immortality of the Soul', in M.J. Osler and P.L. Farber, ed., *Religion, Science, and*

Whereas the mechanical philosphy of Digby was intended to bolster belief in the spiritual side of humanity, the mechanical philosophy of Hobbes was a form of materialism. Even mental and spiritual phenomena, according to Hobbes, had to be extended, if they were real at all, and hence were manifestations of matter in motion. There was no such thing as an immaterial soul or free will. At one point, Hobbes even referred to God as a 'simple corporeal spirit' with magnitude and extension.[339]

Leviathan, Hobbes's famous political treatise, was written during his exile (as a royalist) in Paris and published in 1651. Hobbes not only advocated the absolute power of kings but questioned the authenticity of biblical narratives about the activity of spirits and the occurrence of miracles.[340] Common as such outright skepticism is today, it was nearly unheard of in the mid-seventeenth century, at least in print.

Hobbes made no significant contribution to science as such, but he had a tremendous impact on the development of the mechanical philosophy. His writings were taken by many as evidence of an anti-Christian bias in the mechanical philosophy as a whole. During his years in Paris (the 1630s and '40s) he was known to have frequented the circle of Mersenne and as a close friend of Gassendi. Some of Gassendi's distinctive ideas on atoms and the void were included in Hobbes's treatise *On Bodies*, published in 1655.[341]

We have seen that theological ideas can provide targets for reaction as well as positive influences in the development of science. Much of seventeenth-century mechanical philosophy was a reaction to the perceived dangers of the hermetic and spiritualist traditions. Much of late seventeenth-century English thought was also a reaction to Hobbes's materialism and the perceived dangers of the mechanical philosophy. The Platonist tradition we shall discuss later departed radically from the mechanical model in its attempt to restore a sense of the direct involvement of God in natural events. There were also those who continued to work

---

Worldview, (Cambridge: Cambridge Univ. Press, 1985), pp. 163-83.

Thomas Browne had defended the ideas of Epicurus already in his *Religio Medici* of 1642 (as well as later in his *Vulgar Errors* of 1646, 1650). But, in this famous work, Browne also discussed his earlier sympathy (on the basis of reason alone) with the heresy of the 'Arabians' (and peripatetics like Pomponazzi), 'that the souls of men perished with their bodies'. Digby's initial response was to defend the philosophical credentials of belief in the immortality of the soul (in his *Observations on Religio Medici* of 1642 or 1643). Subsequently, the Cartesian mechanism of his *Two Treatises* (1644) offered Digby a way to separate Epicurean atomism from heretical mortalism and thus reunite faith and reason. Hence, the more conservative (Christian Aristotelian) Alexander Ross referred to Digby's 'erroneous Paradoxes' in his *Philosophicall Touch-stone* of 1645, which was also directed against the mortalism of Richard Overton. See George Williamson, *Seventeenth Century Contexts* (Chicago: Univ. of Chicago Press, 1961), pp. 149ff.

[339]*Leviathan*, Part I, chap. 34; Copleston, *History* 5/1:15-18, 32f.

[340]Rosalie L. Colie, *Light and Enlightenment* (Cambridge: Cambridge Univ. Press, 1957), p. 62; Samuel I. Mintz, 'Hobbes', *DSB* 6:446.

[341]Copleston, *History* 5/1:12; Colie, *Light*, p. 59; Kargon, 'Atomism', p. 136a; Riddell, ed., *Lives*, p. 210; Samuel I. Mintz, 'Hobbes', pp. 445f.

within the mechanical philosophy while modifying it and interpreting it within a Christian context. We shall discuss two of these 'loyalists': Walter Charleton and Robert Boyle.

The fascinating thing about Charleton and Boyle is that they were both adherents of the spiritualist views of Helmont in the late 1640s and early '50s but then adopted the mechanical philosphy in the mid-1650s. In fact, Charleton's *A Ternary of Paradoxes*, published soon after the execution of Charles I (1649), was the first translation of any of Helmont's works into English. Charleton had served as a physician to Charles I since 1643. According to the preface to his translation (1650), he originally found Helmont's elevation of understanding over scholastic reason attractive as an aid to peace in a time of civil war and ecclesiastical strife.[342]

By 1652 Charleton was shifting towards the mechanical philosophy of Descartes and Hobbes, and in 1654 he published a translation and commentary on one of Gassendi's Epicurean treatises[343] under the title, *The Epicurean-Gassendian-Charletonian Physiology* ('physiology' meaning the physics of all material substances). Reversing his former preference, Charleton now rejected the spiritualist idea of a world soul responsible for correspondences and sympathies between spatially separate bodies.[344] He was influenced in this respect by the attack on Robert Fludd's Hermeticism launched by Mersenne and Gassendi.[345] Moreover, spiritualist ideas had by this time become closely associated with sectarian religion and anarchy in England. In place of the spiritualist world of correspondences, Charleton

---

[342]P.M. Rattansi, 'Paracelsus and the Puritan Revolution', *Ambix*, vol. xi (Feb. 1963), p. 26. The three works of Helmont translated by Charleton were *Magnetick Cure of Wounds*; *Nativity of Tartar in Wine*; and *Image of God in Man*. According to Lotte Mulligan, Charleton's Prolegomenon was already critical of Helmont's illuminist epistemology; Mulligan, '"Reason," "Right Reason," and "Revelation" in Mid-Seventeenth Century England', in Brian Vickers, ed., *Occult and Scientific Mentalities in the Renaissance* (Cambridge: Cambridge Univ. Press, 1984), p. 381. However, Mulligan overstates the contrast between the two by comparing Helmont's critical statements about natural reason with Charleton's advocacy of illumined, 'right reason'.

[343]Gassendi's *Animadversions* (1649).

[344]*The Darkness of Atheism Dispelled by the Light of Nature* (1652); *Physiologia Epicuro-Gassendo-Charltoniana* (1654); Gelbart, 'Intellectual Development', pp. 162ff. Antonio Clericuzio overlooks the sweeping nature of Charleton's critique when he states that 'Charleton's criticisms of Helmont in the *Physiologia* (1654)—which have often been considered the proof of Charleton's "conversion" from Helmontianism to the mechanical philosophy—were limited to the Helmontian doctrine of the magnetic cure of wounds'; 'From van Helmont to Boyle: A Study of the Transmission of Helmontian Chemical and Medical Theories in Seventeenth-Century England', *BJHS*, vol. xxvi (June 1993), p. 306.

[345]Gelbart suggests that this influence was mediated by Hobbes, who had arrived back in England in 1650. Charleton quoted Hobbes for the first time in the introduction to his translation of Helmont's *Deliramenta catarrhi* of late 1650. Gelbart also suggests that Hobbes first informed Charleton of Isaac Casaubon's discovery (1614) that the Hermetic corpus was only a late antique composition, not a pre-Mosaic writing as previously believed; Gelbart, 'Intellectual Development', pp. 158f.

now advocated Gassendi's mechanical world of atoms moving in an infinite void.[346]

In the wake of the publication of *Leviathan* (1651), Charleton had to be careful to dissociate Gassendi's version of Epicureanism from the mechanistic materialism of Hobbes as well as the spiritualism of Fludd and Helmont. Thus he stressed, as Gassendi had, the idea that the atoms were originally created and endowed with motion by God.[347] But he went a step further than Gassendi in order to avoid any suggestion of naturalism: even though the atoms moved in accordance with laws dictated by God, they moved blindly and were incapable of either producing or maintaining an ordered cosmos without God's continued providence.

So, according to Charleton, atomism actually provided a proof for the existence and continual providence of God. Without God there would be nothing; or, if there were anything, it would only be a chaos.[348] Moreover, unlike the French mechanical philosophers, Charleton held that living beings could not be explained in mechanical terms: the organisation of plants and animals also required the special providence of God.[349] The mechanistic account of the world had gaps, in effect, and required God to fill them. This was one of the first modern instances of what is known as 'natural theology', the attempt to demonstrate the existence and activity of God from the phenomena of nature.[350]

---

[346]*Physiologia* (1654); Rattansi, 'Paracelsus', pp. 30f.; Kargon, *Atomism*, pp. 84, 87; Gelbart, 'Intellectual Development', pp. 149ff.; Debus, *Chemical Philosophy*, 2:472; Klaaren, *Religious Origins*, p. 210 note 55.

John Henry argues (against the 'mistaken historiography' of Rattansi, Gelbart, and others) that Charleton did not abandon Helmontianism but rather reconciled it with atomism; 'Occult Qualities and the Experimental Philosophy', *History of Science*, vol. xxiv (1986), p. 341 with note 19; cf. Mulligan, 'Reason', p. 381-82. However, Charleton did become increasingly critical of Helmont: first, for failing to provide 'a more substantial and durable Structure of his own' to replace the 'Doctrines of the Ancient Pillars or our Art', which he had demolished (intro. to his trans. of Helmont's *Deliramenta catarrhi*, 1650); second, for his anti-Galenic conception of disease (*Darkness of Atheism*, 1652); and, finally, for holding that there is something in air that burns, for his theory of the rainbow, and for his belief in the effectiveness of the weapon-salve cure—for all of which which he deserves to be called 'Hair-brain'd and Contentious' (*Physiologia*, 1654); Gelbart, 'Intellectual Development', pp. 158, 163, 164f.

[347]That deistic implications could be drawn from the Gassendian-Charletonian view is clear from the latter's statement that God's decrees cannot 'work any the least mutation at all in the natures of his Creatures, or by violence pervert their Virtues to the production of any Effects, to which, by their primitive Constitution and individuation, they were not precisely adapated and accommodated', *Darkness of Atheism* (1652); Williamson, *Seventeenth-Century Contexts*, p. 169.

[348]*Darkness of Atheism*; cf. Westfall, *Science and Religion*, pp. 54, 81ff.; Kargon, *Atomism*, pp. 85f.

[349]Debus, *Chemical Philosophy*, 2:472. In his 1652 *Darkness of Atheism*, Charleton still appealed to the Helmontian *archeus* or 'Formative Spirit' in order to account for the procreation of plants and animals; Gelbart, 'Intellectual Development', p. 161.

[350]Westfall, *Science and Religion*, p. 118.

228   CHAPTER THREE

Robert Boyle experienced a conversion from spiritualist to mechanist leanings similar to Charleton's, though it seems to have taken a longer period of time. The historical reconstruction of the evolution of Boyle's thought is complicated by the fact that many of his writings remain unpublished, and even those published during his lifetime contain sections written years prior to their publication. We shall offer a plausible account of Boyle's conversion, the reasons for it, and the impact it had on his natural philosophy.

In the 1640s and early '50s, Boyle was closely associated with the utopian projects and alchemical research of Samuel Hartlib, as we have noted.[351] During this period he first examined the atomist ideas of Descartes, Gassendi, and Digby, but he remained uneasy about their acceptability as late as 1653 and specified that his papers on the subject were to be burnt.[352] In the early sections of *Some Considerations touching the Usefulnesse of Experimentall Naturall Philosophy* (written in the late 1640s), when he first discussed the new atomical philosophy, Boyle retained many elements of the Helmontian outlook.[353] By the late 1650s, however, when Boyle wrote the later sections of *Usefulnesse* and *The Sceptical Chymist*, he had repudiated what he then saw as Helmont's confusion of the natural and supernatural.[354] In particular, Boyle criticised the reliance of

---

[351]P.M. Rattansi, 'The Social Interpretation of Science in the Seventeenth Century', Peter Mathias, ed., *Science and Society, 1600-1900* (Cambridge: Cambridge Univ. Press, 1972), p. 20; Jacob, *Robert Boyle*, pp. 31-36.

[352]'Of ye Atomicall Philosophy' (compl. *c.* 1653); cf. Kargon, 'Walter Charleton', pp. 187f. For a possible earlier dating of this ms. to 1650, see Richard S. Westfall, 'Unpublished Boyle Papers Relating to Scientific Method', *Annals of Science*, vol. xii (1956), pp. 103-117.

[353]Harold Fisch, 'The Scientist as Priest: A Note on Robert Boyle's Natural Philosophy', *Isis*, vol. xliv (1953), pp. 253ff.; Westfall, *Science and Religion*, pp. 111f.; Antonio Clericuzio, 'Robert Boyle and the English Helmontians', in Z.R.W.M. von Martels, ed., *Alchemy Revisited* (Leiden: Brill, 1990), p. 193f. As Klaaren points out (*Religious Origins*, pp. 156f.), this section of *Usefulnesse* was written before Boyle's 'conversion' to the newer outlook, though not published until 1663.

[354]Klaaren, *Religious Origins*, p. 180. According to Fisch ('Scientist', p. 252) and Rattansi ('Paracelsus', p. 32), Boyle was still sympathetic to Helmont in *The Sceptical Chymist* (published in 1661). However, as Klaaren points out, these passages are included from an earlier draft, entitled 'Reflexions on the Experiments vulgarly alledged to evince the 4 Peripatetique Elements, or ye 3 Chymicall Principles of Mixt Bodies', written not later than 1658. Even so, the passages in *Sceptical Chymist* are far less Helmontian than the corresponding ones in 'Reflexions'; see A. Clericuzio, 'A Redefinition of Boyle's Chemistry and Corpuscular Philosophy', *Annals of Science*, vol. xlvii (1990), p. 567. The text of 'Reflexions' is available in Mary Boas [Hall], 'An Early Version of Boyle's *Sceptical Chymist*', *Isis*, vol. xlv (1954), pp. 153-68.

Antonio Clericuzio points out that Helmontians like George Starkey could easily interpret *The Sceptical Chymist* along Helmontian lines to reinforce their own arguments against the Galenists. Boyle never disavowed this usage, but he did publish an appendix entitled, 'The Producibleness of Chymical Principles', to the 1680 edition, in which he restated his earlier objections to the Paracelsian *tria prima* scheme and criticised the Helmontian theory of water as the material principle in vegetation; Clericuzio, 'Carneades and the Chemists: A Study of

Helmontians on direct revelations of chemical secrets from God (possibly referring to Thomas Vaughan of the Hartlib circle).355

Various reasons have been suggested for Boyle's change of mind. In the background was the rising sectarian activity associated by many with spiritualist views. Some historians have argued that the anarchistic movements of 1646-48 were particularly alarming to Boyle at a time when he and his brother Broghill were trying to secure their inheritance to family property in Ireland. Accordingly, Boyle left behind his earlier pursuit of virtue for virtue's sake and dependence on the unmerited grace of God, due to their association with the antinomianism of radical Puritans and sectarians like the Familists. In its place, he developed an ethic of self-reliance in which God helped those who helped themselves.356 If so, this shift should not be taken as a sign of mere temporising on Boyle's part: as his circumstances changed, his assessment of the relative merits of utopian and establishmentarian strategies for improving society naturally changed too. As an educated person and, potentially at least, a man of property, Boyle abandoned utopian schemes as he began to see the value of utilising the resources at his disposal 'as means to do hansom things with'.357

In any case, the final steps in Boyle's conversion to the mechanical philosophy took place in the late 1650s. An important step here was his move to Oxford where, in 1656, he joined a group of scientists (the 'Oxford Experimental Club') who met in the lodgings of the moderate Puritan divine, John Wilkins, at Wadham College.358 This 'Oxford group' included William Petty, Christopher Wren, John Wallis, and Seth Ward, the nucleus

---

The Sceptical Chymist and Its Impact on Seventeenth-Century Chemistry,' in Michael Hunter, ed., *Robert Boyle Reconsidered* (Cambridge: Cambridge Univ. Press, 1994), pp. 85ff.

355James R. Jacob, 'The Ideological Origins of Robert Boyle's Natural Philosophy', *Journal of European Studies*, vol. ii (March 1972), pp. 15f.

356Rattansi, 'Social Interpretation', p. 21; Jacob, 'Ideological Origins', pp. 15f.; *idem, Robert Boyle*, pp. 37, 85ff. Michael Hunter argues that Boyle's later polemical writings against 'atheists' have a broader target than the political radicals singled out by James and Margaret Jacob; M. Hunter, 'Science and Heterodoxy: An Early Modern Problem Reconsidered', in D.C. Lindberg and R.S. Westman, eds., *Reappraisals of the Scientific Revolution* (Cambridge: Cambridge Univ. Press, 1990), chap. 11; M. Hunter, ed., *Robert Boyle Reconsidered*, p. 3. I accept Hunter's point about the breadth of Boyle's concerns but do not see an incompatibility between these two interpretations, particularly when applied to different aspects of Boyle's development. Here we are concerned only with the transition from spiritualist-alchemical to mechanical principles in the 1650s. Hunter also acknowledges the importance of the common view that alchemists were politically subversive in *Robert Boyle Reconsidered*, p. 14.

357James, *Robert Boyle*, p. 37.

358Rattansi, 'Paracelsus', p. 32; M.B. Hall, 'Boyle', *DSB* 2:373b. Clericuzio is quite guarded in his judgement: 'If one can hardly affirm that at the end of the 1650s Boyle had rejected Helmontianism *in toto*, a shift in his views on the Belgian physician's iatrochemistry is nevertheless to be recorded'; 'Robert Boyle', p. 194; cf. p. 199.

of the later Royal Society of London.[359] Since the group included both Puritans and Anglicans, both supporters of the Commonwealth and Royalists, the principles on which it operated were latitudinarian in religion and pragmatic in politics. Its two great fears were enthusiasm in religion and anarchism in politics.[360]

In addition to these social and theological factors in Boyle's development, there were important advances in technology and experimental science to be considered. In the mid-1650s, Otto von Guericke (1602-86) had publically demonstrated the power of a vacuum in a cavity using a suction pump he had designed.[361] Boyle immediately saw the potential of such experiments as a means of demonstrating the mechanical concept of air pressure. Using a pump built by his assistant Robert Hooke, he performed a variety of experiments and concluded that the apparent power of the vacuum was due to the difference in air pressure between the evacuated cavity and the surrounding atmosphere.[362] There was no longer any need, according to Boyle, to appeal to final causes like nature's abhorrence of a vacuum.[363]

Boyle also investigated the compression of pockets of gas under varying pressures and concluded that the volume of a gas decreased in proportion to the mechanical pressure exerted on it. In other words, the pressure of the gas was inversely proportional to the volume it occupied—a quantitative

---

[359] Thomas Sprat, *History of the Royal Society* (1667); cf. Yates, *Rosicrucian Enlightenment*, pp. 184-87; Barbara J. Shapiro, 'Latitudinarianism and Science in Seventeenth-Century England', *Past and Present*, vol. xl (July 1968), pp. 21-25. Allen Debus has argued that the establishment of academically credentialed academies and journals in the late seventeenth century was a key factor in the developing hegemony of the mechanical philosophy; 'Science vs. Pseudo-Science' (1979), in idem, *Chemistry, Alchemy and the New Philosophy, 1500-1700* (London: Variorum Reprints, 1987), Paper I, pp. 7f., 16.

[360] Steven Shapin and Simon Schaffer, *Leviathan and the Air-Pump* (Princeton, N.J.: Princeton Univ. Press, 1985), chap. 7, esp. p. 300. Latitudinarianism applied to theological doctrine, not to ecclesiastical practise. Latitudinarians placed a high value on conformity to the rites of the Church of England—another reason for their opposition to spiritualists, who were generally nonconformist and often irenic in matters of both faith and practise; Elmer, 'Medicine, Science, and the Quakers', pp. 281-85; idem, 'Medicine, Religion and the Puritan Revolution', pp. 34-45.

[361] Fritz Krafft, 'Guericke', *DSB* 5:574f.

[362] An equivalent interpretation of the vacuum, using the idea of atmospheric weight, had been put forward by Baliani (1630) and 'demonstrated' experimentally by Toricelli (1644), Mersenne (1646), Magni (1646), Baliani (1647), Perier (1648), and Pascal (1648) using a barometric tube. But the same experiments were given anti-void interpretations by Jesuits like Fabri and Zucchi; Redondi, *Galileo Heretic*, pp. 292-97.

[363] *New Experiments Physico-mechanical, touching the Spring of the Air and Its Effects* (1660; Leicester, *Historical Background*, pp. 113f.); *Usefulnesse* (pub. 1663; Thomas Birch, ed., *Works of the Honourable Robert Boyle*, 6 vols. [London, 1772], 2:37f.); cf. James R. Jacob, 'Boyle's Atomism and the Restoration: Assault on Pagan Naturalism', *Social Studies of Science*, vol. viii (May 1978), pp. 213f.

relation known to all students of physics and chemistry as 'Boyle's law'.[364] Moreover, Boyle concluded that gases were composed of discrete particles moving in a vacuum. Compression merely crowded the particles together; conversely, rarefaction allowed them to stretch out. This aspect of nature could be explained by mechanical principles.[365]

From his new perspective, the mechanical philosophy gradually took on greater plausibility in Boyle's mind. He saw the immense importance of moral self-determination in life. He noted, as Digby had already, that the sectarians who were threatening the social order also tended to be mortalists: they did not accept the church's teaching of the immortality of the human soul. Like Digby, Boyle viewed the mortalist heresy as a threat to the uniqueness of human nature and to belief in moral freedom. In order to counter the mortalist threat, Boyle opted for the matter-spirit duality of the mechanical philosophy, which stressed the incapability of inanimate matter to follow God's laws in the way humans were expected to do.[366]

Boyle thus adopted Descartes's notion of matter as completely passive.[367] He did not regard each instant as a new creation, and he did not treat plants and animals as machines, without a teleology, as Descartes did. But he viewed the ability of the atoms to behave in lawful ways as entirely due to the power of God's original creation and continuing concourse. This view is best expressed in a classic passage of *The Usefulnesse of Experimentall Naturall Philosophy* (published in 1663):[368]

---

[364]'Boyle's law' was first stated in the appendix to the 1662 edition of *New Experiments*, but it was only offered as an hypothesis; *Works*, 1:151; cf. M.B. Hall, 'Boyle', *DSB* 2:373b; Milton, 'Origin', p. 182.

[365]Like Robert Hooke, Boyle viewed the corpuscles as tiny compressed springs that naturally expanded when the pressure was reduced; 'Of the Admirably Differing Extension of the Same Quantity of Air Rarefied and Compressed'; *Works*, 3:509. Regardless of the difficulties involved in this theory (Henry, 'Occult Qualities', pp. 348ff.), Boyle's programme was clearly mechanical in intent.

In 1674, Boyle published his *Tracts Containing Suspicions about Some Hidden Qualities of the Air*, in which he attributed the apparently nonmechanical properties of air (the ability to sustain fire and life) to a 'vital substance', possibly originating from the sun or stars; *Works*, 4:85, 90-91; Thomas S. Hall, *Ideas of Life and Matter*, 2 vols. (Chicago: Univ. of Chicago Press, 1969), 1:291f.; Henry, 'Occult Qualities', pp. 344-45. In his *Tracts about the Cosmicall Qualities of Things* (pub. 1671), Boyle attributed such cosmic effects to the impressions left by peculiar sorts of corpuscles originating from the interior of the earth, the atmosphere, and outer space; *Works*, 3:306-8, 316-18; John Henry, 'Boyle and the Cosmical Qualities,' in M. Hunter, ed., *Boyle Reconsidered*, pp. 120-27.

[366]*Usefulnesse* (pub. 1663; *Works*, 2:38ff.); cf. Jacob, 'Ideological Origins', pp. 17f.; idem, 'Boyle's Atomism', p. 214.

[367]Thus God created a 'great Mass of lazy Matter', *Usefulnesse* (*Works*, 2:43); cf. Clericuzio, 'Redefinition', 571-73.

[368]*Usefulnesse* (*Works*, 2:39); cf. *A Free Enquiry into the Vulgarly Received Notion of Nature* Section II (*Works*, 5:170f.), where Boyle takes issue with Helmont and other 'learned men'.

> ...methinks we may, without absurdity, conceive that God...did divide...that matter, which he had provided, into an innumerable multitude of very variously figured corpuscles, and both connected those particles into such textures or particular bodies, and placed them in such situations, and put them into such motions, that by the assistance of his ordinary preserving concourse, the phaenomena which he intended should appear in the universe must as orderly follow, and be exhibited by the bodies necessarily acting according to those impressions or laws, though they understand them not at all, as if each of those creatures had a design of self-preservation, and were furnished with knowledge and industry to prosecute it....

In other words, the lawful behaviour of the atoms was not due to any intelligence on their own part and did not detract from the uniqueness of humanity in possessing the capacity for real self-preservation and industry.

Like Charleton, Boyle had to face a second threat—that of Hobbes's materialism.[369] Consequently, he followed Charleton in arguing that mere matter, even with the laws God had granted it, could not be expected to produce the kind of organisation one observes in living beings. In these cases, seminal principles must be involved (as Paracelsians and spiritualists held), and these in turn pointed to the design and generative activity of God. Thus Boyle was critical of Descartes's version of the mechanical philosophy for not providing any evidence of the operation of God in the physical world.[370] For similar reasons he was not satisfied with Gassendi's Epicurean atomism, in which God merely endowed the atoms with motion and then retired from the scene.[371] Moreover, he rejected Gassendi's idea of absolute space and time as an infringement on the free creative power of God.[372]

So Boyle tried to steer a middle course between the spiritualism associated with the sectaries and the mechanistic materialism associated with

---

[369] Fisch, 'Scientist', p. 252.

[370] Fisch, 'Scientist', pp. 261f.; cf. Dillenberger, *Protestant Thought*, pp. 114f. In one of his unfinished dialogues (early 1670s) and in his later conversations with Gilbert Burnet (late 1680s), Boyle associated communication with spirits with the alchemical preparation and use of the Philosophers' Stone; M. Hunter, 'Alchemy, Magic and Moralism in the Thought of Robert Boyle', *BJHS*, vol. xxiii (1990), pp. 388-91, 397-99; Lawrence M. Principe, 'Robert Boyle's *Dialogue on Transmutation*', a paper presented at the History of Science Society, Madison Wis., 3 November 1991, pp. 17-18; *idem*, 'Boyle's Alchemical Pursuits', in M. Hunter, ed., *Robert Boyle Reconsidered*, pp. 100-101. Hunter and Principe both see Boyle's general purpose as the defence of the mechanical philosophy (probably in its popular, Cartesian form) against the charge of atheism; M. Hunter, 'Alchemy', p. 395; Principe, 'Boyle's Alchemical Pursuits', pp. 101-2.

[371] *About the Excellency and Grounds of the Mechanical Hypothesis* (1674; *Works*, 4:68); cf. Kargon, 'Walter Charleton', p. 191.

[372] Appendix to *Christian Virtuoso*, Part I (*Works*, 6:684); cf. Kargon, 'Walter Charleton', p. 191.

Hobbes.[373] The result was a stratified view of matter in which the simplest particles operated along completely mechanical lines (though not without God's 'preserving concourse'), and the higher levels of living beings were definitely supramechanical (requiring special providence).[374] Intermediate levels, like those of crystals and metals, were subject to question.[375]

In his earlier writings, Boyle viewed the growth of crystals and gems as exhibiting seminal principles (and hence special, generative providence) just as the growth of living creatures did.[376] But later (in 1672), he treated the structure of gems as reducible to the underlying geometry of the arrangements of atoms.[377] In effect, like Charleton, Boyle was appealing to gaps in the mechanical account as hard evidence for the design and immediate providence of God. But, if crystals could be explained in terms of atomic structures, why couldn't living creatures? Once having made a clear distinction between the generative activity of God (in living beings) and the ordinary mechanisms imposed on matter, Boyle was bound to reduce the scope of evidence for divine providence even if he did not intend to do so.[378]

The mechanical philosophy, as we have seen, had its roots in late medieval scholasticism and the idea of God's *potentia ordinata*, the ordinary

---

[373] Michael Hunter describes Boyle's combination of spiritualist and mechanist ideas as 'electicism'; *Robert Boyle Reconsidered*, p. 13. However, there is a clear asymmetry to Boyle's peculiar synthesis: the mechanical philosophy is espoused openly, while the spiritualist and alchemical ideas are veiled in obscurity. Hunter (*ibid.*, p. 14) argues that Boyle's inhibitions about openly practising alchemy were partly due to his association of it with the subversive enthusiasts of the Interregnum.

[374] The need for particular providence (or 'particular contrivance') in the design of plants and animals is explicitly stated in *Usefulnesse* (pub. 1663; *Works*, 2:43; Clericuzio, 'Redefinition', p. 585); *The Origine of Formes and Qualities* (1666; *Works*, 3:48; W.B. Hunter, 'Seventeenth Century Doctrine', p. 206.

[375] See Clericuzio, 'Redefinition', pp. 579-87, for a good overview of Boyle's hierarchical classification of corpuscles, compounds, and animated bodies.

[376] 'Reflexions' (1658?; M. Boas [Hall], 'An Early Version', p. 168); *History of Fluidity and Firmness* (written in 1659, pub. in *Certain Physiological Essays*, London, 1661: 'a plastic Principle implanted by the most wise Creator'; in *Works*, 1:434; W.B. Hunter, 'Seventeenth Century Doctrine', p. 208); *Sceptical Chymist* (1661: 'perhaps'; *Works*, 1:571; Clericuzio, 'Robert Boyle', 195-96); and *Usefulnesse* (pub. 1663: 'something analogous to seminal Principles'; *Works*, 2:44; Debus, *Chemical Philosophy*, 2:477); cf. Clericuzio, 'Redefinition', pp. 567, 584-85.

[377] *An Essay about the Origine & Virtue of Gems* (London, 1672, pp. 55, 71; W.B. Hunter, 'Seventeenth Century Doctrine', p. 208. The idea that precious gems may have 'some Principle of growth' in them recurs, however, in *An Essay of the Great Effects of Even Languid and Unheeded Motions* (1685; *Works*, 5:26-27; Henry, 'Occult Qualities', p. 345). Lawrence Principe points out that Boyle's alchemical interests appear to have slackened in the late-1650s, as he began to advocate the mechanical philosophy, and revived in the 1670s, as he became more concerned about the possible atheistic consequences of that philosophy; 'Boyle's Alchemical Pursuits', pp. 98-102.

[378] Thus Lawrence Principe notes that Boyle's success as a mechanical philosopher created 'potential dilemma' and 'resulted in a tension with his theological commitments', which led to his later interest in the possible role of spirits in alchemical processes; 'Boyle's Alchemical Pursuits', 101-2.

exercise of God's providence. It gave an account of the origins and maintenance of motion that replaced the outdated Aristotelian worldview. The source of the world's dynamism was no longer the *primum mobile* at the outer boundary of the cosmos, but the creation and energising of numerous atoms or corpuscles of matter. The new mechanics could be cited as evidence of the existence and activity of God just as well as the medieval version could. Only now God was viewed as the Designer and Lawgiver, whereas before he was the Prime Mover and Desire of all things.

The mechanical philosophy was not the only candidate to replace Aristotle as the dominant paradigm in the seventeenth century. Its principal competitor was the spiritualist tradition. In the second half of the century, a decision in favour of the mechanical philosophy was made by leading English scientists and clergy. Their reasons were partly theological (opposition to sectarian millenarianism and mortalism) and partly social (fear of anarchy), as well as strictly experimental (quantitative treatment of air pressure).

The effects of this decision were enormous. The mechanical philosophy was to become the dominant paradigm of Western science for the next two centuries and an integral feature of Western technology and industrial manufacture to this day.[379] Though challenged by the Platonic and Newtonian philosophies we shall discuss next, it incorporated the scientific content of the latter in the late eighteenth century and maintained its hegemony in Western thought through most of the nineteenth century. The benefit of the decision for mechanism was a viable programme for understanding and controlling nature at its most fundamental, material level. The weakness was an isolation of the material, mechanical aspects of nature from the aesthetic, moral, and religious dimensions.

This separation of material and spiritual was not an accident. It was programmed into modern science and culture in reaction to the perceived dangers of the more wholistic science and culture of the spiritualists. The mechanical philosophy that emerged in late seventeenth-century England was not just a scientific programme for investigating the mechanical aspects of nature. After all, spiritualists like Hartlib and Leibniz were also interested in mechanical problems. The new paradigm for physical science was intentionally designed as means of establishing a clear antithesis between the material and the spiritual.

In hindsight, we may say that there was no available alternative consistent with the progress of science and human welfare. Given the theological inheritance of the Middle Ages, in which God's direct action and

---

[379] For the effect of the mechanical philosophy on modern consciousness, see Peter L. Berger et al., *The Homeless Mind* (Harmondsworth: Penguin, 1973).

the normal causal connections of nature were mutually exclusive, the only way to isolate the mechanical aspects of nature was to bracket out the spiritual.[380] The decision in favour of the mechanical philosophy was not made until the mid-seventeenth century, but the theology that required it had been worked out as early as the twelfth (chap. 1, sec. D).

3 The Platonist Tradition to Isaac Newton

The third seventeenth-century tradition we shall examine was centred at Cambridge University. In a sense it is a much narrower movement than the international spiritualist and mechanist traditions we have studied. Its importance stems from the fact that it culminated in the development of a new mathematical physics by Isaac Newton. Newton's ability to conceptualise principles like universal gravitation that transcended the mechanical philosophy of the seventeenth century was facilitated by the ideas of a group of Cambridge Platonists led by John Smith and Henry More, partly mediated through the teachings of the mathematician Isaac Barrow. First, we shall look at the Cambridge Platonists in order to discern the theological roots of Newton's physics. Then we shall look at the Cambridge mathematicians, Isaac Barrow and Isaac Newton.

3.1 The Cambridge Platonists: Smith and More

The Cambridge Platonists were a group of academicians who developed a rationalised form of Neoplatonism as an alternative to the mechanical philosophy of Gassendi and Hobbes. At first they were sympathetic to the rationalism of Descartes, but during the 1650s and early '60s they turned against mechanical philosphers like Descartes and Boyle.

The alarm against mechanist ideas was first sounded in the 1640s by John Smith (1618-52). Smith had been a pupil of the Cambridge Puritan Benjamin Whichcote, sometimes regarded as the founder of the Cambridge

---

[380]The spiritual did not need to be bracketed out in medieval Aristotelian natural philosophy because God's everyday action was so remote (*de potentia ordinata*) to being with (chap. 2, sec. B). The replacement of this spatial separation by modern matter-spirit dualism was only necessitated by the gradual elimination of the center-circumference structure of the cosmos by Cusa (chap. 2, sec. C), Copernicus, Digges, and Bruno (chap. 3, sec. B). In other words, medieval God-nature dualism plus Renaissance decentralisation led directly to modern matter-spirit dualism. So the import of the Copernican revolution was not that it removed humanity from the centre of the cosmos (as so often stated), but that it removed God (normal activity) from the periphery. Descartes set the pattern by viewing God's ordinary activity as ubiquitous, on the one hand, but by separating matter and spirit, on the other. In order to uphold the dualism of God and nature, the ubiquity of God's immediate activity had to be limited to the task of sustaining mechanical processes in the material realm (and possibly influencing souls in the spiritual).

Platonist school, though he himself was not a philosopher.[381] Smith's views best represent the philosophical stance of the school in the 1640s. In reaction to the heated doctrinal controversies of Protestant scholasticism, he had turned to mystical philosophy of the third-century Neoplatonist, Plotinus. Smith saw significant points of contact between the Platonic psychology of Plotinus and the introspective rationalism and matter-spirit dualism of Descartes.[382] Indeed, his directive to 'seek for God within thine own soul' had a striking similarity to Descartes's retreat within to establish his own existence and the existence of God. On the other hand, Smith rejected the Epicurean atomism of Gassendi as a masked form of atheism. In particular, he objected to the idea that motion was inherent in matter as militating against the idea of divine creation and providence.[383]

Henry More (1614-87) became the principal advocate of Cartesian ideas in England after the premature death of Smith. He also took over Smith's role as defender of the faith against Epicureanism, materialism, and atheism. To Smith's critique of the Epicureanism of Gassendi, More added his own campaign against the materialism of Hobbes in his *Antidote against Atheisme* (1653). At this stage, More found Descartes's idea of the sheer passivity of matter useful, since it provided evidence for the independent existence of spiritual beings like God and human souls.[384] He even found evidence of the original version of atomism in Genesis 1—an irrefutable proof to 'atheisticall wits' and 'mere naturalists' that Moses was a 'Master of Natural Philosophy'.[385]

Gradually More came to see the ideas of Descartes as unsuited for his campaign against materialism. His disaffection with Descartes's mechanical philosophy began in the late 1640s[386] and led to a final repudiation of

---

[381]C.A. Staudenbauer argues that Whichcote was influenced by More, rather than the other was around: 'Platonism, Theosophy, and Immaterialism: Recent Views of the Cambridge Platonists', *JHI*, vol. xxxv (Jan. 1974), pp. 159-63.

[382]J.E. Saveson, 'Descartes' Influence on John Smith, Cambridge Platonist', *JHI*, vol. xx (April 1959), pp. 258-63; *idem*, 'Differing Reactions to Descartes Among the Cambridge Platonists', *JHI*, vol. xxi (Oct. 1960), p. 567. Charles Webster disputes the extent of Descartes's influence on Smith: 'Henry More and Descartes: Some New Sources', *BJHS*, vol. iv (1969), p. 361.

[383]*Select Discourses* (pub. 1660), pp. 41, 48; cf. Kargon, 'Walter Charleton', p. 185.

[384]Kargon, *Atomism*, p. 83; Rattansi, 'Social Interpretation', p. 24; William H. Austin, 'More', *DSB* 9:509b. Rattansi and Austin give the date of *Antidote* as 1652, but *DNB* 13:868a gives it as 1653.

[385]*Conjectura Cabbalistica* (1653); Brian P. Copenhaver, 'Jewish Theologies of Space in the Scientific Revolution', *Annals of Science*, vol. xxxvii (Sept. 1980), pp. 516ff.; Joseph M. Levine, 'Latitudinarians, Neoplatonists, and the Ancient Wisdom', in Richard Kroll, Richard Ashcraft, and Perez Zagorin, eds., *Philosophy, Science, and Religion in England, 1640-1700*, (Cambridge: Cambridge Univ. Press, 1992), p. 99.

[386]Already in the late 1640s, More was critical of Descartes's attempt to explain all natural phenomena *ex rationibus mechanicis*; Alan Gabbey, '*Philosophia Cartesiana Triumphata*: Henry More (1646-1671)', in Thomas M. Lennon et al., eds., *Problems of*

Descartes in the mid 1660s.[387] The first clear statement of his new views was made in *The Immortality of the Soul* (1659), a work that had a formative influence on Newton during his undergraduate years at Cambridge.[388] Interestingly, it was written at the same time that Boyle underwent the final stage of his conversion from spiritualism to the mechanical philosophy. In fact, More's final rejection of Descartes's philosophy was partly the result of his antipathy to Boyle's new explanation of the power of a vacuum (1660).

For Boyle, the power of a vacuum could be explained in terms of purely mechanical concepts like atmospheric pressure. More was not convinced. If the pressure or weight of the atmosphere was so great, he asked, why doesn't it flatten a lump of soft butter? Unlike the Aristotelians and Cartesians, More accepted the existence of a vacuum, but he reasoned that its properties could best be explained on the basis of an active spirit that was present even in the absence of matter. Beginning in 1662, More argued his point with Boyle for twenty years.[389] As we saw in our discussion of Boyle, scientific theories in the seventeenth century were governed by theological (and social) views as much as by information about the properties of material substances.

The major problem with Descartes's philosophy for More was its identification of extension with matter and the consequent denial of extension to spirit. If so, More concluded, the spiritual would be incapable of having contact with matter and might as well be ignored, as Hobbes had concluded.[390] In order to prove the materialists wrong, therefore, it would be necessary to show that spiritual substance was extended so as to permeate and influence matter.[391] In other words, More revived the ancient Stoic and Neoplatonic idea of a world soul. He called it the 'Spirit of Nature' or 'hylarchic (matter-ruling) principle'. To demonstrate the need of such a spirit, More appealed to the many phenomena which could not easily be accounted for in terms of mechanical concepts: e.g., the sympathetic vibrations of strings, magnetism, gravitation, and the generation of plants and animals. The Spirit of Nature was unconscious and impersonal, but it

Cartesianism, (Kingston Ont., 1982), pp. 190-93; *idem*, 'Henry More and the Limits of Mechanism', in Sarah Hutton, ed., *Henry More (1614-1687): Tercentenary Studies* (Dordrecht: Kluwer, 1990), p. 21.

[387]Colie, *Light*, pp. 49f.; Webster, 'Henry More', pp. 359f., 365, 376. The subtitle of More's 1671 *Enchiridion metaphysicum* was 'the vanity and falsity of Descartes' philosophy, and in truth the philosophy of all others who suppppose the phenomena of the world can be resolved into purely mechanical causes'; Gabbey, 'Henry More', p. 25.

[388]Dobbs, *Foundations*, p. 103.

[389]Beginning with the 1662 edition of *Antidote against Atheisme*; Robert A. Greene, 'Henry More and Robert Boyle on the Spirit of Nature', *JHI*, vol. xxiii (Oct. 1962), pp. 452, 464f.; Gabbey, 'Henry More', pp. 22f.

[390]*Immortality*; cf. J.E. Power, 'Henry More and Isaac Newton on Absolute Space', *JHI*, vol. xxxi (April 1970), p. 289.

[391]*Immortality*; cf. Jammer, *Space*, p. 42; Austin, 'More', p. 510a.

engineered these supramechanical phenomena by directing the motion of the particles of matter in accordance with the designs of God.[392]

More's mature Platonism can be contrasted with Boyle's version of the mechanical philosophy at this point. Boyle, too, recognised that some phenomena, particularly biological ones, could not be explained in purely mechanical terms. However, for Boyle, such supramechanical phenomena were the exception, not the rule. For More, on the other hand, there was 'no purely Mechanicall Phaenomenon in the whole Universe'.[393]

In opposition to Descartes's identification of extension with matter, More concluded that matter and spirit were both extended in space. What distinguished them was that spirit was active and subtle while matter was passive, gross, and impenetrable.[394] Against Descartes, More concluded that space could exist even without matter and that matter was located *in* space. Space was thus independent of matter—infinite, absolute, and uncreated.[395] According to More, space was a representation or 'shadow' of the divine essence in nature.[396] In effect, More substituted a hierarchical view of reality for the radical dualism of the mechanical philosophy. Between God and matter, in descending order, were absolute space (a direct manifestation of God) and the Spirit of Nature (a separate subordinate entity).

As Max Jammer pointed out in 1954, More's concept of absolute space had roots in the Jewish and Christian cabalist traditions, which viewed God as the *maqôm*, or place of the world (chap. 3, sec. A).[397] As early as 1653 More had discussed cabbalistic ideas with his friend and patroness, Lady Anne Conway,[398] and he was initially attracted to them. But More did not

---

[392] Among the writings in which these ideas appear see *Antidote* (1653 and 1662 eds.); cf. Rattansi, 'Social Interpretation', p. 24; Austin, 'More', p. 509b; Gabbey, 'Henry More', pp. 22f. *Immortality*; cf. Greene, 'Henry More', pp. 453ff., 460; Dobbs, *Foundations*, pp. 104f. *Enchiridion metaphysicum* (1671) also cites the tides, the nature of light and colour, and the cohesion of slabs of polished marble; cf. Greene, 'Henry More', p. 471; Gabbey, 'Henry More', p. 25.

[393] *Divine Dialogues* (1668); cf. Greene, 'Henry More', p. 468; Webster, 'Henry More', p. 376. For an attempt to define More's view more precisely (consistently antimechanist or mixed mechanist), see Gabbey, 'Henry More', pp. 24-29.

[394] Jammer, *Space*, pp. 42f.

[395] Jammer, *Space*, pp. 43-47.

[396] Letter to Descartes (1648); cf. Copenhaver, 'Jewish Theologies', p. 519. *Enchiridion* (1671); cf. Power, 'Henry More', p. 290.

[397] Jammer, *Concepts of Space*, pp. 41f., 48; Copenhaver, 'Jewish Theology', pp. 519-28.

[398] Alic, *Hypatia's Heritage*, p. 6. In 1653, More published his *Conjectura Cabbalistica*, dedicated to Ralph Cudworth, but begun at the request of Lady Conway; Richard H. Popkin, 'The Third Force in Seventeenth Century Philosophy: Scepticism, Science and Biblical Prophecy', *Nouvelles de la république des lettres*, vol. i (1983) p. 56n.; Levine, 'Latitudinarians', p. 99.

The philosophy of Anne Conway was a monistic vitalism that belongs in the category of spiritualism along with that of her physician and friend Francis Mercury van Helmont; Carolyn Merchant, *The Death of Nature* (San Francisco: Harper & Row, 1980), pp. 258-64; Popkin, 'The Third Force', p. 60; Alison Coudert, 'Henry More, the Kabbalah, and the

see himself as a cabalist. In fact, he concluded in the 1670s that Cabbalism was inconsistent at points with Christian doctrine. The cabbalistic notion of a series of ten emanations of God giving rise to the phenomenal world, for instance, contradicted the doctrines of the Trinity and creation *ex nihilo*.[399]

Similarly, More's concept of the 'Spirit of Nature' or 'hylarchic principle' may have been inspired by the Paracelsian-Helmontian idea of an 'archeus' ruling over each organic entity.[400] Yet More was not a spiritualist either. Though he gradually parted company with the mechanical philosophy in the 1650s, he retained much of the rationalism of Descartes in his philosophy. In fact, More criticised Paracelsus, Boehme, Francis Mercury van Helmont, and Thomas Vaughan for relying on divine illumination, rather than on illuminated human reason, in their scientific work.[401] In addition, More lacked the pragmatic interests of the spiritualists and pansophists in experimental research and social reform.[402]

3.2 The Cambridge Mathematicians: Barrow and Newton

Barrow and Newton were not primarily philosophers like the Cambridge Platonists: they were mathematicians and biblical scholars. In 1663, Barrow became the first Lucasian Professor of Mathematics at Cambridge. In 1669, he gave up the Lucasian Chair in order to devote himself to the ministry of the Word and sacraments, and Newton was appointed to succeed him.

Though Barrow was not a Platonist in the proper sense, he was deeply influenced by the Platonic ideas of Henry More, and he developed a more precise version of these in his mathematical lectures. As early as 1652,

---

Quakers', in Richard Kroll et al., eds., *Philosophy, Science, and Religion in England, 1640-1700*, pp. 37f.

[399]'Scholium'; cf. Staudenbauer, 'Platonism', p. 168; Copenhaver, 'Jewish Theology', pp. 523-27; Coudert, 'A Cambridge Platonist's Kabbalist Nightmare', *JHI*, vol. xxxvi (1975), pp. 647f. More also attempted to interpret the ten *sefirôth* in terms of the three Neoplatonic hypostases; Coudert, 'Henry More', pp. 51, 65 note 96.

[400]Greene, 'Henry More', p. 454.

[401]On Paracelsus, *Enthusiasmus triumphatus* (1656); cf. Rattansi, 'Paracelsus', pp. 29f. On Boehme, *Philosophiae Tertonicae censura* (1670); cf. Stephen Hobhouse, *Selected Mystical Writings of William Law* (London: Rockliff, 1948), p. 419. On Helmont, *Fundamenta philosophiae*; cf. Staudenbauer, 'Platonism', pp. 166f. On the pamphlet war with Vaughan in 1650-51, see Mulligan, 'Reason', 384-91; Greene, 'Henry More', pp. 456ff.; Dobbs, *Foundations*, pp. 116f.; Robert Crocker, 'Mysticism and Enthusiasm in Henry More', in S. Hutton, ed., *Henry More*, pp. 144-47.

According to More, 'an *Enthusiast* is a *Poet in good earnest*; Melancholy prevailing so much with him, that he takes his no better then Poeticall fits, and figments for divine inspiration and reall truth'; *Enthusiasmus triumphatus* (Williamson, *Seventeenth Century Contexts*, p. 220).

[402]Webster, 'Henry More', pp. 363, 372f. More has also been cited as a member of a 'third school' of thought in seventeenth-century natural philosophy in a slightly different way—as a theosophical-millenarian alternative to rationalism and empiricism in reaction to the sceptical crisis of the seventeenth century; Popkin, 'Third Force', pp. 38f., 41f., 55ff.; B.C. Southgate, '"Forgotten and Lost": Some Reactions to Autonomous Science in the Seventeenth Century', *JHI*, vol. l (April 1989), pp. 265f.

Barrow was using Descartes's idea of the passivity of matter to argue for the existence of spirit. Like More, he criticised Descartes for separating spirit so surgically from matter, thus making nature 'blockish and inanimate'. The mechanical laws of nature, he concluded, could not account for such phenomena as magnetism and the growth of living beings.[403]

In 1665, Barrow gave a lecture, 'Of Space and Impenetrability', which provided the young Newton with the starting point for his own natural philosophy. Barrow criticised Descartes's teachings for implying that matter was infinite and eternal (though Descartes did not draw these implications himself). Since a God who creates could increase or decrease the amount of matter in the universe, one could not regard matter as sharing God's attributes without running the risk of denying creation. Space, on the other hand, could neither be augmented nor diminished. Hence, it was immutable as More had taught. Space must also be infinite, or else God would be bounded. Similarly it must be eternal, or else God would once have been nowhere. Barrow concluded, therefore, that space and time were a mathematical representation of the divine omnipresence and eternity.[404]

Newton's new physics was essentially a mathematical version of the mechanist idea of matter supplemented by the ideas of absolute space and the operation of supramechanical principles like gravitation to mediate between bodies. All of these ingredients were provided by the mechanist and Platonic traditions as we have seen: the passivity of matter, by Descartes, Boyle, and More; absolute space, by Gassendi, Charleton, More, and Barrow; the idea of universal active principles, by More and Barrow.[405] But it required the mathematical skill and the ascetic discipline of Newton to combine these ingredients into a workable mathematical physics.

---

[403]'Cartesiana hypothesis de materia et motu' (*Theological Works*, ed. Alexander Napier, [Cambridge, 1859], 9:79-104); cf. Edward W. Strong, 'Barrow and Newton', *JHP*, vol. viii (April 1970), pp. 156f.; Webster, 'Henry More', pp. 361f.; Dobbs, *Foundations*, pp. 100f.

[404]Jammer, *Space*, p. 111; Strong, 'Barrow and Newton', pp. 156, 160; Whitrow, *Time*, 128.

[405]For the idea of absolute (evenly flowing) time, Newton was indebted to the tradition of Telesio, Suarez (by analogy with the imaginary, potentially infinite space postulated in the late Middle Ages; cf. chap. 2, sec. C), Bruno (in relation to his pluralist cosmology; chap. 3, sec. B), Gassendi, and Barrow; Milic Capek, 'The Conflict between the Absolutist and the Relational Theory of Time before Newton', *JHI*, vol. xlviii (Oct. 1987), pp. 595-608; Richard T.W. Arthur, 'Newton's Fluxions and Equably Flowing Time', *SHPMP*, vol. xxvi (June 1995), pp. 328-33. Betty Jo Dobbs argues that Newton's interest in alchemy was the source of his notion of supramechanical principles, though his alchemical sources were distinctly Neoplatonic in tone; *Newton and the Culture of Newtonianism* (coauthored by Margaret Jacob [Atlantic Highlands, N.J.: Humanities Press, 1995]), pp. 25-31. This may be so, but alchemical spirits were responsible for the formation of all matter, whereas Newton's active principles were only responsible for certain supramechanical phenomena like gravitation, fermentation, and cohesion; *Opticks* (New York: Dover, 1952), pp. 400f.

Newton first developed his ideas on God and nature in a paper, 'On the Gravity and Equilibrium of Fluids' (probably written in the late 1660s).[406] The characteristic features of this early work that place Newton squarely in the tradition of More and Barrow are the rejection of Descartes's identification of extension with matter as leading to atheism and the postulation of absolute space and time as 'emanent effects' of God and 'dispositions of all being'.[407] Following More and Barrow, Newton viewed the attribution to God of extension and duration as the only way to preserve the biblical doctrines of monotheism and creation *ex nihilo*:[408]

> If we say with Descartes that extension is body, do we not manifestly offer a path to Atheism, both because extension is not created but has existed eternally, and because we have an absolute idea of it without any relationship to God,...Moreover, if the distinction [made by Descartes] of substances between *thinking* and *extended* is legitimate and complete, God does not eminently contain extension within himself and therefore cannot create it; but God and extension will be two substances separately complete, absolute, and having the same significance.

In order to maintain a clear distinction between the created and uncreated, Newton felt, it was necessary to separate space (and time) from the realm of matter and place it on the side of the divine. So the basic principles of Newton's natural philosophy were rooted in the creationist tradition as mediated by Henry More's critique of the mechanical philosphy of Descartes.[409]

The development of Newton's theological interpretation of space and matter can be traced in the mature physics of his *Principia mathematica* ('Mathematical Principles of Natural Philosophy': first edition, 1687) and *Opticks* (first edition, 1704). Newton's thinking on specific issues changed over the years, and some historians have developed complex accounts of the

---

[406]*De gravitatione et aequipondio fluidorum*, (A.R. and M.B. Hall, eds., *Unpublished Scientific Papers of Isaac Newton* [Cambridge: Cambridge Univ. Press, 1962], pp. 89-156); cf. Strong, 'Newton', pp. 155ff. Betty Jo Dobbs places this work in the mid 1680s; *The Janus Faces of Genius* (Cambridge: Cambridge Univ. Press, 1991), pp. 143f.

[407]Hall and Hall, eds., *Scientific Papers*, pp. 132, 136f., 142f.

[408]Hall and Hall, eds., *Scientific Papers*, p. 142f.

[409]In support of More's influence: Burtt, *Metaphysical Foundations*, pp. 150f., *passim*; Jammer, *Space*, pp. 110f.; Koryé, *From the Closed World*, pp. 125-89; R.S. Westfall, 'The role of Alchemy in Newton's Career', in M.L. Righini Borelli and W.R. Shea, eds., *Reason, Experiment, and Mysticism* (New York: Science History Publications, 1975), pp. 216f.; *idem, Never at Rest*, pp. 301, 304, 321, 348f.; J.E. McGuire, 'Force, Active Principles, and Newton's Invisible Realm', *Ambix*, vol. xv (Oct. 1968), pp. 184f.; *idem*, 'Neoplatonism and Active Principles', in R.S. Westman and J.E. McGuire, eds., *Hermeticism and the Scientific Revolution* (Los Angeles: William Andrews Clark Memorial Library, 1977), pp. 45, 102, 132. McGuire qualifies his earlier assessment of More's influence in 'Existence, Actuality and Necessity: Newton on Space and Time', *Annals of Science*, vol. xxxv (1978), pp. 463ff., 505f.

ebb and flow of the differing emphases. There was a period in the mid-1670s, for instance, when Newton tried to account for phenomena like gravity and the diffraction of light in mechanical terms. He postulated an etherial fluid of tiny particles that transmitted forces by streaming from one body to another in a manner reminiscent of Descartes's mechanistic explanations.[410] By the mid-1680s, however, Newton had abandoned the attempt to find a mechanical explanation for gravitation. Though he continued to speculate on the existence of a variety of ethers, he never embraced the mechanical model again.[411] Here, for simplicity, we shall treat Newton's basic principles as firm and unchanging after the early 1680s. What were those principles?

As Newton later explained to Richard Bentley (1692), he wrote the *Principia Mathematica* with 'an eye upon such Principles as might work with considering men, for the belief of a Deity'.[412] In the context of his concern to refute atheism, Newton followed More's strategy of postulating supramechanical forces like gravity to account for the interactions between

---

[410]'An Hypothesis to Explain the Properties of Light', paper presented to the Royal Society and published in the *Philosophical Transations* in 1675; cf. Richard S. Westfall, *Force in Newton's Physics* (London: Macdonald, 1971), pp. 323-423; *idem, Construction*, p. 140.

In *Never at Rest* (1975, pp. 307f.), Westfall sees this treatise relating to the alchemical as much as to the mechanical tradition, but, in 'The Influence of Alchemy on Newton' (J. Chance and R.O. Wells, eds., *Mapping the Cosmos* [Houston: Rice Univ. Press] p. 102), he describes it as 'a total, mechanistic system of nature'. See *Never at Rest*, p. 508, on Newton's rejection of the aetherial mechanisms proposed by Huyghens and Leibniz, and Henry, 'Occult Qualities', p. 344, on his second thoughts about the 'Hypothesis' itself.

[411]*De aere et aethere* (*c*. 1679); cf. Westfall, *Force*, pp. 373-78; *Construction*, pp. 141f.; Dobbs, *Foundations*, pp. 211f. Dobbs locates the shift in 1684, when Newton derived Kepler's second law of planetary motion from the inverse square law of gravitation; 'Newton's Alchemy and His "Active Principle"', in P.B. Scheurer and G. Debrock, eds. *Newton's Scientific and Philosophical Legacy* (Dordrecht: Kluwer, 1988), p. 55; *idem, Janus Faces*, pp. 128ff.; *Newton.* pp. 36f.

The aether hypothesised in the 1717-18 *Opticks* was not strictly mechanical, according to Westfall (*Construction*, p. 157), since its parts did not act on each other by contact. According to Kargon (*Atomism*, p. 138), David Kubrin ('Newton and the Cyclical Cosmos: Providence and the Mechanical Philosophy', *JHI*, vol. xxviii [July 1967], p. 339), and McGuire ('Force', pp. 155, 174, 186f.), however, it was quasimechanical. McGuire points out that the aethereal spirit of the *Opticks* was devised more for optical than for gravitational phenomena ('Force', pp. 179ff.). R.W. Home argues that Newton stuck to quasimechanical models for the phenomena of electricity, magnetism, and the vegetative powers of life (nutrition, growth, and generation); 'Force, Electricity, and the Powers of Living Matter in Newton's Mature Philosophy of Nature' in M.J. Osler and P.L. Farber. eds., *Religion, Science, and Worldview*, Cambridge: Cambridge Univ. Press (1985), pp. 101-117.

[412]Letter to Bentley, 10 Dec. 1692; H.W. Turnbull et al., eds., *The Correspondence of Isaac Newton*, 7 vols. (Cambridge: Cambridge Univ. Press, 1959-77), 3:233; D.C. Goodman, ed., *Science and Religious Belief, 1600-1900* (Open University, 1973), p. 131. The context was Newton's encouragement of Bentley's refutation of atheism in the 1693 Boyle Lectures; cf. G.N. Clark, *Science and Social Welfare in the Age of Newton* (Oxford: Clarendon Press, 1949), pp. 82f.

isolated bodies of matter.[413] Since such forces were not explicable as properties of matter, the latter being passive, and could not be accounted for in terms of mechanical contact between bodies, they were evidence of God's imposition and maintenance of control on the stuff of the universe.[414] In effect, Newton's active forces were a mathematical version of More's concept of a Spirit of Nature, similarly adopted in an effort to refute atheism.[415] In place of More's hierarchical model of God, space, the Spirit of Nature, and passive matter, Newton substituted his own hierarchy of God, space, active principles (like gravitation), and passive matter.[416]

A comparison with the spiritualist and mechanist traditions will clarify the distinctive place of these ideas. Newton agreed with the spiritualists that God had prescribed laws for nature, but he did not view these as immanent in matter and specific to individual creatures as the spiritualists did. The principle of gravitation was imposed on matter and universal in scope, like More's Spirit of Nature. On the other hand, Newton agreed with mechanists like Descartes and Boyle that matter by itself was entirely passive, but he did not view the inertial motion and collisions of material bodies as an adequate explanation of inanimate phenomena like gravitation, the diffraction of light, and cohesion. The principles involved in phenomena like these were active and supramechanical.

Newton's supramechanical principles accounted for the source of motion, not just its transmission from one body to another. Thus they played a role

---

[413]The causal role of Newton's theism is stressed by Westfall: 'The Career of Isaac Newton', *American Scholar*, vol. 1 (Summer 1981), pp. 348f.; *idem,* 'Newton and Alchemy', in B. Vickers, ed., *Occult and Scientific Mentalities* (Cambridge: Cambridge Univ. Press, 1984), pp. 321f., 328ff.

[414]*De aere* (Hall and Hall, eds., *Scientific Papers*, p. 223). Second and Third Letters to Bentley, 1693 (I. Bernard Cohen, ed., *Isaac Newton's Papers and Letters on Natural Philosophy* [Cambridge, Mass.: Harvard Univ. Press, 1958] pp. 303, 344); cf. Kargon, *Atomism*, pp. 135ff.; Westfall, *Never at Rest*, p. 505. Query 28 (*Opticks*, p. 369). Letter to Clarke, May 1712 (McGuire, 'Force', pp. 202f.).

Other examples of supramechanical principles are the forces that cause the refection, refraction, and diffraction ('inflecting') of light, electricity, magnetism, fermentation, and cohesion: Query 31 (*Opticks*, pp. 375f., 401). John Henry notes that the supramechanical nature of fermentation and its role in sustaining life processes were common ideas in medical literature of the time, e.g., in Thomas Willis's *A Medico-Philosophical Discourse of Fermentation*, published in 1684; Henry, 'Occult Qualities', pp. 343f.

[415]Note the wording of Newton's Letter to Oldenburg, 1675: 'the power of Nature, wch by vertue of the command Increase & Multiply [Gen. 1:22, 28], became a complete Imitator of the copies sett her by the Protoplast' (Turnbull et al., eds., *Correspondence*, 1:364) and the proposed Corollary 9 to Proposition VI, Book III of the *Principia* (1693): 'There exists an infinite and omnipresent spirit in which matter is moved according to mathematical laws' (Westfall, *Never at Rest*, pp. 308, 509). In 'An Hypothesis...of Light' (1675), Newton introduced the aether as an active spirit of nature (Westfall, 'Influence', p. 111).

[416]McGuire, 'Force', pp. 184ff. According to McGuire ('Force', pp. 161ff., 195ff.), the mediation of God's rule in nature was a new emphasis beginning aroung 1706, but the citations of the previous note show the idea to have been in Newton's mind all along.

analogous to that of the *primum mobile* in Aristotle's cosmology. On one hand, they were directly subject to divine influence; on the other, they operated in accordance with regular natural laws. Like the *primum mobile*, therefore, they provided a link between the natural and the supernatural worlds.

So, even though they depended directly on God, Newton's supramechanical principles were not 'miraculous' or supernatural in the sense of being unpredictable, and they in no way interfered with the ability of scientists to explain phenomena in terms of second causes. Gravity, for instance, operated according to a precise mathematical law: its strength was proportional to the product of the masses of, and inversely proportional to the square of the distance between, the bodies between which it operated. Even comets, traditionally thought to be special agents of God, obeyed this universal law and were periodic phenomena according to Newton.[417] The value of gravitation as evidence of God's activity was due to the fact that it could not be derived from intrinsic (primordial) properties of matter such as mass and extension: hence, it required the imposition of a supramechanical law on matter.[418] An important consequence was that laws like that of universal gravitation could not be deduced from first principles by pure reason as mechanical philosophers like Descartes had hoped. They could only be inferred from an empirical study of the phenomena themselves.[419]

As we have seen, the concept of a law of nature was derived from the ancient Near Eastern (biblical) belief that God ordained laws for all his subjects, even inanimate ones. This idea was transmitted to modern western Europe by the medieval scholastics. It was developed in the late sixteenth century by Suarez and in the early seventeenth by Descartes, whose writings influenced the early members of the Royal Society and Newton.[420] Thus there is a clear line of transmission in the context of the creationist tradition through the seventeenth century before the idea of laws of nature was secularised in the eighteenth century.

Newton did not entirely eliminate the need for the extraordinary exercise of God's power, however. There were at least two points, as he saw it, where the laws of nature were inadequate. One, of course, was the creation of the world system itself. This was not just a matter of faith: it was a direct implication of Newton's physics. One of the basic laws of mechanics

---

[417] J.E. McGuire, 'Transmutation and Immutability: Newton's Doctrine of Physical Qualities', *Ambix*, vol. xiv (June 1967), p. 86.

[418] Letter to Clarke, May 1712 (McGuire, 'Force', pp. 202ff.). But cf. Newton's comparison of gravitation with the primary qualities of matter in his reply to *Memoirs of Literature* (c, 1712); Turnbull et al., eds., *Correspondence*, 5:299f.

[419] Crombie, *Augustine*, vol. II, pp. 323ff.

[420] Zilsel, 'Physical Law', pp. 269-75.

(Newton's third law of motion) was that for every action there was an equal and opposite reaction. Clearly this law did not apply to the moment of creation, however. God set all things in motion, but he was not set in motion himself. Therefore, as Newton put it, the First Cause of all things was certainly not mechanical.[421] Furthermore, the detailed structure of the solar system could not be explained in mechanical terms either, according to Newton. For example, the corotation of the planets and coplanarity of their orbits could only be accounted for by the deliberate design and direct action of God in creation.[422]

Once the bodies of the universe were set in motion and the supramechanical laws of nature were imposed, the world could 'continue by those laws for many ages', as Newton put it in the 1706 (Latin) edition of his *Opticks*. It would only require the willing concurrence, or general providence, of God for its continuance.[423] However, at least two problems would eventually arise in such a way as to require divine intervention. For one thing, irregularities would appear in the orbits of the planets due to the gravitational interaction with passing comets as well as with each other. At some point these irregularities would become so great that the planetary system would require a reformation.[424]

Secondly, the amount of motion (i.e., velocity) in the universe would gradually decrease through friction and inelastic collisions, and periodic

---

[421] Queries 28, 31 (1706; *Opticks*, pp. 369, 397); cf. Stephen Hobhouse, 'Isaac Newton and Jacob Boehme', *Selected Mystical Writings of William Law*, pp. 417f. In the General Scholium to the 1713 *Principia*, Newton argued that the omnipresence of God continually contained all things without offering any resistance to motion (Florian Cajori, ed., *Sir Isaac Newton's Mathematical Principles* [Berkeley: Univ. of California Press, 1962], p. 545).

[422] Letters to Bentley, 10 Dec. 1692; 17 Jan. 1693; 11 Feb. 1693; 20 Feb. 1693; Turnbull et al., eds., *Correspondence* 3:234f., 240, 244, 254f.; Goodman, ed., *Science*, pp. 132ff., 136.

[423] Query 31 (*Opticks*, p. 402). Newton was recorded by David Gregory as saying (May 1694) that the sun and stars would gravitate towards their common center of gravity unless God continually intervened through a 'continual miracle' to prevent their collapse; Turnbull et al., eds., *Correspondence*, 3:334 (Latin), 336 (ET); cf. Letters to Bentley, 10 Dec. 1692; 25 Feb. 1693; Turnbull et al., eds., *Correspondence* 3:234, 254f.; Michael Hoskin, *Stellar Astronomy* (Chalfront St Giles: Science History Publications, 1982), pp. 71, 87f. This thought also appeared in Query 28 (1706; *Opticks*, p. 369). This put God in the unenviable position of having to prevent some of the consequences of the active principles he normally sustained! In the 1713 and 1726 *Principia*, however, Newton suggested that the stars were far enough apart to make their mutual gravitation negligible and hence that the original design of God was sufficient to account for the relative stability of the stellar realm (Proposition XIV, Corollary 2 and General Scholium; Cajori, ed., *Mathematical Principles*, p. 422, 544). This may have been in response to Berkeley's citation of the stability of the fixed stars as an exception to universal gravitation: *Principles of Human Knowledge* CVI. According to Hoskin (*Stellar Astronomy*, pp. 87f.), continual divine intervention was still required due to small deviations from spherical symmetry in the stellar system.

[424] Query 31 (*Opticks*, p. 402).

reformations would be required to restore it.[425] Newton was never quite sure just how God would accomplish such a reformation. He made an exhaustive study of biblical prophesies like 2 Peter 3:7-13 to find answers.[426] With Edmond Halley and others, he explored the possibility of a periodic influx of cometary exhalations.[427] But at some time, in some way, a supernatural intervention by God would be necessary. Indeed, God was capable of intervening at any time, since the very framework of space and time was his emanent effect. Thus, God was...[428]

> ...a powerful ever-living Agent, who being in all Places, is more able by his Will to move the Bodies within his boundless uniform Sensorium [space], and thereby to form and reform the Parts of the Universe, than we are by our Will to move the Parts of our own Bodies.

The problem of the dissipation of motion in Newton's cosmology was due to the underlying dualism of active principles and passive matter. Active principles like gravitation were superimposed on a recalcitrant matter in the form of fundamental particles. The particles were infinitely hard, immutable and inert. Though created by God in the beginning, they did not manifest his continued presence as space did or his immediate activity as active principles did.[429]

It was at this point that Leibniz took issue with Newton. A brief comparison between the two is instructive for the differing ways in which the theological ideas of the creationist tradition could be developed. There were two fundamental issues: the autonomy of nature and the relationship of God to space.

Leibniz, as we have seen, held that God had endowed brute matter with the ability to execute his laws perpetually. It was partly his belief in the indefatigability of creaturely motion that led him to postulate a *vis viva*

---

[425]Query 31 (*Opticks*, pp. 397ff.). Ted Davis states that Newton 'captured the essential thrust of the law of entropy', discovered a century and a half later by Carnot and Kelvin; Edward B. Davis, 'Newton's Rejection of the "Newtonian World View"', *Fides et Historia* vol. xxii (Summer 1990), p. 13. The analogy is worth developing. Newton's 'quantity of motion' could be 'conserved and recruited' by active principles like gravitation and fermentation (Query 31; *Opticks*, pp. 399ff.), i.e., by sources of energy given in the initial state of a closed system. However, the Leibnizian concept of energy conservation (*vis viva*) had to be developed in a mathematical form before these ideas could be given precision in nineteenth-century thermodynamics.

[426]Kubrin, 'Newton', p. 332.

[427]Simon Schaffer, 'Newton's Comets and the Transformation of Astrology', in P. Curry, ed., *Astrology*, pp. 233-37; Sara Schechner Genuth, 'Newton and the Ongoing Teleological Role of Comets', in Norman J.W. Thrower, ed., *Standing of the Shoulders of Giants* (Berkeley: Univ. of California Press, 1990), pp. 301f.

[428]Query 31 (*Opticks*, p. 403); cf. *De gravitatione* (Hall and Hall, eds., *Scientific Papers*, pp. 138f.). Comparison of these two passages shows the remarkable continuity of Newton's thought from *c.* 1670 to 1706.

[429]McGuire, 'Transmutation', pp. 82ff.

(equivalent to the later idea of energy) that was conserved even in inelastic collisions. According to Leibniz, the *vis viva* was absorbed into the internal dynamics of the constituent particles of the bodies even when the latter ceased to move visibly. Newton, on the other hand, refused to allow a reduction of the inner dynamics of bodies to mechanical terms due to his own concern to preserve the role of God in the cosmos. As a result, God was required to restore the amount of motion supernaturally at various intervals. To Newton, a lack of complete autonomy in nature was consistent with the omnipotence of God. For Leibniz, on the other hand, it was a denial of the perfection of the original creation and, hence, inconsistent with the omnipotence of God.

Similarly, Leibniz could only see Newton's postulation of absolute space as the elevation of a creature to quasidivine status. Corresponding to Leibniz's view of nature as nearly autonomous, was a view of God as highly transcendent, beyond all space and time. Consequently, Leibniz advocated a concept of relational space which had similarities to the post-Newtonian view of space-time developed by Albert Einstein. Newton, on the other hand, viewed space as God's means of controlling all nature. Absolute space, for him, like the decaying amount of motion in the universe, was required by the notion of an omnipotent God, whereas just the opposite conclusions were reached by Leibniz.

In his theology, Newton was an 'Arian', that is, he believed that the pre-existent Christ was the first of God's creatures.[430] His views were not widely known during his lifetime and did not influence the teaching of the Church of England in any way. However, they were influential for an important group of his disciples, including Samuel Clarke and William Whiston, whom we shall discuss in the next chapter.

Indeed, there was an organic (if not necessary) relation between Newton's natural philosophy and his Arianism. This can be seen by looking again at his concept of the relationship between God, space, and time. For Newton, God was not outside space and time. God's infinity was an infinite extension or spatial infinity and his eternity was a limitless duration or unending time. God's extension and duration constituted space and time as we know them.[431] When challenged on this point by Berkeley and

---

[430]Newton clearly rejected the theology of Athanasius (largely based on the 'Athanasian Creed' of later date), but he could also be critical of Arius for introducing 'metaphysical opinions...expressed in a novel language not warranted by Scripture'; Yahuda Ms. 15.7, in Frank E. Manuel, *The Religion of Isaac Newton* (Oxford: Clarendon Press, 1974), p. 58; cf. Rob Iliffe, "'Making a Shew": Apocalyptic Hermeneutics and the Sociology of Christian Idolatry in the Work of Isaac Newton and Henry More', in James E. Force and Richard H. Popkin, *The Books of Nature and Scripture* (Dordrecht: Kluwer, 1994), p. 65.

[431]McGuire, 'Force', p. 200.

Leibniz,[432] Newton added a note to the second (1713) edition of the *Principia* which differentiated his view from a pantheistic identification of God with space and time but did not alter his essential position:[433]

> He is not eternity and infinity, but eternal and infinite; he is not duration or space, but he endures and is present. He endures forever, and is everywhere present; and by existing always and everywhere, he constitutes duration and space....He is omnipresent not *virtually* only, but also *substantially*....

This correlation of God's eternity with unending time meant that all of God's acts, internal and external, could be placed on a time line. Newton saw God as 'He that was and is and is to come' (Rev. 1:4, 8, altered to suggest a temporal sequence of past, present, and future for God), rather than the mysteriously transcendent God of scholasticism for whom all time was a single point (*totum simul*).[434]

Therefore, Newton could not differentiate between God's internal operations (*opera ad intra*), like the generation of the Son and the procession of the Spirit, and his external operations (*opera ad extra*) by appealing to a qualitative difference between eternity and time. Christ was, for Newton, the pre-existent Son of God, begotten before all worlds, but his generation was an event on the same time line as the creation of the world. The only difference between the generation of the Son and the creation of the world was that the former was invisible and temporally prior to the latter. Before the creation of this visible world, God exercised his omnipotence by creating an invisible world, as Origen and Basil had held (chap. 1, sec. B).[435] But, for Newton, this invisible creation included the Son of God as well as the angels. The Son was, therefore, a perfect creature, the first and greatest of all God's creatures, the instrument through which the material world was subsequently created, but neither coeternal nor consubstantial with the

---

[432]Berkeley, *Principles of Human Knowledge* (1710); Leibniz, Letter to Hartsoeker, 10 Feb. 1711 (pub. in *Memoirs of Literature*, 5 May 1712); cf. Cajori, ed., *Mathematical Principles*, pp. 668f.; Turnbull et al., eds., *Correspondence*, 5:300.
[433]*Principia*, General Scholium to Book III (1713 ed.; Cajori, ed., p. 545).
[434]Additional ms. 3965, draft revision for 2nd ed. of the *Principia* written in the early 1690s [J.E. McGuire, 'Newton on Place, Time, and God: An Unpublished Source', *BJHS*, vol. xi (1978), pp. 121; Copenhaver, 'Jewish Theologies', p. 543].
[435]Additional ms. 3965 (McGuire, 'Newton', p. 143).

Father.[436] Thus the notion of God's eternity as limitless duration or unending time made Arianism a plausible Christology for Newton.

The overall consistency of Newton's Arian Christology with his natural philosophy can also be seen from his view of space as an 'emanent effect' of God. Space played the role, in Newton's mind, of the eternal Son of God in traditional theology. It was eternal; it was an image of God's substance; it was the primary mediator between God and his creation; it was that without which God would not be the God of Scripture.[437] However, space was entirely impersonal and could not be regarded as a second hypostasis or personal companion of God. Newton's God was, therefore, alone in infinite space and endless time, having only a finite creature to share with (after creation) for a limited time. Again his natural philosophy made the Arian view of God and Christ quite plausible.

In order to assess Newton's contribution to our understanding of God and nature from a historic creational perspective, one must recall the situation in theology and science as he found it. The dominant paradigm of natural philosophy was the mechanist tradition which was predicated on a dualism of spirit and matter, God and nature. Newton rightly sensed that such a dualism was unbiblical and would lead to deism, or even atheism, if allowed to persist unchallenged. Positively, then, we may view Newton as a champion of biblical theism, a view in which God is active in nature, and nature is only relatively autonomous. The affirmation of a First Cause was,

---

[436]Newton's first explicit statements of an Arian Christology occur in his theological mss. of the early 1670s (Westfall, *Never at Rest*, pp. 310-18, 331f.). An Arian stance is indicated already in the *De gravitatione* of the late 1660s (Hall and Hall eds., *Scientific Papers*, p. 142), although Newton did affirm the Thirty-Nine Articles of the Church of England in 1665 and 1668 (Westfall, *Never at Rest*, pp. 330f.). Note that Betty Jo Dobbs places *De gravitatione* in the mid 1680s; *Janus Faces*, pp. 143f.

Thomas Pfizenmaier has shown that between 1675 and the writing of the Mint papers (after 1696) Newton developed a sympathy with the moderate Post-Nicene theologians who viewed Christ as similar to the Father in substance (*homoiousios*) rather than identical (*homoousios*); 'Was Isaac Newton an Arian?' *JHI*, vol. lviii (Jan. 1997), pp. 69-79. In the absence of any clear affirmation of this position, however, the consensus that Newton was an Arian will stand.

[437]'If ever space had not existed, God at that time would have been nowhere...' (*De gravitatione*, Hall and Hall, eds., *Scientific Papers*, p. 137). Compare Origen, Alexander of Alexandria, and Athanasius on the necessity of having an eternal Son to the being of God as Father.

Betty Jo Dobbs has argued that Newton equated the pre-existent Christ with the alchemical spirit symbolised by Hermes; 'Newton's *Commentary* on the *Emerald Tablet* of Hermes Trismegistus', in Merkel and Debus, eds., *Hermeticism*, pp. 187ff. For the background of this idea in sixteenth- and seventeenth-century alchemy, see also Dobbs, *Alchemical Death and Resurrection* (Washington D.C.: Smithsonian Institution Libraries, 1990), pp. 11-13, 16-17. Unfortunately, the active (or vegetable) spirit in matter is hardly personal or volitional enough—even if it is symbolised by Hermes—to allow a robust christology. But whether this observation would have troubled Newton or not is difficult to say.

for Newton, a legitimate part of natural philosophy, not an import from some other mode of knowledge that transcends science.[438]

On the negative side, however, one must recognise that Newton's doctrine of God was far from biblical. In order to defend the faith, he rightly insisted that it be 'as agreeable to reason as possible'.[439] But, for many, orthodox Christianity appeared far from reasonable in Newton's day.[440] As Newton saw it, the traditional idea of divine transcendence, that of the 'Athanasian Creed', for instance, made God appear static and uninvolved in history, effectively leading to atheism. (So, to be fair, we might say that the scholarly trinitarianism of Newton's time was not quite biblical either.) Like many of his fellow Anglicans, Newton viewed the doctrine of the Trinity as a mystery beyond reason.[441] But, in contrast to to his coreligionists, he also regarded it as being tantamount to polytheism and lacking clear biblical support.[442] Consequently, Newton turned his back on some of the most important truths of biblical theism in order not to 'render it suspect, and exclude it from the nature of things'.[443]

One must also be concerned about the way in which Newton made room for God in his physics. Newton's intent was clearly to formulate a natural philosophy that would vindicate his faith and refute atheism. The need for supramechanical, active principles and for periodic supernatural interventions were his two principal ways of securing God's participation in nature. But supramechanical principles could be regarded as perfectly natural and even

---

[438] General Scholium (1726; Cajori, ed., *Mathematical Principles*, p. 546); cf. J.E. McGuire and P.M. Rattansi, 'Newton and the Pipes of Pan', *Notes and Records of the Royal Society of London*, vol. xxi (1966), pp. 121f., 126, 138.

[439] Additional ms. 3965 (McGuire, 'Newton', p. 121).

[440] According to William S. Babcock, the transformation of the patristic doctrine of the Trinity into a logical puzzle began with Boethius and culminated in the time of Newton with the controversy between Edward Stillingfleet and John Locke; Babcock, 'A Changing of the Christian God: The Doctrine of the Trinity in the Seventeenth Century', *Interpretation*, vol. xliv (April 1991), 135-42.

[441] Isaac Barrow, for example, viewed the doctrine of the Trinity as beyond the capacity of human reason to demonstrate or to assess. See his sermon, 'A Defence of the B[lessed] Trinity', Alexander Napier, ed., *Theological Works of Isaac Barrow*, 9 vols. (Cambridge, 1859), 4:492-523, cited in Irene Simon, 'The Preacher', in Mordechai Feingold, ed., *Before Newton: The Life and Times of Isaac Barrow* (Cambridge: Cambridge Univ. Press, 1990), p. 316.

[442] Yahuda ms. 2.2, 11, 15.5 (Westfall, *Never at Rest*, pp. 344, 823). In his personal notes on Ralph Cudworth's *True Intellectual System of the Universe* (1678), Newton stated that 'the idea of the Trinity was everywhere to be found' in pagan theology, and he included himself among those 'who Boggle so much at the Trinity, and look upon it as the Choak-Pear of Christianity'; Danton B. Sailor, 'Newton's Debt to Cudworth', *JHI*, vol. xlix (July 1988), p. 516. For example, John Sherlock's *Vindication of the Doctrine of the Trinity* (1690) tried to explain the Trinity in terms of three centers of self-consciousness. Many interpreted this as a move toward tritheism; Babcock, 'Changing of the Christian God', p. 142.

[443] Additional ms. 3965 (McGuire, 'Newton', p. 123).

'mechanical' in a more generalised sense of the term.[444] This was how Newton's physics was to be interpreted by materialists and mechanists in the later eighteenth century.[445] In effect, belief in the regular activity of God (*potentia ordinata*) was to be swallowed up by the idea of the autonomy of nature in the eighteenth century as it had in the twelfth. And, paradoxically, as Richard Westfall has pointed out, Newton himself could regard his work as the perfection of the mechanical philosophy rather than its denial.[446]

In fact, Newton's physics was better than he realised. Once mathematical physicists of the eighteenth century accepted his 'supramechanical' principles, they were able to show that the decay of motion in the solar system was negligible and that there was no need for periodic reformations of the planetary orbits. This left God without much to do as far as the physical world was concerned except concur with the laws of mathematical physics. Nature finally became viewed as entirely autonomous.

---

[444]Newton himself referred to gravitation as 'a power seated in the frame of nature by the will of God'; Undated Reply (*c*. 1712) to *Memoirs of Literature*; Turnbull et al., eds., *Correspondence* 5:299f. In the 1717 *Opticks* (Queries 17-24), Newton reintroduced the notion of a subtle (nonmechanical) ether as the material substratum of all supramechanical forces; Schofield, *Mechanism*, pp. 12-15; Thackray, *Atoms*, pp. 26-30.

[445]Peter M. Heimann, '"Nature is a Perpetual Worker": Newton's Aether and Eighteenth-Century Natural Philosophy', *Ambix*, vol. xx (March 1973), pp. 2f., 6, 8f.

[446]Westfall, 'Newton', p. 332; *idem*, 'The Rise of Science and the Decline of Orthodox Christianity', in David C. Lindberg and Ronald L. Numbers, eds., *God and Nature* (Berkeley: Univ. of California Press, 1986), p. 233.

CHAPTER FOUR

THE HERITAGE OF ISAAC NEWTON:
FROM NATURAL THEOLOGY TO NATURALISM
(The Eighteenth Century)

A. The Newtonian Tradition from Newton to Hutton

The eighteenth-century or Enlightenment period is often viewed as a radically new departure in Western thought. It can be portrayed, for instance, as a cultural transition from faith to reason, or from the hegemony of theology to the dominance of natural science. It should be apparent from the material covered in earlier chapters, however, that such transitions had been alternately proffered and resisted many times since the eleventh century. In the eighteenth century, the names were different, but the issues were fundamentally the same as in earlier periods. Newton's natural philosophy could be interpreted in such a way as to stress either the natural or the supernatural side, just as the Aristotelian natural philosophy could in the late Middle Ages or the hermetic could in the Renaissance. If there was a significant shift in the eighteenth century, it came from the fact that science and technology had placed Western Europeans in a position of dominance over the peoples they colonised and over nature itself.

Theological beliefs continued to play an important role in the development of the natural sciences throughout the eighteenth, and well into the nineteenth century. The main difference from earlier periods was that there was less orthodoxy and more variety—ranging from orthodox trinitarianism to monistic materialism—in the theological stances assumed. The fact that unanimity in theology seemed more remote, or even impossible, undoubtedly contributed to the later (nineteenth- and twentieth-century) tendency to suppress personal convictions in science—scientists no longer needed theological legitimation for their work anyway—but this trend was only dimly evident in the eighteenth century itself.

If there is any unity to the eighteenth century as far as science and theology are concerned, that unity stems from the person and work of Isaac Newton. Newton had a very distinctive view of the relationship between God and nature, and his ideas were quite well known and understood in the eighteenth century. Still, they may seem strange to those today who are accustomed to associate the name of Newton with nineteenth-century mechanistic and reductionist ideas.

For Newton, laws of nature like universal gravitation depended on God's immediate presence and activity as much as the breathing of an organism depends on the life-principle within. Like breathing, the operations of nature were regular and natural. Yet there was no possibility, in Newton's view, of taking them to be self-explanatory. Like breathing, they gave no indication as to how the present system originated or how its laws were sustained. Moreover, like breathing, they were subject to modification (e.g., suspension or acceleration) by the free self-determination of the life within. God and nature were thus in a symbiotic relationship, according to Newton, each pointing to the reality and reliability of the other.

Newton's position provided the standard against which all others in the eighteenth century were measured and adjusted. For pedagogical purposes, we may categorise the various thinkers of the period in terms of four basic responses to Newton:

(1) the Newtonians themselves, who followed the master in affirming a basically stable, steady-state symbiosis of God and nature, and, within nature, of matter and spirit;

(2) monists and materialists who went beyond (or behind) Newton by granting energy, life, and even thought to matter itself (not just to nature) and thus affirmed a deeper unity of matter and spirit;

(3) anti-Newtonians and antimaterialists motivated by the desire to reaffirm a more traditional theology (some more dualist and others more monistic than Newton's view);

(4) proponents of analytical mechanics (and chemistry) who went beyond the master by attempting to account for the very origin and stability of the present system of nature in strictly mechanical terms without reference to divine presence or activity.

Theologically minded readers with a homiletical bent might refer to these four responses as:

(1) God required (natural theology);

(2) God reduced (hylozoic monism);

(3) God reaffirmed (conservative anti-Newtonianism);

(4) God retired (neomechanism).

The purpose of the categorisation is to ensure as much as possible the comparison of like thinkers with like. However, the lines are not sharply drawn and, as in any historical taxonomy, there are many intermediate cases and cross-influences.

So Newton's views provide unity to the eighteenth century only as a point of reference for agreement and disagreement. In fact, Newton's God-world symbiosis was satisfactory to few other than his most loyal followers. His assimilation of divine providence to cosmic functions like upholding gravitation and replenishing motion suggested a trend towards deism or materialism to the anti-Newtonians, many of whom were also concerned about the antitrinitarian theologies of Newtonians like Clarke and Whiston. On the other hand, Newton's refusal to speculate concerning the origin of the cosmic system raised a red flag for the more bullish and mathematically inclined of his followers, the neomechanists. And his appeal to divine providence as the source of continued motion was regarded by monistic naturalists as detracting from the integrity and self-determination of nature. None of these emphases were entirely new: in fact, all of them can be seen as truncated forms of the historic creationist tradition. But the coherence of that tradition had gradually been lost and with it the possibility of providing a theological and ethical framework for scientific endeavour as a whole in the Western world.

The characteristic idea of the Newtonians was a symbiosis of God and nature, and of theology and science. A God-nature symbiosis meant, in this instance, a somewhat simplified doctrine of God, tending in some cases to unitarianism and deism. In the case of strict Newtonians, it also meant a low view of matter as absolutely inert and passive. The inertness of matter was used as evidence for the all-encompassing activity of God, and the omnipotence of God was cited in support of the stability of the present order of nature. In some cases, God's control of nature through active principles was also taken to legitimate a particular political order (based on a strong monarchy). This latter, political form of Newtonianism is particularly evident in writers like Samuel Clarke in England and Voltaire in France.[1]

Even though the Newtonian tradition was just one of several streams of thought in the eighteenth century, it was far from homogeneous in itself. In this section we shall review a variety of options within that tradition. For one thing, we must allow for regional variation among the Newtonians of England, America, continental Europe, and Scotland. In England and America, it is useful to differentiate further between three groups of Newtonians: the antitrinitarian Anglicans, Samuel Clarke and William Whiston; English matter theorists and astronomers like John Keill and Roger Cotes; and English and American Independents (Congregationalists) like Isaac Watts and Cotton Mather. Even within these subgroupings, we shall find considerable variation of thought, with most Newtonians leaning

---

[1] On Clarke: Margaret C. Jacob, 'Newtonianism and the Origins of the Enlightenment', *ECS*, vol. xi (1977), p. 9; Steven Shapin, 'Of Gods and Kings: Natural Philosophy and Politics in the Leibniz-Clarke Disputes', *Isis*, vol. lxxii (1981), pp. 210f. On Voltaire: Margaret Jacob, *The Radical Enlightenment* (London: Allen & Unwin, 1982), pp. 102ff. Also, on William Derham: M.C. Jacob, 'Science and Social Passion: The Case of Seventeenth-Century England', *JHI*, vol. xliii (1982), p. 337. On Thomas Prince: John E. Van de Wetering, 'God, Science and the Puritan Dilemma', *New England Quarterly*, vol. xxxviii (1965), p. 500. On Joseph Butler: Colin A. Russell, *Science and Social Change in Britain and Europe, 1700-1900* (London: St. Martin's Press, 1983), pp. 49f. On Jacques Turgot: J.H. Randall, Jr., *The Making of the Modern Mind* (New York: Columbia Univ. Press, 1926, 1976), pp. 323f.

either to the side of hylozoic materialism or to that of neomechanism.

1 Anti-Trinitarian Anglicans: Samuel Clarke and William Whiston

The two most prominent Newtonians in the early eighteenth century were Samuel Clarke and William Whiston. We consider them first as the standard bearers for the Newtonian tradition before we glance at a variety of lesser lights who modified Newton's ideas in significant ways.

1.1 Samuel Clarke

Samuel Clarke (1675-1729), an ordained Anglican clergyman, was noted among other things for his Boyle Lectures on Newtonian physics and natural theology (1704-5),[2] his skepticism about the doctrine of the Trinity (1705, 1712),[3] and his epistolary debate with Leibniz concerning the relationship of God and nature (1715-16).[4] In all three endeavours, he was the chief spokesman for Newton in science and theology.

Like Newton, Clarke held that the important principles of physics like gravitation were active and supramechanical; that is, they could not be accounted for by the motion of passive material corpuscles. The opponents he had in mind here were three: the mechanical philosophy of Descartes; the hylozoic monism of Hobbes, Spinoza, Toland, and Collins; and the anti-Newtonianism of Leibniz.

According to the mechanical philosophy, all nature could be explained in terms of matter in motion, and God was required only at a metaphysical level as the presupposition of science. The strongest scientific argument against consistent mechanism, according to Clarke, was the existence of supramechanical principles like gravitation that were functions of the total mass of the bodies involved, not of the surface areas (or cross-sections) in proportion to which some impulse might be impressed by the surrounding medium. Hence, active principles could be taken as indications of the immediate presence and activity of God—albeit under the constraint of God's self-imposed laws—in the same way that miracles were used as evidence by other theologians. But active principles were regular and predictable; they

---

[2] *A Demonstration of the Being and Attributes of God* (pub. 1705); *A Discourse concerning the unchangeable Obligations of Natural Religion and the Truth, and Certainty of the Christian Revelation* (1706). For bibliographic details on Clarke, see J.P. Ferguson, *An Eighteenth-Century Heretic: Dr Samuel Clarke* (Kineton: Roundwood Press, 1976), pp. 230ff.

[3] *The Scripture Doctrine of the Trinity* (1712).

[4] *A Collection of Papers which passed between the late learned Mr Leibniz, and Dr Clarke...relating to the Principles of Natural Philosophy and Religion* (1717).

were not themselves 'miracles' in the sense of being unique or unrepeatable.[5]

According to hylozoic monists, on the other hand, matter was invested with life and movement (*conatus*) of its own. Hylozoic monists effectively absorbed God into nature, and spirit into matter. The strongest scientific argument against the monists, according to Clarke, was that there was nothing in the Newton's laws themselves to determine the direction in which a supposed inherent motion should occur: directionality had to be imposed by an externally applied force. Accordingly, without a supramechanical First Cause to set things in motion with a particular initial direction (in an absolute space), all things would have remained eternally at rest.[6]

Like a good theologian, Clarke made a point of affirming the activity of God equally in all events, whether natural (occurring through the agency of active principles) or miraculous. The only difference between the two was that miracles were unusual, whereas active principles (like mutual gravitation) were not. From God's point of view, in fact, the natural and the miraculous were effectively the same.[7] Clearly Clarke's intention, like that of pious Christians as far back as Augustine, was to avoid the suggestion that the 'natural' was in any sense independent of God. The suggestion of independence was a conclusion that many had drawn from the underlying dialectic of God's ordinary and absolute powers. But, intentions aside, Clarke stressed the natural (though not just the material or mechanical) and tried to refute Leibniz's charge (1710) that Newtonians appealed to miracles in their science.[8]

In response to Leibniz, Clarke argued that, even though they depended directly and continuously on the action of God, forces like gravitation and cohesion were not 'perpetual miracles'.[9] Clarke was right to make this distinction, for active principles were universal in scope and hence played a fundamentally different role in Newtonian cosmology from that of the

---

[5] Koryé, *From the Closed World* (1957), pp. 258, 271; Kargon, *Atomism* (1966), pp. 137f.; Ezio Vailati, 'Leibniz and Clarke on Miracles', *JHP*, vol. xxxiii (1990), pp. 588f.

[6] M.C. Jacob, 'John Toland and the Newtonian Ideology', *Journal of the Warburg and Courtauld Inst.*, vol. xxxii (1969), p. 322; *The Newtonians and the English Revolution*, (Ithaca: Cornell Univ. Press, 1976), pp. 238-41; John J. Dahm, 'Science and Apologetics in the Early Boyle Lectures', *Church History*, vol. xxxix (June 1970), pp. 181-84; Vailati, 'Leibniz and Clarke', p. 568.

[7] H.G. Alexander, *Leibniz-Clarke*, p. 24; McGuire, 'Force' (1968), pp. 202f.; F.E.L. Priestley, 'The Clarke-Leibniz Controversy', in R.E. Butts and J.W. Davis, eds., *The Methodological Heritage of Newton* (Toronto: Univ. of Toronto Press, 1970), p. 53; Peter M. Heimann, 'Voluntarism and Immanence: Conceptions of Nature in Eighteenth-Century Thought', *JHI*, vol. xxxix (1978), p. 274.

[8] A. Rupert Hall, *From Galileo to Newton* (New York: Harper & Bros., 1963; Dover, 1981), p. 317.

[9] Koryé, *From the Closed World* (1957), p. 271.

instances of particular providence cited earlier by Luther and Calvin (in the context of an Aristotelian cosmology). In other words, the naturalistic tendency in the Newtonians was not just the result of an increasingly comprehensive natural science—there were gaps in Newton's cosmology as great as those in Aristotle's. The tendency to naturalism was a philosophical or theological bent with a history of its own, going back at least as far as the twelfth century.

Part of the apologetic value of affirming the dependence of active principles on providence for Clarke was that the origins of the system of the world were not then amenable to naturalistic accounts. If God were not needed to account for the maintenance of the world (by supramechanical active principles), Clarke reasoned, God would not be needed to account for its formation either. In that case, everything could be accounted for by natural reason, and belief in a creator God would become as redundant as it appeared (to Newtonians) to be in Descartes's cosmology. On Newton's accounting, however, the world could be seen to be the purposeful result of God's free choice and design.[10]

In what we have considered thus far, Clarke was merely the mouthpiece of Newton. Newton too had affirmed the irreducible role of God in both the formation and the maintenance of the cosmos. But he also invoked divine providence to account for the periodic adjustments needed to provide the solar system with a stability that he could not see resulting from the laws of gravitation and motion alone. Here, under pressure from Leibniz (and possibly with Newton's approval), Clarke was willing to retreat a step. Leibniz required that there be a 'sufficient reason' for every phenomenon, that is, a reason good enough to account for its occurrence at one given place and time rather than another. Otherwise God would appear to be capricious. Clarke responded by grounding the periodic adjustments required to preserve the stability of the planetary system in the eternal decrees of God. In other words, these occasional suspensions and amendments of the laws of nature were not capricious or random; they were just as much a part of God's ordained will (*potentia ordinata*) as the laws of nature themselves.[11]

Here again (as in the case of Descartes), we have the paradoxical result that an emphasis on the omni-activity of God can lead in the direction of consistent (though not monistic) naturalism. There were six commonly cited ways in which God's actions could play a role in nature: (1) the creation of matter and setting it in motion in accordance with certain prescribed laws; (2) the formation of the present world system; (3) its continued operation; (4) its occasional reformation; (5) occasional spiritual

---

[10] Koryé, *From the Closed World*, pp. 237-41.
[11] Koryé, *From the Closed World*, p. 241ff.; Shapin, 'Of Gods and Kings', pp. 194f.; Priestley, 'The Clarke-Leibniz Controversy', pp. 51f.

intrusions in human affairs through the agency of natural phenomena (e.g., comets and epidemics); and (6) miracles. Three of these ways—primary creation, spirit intrusions, and miracles—were not major issues prior to the philosophical critique of David Hume (mid-eighteenth century): they belonged to the realm of *potentia Dei absoluta*, which no one could positively deny (though the Deists openly questioned biblical miracles, Priestley still believed in them in the late-eighteenth century)[12] but everyone could conveniently ignore for the purposes of natural science. Newton himself accepted all three, but he stressed the role of God in the three other ways we have noted: the formation of the solar system, its continued operation by active principles, and its occasional reformation. In Clarke's writing, here really an extension of Newton's, it became clear that operation and even reformation were completely lawful and subject to scientific investigation. Only the formation of the solar system remained, for the time being, entirely beyond the province of human science. The programme of the neomechanists would later reduce the original formation as well as the occasional reformations to the laws of mechanics, thus making God appear entirely redundant.

In theology, Clarke did more than any other scientifically oriented figure to legitimate and reinforce three tendencies among latitudinarian (low-church) Anglicans and Nonconformists or Dissenters. First there was the tendency we have already noted to limit the present-day actions of God to the (self-imposed) laws of nature.

The second tendency Clarke reinforced was a movement towards equating Christian dogma with the conclusions of 'right reason'.[13] In theory, this move was intended to commend Christian doctrine to scientists and other intellectuals. In fact, the God-givenness of human reason was a venerable idea with roots in the creationist tradition. In practise, however, it often meant limiting religious faith to those doctrines that could command the assent of all reasonable people.

The third tendency reinforced by Clarke is a case in point: a growing scepticism concerning the doctrine of the Trinity. For one thing, Clarke wrote, the idea of three coequal persons in one individual substance is a

---

[12] Colin Brown, *Miracles and the Critical Mind* (Grand Rapids, Mich.: Eerdmans, 1984), chaps. 2-4.

[13] For his D.D. at Cambridge (1709), Clarke defended the thesis that 'No Article of the Christian Faith delivered in the Holy Scriptures is Disagreeable to Right Reason', Roland N. Stromberg, *Religious Liberalism in Eighteenth-Century England* (Oxford: Oxford Univ. Press, 1954), p. 43; John Gascoigne, *Cambridge in the Age of the Enlightenment* (Cambridge: Cambridge University Press, 1989), p. 117. Gascoigne documents Clarke's influence at Cambridge University; *ibid.*, pp. 119f., 126f.

logical self-contradiction.[14] By reason, Clarke argued, there must exist one single being who is immaterial, simple (without parts), and self-existent.[15] However, Scripture describes Christ as the 'Son of God' and as 'begotten' by God. Terms like these clearly imply dependence on the will of the Father who alone is self-existent or unoriginate and hence 'God' in the proper sense.[16] In fact, since the eternity of God is infinite duration or everlastingness (after Newton), rather than the transcendence of time, the 'begetting' of the Son must be a temporal act of God like the creation of the world, only indefinitely earlier in time.[17]

As in the case of Newton (chap. 3, sec. C), one can see here a strong correlation between the notion of absolute space (and time) as an emanent effect of God, the interpretation of God's eternity as everlasting time, and the implausibility of the triunity of God—a cluster of ideas that made possible a complete reinterpretation of Christian doctrine in progressive circles of the eighteenth century.

Clarke's importance for English theology stems from the fact that he backed the new liberalism with the authority of Newtonian science. He personally omitted recitation of the 'Athanasian Creed' (the *Quicunque vult*) and advocated nontrinitarian modifications of the liturgy and of subscription to the Thirty-Nine Articles of the Church of England. His early questions (1705) about the authority of the 'Athanasian Creed' encouraged William Whiston to do his pioneering study (1707ff.) of the the doctrine of the Trinity in the early church.[18] Clarke's own full-length critical study of the biblical basis of the Trinity (1712) made such an impact on several Anglican clergymen that they ceased to read the Athanasian Creed in church services.[19] And his proposal to revise subscription to the Thirty-Nine Articles was revived twice in the eighteenth century though it was never carried.[20] But Clarke's influence was, along with that of Whiston, greatest among the dissenting churches, as we shall see momentarily.

---

[14] Letter to John Jackson, 23 Oct. 1714; Larry Stewart, 'Samuel Clarke, Newtonianism, and the Factions of Post-Revolutinary England', *JHI*, vol. xlii (1981), p. 58.

[15] *Demonstration* (1705); J.P. Ferguson, *The Philosophy of Dr Samuel Clarke and Its Critics* (New York: Vantage Press, 1974), pp. 23-26; Stewart, 'Samuel Clarke', p. 56.

[16] *Scripture Doctrine* (1712); Stewart, 'Samuel Clarke', p. 57; Stromberg, *Religious Liberalism*, p. 45.

[17] Correspondence with Joseph Butler (at that time a scholar at the Dissenting Academy at Tewkesbury); Stewart, 'Samuel Clarke', pp. 58f.

[18] Whiston, *Historical Memoirs of the Life of Dr Samuel Clarke* (London, 1730), pp. 12f.; J. Hay Colligan, *The Arian Movement in England* (Manchester: Manchester Univ. Press, 1913), pp. 32, 105f.; Eamon Duffy, '"Whiston's Affair": The Trials of a Primitive Christian, 1709-14', *Journal of Ecclesiastical History*, vol. xvii (1976), p. 134; James E. Force, *William Whiston* (Cambridge: Cambridge Univ. Press, 1985), pp. 14f.

[19] Colligan, *Arian Movement*, pp. 109ff.; Stromberg, *Religious Liberalism*, pp. 44ff.; Stewart, 'Samuel Clarke', pp. 57f.

[20] Stromberg, *Religious Liberalism*, p. 48; Gascoigne, *Cambridge*, pp. 131-34, 194-97.

1.2 William Whiston

William Whiston (1667-1752) was a professional mathematician who succeeded Newton as the Lucasian Professor at Cambridge in 1701. In 1710, he was dismissed on the grounds of his published disagreement with orthodox Christology.[21] Whiston was widely regarded as an Arian (one who views Christ as a primordial angelic creature). Actually, he was more of an Origenist or Eusebian than an Arian: he held that Christ was a 'second god' in the Neoplatonic sense—divine, yet subordinate to the supreme God.[22]

Origenist, Eusebian, and Arian leanings were all tolerated in the Church of England under Latitudinarian (Whig) policies, so long as the advocates of such unorthodox views did not speak out against subscription to the Thirty Nine Articles. Newton himself was an Arian, though he only divulged his heterodox views to a small circle of friends and they were not widely known at the time. However, Whiston argued his case publicly and paid the price of dismissal from the chair Newton had once held uncontested. Later in his life (1747), Whiston joined the General Baptists, who had officially broadened their standards to embrace a greater variety of theological views. Significantly, Whiston's father had been one of those ministers of Presbyterian-Puritan background who had elected to conform to the standards of the Church of England at the Restoration of the monarchy in 1660.[23] This generational slide from Calvinism to latitudinarianism and unitarianism was characteristic of many Christians in the late eighteenth century. We shall see one of its many fruits in the monistic materialism of Joseph Priestley.

Whiston's criticisms of the orthodox doctrine of the Trinity were similar to those of Newton and Clarke, but even more influential. The two main points that counted against the doctrine in Whiston's view were (1) that it implied the existence of three coequal gods (if Christ were granted self-existence like that of the Father)[24] and (2) that the notion of a plurality of divine hypostases came from pagan thought rather than from the simple teachings of Jesus.[25]

Although he was shunned by the Anglican establishment, Whiston's theological views, along with Clarke's, were influential among English Presbyterians like James Pierce and Joseph Hallett, who precipitated the 'Non-Subscription Controversy' at the Salters' Hall Conference of Dissenters in 1719. Not only did the failure to enforce subscription to the doctrine of the Trinity bring about the demise of the fledgling union of

---

[21] Force, *William Whiston*, p. 14.
[22] Whiston, *Historical Memoirs*, pp. 15, 18, 20; Duffy, 'Whiston's Affair', pp. 141f.
[23] Duffy, 'Whiston's Affair', p. 130.
[24] John Redwood, *Reason, Ridicule and Religion* (London: Thames & Hudson, 1976), p. 167.
[25] Duffy, 'Whiston's Affair', p. 134.

London Presbyterians, Independents (Congregationalists), and Baptists, but it established the respectability of Arian and Unitarian views within Dissenting Academies for the duration of the century.[26] One such academy was Daventry, where Joseph Priestley received his early training.

Like Newton and Clarke, Whiston treated matter as entirely passive and dependent on the continued exercise of divine power for its behaviour in accordance with the laws of nature. Whiston pressed this aspect of Newtonianism to its logical conclusion, however: the normal operation of God was consistent and universal, 'acting by fixed and constant rules'.[27] Therefore, Whiston argued, whenever possible, events should be accounted for in terms of those rules and not ascribed to miracles unless there was sufficient reason to do so.[28]

The desire to explain as much as possible in naturalistic terms had been, as we have seen, a persistent theme in Western thought at least since the twelfth century (Adelard of Bath, Thierry of Chartres, William of Conches). Whiston stood firmly within this tradition in spite of his invocation of divine providence to sustain the processes of nature. In fact, now that comets were understood to be lawful, periodic phenomena, they too could be pressed into service as the natural causes of extraordinary events. For example, Whiston appealed to an ancient appearance of the comet of 1680 to account for the Great Flood (enabling him to date it to 2349 BC),[29] the rotation of the earth about its axis, and the twenty-three-degree inclination of the earth's equator to the ecliptic plane.[30] So even idiosyncracies of the present structure of the solar system could be accounted for in naturalistic terms. By treating significant aspects of the system of nature as in process in historical (albeit biblical) time, Whiston thus opened the way for later speculations concerning the origin of the cosmic system as a whole.

On the other hand, Whiston was still a great believer in the supernatural, that is, the presence and power of immaterial agencies. He occasionally experienced prophetic trances,[31] and, like most of his contemporaries, he could see the hand of Satan and his demons in untoward events like mental

---

[26] Stromberg, *Religious Liberalism*, p. 47; Peter Toon, *The Emergence of Hyper-Calvinism in English Non-Conformity* (London: Olive Tree, 1967), pp. 37ff.; Roger Thomas, 'Presbyterians in Transition', in C.G. Bolman et al., *The English Presbyterians* (London: Allen & Unwin, 1968), pp. 155f., 160-63; Ferguson, *Eighteenth-Century Heretic*, pp. 148f.

[27] F.E.L. Priestley, 'Newton and the Romantic Concept of Nature', *Univ. of Toronto Quarterly*, vol. xvii (1948), p. 333.

[28] *New Theory of the Earth* (1696); Colin A. Russell, *Cross-Currents: Interactions between Science and Faith* (Leicester: Inter-Varsity, 1985), p. 133.

[29] Toulmin and Goodfield, *Discovery*, pp. 114f.

[30] Whiston, *New Theory* (1696); Webster, *From Paracelsus to Newton*, p. 41; Jacques Roger, 'William Whiston', *DSB* 14:296a.

[31] Roger, 'Whiston', p. 295b.

disturbance, pestilence, and famine. But even the Devil operated through the medium of second causes like meteors, which could be investigated scientifically.[32] The Newtonianism of Whiston, like that of Clarke, portrayed God as functioning in mechanically predictable ways. The 'God' later neomechanists like Laplace were explicitly to exclude from natural science was fast becoming as monotonous as he was redundant.

2 English Newtonian Matter Theorists from Keill to Knight

We consider here a group of minor, yet significant, Newtonians who were primarily matter theorists. That is, they speculated about the properties of matter as functions of the underlying corpuscular structure and the active principles of Newton. It was characteristic of Newtonian matter theorists that they regarded the manifest properties of matter as epiphenomena, or secondary qualities, compared with the primary qualities and forces of the unmeasurably minute atoms.

The Newtonians at hand were primarily scientists, but they were also theologically informed and religiously motivated. Some, like Derham and Cotes, were ordained as priests. Paradoxically, it was Derham and Cotes who most toned down Newton's emphasis on the omniactivity of God.

Here we shall consider three slightly different directions in which matter theory developed within the Newtonian tradition:

(1) the matter minimalism of Oxford matter theorists like Keill, Freind, and Desaguliers;

(2) the tendancy towards matter maximalism of Derham and Cotes;

(3) the two matter types of Gowin Knight.

2.1 Oxford Matter Minimalists: Keill and Freind

John Keill (1671-1721) was lecturer in natural philosophy (1694) and professor of astronomy (1712) at Oxford.[33] He was the first to formulate the so-called 'nutshell theory' of matter (1708) and was thus partly responsible for instigating the later Leibniz-Clarke debate (1715-16) concerning the relation of God and nature.[34] Prior to that he had coined the term 'momentum' (1700), one of the most important concepts in modern physics, thus instigating the so-called *vis viva* debate between Newtonians and Leibnizians. Although they were distinct issues, the nutshell theory and opposition to *vis viva* were coordinated thrusts of an anti-Lebnizian

---

[32] *An Account of the Demoniacks and of the Power of Casting Out Demons* (1737); Webster, *From Paracelsus to Newton*, p. 98.

[33] David Kubrin, 'Keill', *DSB* 7:275.

[34] Arnold Thackray, 'Matter in a Nut-Shell: Newton's *Opticks* and Eighteenth-Century Chemistry', *Ambix*, vol. xv (1968), pp. 38f.

campaign. All three elements—the nutshell matter theory, opposition to *vis viva*, and criticism of Leibniz's view of matter as self-active—were clearly stated in the 'authorised' popularisation of Newton's ideas, Henry Pemberton's *View of Sir Isaac Newton's Philosophy* (1728).[35]

As Keill defined it, the momentum of a moving object is the product of its mass and its velocity. Like velocity, it is a vector quantity; that is, it has direction in space. The momentum of a body is the impulse that it would communicate to another body at rest. Thus it could be taken as a measure of the force exerted by a falling body on the earth where it impacts. Leibniz (a matter maximalist, if you will) argued that the proper measure of such force should be the product of the mass and the *square* of the velocity (chap. 3, sec. C), and a philosphical-scientific-theological debate ensued which was never quite resolved. Eventually, second-generation Newtonians like Desaguliers and Boscovich came to the conclusion that both mv and mv^2 were legitimate measures of force.[36]

The *vis viva* debate is interesting for the way in which theological and apologetic (and even political) concerns interacted with experimental physics. The importance for Newtonians of the vectorial nature of force lay, as we have seen, in the inferred necessity for a prime mover to start things out with a particular initial direction in absolute space. In contrast, scalar quantities, which could be defined without reference to direction, were preferred by Leibnizians partly because they saw motion as inhering in matter by virtue of God's creation and providence. The ostensibly scientific argument about *vis viva* thus involved two totally different philosophies and theologies of nature. Each side feared (or at least claimed) that the other would lead to atheism.[37]

This attempt by Newtonians to empty matter of qualities like inherent motion was paralleled by a reduction of matter to a nutshell in volume and the consequent emphasis on the role of impressed forces in physics and chemistry. The nutshell theory developed one of the most astounding thoughts of Isaac Newton—the notion, first adumbrated in the early editions of his *Opticks* (1704, 1706), that all the atoms in a solid would, if packed tightly together, fit neatly inside a much smaller volume. The scientific rationale for this paradox was that light (consisting of fine material particles for Newton) could pass unhindered through great quantities of transparent substances like glass and water.[38] In other words, matter was highly porous,

---

[35] Thackray, 'Matter', p. 42; Carolyn Iltis, 'The Leibnizian-Newtonian Debates', *BJHS*, vol. vi (1973), pp. 363ff.

[36] Iltis, 'Leibnizian-Newtonian', pp. 366ff.

[37] E.g., Desaguliers, *Lectures of Experimental Philosophy* (2nd ed., 1719); Iltis, 'Leibnizian-Newtonian', p. 367.

[38] Thackray, 'Matter', pp. 30ff.; *idem, Atoms and Powers*, pp. 22f.

and what appeared to be a solid to human senses was really mostly empty space between indefinitely small atoms. The responsibility for maintaining the apparently stable structure of the macroscopic world thus lay with supramechanical forces (and hence with God) rather than with matter itself.

The philosophical effect of Newton's idea was to undermine the very foundation of mechanism and materialism, the notion of the primacy of matter in physics. In fact, the context in which Keill first made the nutshell theory explicit was a series of attacks on the matter theory of Leibniz in 1708, 1714, and 1715. As Keill put it in his 1715 textbook, the proportion between the space occupied by the constituent atoms of a solid and the total volume of that solid is comparable to that of a grain of sand to the whole earth.[39]

Since the amount of matter in a substance was so insignificant, it made sense to describe its macroscopic (chemical) properties in terms of forces acting at a distance between the microscopic atoms. Keill himself attempted (1708) to develop an analogue to Newton's law of gravitational attraction for the attractive forces between the atoms of a chemical compound.[40] John Freind (1675-1728), who studied under Keill at Oxford, described (1709) the matter-minimalist programme as one in which 'almost all the Operations of Chymistry are reduced to their true [i.e., Newtonian] Principles, and the Laws of Nature'.[41] Thus the minimisation of matter did not put natural processes beyond the reach of mathematical science. To the contrary, a subordination of matter to immaterial principles went together with a high degree of mathematical formalism, as had often been the case earlier in the Neoplatonic tradition of the Middle Ages and the Renaissance.

2.2 Newtonian Matter Maximalists: Derham and Cotes

If the Newtonian followers of Keill tended to minimise the extent and power of matter, Derham and Cotes tended in a very different direction.

William Derham (1657-1735) is best known for the 'Physico-theology' of his Boyle Lectures (1711-12),[42] which set a standard for the natural theology tradition that culminated in the turn-of-the-century writings of

---

[39] Thackray, 'Matter', pp. 34-41.
[40] 'On the Laws of Attraction and Other Physical Properties' (1708); A.R. Hall, *From Galileo*, p. 324; Robert E. Schofield, 'Joseph Priestley and the Physicalist Tradition in British Chemistry', in L. Kieft and B.R. Willeford, Jr., eds., *Joseph Priestley* (Lewisburg, Penn.: Bucknell Univ. Press, 1980), pp. 97f.
[41] *Praelectiones Chymiae* (*Chymical Lectures*, 1709); Hall, *From Galileo*, p. 325; Schofield, 'Joseph Priestley', p. 98.
[42] *Physico-Theology, or a Demonstration of the Being and Attributes of God from His Works of Creation* (pub. 1713). Derham also published a new edition of John Ray's *Physico-Theological Discourses* in 1713; D.M. Knight, 'William Derham', *DSB* 4:40b.

William Paley (1795, 1802).[43]

Derham continued the patristic and Reformed tradition of valuing secular skills as divine gifts. The basic idea is now familiar from figures like Basil, Hugh of St Victor, Paracelsus, Luther, and Calvin. Derham went beyond his forebears, however, in including nonproductive, entrepreneurial skills like commerce and management among divine gifts along with more traditionally valued trades like craftsmanship and medicine. He also concluded that Africans and American Indians were, by God's design, less gifted as entrepreneurs than Europeans, and that the latter were thereby entitled to 'ransack the whole globe'.[44] At this point in history, western Europeans were clearly achieving a degree of power that made their science and technology as oppressive to nations they colonised as foreign technologies had been to the Israelites in the Old Testament and intertestamental periods (chap. 1, sec. E).

Just as God implanted spiritual inclinations in all humans (albeit unequally) at birth, he also impressed active principles like gravitation on all matter at creation. The positive qualities of matter were, for Derham, not intrinsic to matter, but implanted, just like those of gifted people.[45] There was thus a direct analogy here between the sociopolitical and the natural worlds.[46]

The analogy was powerful, but it came at a price. Just as gifts and inclinations once implanted within humans become part of their nature, so active principles like gravitation and cohesion once impressed on material bodies become properties of their nature. This was a significant departure from Newton's view that active principles represented the continued immediate, though lawful, activity of God.

In fact, Derham was shying away from the suggestion that God might be viewed as a world-soul like Henry More's Spirit of Nature. That such a notion could be inferred from Newton's symbiotic model of God and nature had already been argued by Leibniz in support of his charge that Newton's ideas would lead in the direction of pantheistic materialism.[47] Derham was thus pulled away from the matter-minimalism of Newton and Keill in the direction of another sun, the matter maximalism of Leibniz. He was also

---

[43] D.L. LeMahieu, *The Mind of William Paley* (Lincoln: Univ. of Nebraska Press, 1976), pp. 76f.

[44] William Coleman, 'Providence, Capitalism, and Environmental Degradation', *JHI*, vol. xxxvii (1976), pp. 33ff.; C.A. Russell, *Cross-Currents*, p. 232.

[45] *Astro-Theology, or a Demonstration of the Being and Attributes of God from a Survey of the Heavens* (1715); Jammer, *Concepts of Force*, p. 155; Schofield, *Mechanism*, p. 23.

[46] M.C. Jacob, 'Science and Social Passion: The Case of Seventeenth-Century England', *JHI*, vol. xliii (1982), p. 337.

[47] C.B. Wilde, 'Matter and Spirit as Natural Symbols in Eighteenth-Century British Natural Philosophy', *BJHS*, vol. xv (1982), pp. 102ff.

following a variant of the mechanical philosophy developed by Robert Boyle and John Ray according to which the best evidence of the work of God was to be found in the structure and function of living organisms, particularly the instincts imprinted in animals.[48] Whether by the anti-Newtonian pull of Leibniz or the pre-Newtonian push of Boyle and Ray, Derham was inclined to deviate from the strictly Newtonian line.

Another follower of Newton who deviated from the strict Newtonian line was Roger Cotes (1682-1716), first Plumian professor of natural philosophy at Cambridge. A trusted collaborator of both Newton and Whiston, Cotes was given the responsibility of editing and writing a preface for the second edition of Newton's *Principia* (1713). The new preface read like a counterattack against the arguments Leibniz had mounted against Newton's hypothesis of forces acting at a distance without material mediation.[49]

In order to defend Newton, Cotes made two moves that were to have an impact on subsequent matter theory. First, he stated that Leibniz's requirement for a material substratum of all forces would lead to the conclusion that the world was driven to its present state of order by qualities inherent in its material nature rather than by the sheer will of God. In fact, any matter imbued with such qualities must always have existed, thus ruling out the notion of an original creation as well as the formation of matter.[50] In this respect, Cotes spoke rather prophetically of much of later eighteenth-century natural philosophy.

Cotes's second move against Leibniz was to forestall the charge that active principles were occult phenomena beyond the scope of rigorous science. In order to do so, however, he was forced to classify gravitation 'among the primary qualities of all bodies'.[51] Prior to this, Newton had specified the primary qualities of matter to be those, like spatial extension, impenetrability, mobility, and inertia, which could not be denied as attributes of a portion of matter without denying its very existence.[52] The status of gravitation as a manifestation of God's regular activity in nature seemed to place it on an entirely different footing from the primary qualities, even though the law governing it was perfectly mathematical. Newton

---

[48] Dahm, 'Science and Apologetics', pp. 184ff.
[49] Leibniz referred to Cotes's preface as 'pleine d'aigreur'; J.M. Dubbey, 'Roger Cotes', *DSB* 3:432b. Newton refused to read it before publication in order not to implicate himself too directly; Hall, *From Galileo*, p. 312.
[50] Priestley, 'Clarke-Leibniz', p. 34.
[51] 1713 Preface to the *Principia* (Cajori, ed., p. xxi); Jammer, *Concepts of Force*, pp. 147f., 201. However, in writing to Clarke on 25 June 1713, Cotes denied that he had said gravitation was essential to matter; Turnbull et al., eds., *Correspondence*, 5:412f.
[52] As Newton put it in his third 'Rule of Reasoning in Philosophy', the universal, primary qualities of bodies were those 'which would admit neither intensification nor remission of degrees' (*Principia*, Cajori, ed., p. 398).

himself had tried to parry Leibniz's blow by arguing that gravitation was 'seated in the frame of nature' just like the primordial qualities of matter (c. 1712),[53] and by grounding the active principles in a subtle material ether (1717).[54] But Cotes's solution was far more radical and ran directly counter to Newton's earlier denial (1692-93) of the notion that gravitation was in any sense a property of matter.[55]

With or without his consent, Newton's foremost mental offspring, the immaterial and supramechanical principle of gravitation, had thus been wedded to brute matter, or so Cotes's new preface could be read. And so it was read, particularly among post-Newtonians of the mid-eighteenth century who tended towards pantheistic materialism and neomechanism. Cotes thus contributed to the fulfillment of his own dire prediction about the possibility of a purely naturalistic science without God.

2.3 Two Matter Types of Gowin Knight

Gowin Knight (1713-72) was born the same year Cotes published the new edition of the *Principia*. Knight is important primarily for his generalisation of active principles to account for repulsive as well as attractive forces. Interestingly, it was Cotes who first drew the attention of Newtonians to the role of repulsive forces, in his account of the elasticity of air.[56] Knight, however, went further and attempted 'to demonstrate that all the Phaenomena in Nature May be explained by Two simple active Principles, Attraction and Repulsion' (1754).[57] Part of the reason for his appreciation

---

[53] Undated letter to the editor of *Memoirs of Literature* (c. 1712). Cotes himself had brought Leibniz's letter to Hartsoeker (Feb. 1711), published by the *Memoirs* on 5 May 1712, to Newton's attention; Turnbull et al., eds., *Correspondence*, 5:300ff.; Edward B. Davis, 'Creation, Contingency, and Early Modern Science' (Ph.D. diss., Indiana Univ., 1984), pp. 224f.

[54] Peter M. Heimann (Harman), 'Concepts of Inertia: Newton to Kant', in M.J. Osler and P.L. Farber, eds., *Religion, Science, and Worldview* (Cambridge: Cambridge Univ. Press, 1985), p. 124.

[55] Letters to Bentley (17 Jan. 1693; 25 Feb. 1693), who was then preparing the Boyle Lectures for 1693; Turnbull et al., eds., *Correspondence*, 3:240, 253f.; Hall, *From Galileo*, pp. 314f. John Henry's argument that Cotes was 'drawing directly upon his understanding of Newton's opinion' is plausible enough but must be balanced by Henry's main point about Newton's 'voluntarist theology'; '"Pray Do Not Ascribe that Notion to Me": God and Newton's Gravity', in J.E. Force and R.H. Popkin, eds., *Books of Nature and Scripture*, pp. 123, 140f.

[56] Cotes, *Hydrostatical and Pneumatical Lectures* (posth. 1738); F.E.L. Priestley, 'Joseph Priestley', p. 99. The expansive power of air was also analysed in Stephen Hales's *Vegetable Staticks* (1727); Thackray, *Atoms*, pp. 122f.

[57] *Attempt to Demonstrate* (1754); Schofield, *Mechanism*, pp. 175f. Cf. Robert Greene's *Principles of the Philosophy of the Expansive and Contractive Forces* (1727) and Desaguliers's contention that there are only two forces 'in all the Phaenomena and Changes in Nature'; *Course of Experimental Philosophy* (1734); Heimann and McGuire, 'Newton's Forces', p. 255, 259; A.R. Hall, 'Desaguliers', *DSB* 4:44b.

for the duality of forces was Knight's early production of and experimentation with artificial magnets, which, of course, exhibit clear polarity.58

For Knight, as for all Newtonians, nature depended directly on the presence and activity of God. But Knight took this principle to its logical conclusion. Even the immutability of the primary qualities of matter merely expressed the immutability of God. And the irresistibility and inexorability of natural causation expressed the immutability of his will (as for Descartes).59 It appears that Cotes's wedding of active principles to matter had had the effect for Knight of eliminating Newton's duality of matter and force. But rather than force being reduced to a property of matter, as it was for materialists, matter for Knight was reduced to a function of force and divine power.60 A similar challenge to the notion of primary qualities was developed by Jonathan Edwards, whom we consider below, as a counter to materialistic monism.

The reduction of matter to a function of force had the effect of requiring two distinct types of matter for Knight, who prefigured in this respect the later work of James Hutton. Unlike gravitation which was universally attractive, electric and magnetic forces could be either attractive or repulsive. Therefore, there must be two types of primary particle, Knight reasoned. Clearly one type of matter consisted of particles that attracted all other particles indiscriminately—this was ordinary Newtonian gravitational matter. In addition, Knight hypothesised another type of matter consisting of particles that attracted gravitational matter but repelled other particles of their own kind. This new, repellant matter accounted for the phenomena of light, heat, electricity, and magnetism.61 The signficant feature of Knight's speculations was the rejection of the homogeneity of matter assumed by the mechanical philosophy and affirmed in most of Newton's writings.62 Aristotle's five types of matter (four gross, one subtle) had been reduced to one by the mechanical philosophers, but that one had now increased back to two, even within the Newtonian tradition.

In a sense, Knight's division of the kingdom of matter complemented his elimination of Newton's duality of passive matter and active force. Both moves had the effect of weakening the primacy of matter as an irreducible

---

58 Schofield, *Mechanism*, pp. 175f.
59 P.M. Heimann and J.E. McGuire, 'Newtonian Forces and Lockean Powers', in Russell McCormmach, ed., *Historical Studies in the Physical Sciences*, vol. iii (Philadelphia, 1971), p. 296.
60 Heimann and McGuire, 'Newtonian Forces', p. 298.
61 Heimann and McGuire, 'Newtonian Forces', p. 297; Schofield, *Mechanism*, p. 177.
62 Newton allowed for two types of matter, one ponderable and the other not, in a ms. revision of the *Principia* dating from the early 1690s; J.E. McGuire, 'Transmutation and Immutability: Newton's Doctrine of Physical Qualities', *Ambix*, vol. xiv (1987), pp. 72f.; Thackray, *Atoms*, pp. 29f.

given in physics and thus paralleled the notion of the porosity of matter with the 'nutshell' matter theorists. However, Knight's differentiation of two types of primary particle also had the effect of making mutual attraction and mutual repulsion intrinsic properties of those types of matter over and above the mechanical primary qualities common to all particles. As we shall see again in the case of Boscovich and Priestley, the reduction of matter to force could lead to a higher view of matter itself.

3 English and American Independents: Isaac Watts and Cotton Mather

Newtonian ideas were particularly popular among moderate Independents (Congregationalists) in England and America who were sympathetic to Whig policies of toleration for Nonconformists. Here we shall briefly consider two representative figures: Isaac Watts and Cotton Mather.

3.1 Isaac Watts

Isaac Watts (1674-1748) typifies the concurrence in many early eighteenth-century Christians of warm-hearted piety with hard-headed natural philosophy, and of liberal churchmanship with speculative theology. The common denominator to these seemingly contradictory tendencies was the currently popular attempt to differentiate between unambiguous data upon which all reasonable persons should agree and individual speculation consistent with the data. Watts was sparing in his listing of data and, at the same time, expansive in his personal theology.

Though opposed to the excesses of antinomian 'enthusiasm' and ambivalent towards the Great Awakening in America, Watts was concerned to bring a degree of warmth to Calvinist worship through his sermons and hymns.[63] The words and music of hymns like 'Our God, Our Help in Ages Past', 'When I Survey the Wondrous Cross', and 'Joy to the World' were enough to move even the strictest Calvinists and most hardened agnostics to genuine religious experience. But deep Christian piety did not prevent Watts from affirming that nature was entirely mechanical: so mechanical, in fact, that, once God had imposed the laws of motion (i.e., Newton's laws) on matter at creation, the subsequent formation and history of all things, including living organisms, took place strictly as a consequence of those laws.

Of course, Watts did not deny the complete freedom of God in choosing the laws of nature in the first place, or the complete dependence of both matter and motion on continued divine preservation: this much was in accordance with Newton. However, all the active powers arising from

---

[63] Paul Ramsey, 'Editor's Introduction', *The Works of Jonathan Edwards*, vol. 1 (New Haven: Yale Univ. Press, 1985), pp. 91, 96f.

divinely imposed laws could, according to Watts, be included within the 'Mechanical Motions and Powers of Matter' by definition.[64] Here, as with Samuel Clarke and William Derham, Newton's divinely imposed active principles were absorbed into the very mechanical philosophy which they had been designed by Newton to refute.

As to churchmanship and theology: according to Watts, only the clear statements of Scripture should be taken as binding on Christian consciences. One ought not, therefore, to require subscription to human doctrines, not even venerable doctrines like the view of the Trinity enshrined in the 'Athanasian Creed'.[65] Watts, therefore, supported the majority of his Dissenting colleagues in opposing the motion to require subscription to 'Athanasian' trinitarian standards at the Salters' Hall conference of Dissenting ministers in 1719.[66] One sees here again the impact on Dissenting groups of current Latitudinarianism and Newtonian natural theology as mediated by Clarke and Whiston.

On the other hand, Watts was by no means a minimalist in his personal theology. He did not reject the Trinity or the deity of Christ himself, but rather sought to reconcile those leaning towards Arianism and Unitarianism. Accordingly, he attempted (1722-25) to reformulate the traditional doctrine in order 'to vindicate the true and proper deity of Christ and the Holy Spirit' against the charge that it implied 'three distinct conscious minds' in the godhead.[67] For example, he developed a speculative notion, going back to the seventeenth-century Independent, Thomas Goodwin (also Origen), that Christ had possessed a human soul or conscious mind since the begining of time.[68] On the basis of this theory, Watts could argue that the coexistence with God of a second (divine-human) person did not necessarily imply a plurality of conscious minds in the godhead itself and thus avoid both Arianism and tritheism.[69] Note that the pre-existence of the human soul in the case of Jesus also had the effect of reinforcing the notion of a strict

---

[64] *The Works of Isaac Watts* (D. Jennings and P. Doddridge, eds., 6 vols., London, 1753), 5:594; G.N. Cantor and M.J.S. Hodge, 'Introduction', *Conceptions of Ether* (Cambridge: Cambridge Univ. Press, 1981), pp. 43f. On the other hand, Watts held to the passive nature of matter, in contrast to that of soul, in his *Essay on the Freedom of the Will in God and in Creatures* (1732); Ramsey, *Jonathan Edwards*, p. 99.

[65] *Useful and Important Questions concerning Jesus the Son of God* (1746; *Works* 6:713ff); Arthur Paul Davis, *Isaac Watts* (New York: Dryden Press, 1943), p. 118.

[66] *Oxford Dictionary of the Christian Church* (1957), p. 1441b.

[67] Davis, *Isaac Watts*, pp. 112, 114f.

[68] *The Glory of Christ as God-Man Displayed* (1746); Part III is entitled: 'An Argument tracing out the early Existence of the Human Soul of Christ, even before the Creation of the world, With an Appendix containing an Abridgement of Doctor Thomas Goodwin's Discourse of the "Glories and Royalties of Christ"'; Davis, *Isaac Watts*, pp. 118f., 278. Watts held this view as early as his 1725 *Four Dissertations relating to the Christian Doctrine of the Trinity*; Davis, *Isaac Watts*, p. 114; Toon, *Hyper-Calvinism*, pp. 43f.

[69] Davis, *Isaac Watts*, p. 112.

separability of matter and spirit that was advocated by many British moderates.

Watts's Christology was more than just a curious vagary in the history of Christian doctrine. When Joseph Priestley later decided against the traditional doctrines of the Trinity and matter-spirit dualism on the grounds that they were pagan intrusions into biblical faith, he had in mind teachings then current in Dissenting academies influenced by Watts, which associated the deity of Christ and the duality of matter and spirit with the idea of the pre-existence of the human soul as found in Plato and Origen.[70] Thus, a Dissenting minister known today primarily for his moving hymns could play a vital, if minor, role in the development of eighteenth-century natural philosophy towards materialism and neomechanism.

3.2 Cotton Mather

Cotton Mather (1663-1728), a leading churchman of colonial New England, was an even more complex figure than Watts. His modern biographers invariably find outright contradictions:[71] e.g., between his angel-mysticism and his mechanistic worldview; between his empirical medicine and his belief in the existence of a Nishmath-Chajim that governs all bodily functions; and between his concern for rules of evidence and his gullibility concerning the confessions used as evidence in the Salem witchcraft trials. Clearly Mather reflected the confluence of a variety of intellectual and spiritual themes, but, if more eclectic, he was no more inconsistent that many of his contemporaries. Like most supposed traits of human nature, consistency is a construct that is meaningful for certain eras and contexts, but not for others.

Our concern here is with the degree and nature of Mather's response to the new science and particularly to Newtonianism. Though more conservative theologically than their English colleagues, the American Puritans had most of the latest books and were well read in current science. Cotton was familiar with the writings of mechanical philosophers like Descartes, Gassendi, and Boyle,[72] declaring in 1689 that corpuscularianism was the only right philosophy.[73] Like Charleton and Boyle, he argued (1702) that the atomistic concept of matter required the impression of order by an

---

[70] Erwin N. Hiebert, 'The Integration of Revealed Religion and Scientific Materialism in the Thought of Joseph Priestley', in Kieft and Willeford, eds., *Joseph Priestley*, p. 37.

[71] E.g., the excellent biography by Kenneth Silverman upon which I rely here: *The Life and Times of Cotton Mather* (New York: Columbia Univ. Press, 1985).

[72] Silverman, *Life*, pp. 40ff., 262; Jeffrey Jeske, 'Cotton Mather: Physico-Theologian', *JHI*, vol. xlvii (Oct. 1986), p. 585.

[73] *Early Piety Exemplified* (1689); Silverman, *Life*, p. 443.

external agent and thus furnished convincing proof of the existence of God.[74]

Cotton's father, Increase Mather, had been inspired by the appearance of the comet of 1680[75]—and by the naturalistic explanations of the comet by Pierre Bayle, whose *Pensées* were published in 1682[76]—to read up on the latest European astronomical studies. His *Kometographia* (1683) recognised that comets, like planets, move in elliptical orbits.[77] Even though he did not fully understand the periodicity of cometary motion[78] and he still took comets to be direct signs of judgement from God (*pace* Bayle),[79] Increase suggested that their appearance could be predicted scientifically from conjunctions of the superior planets.[80]

Cotton Mather also observed the comet of 1680 and published a technical description of it in 1683. The deep impression made on his mind by this new messenger from the heavens can be sensed from a statement he made in one of his sermons that year: 'Every *Wheel* in this *huge clock* moves just according to the *Rule* which the *All-wise Artist* gave it at the first'.[81] The cosmos thus appeared to be as mechanical in the late seventeenth century on one side of the Atlantic as on the other. Even thunder and lightning, according to Mather (1694), were caused by the mechanical clashing of clouds containing explosive vapours from decaying vegetable matter.[82]

Cotton's mechanistic view of thunder did not exclude the active role of evil spirits, however.[83] According to an older cosmology surviving in passages of Scripture, the Devil was the 'Prince of the power of the air' (Eph. 2:2). In accordance with God's prior design and permission, Satan could use his power to set the cosmic wheels in motion and thus wreak

---

[74] Illustration (dated 1702) of Genesis 1 for 'Biblia Americana'; Silverman, *Life*, pp. 167f.

[75] Thomas Brattle also observed the comet of 1680 and sent data to the Royal Observatory of Greenwich which were later used by Newton in the *Principia* (III, prop. 41, prob. 21) to confirm his theory of elliptical orbits; Wallace E. Anderson, 'Editor's Introduction', *Jonathan Edwards: Scientific and Philosophical Writings* (*Works*, vol. 6, New Haven: Yale Univ. Press, 1980), pp. 19, 38.

[76] Bayle, *Pensées diverses...a l'occasion de la comète* (1682).

[77] Silverman, *Life*, p. 40.

[78] Newton did not accept the periodic motion of comets himself until the mid 1680s; McGuire, 'Transmutation', note on p. 86.

[79] Silverman, *Life*, pp. 56f.

[80] Jeske, 'Cotton Mather', p. 590 and note.

[81] *Elegy on...Collins*, p. 5; Silverman, *Life*, pp. 42, 434.

[82] Sermon of 1694, pub. in *Brontologia sacra* (1695); Silverman, *Life*, pp. 93f. Cf. *The Christian Philosopher* (London, 1721), Essay 16, pp. 61ff.; also in Winton U. Solberg, ed. (Urbana: Univ. of Illinois Press, 1994), pp. 70f.; Dillenberger, *Protestant Thought*, p. 161.

[83] Angels and demons could also cause meteors, tempests, and thunders; *A Voice from Heaven* (1719); Margaret Humphreys Warner, 'Vindicating the Minister's Medical Role: Cotton Mather's Concept of the Nishmath-Chajim', *JHMAS*, vol. xxxvi (July 1981), p. 292.

havoc and cause plagues on earth.[84] In context, there was no real 'inconsistency' here. Most early modern machines were operated by humans; wind and water power were ancient but imperfectly harnessed; steam engines and automation came later.[85] Correspondingly, the course of nature was thought to be governed by mathematical laws and mechanisms but also subject to alteration—all the more so as the Millennium approached and Satan stepped up his attacks.[86]

Mediating between intelligences like demons and inert matter was a 'Plastic Spirit'. Though lacking intelligence and will, the Plastic Spirit permeated the world and governed the dynamics of matter in accordance with mathematical laws.[87] In individual bodies, the spirit took the form of the Helmontian *archeus*—Mather called it the 'Nishmath-Chajim' ('breath of life'; cf. Gen. 2:7)—which presided over the metabolism and reproduction of each organism.[88] Mather found clear scientific evidence for the existence of such a spirit in the ability of plants to regenerate themselves from mere cuttings.[89] This subtle material substance perfectly expressed Mather's ambivalence regarding the relation of nature to God; it was both natural and spiritual, both autonomous and dependent on God, at the same time.[90] Like the Cambridge Platonists, Mather saw in the Plastic Spirit evidence that

---

[84] Sermon of 1694; Silverman, *Life*, p. 94. As the Millennium approached, Satan would be confined closer to the centre of the earth (Rev. 12:9-12) and could bewitch people as well as cause earthquakes like that in Jamaica of 1692, the year of the outbreak of demon possession in Danvers, Salem, and Andover; sermon printed in *Wonders of the Invisible World* (1692); Silverman, *Life*, pp. 103, 107f. In his 'Biblia Americana', Mather still held on to aspects of Aristotelian cosmology: the cosmos had a geometric centre towards which heavy objects naturally gravitated (Beall and Shyrock, *Cotton Mather*, pp. 40f.) and an outermost celestial sphere beyond which God's heaven was located (Jeske, 'Cotton Mather', p. 586). However, Mather knew and promoted Copernican ideas generally; *Christian Philosopher*, Essay 20, pp. 75f.; Solberg, ed., p. 84; Silverman, *Life*, pp. 22, 253.

[85] Reports of Newcomen's early version of the steam engine were published by Desaguliers (*Course*) in 1734; G.S. Rousseau, 'Science', in Pat Rogers, ed., *The Eighteenth Century* (London: Methuen, 1978), p. 170.

[86] Increase Mather's eschatological sermons preserved in Cotton's notebooks; Cotton Mather, 'Problemata theologicum' (comp. 1703); Silverman, *Life*, p. 171.

[87] Silverman, *Life*, pp. 122, 133. The idea of a 'Plastic Spirit' reminds us of the 'Plastick Nature' described in Ralph Cudworth's *True Intellectual System of the Universe* I.iii.28; Hunter, 'Seventeenth-Century Doctrine of Plastic Nature', 201f.; Klaaren, *Religious Origins*, p. 227 note 71.

[88] Warner, 'Vindication', pp. 278, 285f. (Nishmath-Chajim).

[89] 'Monstrous Impregnations' (1716); Silverman, *Life*, p. 252. According to Otho Beall and Richard Shyrock, Mather inclined more to the iatro-mechanical (preformationist) than to the iatro-chemical theory of disease; *Cotton Mather: First Significant Figure in American Medicine* (Baltimore: Johns Hopkins Press, 1954), pp. 24, 32f. That is, he understood regeneration, heredity, and disease to be controlled by 'animalcules'. But the animalcules themselves were guided by Plastic Spirit as the regenerative ability of plants and animals indicated; Silverman, *Life*, pp. 252, 343f.

[90] As Warner states, the Nishmath-Chajim allowed a naturalistic account of spiritual healing; 'Vindication', p. 292.

there was something more than mechanics in the operations of nature.[91] The operation of Plastic Spirit was no more unintelligible or 'occult' than a strictly mechanical account would be.[92]

The impression made by the idea of mechanism was a lasting one, but the idea took on a different meaning in Mather's later work as the result of the newer, Newtonian science. The mechanical ideal was reinforced by his reading of John Ray and William Derham[93] and was clearly stated in his influential *Christian Philosopher* (pub. 1720-21), the first textbook on science to be written by an American:[94]

> The Great God has contrived a mighty *Engine*, of an Extent that cannot be measured, and there is in it a Contrivance of *Motions* that cannot be numbered. He is infinitely gratified with the View of this *Engine* in all its *Motions*....

Comets, of course, were part of the 'mighty Engine'. That in itself would not prevent their serving (as they did for Whiston) as the instruments of spiritual powers and hence portending untoward events like war and pestilence. However, by the time he wrote the *Christian Philosopher*, Mather had also read 'the admirable Sir Isaac Newton, whom we now venture to call the Perpetual Dictator of the learned World in the *Principles of Natural Philosophy*'.[95] Among other things, Mather picked up on Newton's speculations in the *Principia* about the vapours of comets replenishing the water supply on earth. Newton's speculations had been

---

[91] 'The more progress we make in *Experimental philosophy*, the oftener we shall find ourselves driven to something so much beyond *mechanical principles*'; 'Monstrous Impregnations'; Silverman, *Life*, p. 252. Features of human metabolism that could not be explained in strictly mechanical terms included foetal formation, emotional disorders, and the body's capacity for fighting disease. These depended on the Nishmath-Chajim; *Angel of Bethesda* (1722); Pershing Vartanian, 'Cotton Mather and the Puritan Transition to the Enlightenment', *Early American Literature*, vol. vii (Winter 1973), pp. 218f.

[92] '...if I call [the Plastic Vertue] *unintelligible*, I must confess my Mechanism to be so too'; 'Monstrous Impregnations' (1716); Silverman, *Life*, p. 252.

[93] *Christian Philosopher*, p. 3; Solberg, ed., p. 10; Beall and Shyrock, *Cotton Mather*, pp. 50f.; Silverman, *Life*, pp. 249f.

[94] *Christian Philosopher*; Silverman, *Life*, pp. 250f. Silverman comments that 'Such a predictable universe seems inconsistent with inflows of supernatural grace...and other unscheduled wonders of the invisible world'. Silverman then has to explain how Mather 'made room for pneumatological phenomena in a mechanical universe'. But the same apparent inconsistencies appear in other figures of the time like Whiston. In any case, psychic phenomena were never excluded by cosmic mechanism. In fact, the mechanisation of the external, physical world may have encouraged efforts towards transcendence by inner, spiritual means.

[95] *Thoughts for the Day of Rain* (1712); W.E. Anderson, *Jonathan Edwards*, p. 19; *Christian Philosopher*, p. 56; Solberg, ed., p. 65; Dillenberger, *Protestant Thought*, p. 159; Silverman, *Life*, p. 249. Mather's puzzlement over the nature and extent of the atom (Silverman, *Life*, p. 251) is probably a reflection of Newton's in the 1713 Query 31 (Thackray, 'Matter', pp. 37f.).

motivated by a concern for the stability of the system of nature (mirrored always in the stability of early modern society).[96] The facts were these: comets periodically spewed vapours into space, and the water supply on earth was was gradually being converted into solid vegetable matter and earth. Therefore, Newton had reasoned, a balance would be maintained if the excess caused by one made up for the deficiency caused by the other.[97]

'If this be so', wrote Mather, 'the Appearance of *Comets* is not so dreadful a thing, as the *Cometomantia*, generally prevailing, has represented it.'[98] Mather's concern was apparently the pastoral one of delivering people from fear of the unknown rather than the more scientific one of balancing the processes of nature. Still, he was hedging his bets: his citation of Newton was followed by quotations from other authorities who still held that cometary vapours were potentially noxious and could be instruments of divine judgement.[99]

There was more at stake here than just an emphasis on mechanisms and cycles in nature: a fundamental shift was taking place in the meaning of a numinous natural phenomenon. Since ancient times comets had been omenous clues to the meaning of history, cosmic copulas that bound the heavens together with human life in mutual sympathy. Newtonian comets were life-supporting but silent; they ensured that the water supply would not run out, but they did nothing to make human life more bearable or history more meaningful.

However, comets were just one of many traditional links between heaven and earth. Mather could afford to trade them away partly because he still had meteors, tempests, and thunder to ensure the continued activity of spirits in the everyday world: '...the *Heavens do Rule*', he said (1719), 'and the *Invisible World* has an astonishing share in the Government of *Ours*.'[100]

Clearly, Cotton Mather was an eclectic more than a Newtonian in the strict sense. His distance intellectually from the mainstream of English

---

[96] On Mather's concern for the stability of New England society, see Silverman, *Life*, pp. 55f., 106ff., 115ff.

[97] *Principia* Book III (1687, 1713; Cajori, ed., 2:529f.); McGuire, 'Transmutation', pp. 87f. The notion that water was transmuted into earth by the growth and decay of vegetation goes back to the alchemical tradition via Helmont and Boyle; A.G. Debus, 'Becher', *DSB* 1:547b; Sara Schechner Genuth, 'Comets, Teleology, and the Relationship of Chemistry to Cosmology in Newton's Thought', *Annali dell' Instituto e Museo di Storia della Scienza di Firenze*, vol. x (1985), pp. 42-52. This idea was not refuted until Lavoisier; Thackray, *Atoms*, p. 196. Newton's speculations on the occasionally destructive effects of comets were never published; Genuth, 'Comets', pp. 54ff., 63f.

[98] *Christian Philosopher*, p. 43; Solberg, ed., p. 52; Silverman, *Life*, p. 251.

[99] Particularly George Cheyne; *Christian Philosopher*, pp. 44f.; Solberg, ed., p. 53.

[100] *A Voice from Heaven* (1719); Warner, 'Vindicating', p. 292. A key biblical text for the influence of the heavens on terrestrial affairs was Job 38:33b. Eclipses had been naturalised already by Pierre Bayle; *Pensées* (1682, Herbert H. Rowen, ed., *From Absolutism to Revolution, 1648-1848* [New York: Macmillan, 1968], p. 9).

Newtonian thought may help to account for the fact that, while adopting the new cosmology, he did not follow the theological trend towards Arianism and Unitarianism. Like many English Latitudinarians and Dissenters of his time, Mather emphasised the reasonableness of Christian faith.[101] He even offered a simplified formulation of doctrine (1717) that would allow all genuine believers to join together in a pan-Christian union. His aim again was the pastoral one of avoiding the excesses of theological controversy and easing the consciences of lay Christians uncertain about the finer points of doctrine. But unlike some of his colleagues, Mather included the Trinity as one of the non-negotiable truths common to all true believers.[102]

Mather was aware of Arian leanings among his English colleagues as early as 1699. He became concerned about the challenge of his 'learned friend Whiston' in 1711 and was even assailed by doubts himself. But Mather prayed to God and received 'Sweet Satisfaction...in His Truth, concerning Three Eternal Persons in His infinite Godhead'.[103] Like William Derham, Mather finally gravitated towards another sun besides Newton, and it wasn't Leibniz.

Mather's view of human reason was more theistic than naturalistic.[104] Like most of his colleagues, he viewed reason as a divine gift and was optimistic about the possibility of understanding God's works with its aid (cf. chap. 1, sec. B). But the grounding of reason in the divine image implied for Mather that direct illumination by God was both possible and required. Reason was not autonomous, but a participant in a larger reality and responsive to the voice of God.[105] Hence, there was no 'contradiction' between following reason and receiving divine guidance.

The difference between Mather's stance and that of Isaac Watts became clear in the wake of the Salters' Hall controversy of 1719. Both were motivated by a desire to unite dissenting Christians around the essentials of the faith. But, whereas Watts saw the imposition of subscription to a trinitarian standard as exceeding the clear requirements of Scripture and potentially divisive, Mather saw the lack of doctrinal imposition as refusal to rally around a common, biblical standard and as opening the fellowship of

---

[101] E.g., *Reasonable Religion* (1700); *Reason Sanctified and Faith Established* (1712); Jeske, 'Cotton Mather', pp. 587f.

[102] *Malachi* (1717); Silverman, *Life*, pp. 300f., 331.

[103] Richard F. Lovelace, *The American Pietism of Cotton Mather* (Grand Rapids, Mich.: Christian Univ. Press, 1979), pp. 42ff.; Silverman, *Life*, p. 329.

[104] Vartanian, 'Cotton Mather', p. 221.

[105] 'The Light of Reason is the Work of God; the Law of Reason is the Law of God; the Voice of Reason is the Voice of God'; Diary for 1711; Vartanian, 'Cotton Mather', p. 221. 'By the light of this precious and wondrous Candle, we discern the Connection and Relation of things to one another'; *Man of Reason* (1718). 'Reason, what is it but a Faculty formed by God, in the Mind of Man, enabling him to discern certain Maxims of Truth, which God himself has established'; *Christian Philosopher*, p. 283; Jeske, 'Cotton Mather', p. 588.

the church to heretics.106 The underlying difference seems to have been that Watts, like most English Latitudinarians and Dissenters, took God the creator to be the focus of his piety. Mather's religion, on the other hand, was centred in the immediate lordship of Jesus Christ over the Church, the state, and personal life.107 Any attempt to allow doubts about the divine status of Christ was as much an attack on the New England way of life as witchcraft was.108

The examples of Watts and Mather indicate the pervasiveness of mechanistic and Newtonian ideas in early eighteenth-century English-speaking culture. There is little evidence here of any 'conflict' between science and religon. On the other hand, serious questions were being raised about the impact of Newtonian natural theology on biblical faith and on the doctrine of the Trinity in particular. Accordingly, some of the anti-Newtonian figures we shall consider later (sec. C) made the Trinity foundational to their own distinctive natural philosophies.

4 Continental Newtonians from Boerhaave to Boscovich

Newtonian ideas were first introduced into France in the late seventeenth century by Oldenburg and Malebranche, but it was not until their championing by Boerhaave, Maupertius, and Voltaire in the 1730s that they began to attract many followers in continental Europe.109 As we shall see, the manners in which Newton was interpreted differed greatly, with Boerhaave and Maupertius tending towards the materialist side and Euler, Voltaire, and Boscovich opposing that tendency in various ways. Boerhaave's younger colleague, s'Gravesande, who popularised Newtonian ideas already in 1720, went so far in the materialist direction that we shall treat him in the following section.

4.1 Boerhaave and Maupertius

Herman Boerhaave (1668-1738), the son of a Dutch Reformed minister,

---

[106] Silverman, *Life*, pp. 330ff.

[107] Silverman, *Life*, pp. 51, 108, 331f. Dillenberger's conclusion that Mather was in danger of 'neglect of the Christological centre of Christian understanding' is based on a reading of the *Christian Philosopher* alone; *Protestant Thought*, p. 160; cf. Silverman, *Life*, p. 250. Even so, the concluding pages of the *Christian Philosopher* are devoted to the praise of Christ as 'Lord of all'. And one should not overlook passages like Essay 13, p. 58; Solberg, ed., p. 67: '...how can we forget the Glorious CHRIST, who is our *Head* in the *Covenant*; and about whose *Head* there has been the appearance of a *Rainbow*, in the Visions of his Prophets, betokening our Dependence upon Him for all our Preservations!'

[108] Mather on the demons: 'These Monsters have associated themselves to do no less a thing than, To destroy the kingdom of our Lord Jesus Christ, in these parts of the World'; Silverman, *Life*, p. 108; cf. 106.

[109] Henry Guerlac, *Newton on the Continent* (Ithaca: Cornell Univ. Press, 1981), pp. 62ff.; C.A. Russell, *Science*, p. 61.

began his studies at Leiden in the fields of theology and philosophy before turning to medicine. He eventually became professor of medicine, chemistry, and botany, thus holding three of the five chairs on the Leiden Faculty of Medicine.[110]

Boerhaave is often regarded as the founder of 'rational' medicine and chemistry: his system of medicine (1708)[111] was so influential that the early eighteenth century was known among contemporary physicians as the 'Age of Boerhaave'.[112] The new system's approach to natural science was empirical and eclectic, based on the chemical philosophy of Helmont (iatrochemistry) as well as the mechanical philosophy of Bacon and Boyle (iatrophysics). As early as 1715, Boerhaave incorporated Newton's idea of attraction at a distance into his chemistry.[113] This overall synthesis was later popularised in his *Elements of Chemistry* (1732).

Boerhaave's most important contribution to the Newtonian tradition was his view of elemental fire as the basic active principle that pervades the cosmos. Fire played an instrumental role in chemical reactions by altering the forces of attraction between the atoms of various substances.[114] Boerhaave's fire was material, but not ponderable; that is, it had no weight, offered no resistance, and could penetrate solid matter. It radiated from the sun (an old hermetic idea) and circulated through the solar system revitalising motion and life, particularly on the earth.[115] Though consistent with Newtonian speculations on the ether and on comet vapours, Boerhaave's ideas pointed in the direction of hylozoic materialism and had a deep influence on early vitalists like s'Gravesande, Haller, and Shaw (sec. B).[116]

In medicine, Boerhaave is significant for his notion of a supramechanical 'Aura' or *spiritus rector* which presided over the body and directed its functions. This Aura had been implanted by God in each living creature at creation. Clearly Boerhaave's Aura was related to Helmont's *archeus* and Mather's Nishmath-Chajim.[117] Spiritualist ideas were thus alive and well in the early eighteenth-century Netherlands and America in spite of their association with the radical fringe in seventeenth-century England (chap. 3, sec. C).

---

[110] G.A. Lindeboom, 'Boerhaave', *DSB*, 2:224f.

[111] *Institutiones medicinae* (1708); Dolan and Adams-Smith, eds., *Health and Society*, p. 114.

[112] Peter Gay, *The Enlightenment*, vol. 2 (London: Wildwood House, 1973), p. 18.

[113] Boerhaave, *Oratio de comparando certo in physicis* (1715); Gay, *Enlightenment*, vol. 2, p. 135.

[114] Leicester, *Historical Background*, p. 123.

[115] Heimann, 'Nature', pp. 12f.; Cantor and Hodge, eds., *Conceptions of Ether*, pp. 24ff.

[116] Haller was his student at Leiden; Dolan and Adams-Smith, eds., *Health and Society*, p. 115.

[117] Heimann, 'Nature', note on p. 13.

Pierre Louis Moreau de Maupertius (1698-1759) was as much a Newtonian as Boerhaave and like the latter shifted in a materialistic direction, though with a very different emphasis and far more radical consequences. Whereas Boerhaave followed the antispeculative, empiricist side of Newtonian research, Maupertius developed the mathematical and philosophical side. And, whereas Boerhaave combined Newtonian ideas with the Helmontian chemical philosophy, Maupertius combined them with Leibniz's monadology (chap. 3, sec. C). He was thus transitional from Newtonian natural theology to hylozoic monism and could just as easily be treated in the following section on post-Newtonian materialists.

In 1728, Maupertius visited England and became a convinced exponent of the Newtonian philosophy. He was instrumental in introducing Newtonian ideas into both France and Germany, where he served (after 1746) as president of the Berlin Academy at the invitation of Frederick the Great.[118] When in his later work (1750s) Maupertius turned in a decidedly materialist direction, he was partly responsible for introducing Leibnizian ideas back into France.[119]

The development of Maupertius's thought is therefore a confirmation of a materialist tendency we have already seen in Newtonians like William Derham and Roger Cotes. On the other hand, the steady anti-Leibnizianism (tending towards neomechanism) of his younger colleague at Berlin, Leonhard Euler (1707-83),[120] is further evidence of the variety of possible directions Newtonians could develop. Whereas Maupertius combined Newtonian ideas with Leibniz's pansophism, Euler combined them with the very Cartesian mechanism Maupertius (and Newton) opposed.[121]

Maupertius began as a perfectly respectable Newtonian, however. In fact, he was responsible for 'verifying' (1736-37) one of the key predictions of Newton's *Principia*—the flattening of the earth at the poles—over against the equatorial constriction predicted by the rival Cartesian philosophy then popular in France.[122] With the awesome reputation of 'Flattener of the Earth',[123] Maupertius then campaigned against the Cartesians, arguing that

---

[118] In France as early as 1732: *Discours sur la figure des astres* (1732); Randall, *Career*, vol. I, p. 874; Gay, *Enlightenment*, vol. 2, pp. 136f.

[119] Ernst Cassirer, *Philosophy of the Enlightenment* (Boston: Beacon Press, 1955), p. 86. The Marquise du Châtelet, originally a pupil of Maupertius, had already introduced Leibnizian ideas through her popular *Institutions de physique* (1740), which was based on the *Ontologia* of Christian Wolff (1729); Hankins, *Science*, pp. 35, 40.

[120] Calinger, 'Newtonian-Wolffian', pp. 321ff. Note that Euler's opposition to Leibniz began (1738-45) before Maupertius arrived in Berlin.

[121] Calinger, 'Newtonian-Wolffian', p. 329; Heimann, Concepts', pp. 127ff.

[122] A.R. Hall, *The Scientific Revolution, 1500-1800* (Boston: Beacon Press, 1966), pp. 343f. The test was proposed in Maupertius's *Sur la figure de la terre* (1733); B. Glass, 'Maupertius', *DSB* 9:186b.

[123] Randall, *Career*, vol. I, p. 874.

Newton's concept of action at a distance was perfectly acceptable in physics and need not be replaced by a mechanistic explanation of gravitation.[124] An adequate rationale for Newton's physics could be found instead, Maupertius contended, in a 'Principle of Least Action' which he offered (1740-46)[125] as an antidote to Cartesian mechanism as well as to Leibniz's principle of the conservation of *vis viva*.[126]

'Action', a foundational concept of modern physics, was first defined by Maupertius as the product of the velocity (or momentum) and the distance through which a material body travels at that velocity. From the empiricist point of view, action was clearly a second-order concept, less closely tied to immediate observation than the standard Newtonian concepts of distance and velocity, or even mass, force, and momentum. It turns out to be a more fundamental concept, however, and it has played a role in several major advances in nineteenth- and twentieth-century physics.

The 'Principle of Least Action' was a reformulation of Newton's laws of motion in teleological terms using the concept of action: given the starting and ending points of its path in space, the trajectory of a body (position and velocity at each point) between those points must be such as to make its cumulative action a minimum (in general, an extremum). In other words, the slightest deviation from that trajectory would result in an increase in the cumulative action. This truly amazing result was bound to raise theological questions. How does a bit of matter know enough to move in just the right way to realise such an economy of motion?

In Maupertius's view, the Principle of Least Action was a necessary consequence of the wisdom of God and hence provided a metaphysical foundation for natural philosophy.[127] Conversely, the experimental verification of least action in nature could be used as a proof of the existence and basic attributes of God.[128] Moreover, the gradual decrease of the total amount of action in the universe due to inelastic collisions indicated the need for periodic divine intervention to keep things going.[129]

At this point in his career, Maupertius followed the natural theology of Newton, though he appealed to a teleology inherent in Newton's equations

---

[124] Maupertius, *Essai de cosmologie* (1750); Jammer, *Concepts of Force*, p. 208.

[125] 'Lois du repos des corps' (1740); 'Accord de differentes lois de la nature' (1744); 'Les lois du mouvement et du repos' (1746); Ronald S. Caliger, 'The Newtonian-Wolffian Controversy (1740-1759)' *JHI*, vol. xxx (July 1969), p. 324; Glass, 'Maupertius', p. 187.

[126] Aram Vartanian, *Diderot and Descartes* (Princeton: Princeton Univ. Press, 1953), pp. 95f.

[127] Thomas L. Hankins, *Science and the Enlightenment* (Cambridge: Cambridge Univ. Press, 1985), p. 210.

[128] *Essai de cosmologie* (1750); Dijksterhuis, *Mechanization*, p. 496.

[129] *Lettres* (Dresden, 1752); Herbert H. Odom, 'The Estrangement of Celestial Mechanics and Religion', *JHI*, vol. xxvii (Oct. 1966), p. 545.

rather than the supramechanical nature of active principles like gravitation. On the other hand, he rejected Newton's argument that initial features of the solar system like the coplanarity of the planets could not be explained by mechanical principles and hence had to be referred directly to divine manipulation.[130] In this step towards neomechanism, Maupertius resembled other Newtonians like Whiston and Watts.

An earlier 'minimum principle' for light, using increments of time, had been suggested by Pierre de Fermat already in 1661. Maupertius regarded his Principle of Least Action as more general than Fermat's, applying to both solid bodies and to light (which was regarded as corpuscular by Newtonians in any case). He thus suggested a unification of mechanics and optics that remained a guiding ideal of theoretical physics until it was realised in nineteenth-century Hamiltonian, and twentieth-century quantum, mechanics.[131]

While the Principle of Least Action was a legitimate mathematical development of Newton's mechanics, philosophically it owed much to the pansophism of Newton's antagonist Leibniz. Even brute matter in Leibniz's view displayed a degree of intelligence by virtue of its God-given ability to function in accordance with prescribed laws. In his early work (1740s), Maupertius was influenced mostly by Leibniz's ideas (in opposition to Descartes's) on final causes in nature. In fact, he became embroiled in a priority dispute (1751-52) with one of Leibniz's disciples concerning the origins of the Principle of Least Action. In an alternative form the idea may have been stated by Leibniz as early as 1707.[132]

In the final decade of his life (1750s), Maupertius developed a hylozoic materialism similar to that of Buffon and Diderot (sec. B). He was impressed by recent evidence for spontaneous generation gathered by John Tuberville Needham (late 1740s).[133] He also did studies of his own, demonstrating the inheritance of traits from both maternal and paternal lines of descent.[134] Newton's principles, Maupertius concluded, were adequate (once supplemented by the Principle of Least Action) for the purposes of physics and astronomy, but biological problems like spontaneous generation and heredity required additional principles like appetite, perception, memory, and

---

[130] *Essai de cosmologie* (1750); Roger Hahn, 'Laplace and the Mechanistic Universe', in Lindberg and Numbers, eds., *God and Nature*, p. 265. On the other hand, Maupertius also rejected the Cartesian program of explaining the origin of the world in terms of matter and motion: Stanley L. Jaki, trans., *Immanuel Kant: Universal Natural History and Theory of the Heavens* (Edinburgh: Scottish Academic Press, 1981), pp. 249f, note 26.

[131] W.F. Bynum et al., eds., *Dictionary of the History of Science* (Princeton: Princeton Univ. Press., 1981), p. 146.

[132] Calinger, 'Newtonian-Wolffian', pp. 324ff.

[133] Vartanian, *Diderot*, pp. 258-62. Maupertius opposed the election of Mettrie to the Berlin Academy in 1748; Calinger, 'Newtonian-Wolffian', p. 327.

[134] Kiernan, *Enlightenment*, p. 220; Glass, 'Maupertius', 187-88; Hankins, *Science*, p. 138.

will as properties of atomic matter itself.[135] Mental activity, he concluded, was not a separate substance as Descartes and the mechanical philosophers held, but a fundamental property of matter like extension.[136] Even human thought was the result of the combined activity of the elementary atoms.[137] Various forms of life and intelligence could, therefore, have evolved naturally from mere matter through the inheritance of acquired traits and natural selection.[138]

Maupertius was thus an important forerunner of the evolutionary ideas of Lamarck in the late eighteenth and early nineteenth centuries. Still, there was no conflict between natural selection and theology in his view. Since the laws of nature and the properties of matter were God's own work to begin with and reflected divine wisdom, there was no contradiction between attributing such powers to matter and believing in God.[139]

The ease with which Maupertius shifted from Newtonian natural philosophy to hylozoic materialism is breathtaking! As contributing factors we have noted the tendency of some Newtonians (e.g., Roger Cotes) to anchor active principles in matter, the direct influence of Leibniz, and experimental evidence for spontaneous generation and bilineal descent. Behind all of these partial explanations, however, was the multivalent idea of creation itself. As in the Middle Ages and Renaissance, the idea of creation could inspire awe for the power of the Creator, but it could also evoke faith in the power of God's creature.

4.2 Voltaire and Boscovich

In the face of the mounting threat of neomechanism and materialism, Newtonians in France and Italy rallied around the Newtonian standard. The prestige of Newton was exploited particularly by French social moderates like Voltaire and Boscovich to counter more radical tendencies in philosophy and politics.[140]

---

[135] *Dissertatio inauguralis metaphysica de universali Naturae systemate* (pseud., Erlangen, 1751); Cassirer, *Philosophy*, p. 86.

[136] Baumer, *Modern European Thought*, p. 210.

[137] *Dissertatio* (1751); Émile Bréhier, *The History of Philosophy*, vol. 5 (Chicago: Univ. of Chicago Press, 1967), pp. 135f.

[138] *Essai de cosmologie* (1750); Colm Kiernan, *The Enlightenment and Science in Eighteenth-Century France* (Banbury: Voltaire Foundation, 1973), p. 220; Hahn, 'Laplace and the Mechanistic Universe', p. 265.

[139] Jacques Roger, 'The Mechanistic Concept of Life', in Lindberg and Numbers, eds., *God and Nature*, pp. 288f.

[140] As early as 1739, the Abbé Noel-Antoine Pluche (1688-1761) attacked advocates of the cosmogony of Descartes, charging that the naturalistic explanation of the origins of the world system was designed 'to relieve Providence of its role in the creation of the universe': '...God is as much forgotten as if he had never existed'. But this was impossible, argued Pluche, since matter by itself was blind and devoid of the slightest intelligence. The laws of

Francois-Marie Arouet de Voltaire (1694-1778) was nearly contemporary with Maupertius both in age and in championing the ideas of Newton. After visiting England and consulting with Newtonians like Clarke and Pemberton (1726-29),[141] he published his famous *Philosophical Letters on the English* (1733), in which he contrasted Newton with Descartes to the decided advantage of the former.[142]

Realising the importance of natural philosophy for an integrated picture of the world, Voltaire studied Newton's work for five years (1736-41). Not being trained in mathematics himself, he relied on the help of his mistress, his 'Lady Newton', the Marquise du Châtelet, who had studied Newtonian physics with Maupertius and his pupil Clairaut. Voltaire's *Elements of the Philosophy of Newton* (1738), largely based on the work of Mme Châtelet,[143] was a vibrant declaration of independence from Descartes's necessitarianism and from the church's authority as much as a defense of Newton himself.

Descartes's ideas had been adopted by conservative scholars like the Jesuits in the early eighteenth century as the result of their newly alleged consistency with Catholic dogma. Now Voltaire exulted in Newton's sense of divine sovereignty as a symbol of human freedom from all worldly systems.[144] Whereas English latitudinarians valued their church as comprehensive and socially beneficial, Voltaire and other French *philosophes* abhorred theirs as power hungry and bigoted.[145] One might well ponder the change that had taken place since the eleventh and twelfth centuries when the sovereignty of God had been cited by conservative Catholics as an antidote to the strictures of Aristotelian dialectic and the power of the state (chap. 1, sec. D).

Voltaire was anticlerical, but not irreligious. 'People must have a religion', he quipped, 'and not believe in priests, just as they must have a diet and not believe in physicians.'[146] On the other hand, religion (like diet) was only a part of life for Voltaire and should not be allowed to interfere with secular affairs like investments in the stock market. His comments on the secularity of contemporary business practise perfectly express the way in which moral and religious considerations had been confined to social enclaves of personal preference:[147]

---

mechanics could be used only to explain the operations of a preassembled system, not to account for the origins of the system itself; *Histoire du ciel* (1739); Vartanian, *Diderot*, pp. 90-93. In this respect, Pluche followed the orthodox line of Newton and Clarke.

[141] M.R. Jacob, *Radical Enlightenment*, pp. 102, 104.

[142] *Lettres philosophiques sur les Anglais*, Letters 14-17; Gillispie, *Edge*, pp. 157f.

[143] *Éléments de la philosophie de Newton* (1738 and 1741).

[144] Gillispie, *Edge*, pp. 158f.

[145] A. Robert Caponigri, *A History of Philosophy*, vol. III (Notre Dame: Univ. of Notre Dame Press, 1963), p. 352.

[146] Scribbled in a notebook; Gay, *Enlightenment*, vol. 2, p. 19.

[147] *Lettres philosophiques*, Letter 6; Gay, *Enlightenment*, vol. 2, p. 51.

> Enter the London Stock Exchange, that place more respectable than many a court....There the Jew, the Mohammedan, and the Christian deal with each other as if they were of the same religion and give the name of infidel only to those who go bankrupt. There the Presbyterian trusts the Anabaptist, and the Anglican honors the Quaker's promise. On leaving these peaceful and free assemblies, some go to the synagogue, others to drink;...others go their church to await the inspiration of God, their hats on their heads, and all are content.

In order to appreciate the peace and joy Voltaire experienced in the London Stock Exchange of all places, one must recapture some sense of the pain and disillusionment with doctrinal differences that had been occasioned by religious disputes in Western Europe in the two centuries since the Reformation.

Voltaire's dislike of authoritarianism perhaps accounts as much as anything for his break with Maupertius, since 1746 the autocratic president of the Berlin Academy of Sciences. For reasons that are now hard to reconstruct, the two took opposite sides in the dispute (1751-52) over the authorship of the Principle of Least Action, Voltaire being content to associate the non-Newtonian idea cherished by Maupertius with the name of Leibniz. In a diatribe of 1752, he created a hilarious scene in which Maupertius was himself reduced to 'least action' by a Leibnizian bullet traveling at the square of its speed, thus ridiculing the Principle of Least Action and the conservation of *vis viva* with a single stroke.[148]

But there was a serious issue at stake between the two as well. Maupertius, as we have seen, followed a current of Newtonian thought that led from active principles within nature, but external to matter, to a dynamic view of matter imbued with active principles of its own. Voltaire resisted this tendency. His study of Newton had convinced him that matter was completely passive and that its present order was impressed upon it and maintained by God.[149]

The liberal, cosmopolitan culture so dear to Voltaire depended on the assumption that human beings were basically the same everywhere, regardless of their times, nationality, or religion. But the equal accessibility of reason and faith to all peoples required universal divine providence as its ground. It also required that human nature, and all nature, had always been what they now appeared to be to an enlightened French philosopher![150] The suggestion that there might be a progressive development in revelation, or in nature itself, would only lead back to the authoritarian claims of a

---

[148] Calinger, 'Newtonian-Wolffian', pp. 328f.
[149] *Éléments* (1738); Baumer, *Modern European Thought*, pp. 205f.
[150] *Essai sur les moeurs et les esprits des nations* (1756); Baumer, *Modern European Thought*, pp. 152, 195.

guardian of tradition like the church.[151] In other words, the only binding system could be one that antecedes all process of development, whether in religion or in nature. In religion, such a system was that of the basic truths common to all peoples and known in the West by the name of Deism.[152] In nature it was that of Newton's laws.

In the late 1740s, John Tuberville Needham's research on spontaneous generation was published. Father Needham had found that jars of mutton broth and dishes of grain would putrefy and produce microorganisms even when heated and sealed off from the outer world (twenty years later Spallazani and Pasteur proved otherwise). Buffon, Maupertius, and Diderot were impressed by Needham's work, but Voltaire satirised it (since worms could be produced from rye flour, perhaps humans could be made from wheat flour!) and accused Father Needham (wrongly) of materialism and atheism.[153] Beginning around 1765, Voltaire waged a major campaign against the new materialists, attacking their notion that motion was inherent in matter and insisting that only God could be credited with the generation and maintenance of motion.[154]

The continental Newtonians we have looked at have all had approximate counterparts in England and America. Boerhaave reminds us in some ways of Cotton Mather; Maupertius, in some ways of Whiston and in others of Cotes; and Voltaire, of Clarke. The Jesuit Abbé, Roger Joseph Boscovich (1711-87), a native of Croatia and sometime professor of mathematics at the Collegium Romanum (1740-59), might be said to remind us of John Keill, Gowin Knight, and Isaac Watts.[155] Like Keill, Boscovich stressed the insignificance (in bulk) of matter. Like Knight, he saw force as the most important aspect of physical reality. Like Isaac Watts, he was concerned to ensure the priority of spirit to matter.

Descartes and the mechanical philosophers, it will be recalled, tried to reduce all phenomena in nature to the effect of collisions between solid

---

[151] Voltaire had particularly in mind the case argued by Bossuet in his *Discours sur l'histoire universelle* (1681), against which he wrote his *Essai* of 1756; Baumer, *Modern European Thought*, p. 158. Claims to historic rights by clergy and nobility were the problem leading to the financial crisis of 1774-76 according to Turgot, then appointed controller general by Louis XVI. Turgot's solution: replace historical rights by natural rights common to all subjects of the state; Hankins, *Science*, pp. 158f. Turgot's naturalism was ethnocentric and nationalistic, however, not cosmopolitan and internationalist like Voltaire's; Nisbet, *History*, p. 180.

[152] Art. 'Deism' (later 'Theism') in Diderot's *Encyclopedia* (1751-72); Baumer, *Modern European Thought*, p. 195.

[153] *Dictionaire philosophique* (1764); Vartanian, *Diderot*, pp. 258-62. Holbach argued the possibility of generating a human being from wet flour by analogy with Needham's work; *Système de la nature* (1770); D.C. Goodman, ed., *Science and Religious Belief, 1600-1900* (London: Open University, 1973), 279.

[154] *Des singularitées de la nature* (1768); Baumer, *Modern European Thought*, pp. 212f.

[155] Copleston, *History* 6/1:69.

material particles. Newton allowed for such collisions, but added that the more interesting phenomena were due to forces acting at a distance and only accounted for in part by the intervening ether. Boscovich went a step further away from the mechanical philosophy and hypothesised (1745, 1758) that atoms were merely mathematical points and that all interactions, even seeming collisions, were the result of forces acting at a distance between those points. In complete contrast to what the Cartesians held, there was no such thing as extension in the material sense: there were only forces acting between dimensionless point-masses. Minimising matter even more than Keill and Pemberton did, Boscovich held that all the matter in the universe would not fill even a nutshell.[156]

Since contact forces were repulsive and operated only near the surface of a body, while chemical-bonding forces were attractive at medium range and gravitation was attractive and long-range, Boscovich hypothesised (1758) the existence of a single field with alternating zones of force around each mass point. The force surrounding a particle alternated sign from one zone to the next, being repulsive at small distances from the centre and alternatively attractive and repulsive at larger distances. The force-field also diminished in peak intensity with increasing distance from its centre. Then all the phenomena of physics and chemistry—including cohesion, fermentation, and chemical combination as well as gravitation and collisions—could be reduced to a single, unified force law as Newton himself had once suggested.[157]

Like Voltaire, Boscovich was concerned about the current tendency of many natural philosophers to treat spirit as a property of matter. By reducing matter to extensionless points, he hoped to negate its power in the minds of his contemporaries. But by reducing everything to force, he may have encouraged some, like Priestley, who came to see all matter as inherently dynamic. Boscovich was deeply chagrined at the way in which Priestley later (1777-78) exploited his ideas.[158]

Newtonian principles first successfully challenged the older ideas of Descartes on the Continent in the 1730s. But, by the 1750s, continental Newtonians were themselves on the defensive in the face of a new movement towards materialism. Theologically, the progression was from a clear matter-spirit dualism (Descartes), to matter-spirit symbiosis (Newton), to matter-spirit identity (the later Maupertius and Diderot). The corresponding shift in physics was from a science restricted to geometric quantities

---

[156] Thackray, 'Matter', pp. 49f.
[157] Jammer, *Concepts of Force*, pp. 173-77. Newton's suggestion of force zones was made in Query 31 of the *Opticks* (Query 23 in the 1706 *Optice; Opticks*, pp. 375f.); Heimann and McGuire, 'Newtonian Forces', p. 278.
[158] Thackray, 'Matter', p. 50; *Atoms*, p. 249.

(Descartes), to a more comprehensive one of dynamic quantities (Newton), to a science so comprehensive that it could not be quantified at all (Diderot).[159]

5 Scottish Newtonians: Colin Maclaurin and James Hutton

The last two Newtonians we shall consider are the Scots, Maclaurin and Hutton. As with the Anglo-American and continental Newtonians there is as much diversity as unity among them, with Maclaurin tending towards materialism and Hutton towards neomechanism, though the Scots do exhibit a distinctive common interest in the implications of natural philosopy for moral life.

5.1 Colin Maclaurin

Colin Maclaurin (1698-1746) became professor of mathematics and physics at the University of Edinburgh in 1725 upon the recommendation of Newton. He disseminated Newtonian ideas through his lectures and gave the reading public a faithful rendering of the master in his *Treatise of Fluxions* (1742) and his *Account of Sir Isaac Newton's Philosophical Discoveries* (published posthumously in 1748). As it turned out, the computational methods of Newton were too tied to geometrical constructions ('fluxions') to allow further progress in mathematical physics, and the hegemony of Newton's ideas in England and Scotland led to the neglect of current advances in continental mathematics based on the calculus of Leibniz and Euler.[160]

According to Maclaurin, Newtonian science provided a solid foundation for both natural religion and moral philosophy.[161] It pointed to God as the source of all efficacy in nature, equally present and active everywhere.[162] The scientist could never actually reach the First Cause as the result of scientific demonstration, but the increasing simplicity, generality, and beauty of scientific laws that scientists inferred and their transcendence of mechanical explanations were clear signs of God's presence and attributes.[163]

The publication of Maclaurin's textbook also reflected a distinct growth

---

[159] Aram Vartanian (*Diderot and Descartes*) overstates his case for continuity between Descartes and Diderot but captures the dynamics of the transition very well.
[160] J.F. Scott, 'Maclaurin', p. 611b.; Bynum et al. eds., *Dictionary*, pp. 150f.
[161] Odom, 'Estrangement', p. 541.
[162] *An Account of Sir Isaac Newton's Philosophical Discoveries* (1748) IV.9; Goodman, ed., *Science*, p. 275.
[163] Odom, 'Estrangement', p. 541.

of interest in the ether speculations found in some of Newton's writings.[164] Boerhaave, as we have seen, popularised the idea of a fiery substance permeating the cosmos in the 1730s. In 1740, an anonymous *Examination of the Newtonian Argument for the Emptiness of Space* was published, in which Newton's subtle ether was identified with Boerhaave's fire, and the resulting universal medium was credited with all the phenomena of physics, including gravitation, cohesion, electricity, and magnetism.[165] Maclaurin adapted this quasimaterialistic interpretation of Newton by stipulating that the efficacy of the ethereal medium was solely due to the underlying power and free will of the Deity.[166] In contrast to Boscovich, who tried to account for all physics in terms of force-fields between particles with mass but no extension, these two ether theorists suggested an explanation in terms of subtle matter-fields with extension but little or no mass.

Though active principles and ethers depended on the immediate will of God for their creation and continued general operation, their particular effects, Maclaurin was careful to point out, were not 'immediate volitions' of the Deity. For one thing, the supramechanical powers of nature were subject to the constraints of mechanical principles like Newton's law of action and reaction. But, more than that, as subordinate powers they had their own 'proper force and efficacy' in accordance with the common course of nature.[167] In other words, Maclaurin's Newton was not an occasionalist, but more of a deist: once his God had established the laws of nature, things operated pretty much on their own until a general reformation was required analogous to the original formation of the world-system.[168]

Like many other Newtonians, Maclaurin treated induction as the only proper scientific method in spite of his own bold theorising on subtle ethers. The Newton he presented to his public was 'distinguished for his caution and circumspection'.[169] Maclaurin's concern was to oppose the hypotheses invoked by the mechanical philosophers to account for what Newtonians took to be supramechanical phenomena. His theological

---

[164] Particularly (in order of publication), the General Scholium to the 1713 *Principia*; Queries 17-24 to the 1717 *Opticks*; the 1675-76 'Hypothesis Explaining the Properties of Light' (first published in Thomas Birch's *History of the Royal Society* in 1756-57); and the 1678/9 Letter to Oldenburg and 1678/9 Letter to Boyle (published in Birch's five-volume folio edition of the Works of Boyle in 1744); Thackray, *Atoms*, pp. 26-31; Schofield, *Mechanism*, pp. 13ff., 101ff.

[165] Schofield, *Mechanism*, pp. 107f.

[166] *Account* IV.9; Goodman, ed., *Science*, p. 265; Schofield, *Mechanism*, pp. 105f.

[167] *Account* IV.9; Goodman, ed., *Science*, pp. 263-66.

[168] Evidence for a coming reformation could be found in the material as well as the moral world; looking backwards, therefore, the world could not be eternal; *Account* IV.9; Goodman, ed., *Science*, pp. 263f. One type of physical evidence was the discontinuities and displacements of geological and paleontological remains; *ibid.*, pp. 266f.

[169] *Account* IV.9; Goodman, ed., *Science*, p. 256.

argument was that God's wisdom and ways far transcended human powers, so that humans could not understand nature by speculative reason alone. Such an emphasis on the inscrutability of nature and the limitations of human reason did not, however, negate the underlying faith in the ultimate comprehensibility of nature (chap. 1, sec. B). The comprehension of nature remained an ideal that could and would be attained, if not in this life, then in the next.[170]

Similar inductivist epistemologies had been advocated by Newtonians like Clarke,[171] Boerhaave,[172] and Pemberton[173] and were later to be championed by the 'common sense' philosopher Thomas Reid (1710-96). Newton himself had been highly ambiguous on the question of scientific method. He claimed that he 'feigned no hypotheses',[174] while boldly proposing speculative principles like supramechanical forces and imponderable ethers. Reid, on the other hand, used strict inductivist principles as an argument against speculative hypotheses like Maclaurin's imperceptible ether.[175] He could be said to have reflected the cautious side of Newton's personality. The other, bolder side was better reflected in the hypotheses of Maclaurin and Hutton.[176]

5.2 James Hutton: Uniformitarianism Versus Catastrophism

James Hutton (1726-97) attended the lectures of Maclaurin at University of Edinburgh (early 1740s) and studied medicine at Edinburgh and Leiden (1744-49), coming under the influence of Boerhaavian ideas along the way. He took up farming in Berwickshire where he formulated his basic ideas about the importance of erosion for various soils.[177] Moving back to Edinburgh in 1768, he then turned his attentions to the relatively new field of geology, a field of which he has long since been regarded as one of the founding fathers.[178]

Hutton is best known for his 'principle of uniformity' (1788,

---

[170] *Account* IV.9; Goodman, ed., *Science*, pp. 268f.; Odom, 'Estrangement', pp. 542f.
[171] M.C. Jacob, *Newtonians*, p. 239.
[172] Gay, *Enlightenment*, vol. 2, pp. 18, 135; Schofield, *Mechanism*, pp. 136f.
[173] Larry Laudan, *Science and Hypothesis* (Dordrecht: Reidel, 1981), pp. 103-8, notes 3 and 43.
[174] Query 31 (23) to the 1717 (1706) *Opticks*.
[175] Richard G. Olson, *Scottish Philosophy and British Physics* (Princeton: Princeton Univ. Press, 1975), pp. 39-44, 52f.; Laudan, *Science*, pp. 91-98.
[176] Heimann and McGuire, 'Newtonian Forces', pp. 281-84.
[177] According to Joseph Black (1787), Hutton formulated his ideas in his last years of farming, *c.* 1765-68; V.A. Eyles, 'Hutton', *DSB* 6:579b.
[178] Eyles, 'Hutton', p. 578.

1795)[179]—the supposition that all geological change has always involved the very same natural processes that we find at work today. Hutton proposed his Uniformitarianism as an alternative to 'Catastrophism', the view that significant change had taken place through one-time events like the passing of a huge comet or the Great Flood.[180]

Hutton's principle of uniformity is just the traditional idea of a common course of nature (chap. 1, sec. D) applied to the long-range history of nature.[181] Like Newton, Hutton assumed that the world-system has always been fundamentally the same as it is now: since the day of creation it has been in a steady state, the perfect order of which reflects the perfect wisdom of God.[182] The difference is that Hutton attempted to include processes of geological change like the building of mountains into the system, whereas Newton had only been concerned with the orbits of the planets in space and with processes like chemical change on earth.[183] Moreover, Hutton saw no need for the periodic reformations which Newton had assigned to divine intervention.[184]

In other words, in comparison with Newton's, Hutton's was a higher order concept of the system of nature which included not only the present structure of the world, but the process (or natural history) by which the present structure had come into existence and was maintained. As with Newton, and in contrast to materialists like Buffon and neomechanists like Laplace, the origins of the system were beyond the scope of science for Hutton: in nature itself, he found 'no vestige of a beginning—no prospect of an end'.[185] But Hutton came about as close to being a neomechanist as

---

[179] First defended in a paper 'Concerning the System of the Earth, Its Duration, and Stability' presented to the Royal Society of Edinburgh in 1785 (in *Proceedings of the Royal Society of Edinburgh*, vol. 63B [1950], pp. 380-82); first published in 'Theory of the Earth; or an Investigation of the Laws Observable in the Composition, Dissolution, and Restoration of Land Upon the Globe' in *Transactions of the Royal Society of Edinburgh*, vol. i (1788), pp. 209-304; later, expanded version published as *Theory of the Earth* (2 vols., 1795); Hector MacPherson, *The Intellectual Development of Scotland* (London: Hodder & Stoughton, 1911), pp. 115ff.; Eyles, 'Hutton', pp. 579f., 587b.

[180] Toulmin and Goodfield, *Discovery*, pp. 190f.; Russell, *Cross-Currents*, p. 131. The Catastrophism of de la Pryme should not be confused with the Neptunism of Abraham Werner; Russell, *Cross-Currents*, pp. 132, 134.

[181] Heimann, 'Voluntarism', p. 282; R. Grant, 'Hutton's Theory of the Earth', in L.J. Jordanova and R.S. Porter, eds., *Images of the Earth* (Chalfont St Giles: British Society for the History of Science, 1978), pp. 29ff.

[182] Toulmin and Goodfield, *Discovery*, pp. 189-92; Heimann and McGuire, 'Newtonian Forces', p. 292.

[183] Hutton drew a parallel between the succession of the positions of the planets in the system of the heavens (Newton) and the succession of epochs in the system of natural history of the earth; Toulmin and Goodfield, *Discovery*, pp. 191f.

[184] Stephen Jay Gould, 'Hutton's Purposeful View', *Natural History*, vol. xci (May 1982), p. 9c.

[185] Toulmin and Goodfield, *Discovery*, p. 192.

one possibly could without abandoning the Newtonian framework of God and nature. Only the Newtonian stipulation that God had personally designed the present system of nature stood between natural theology and the retirement of God from science altogether.

Hutton viewed nature as an entirely self-contained system. This was seemingly in contrast to Newton, but quite in line with the tendencies of some of his followers. Like Derham and Cotes, Hutton believed that God had implanted active principles in nature at creation sufficient to account for all its normal functions. From Boerhaave, Hutton took the notion of a fiery substance emanating from the sun, circulating throughout the planetary system, and revitalising life on earth (cf. Newton's cometary vapours).[186] From Gowin Knight, he took the idea of positing two totally different types of matter: one type was Newton's ordinary gravitational (attractive) matter; the other was Boerhaave's solar (expansive) matter and was responsible for heat, light, and electricity (cf. Newton's and Maclaurin's subtle ethers). As with Knight, matter and force were thus interdependent aspects of the system of nature.[187]

In Hutton's geology, for instance, the heat within the earth formed new rock and raised up new continents (expansion) and thus compensated for the continual erosion of land caused by the downhill course of water (gravitation). The perpetual circulation of solar matter and the balance of attractive and repulsive forces in nature thus provided a comprehensive cyclical mechanism that Hutton felt could account for all the operations of nature without the need for divine intervention. Actually, Hutton's system of nature was no more cyclical than Newton's; it was just that God was no longer involved as part of the cycle.[188]

We need to examine two dimensions of Hutton's natural philosophy that are external to the discipline of geology itself: one is technological; the other, theological.

Historically, fundamental shifts in worldview have often been accompanied, and in part occasioned, by changes in technology. In fact, there is a two-way interaction between the two. Cosmologies are ideal systems based on metaphors taken from everyday life, and the metaphors used in new cosmologies are frequently taken from new or newly impressive technologies. Conversely, technology is often a construction in miniature of what a culture takes to be the nature of the cosmos as a whole. The partial

---

[186] Heimann, 'Nature', p. 22. Hutton did his medical training at Leiden, receiving his M.D. in 1749.

[187] *An Investigation of the Principles of Knowledge* (1794); Heimann and McGuire, 'Newtonian Forces', pp. 289f., 299.

[188] *Theory of the Earth* (1795); Heimann and McGuire, 'Newtonian Forces', pp. 291ff.; Heimann, 'Nature', pp. 8f., 21f.; *idem*, 'Voluntarism', pp. 281ff.

dependence of theoretical science on technology does not imply a lessening of the ideological nature of science or a reduction of superstructure to substructure (Marx). Technologies have ideas and ideals of their own and can reflect creationist ideas as well as scientific theories can (chap. 1 sec. E).

In ancient times, hierarchical cosmologies had been reflected in towers with many levels and temples with many antechambers. The development of homocentric cosmologies from Eudoxus and Aristotle to the Arabs was accompanied by the manufacture of armillary spheres and water clocks, all of which required a continuous and variable input of energy in order to function. The late medieval idea of impetus and the increasing sense of the autonomy of the world-machine came together with the development of the first truly mechanical clocks in the fourteenth century. Descartes's mechanical philosophy was really the perfection of the late medieval notion of impetus. Newton's system of supramechanical forces, on the other hand, was based on the realisation that the cosmos was more complex than any machine existing at that time.

The rapid development of technology in the eighteenth century must also be viewed as an important factor in the changing views of the cosmos. Two major innovations were experimental devices that exploited electric and magnetic forces and the steam engine. Whereas invisible forces seemed to transcend the terms of known mechanisms to Newton, they gradually appeared to be as mechanical as nuts and bolts to those like Knight who were familiar with the production and operation of artificial magnets. Similarly, whereas fire and heat had been known primarily for their cataclysmic, destructive effects, they could appear to be controlled and constructive to someone like Hutton, who was familiar with James Watt's new, more efficient steam engines (*c.* 1769-90).[189]

On the theological front, Hutton was motivated by the importance of two things: divinely appointed ends to be realised in this life, and the possibility of moral perfection in the next. In effect, Hutton traded off one form of divine involvement in nature for another, less immediate, one. He completely eliminated the role of God as an efficient cause in nature—e.g., in maintaining active principles or restoring the quantity of motion in the system (Newton's ideas). On the other hand, he heightened the scientific importance of final causes, ends to which the normal processes of nature are fitted as means.

Like any good machine, the system of nature was designed (by God) to accomplish a certain task, in this case, the creation and maintenance of a

---

[189] The connection with Watt's engines has frequently been noted, e.g. by Gould, 'Hutton's Purposeful View', p. 9c.

habitat suitable for human beings.[190] Hutton's early work in agriculture had impressed on him the fact that good soil is produced by erosion. Yet erosion was also a seemingly destructive process and would eventually lead to the destruction of the continents if there were no compensating mechanism. Faith in the wisdom and benevolence of God, however, required that the processes of nature be entirely suited to human good. The apparent contradiction was resolved by Hutton's idea of the continent-raising power of subterranean heat described above.[191]

In keeping with the thinking of his time, Hutton also saw a deep relationship between natural and moral philosophy. A major tenet of the moral philosophy of the time was the notion of a life after death. Without such a belief, it was thought, there would be no motivation even to try to be moral in this life. On the other hand, the best evidence for life after death, Hutton thought, would be evidence for the beginnings of a moral development in this world that could not be completed in one lifetime. The best evidence for such a development in human life, in turn, would be the existence of a purposeful scheme in nature itself. Conversely, if the intentions of God for human good were effectual in nature, then it was reasonable to suppose that they would be effectual in the moral realm as well.[192]

Though not without its problems, Hutton's Uniformitarianism was as legitimate an interpretation of the idea of creation as any of its eighteenth-century alternatives. However, Hutton's ideas were interpreted as divorcing God from nature by most of his contemporaries—supporters (e.g., Toulmin and Playfair)[193] and opponents alike.[194] As a matter of fact, then, Hutton's principle of uniformity was a major stimulus to the development of 'flood geologies' like that of William Buckland in the early nineteenth century.[195]

---

[190] R. Hooykaas, *Natural Law and Divine Miracle* (Leiden: Brill, 1959), pp. 180f.; Gould, 'Hutton's Purposeful View', p. 11a.

[191] Hooykaas, *Natural Law*, pp. 181f.; R. Grant, 'Hutton's Theory', p. 26; Gould, 'Hutton's Purposeful View', p. 11a.

[192] R. Grant, 'Hutton's Theory', pp. 25, 27ff.

[193] George Hoggart Toulmin's *The Antiquity and Duration of the World* (1780) had already concluded (*pace* Hutton) that all nature, including humanity, was eternal and that there were no moral constraints on life; Toulmin and Goodfield, *Discovery*, pp. 193f.; Martin J. S. Rudwick, 'The Shape and Meaning of Earth History' in Lindberg and Numbers, eds., *God and Nature*, p. 307. John Playfair's *Illustrations of the Huttonian Theory of the Earth* (1802) removed Hutton's references to God's design in the raising and lowering of continents; Hankins, *Science*, p. 155.

[194] Hutton was already under attack in the late 1780s or early 1790s when he wrote his (unpublished) 'Memorial Justifying the Present Theory of the Earth from the Suspicion of Impiety'; Eyles, 'Hutton', pp. 586f.

[195] Russell, *Cross-Currents*, pp. 127f.

5.3 Conclusion: The Modernity of Flood Geology

It is as important to understand the underlying dynamics of developments in science and theology as it is in geology itself. In our discussion of Samuel Clarke, we noted six distinct ways in which God was normally understood to be active in the world of nature and history. The three that Newton emphasised as the theological basis of his system (formation of the solar system, active principles, and periodic reformations) all appeared to have been replaced by strictly naturalistic accounts (taking the popular 'atheistic' reading of Hutton) by the close of the eighteenth century. The original creation (*ex nihilo*) had apparently left no trace other than the inscrutable existence and lawful motion of nature itself. The historic occurrence of miracles (transcending natural second causes) was being questioned by both philosophers and biblical scholars: miracles had ceased in the minds of most Protestants, so they could be neglected for most scientific purposes in any case.

That left only occasional intrusions of spiritual beings (through the medium of second causes) in dramatic natural phenomena to signal humans that there was a God in heaven and that the cosmos was somehow tuned to their existence on earth. Even here, strictly naturalistic explanations were beginning to prevail. Events like storms at sea and comets from heaven were not so awesome as they had once seemed. Due to advances in science and technology, nature apparently had been domesticated by the end of the eighteenth century. For those with religious or romantic impulses, that presented a serious problem. Uniformity of natural processes in a closed system of cause and effect proved to be as difficult to live with as the violence-ridden and life-threatening world of an earlier generation. Whereas the earlier generation (lasting through Hutton) needed to be convinced that this earth was a stable environment for humanity, the new one needed to be convinced that it was a meaningful one.

The Great Flood, however, was an event so stupendous, so clearly described in Scripture, and, on the new geological timescale, so recent that it naturally became a *cause célèbre* among theologically interested scientists in late eighteenth- and early nineteenth-century France and England. If it could be shown that God had acted in such a definitive way in Noah's time, it could more readily be believed that he was still in control and that he could act again even in our own. The Flood thus became for many, as it still is for some today, the one remaining scientifically verifiable sign of God's activity and hence of the fundamentally personal character of the world of nature. What the containment of terrestrial waters had been for Luther, the immobility of the earth for Calvin, the active principles and periodic reformations of the world for Newton, and the balance of erosion and continent-building for Hutton, the Universal Deluge became for the early Buckland.[196] The gradual shift over three centuries is seen in the fact that the Reformers cited the normal condition of the absence of flooding as evidence of God's supernatural intervention, whereas the early Buckland (1819, 1823) appealed to geological evidence for the Flood itself as evidence of God's creative intervention, operating strictly as a final cause within the natural framework of cause and effect.[197] Paradoxically, Flood geology was the product of the advance of science just as much as Uniformitarianism was.

We have reviewed a representative sampling of natural philosophers who worked within Newton's framework of a symbiosis between God and nature. The overwhelming variety of types of Newtonianism and the tendency of most of its exponents towards either monistic materialism (Derham, Cotes, Boerhaave, Maupertius, and Maclaurin) or neomechanism (Clarke, Whiston, Watts, Euler, and Hutton) indicate the basic instability of the Newtonian paradigm. We recall that Newton originally developed his notions of absolute space and active principles as alternatives to two other systems: spiritualism, which held to a deeper unity of God and nature; and the mechanical philosophy, which posited a sharper dichotomy. Monistic materialism and neomechanism were the eighteenth-century sequels to the spiritualism and mechanism of the seventeenth. Newton's innovations thus devoured the older alternatives and brought forth two new ones in their place.

B. Post-Newtonian Materialists

In this section, we consider the work of a variety of non-Newtonians who advocated a fundamentally dynamic view of matter, in some cases leading to hylozoic materialism. In general, these figures were reacting against the Newtonian ideas of the passivity of matter and the dualism of matter and spirit. They were deeply influenced by the spiritualism of Helmont and the

---

[196] 'Again, the grand fact of *an universal deluge* at no very remote period is proved on grounds so decisive and incontrovertible...and which are unintelligible without recourse to a deluge exerting its ravages at a period not more ancient than that announced in the Book of Genesis', Buckland, *Vindiciae geologicae* (1820), pp. 23f.; Goodman, ed., *Science*, pp. 354f.; Russell, *Cross-Currents*, p. 127.

[197] Innaugural Address for Readership in Geology at Oxford (1819); *Reliquiae diluvianae* (1823); Walter F. Cannon, 'Buckland', *DSB* 2:568b-569b. After much criticism, Buckland renounced his identification of the geological flood with that described by Moses in Genesis. For the later Buckland (1832ff.), the geological universal deluge, like all other geological epochs, belonged to the period before the creation story in Genesis 1:3ff.; *Geology and Mineralogy Considered with Reference to Natural Theology* (Bridgewater Treatise, written 1832-36); Canon, 'Buckland', p. 570.

pansophism of Leibniz (chap. 3, sec. C). Theologically, they tended toward pantheism rather than biblical theism. The British matter theorists made matter entirely autonomous, thus eliminating the third tenet of the creationist tradition. The French materialists undercut the first tenet—the comprehensibility of the world based on the creation of the world according to the plan of God, and the creation of the human mind in the image of God. However, the other two points of the historic creationist tradition—the unity of nature and the ministry of healing—were retained and developed in a secularised form.

We shall discuss three groupings: the early continental vitalists, British matter theorists (non-Newtonian), and French materialists.

1 Early Continental Vitalists: Stahl and s'Gravesande

We begin our survey with two continental matter theorists who, under the influence of spiritualist and Leibnizian notions, departed from the Newtonian framework in the direction of materialism.

Georg Ernst Stahl (1660-1734), a devout Pietist best known as the author of the phlogiston theory of combustion, clearly demonstrates the continuity of seventeenth-century spiritualism and eighteenth-century vitalism as well as their common antipathy to the mechanical philosophy.

To begin with, Stahl was the pupil of Johann Joachim Becher (1635-82), a German chemist who was himself influenced by Paracelsus and Helmont. Becher revised the Paracelsian *tria prima* (salt, sulfur, and mercury) and developed a classification (1664) of three kinds of earth that played a role in chemical reactions: one vitreous (like salt, gives substance); one fatty (like sulfur, gives colour and combustibility); and one fluid (like mercury, gives odour, form, and weight). Chemical combustion was an organic process made possible by the presence of the second, 'fatty' earth (*terra pinguis*), which Becher also called sulfur *phlogistos* ('combustible sulfur'). Sulfur *phlogistos* was released from a compound through heating (calcination) or combustion ('oxidation' in modern chemistry) and could be restored by heating with charcoal ('reduction').[198] In spiritualist terms, it was the equivalent of the vital principle or soul which animated the body of a living creature.

Stahl took over Becher's idea of different types of earth, combined it (1723) with Newton's notion of attractive forces between the atoms of a chemical compound, and used the result to develop a viable alternative to the

---

[198] Becher, *Oedipus chimicus* (1664); Leicester, *History*, pp. 121f.; A.G. Debus, 'Becher', *DSB* 1:549f.

Newtonian chemistry.[199] To Becher's fatty earth or sulfur *phlogistos* he gave the name, 'phlogiston'. Substances like sulfur and phosphorus were richly endowed with phlogiston by their natural (organic) formation in the earth. The process of combustion released the phlogiston to the air and left behind an acid (vitriol). Air was not an agent in the chemical process itself; it only served to carry away phlogiston so that it could be reabsorbed by plant life and the cycle of nature could continue.[200]

The identification of three fundamentally different kinds of matter implied that atoms had intrinsic properties that qualified them as belonging to one type rather than another. As in the spiritualist tradition, however, such properties were regarded as granted by God at creation.[201]

Though the idea of phlogiston had to be discarded once the existence and chemical action of oxygen were properly understood (Lavoisier), Stahl's theory was an excellent unified theory of the various facts of chemistry known in the early eighteenth century.[202] It remains an important historical witness to the ideal of the unity and activity of nature which has motivated so much of modern science, even though it has gradually been divorced from its creationist roots.

Willem Jacob s'Gravesande (1688-1742) was a professor of mathematics and astronomy and a colleague of Boerhaave at Leiden. His *Mathematical Elements of Physics* (1720) was one of the first publications in continental Europe to advocate the natural philosophy and God-nature symbiosis of Newton.[203] Accordingly, he could have been included in our treatment of continental Newtonians (sec. A, this chap.). However, s'Gravesande was also one of the first Newtonians to convert wholeheartedly to the Leibnizian camp. Already in the Newtonian treatise mentioned, he followed Boerhaave and others in treating fire as a subtle substance different from ordinary matter and responsible for the phenomena of heat, light, and electricity.[204]

Such notions were still compatible with a generally Newtonian stance as we have seen in the cases of Boerhaave, Maclaurin, and Hutton. But when s'Gravesande tried to refute Leibniz's claims for *vis viva*, he concluded (1722) that the Leibnizians were actually correct—force (or what we would

---

[199] Stahl, *Fundamenta chymiae dogmaticae et experimentalis* (1723); Schofield, *Mechanism*, pp. 211f.; Hankins, *Science*, p. 124.

[200] For a good summary, see Gillispie, *Edge*, p. 205; Leicester, *History*, p. 122; L.S. King, 'Stahl', *DSB* 12:604a-605b.

[201] Leicester, *History*, p. 122. According to Thackray's analysis, Stahl was somewhat anti-Newtonian; *Atoms*, p. 194.

[202] Leicester, *History*, p. 123.

[203] *Physices elementa mathematica experimentis confirmata* (1720, 1721); Hall, *From Galileo*, pp. 319f.; idem, 's'Gravesande', *DSB* 5:511a; Iltis, 'Leibnizian-Newtonian', p. 358; Hankins, *Science*, p. 49.

[204] Schofield, *Mechanism*, pp. 140f., 143f.; A.R. Hall, 's'Gravesande', p. 511a.

call energy) was really inherent in a moving body and not just impressed from without.[205] Consequently, according to s'Gravesande, effects proceeded automatically and necessarily from their natural causes.[206]

In developing his new view of matter (1736-37), s'Gravesande had to answer the argument of Samuel Clarke that positing a necessary order in nature would lead to atheism. This he did by appealing to the wisdom of God as manifested in the order of nature, as Leibniz had done, and by arguing against Clarke that tying providence to the maintenance of the properties of matter was an unnecessary constraint on the exercise of divine power.[207] While certainly missing the intent of Newton and Clarke, s'Gravesande had a valid point here. As we have seen, Newton's God was free only in the sense of being ceaselessly occupied with the maintenance of every single particle in the universe in accordance with a freely chosen set of exact mathematical laws.[208] In comparison, the Leibnizian-Gravesandian view suggested a more economical operation of nature and a more relaxed (if not idle) God.

2 British Matter Theorists from Hales to Priestley

The English and Scottish scientists who tended towards materialism were pioneers of the relatively new discipline of chemistry. In contrast to their Newtonian contemporaries (sec. A), they stressed quantitative experimental techniques rather than mathematical models and were less concerned with natural theology.

Prior to the philosophical and theological materialism of David Hartley and Joseph Priestley, the materialist tradition in Great Britain was decidedly nontheological. Even though continental vitalist ideas made an impact, their spiritualist associations were entirely eliminated in keeping with the antispiritualist tenor of moderate British thought since the mid-seventeenth century. Consequently, we mention here only a few of the more important early British matter theorists prior to Hartley and Priestley to give an idea of the progressive secularisation of historically creationist ideas.

2.1 Secular British Matter Theorists from Hales to Black

Stephen Hales (1677-1761) treated air (1727) as a universal agent in chemical reactions somewhat similar to Stahl's phlogiston.[209] Like

---

[205] 'Essai d'une nouvelle theorie du choc des corps' (1722); A.R. Hall, 's'Gravesande', p. 509b; Iltis, 'Leibnizian-Newtonian', pp. 359f.
[206] *Introduction to Philosophy* (1736-37); Iltis, 'Leibnizian-Newtonian', p. 362.
[207] *Ibid.*
[208] 'Gravity must be caused by an agent acting constantly according to certain laws...'; Newton's letter to Bentley, 25 Feb. 1693; Turnbull et al., eds., *Correspondence*, 3:254.
[209] *Vegetable Staticks* (1727).

phlogiston, Hales's air (in modern chemistry, just the carbon dioxide component) could be bound or 'fixed' in compounds and then released by heating or fermentation.

Hales established the discipline of pneumatic chemistry by collecting the 'airs' released by various reactions and analysing their properties.[210] Thus, like Stahl, he abandoned (1733) the Newtonian programme of reducing all chemical phenomena to the interaction of atoms by means of forces acting at a distance.[211] But, in as much as Hales was an Anglican clergyman, his ideas lacked the spiritualist associations of Stahl's.

Peter Shaw (1694-1764) was responsible more than anyone for introducing materialist concepts into England through his translations of Boerhaave (1727, 1741) and Stahl (1730).[212] Shaw started out as a Newtonian and in a sense continued the alchemical side of Newton's own work.[213] But, like Stahl and Hales, he eventually came to the point of rejecting the reduction of chemistry to atoms and forces as the Newtonians required. The concepts and methods of mathematical physics only 'scratched the shell and surface of things and left the kernel untouched'.[214] Shaw's chemistry was instead based on the role of active principles like Boerhaave's fire that were inherently material and observable.[215] But, as in the case of Hales, the spiritualist ideas associated with the work of Boerhaave and Stahl were conspicuous by their absence.

Cadwallader Colden (1688-1776), a colonial American physician trained at Edinburgh, was more explicit in his secularising of Newtonian and spiritualist notions (1746, 1751). Like Gowin Knight and others, Colden allowed for inherent differences in the qualities of matter. In fact, he postulated three species of matter distantly resembling Paracelsus's *tria prima*: one was self-moving (light); one transmitted motion from one body to another (an elastic ether); and one actively resisted motion (ordinary inertial matter).[216] Not only did Colden's scheme attribute distinctive properties to each species of matter over and above the primary mechanical qualities common to all three; it also attributed a peculiar kind of activity to each species, even interpreting inertial resistance to motion as a form of activity.

---

[210] *Vegetable Staticks* (1727); R.E. Schofield, 'Joseph Priestley', pp. 99ff.

[211] *Haemastatics* (1733); R.E. Schofield, 'Joseph Priestley', p. 102; cf. Hall, *From Galileo*, p. 326.

[212] Schofield, *Mechanism*, pp. 147, 154f., 211.

[213] Heimann, 'Nature', p. 12.

[214] *Chemical Lectures* (1734, 1755); Schofield, *Mechanism*, pp. 211f.; Heimann, 'Nature', pp. 11f.

[215] Notes on Shaw's revised translation of Boerhaave's *New Method of Chemistry* (1741); Heimann, 'Nature', p. 12.

[216] *An Explication of the First Causes of Action in Matter* (New York, 1746); *Principles of Action in Matter* (London, 1751); Schofield, *Mechanism*, p. 130.

The theological significance of Colden's ideas is seen from the title of his first work: *An Explication of the First Causes of Action in Matter; and of the Causes of Gravitation* (1746). Whereas earlier scientists had attributed primary causation to God and allowed only secondary causation (if any) to creatures, Colden located the seat of causation, even that of Newton's gravitation, in matter itself. In a second work (1751), Colden explicitly rejected Newton's contention that all motion was directly sustained by God as 'very unphilosophical'.[217] In other words, it was more economical to suppose that the principles of motion Newton had discovered were inherent in matter, even though they were radically different from the properties originally attributed to matter in the mechanical philosophy. Matter, even inert matter, was no longer strictly passive. From the Newtonian standpoint, God was out of a job—or rather, his job had been taken over by matter.

Another example of the secularisation of Newtonian ideas is William Cullen (1710-90). Cullen taught chemistry at Glasgow and Edinburgh—his lectureship at Glasgow (1747) was the first position in chemistry to be established independently of the discipline of medicine.[218] In Glasgow and Edinburgh, he was associated with such giants of Enlightenment thought as David Hume and Adam Smith.[219]

Like Peter Shaw, Cullen was influenced by the vitalism of Stahl to adopt a more materialist interpretation of Newtonian ideas.[220] For example, Cullen explained gravitational attraction in terms of a gradient in the ether density between gross material bodies.[221] Chemical reactions and functions of the nervous system could also be described in terms of the ether.[222] Newton himself had speculated along such lines but never regarded such ether mechanisms as complete explanations that might exclude the active role of God. Cullen, however, cast Newton's ether in the role of primary causal agent in nature, thus eliminating God altogether.[223]

Joseph Black (1728-99) was a pupil of Cullen at Glasgow, completed his M.D. at Edinburgh (1754), and succeeded his mentor in his chairs successively at Glasgow and Edinburgh. Like Cullen, he was associated with major figures of the Scottish Enlightenment—Adam Smith, David

---

[217] *The Principles of Action in Matter* (1751); Heimann and McGuire, 'Newtonian Forces', p. 303.
[218] W.P.D. Wightman, 'Cullen', *DSB* 3:494a.
[219] J.R.R. Christie, 'Ether and the Science of Chemistry, 1740-1790', in G.N. Cantor and M.J.S. Hodge, eds., *Conceptions of Ether*, pp. 86, 91f.
[220] Thackray, *Atoms*, pp. 194f.
[221] Christie, 'Ether', pp. 99f.
[222] Christie, 'Ether', p. 101; Dolan and Smith, *Health*, p. 118.
[223] Christie, 'Ether', pp. 93f.

Hume, and James Hutton.[224] Like Stahl, Hales, and Shaw, he rejected the Newtonian matter theorists' programme of reducing chemical phenomena to forces acting between minute atoms.[225] Black laid the basis for quantitative analysis by improving Stephen Hales's experimental methods for differentiating the various kinds of air given off by chemical reactions.[226]

Like Boerhaave, s'Gravesande, and Cullen, Black regarded heat as an active material substance involved in both physical and chemical change.[227] Noticing that ice remains on the ground long after the temperature rises above freezing, he took a few measurements and discovered that the temperature of the water dripping off the melting ice was still at the freezing point. Black concluded that there must be a 'latent heat' in water that is given up in the process of freezing, and that must be restored in melting before further heat input will raise its temperature above thirty-two degrees Fahrenheit. Black also discovered (following the work of Gabriel Daniel Fahrenheit) that different materials had different capacities for absorbing heat, or different 'specific heats'.[228] For the first time a clear distinction had been made between the temperature of a body and the amount of heat it absorbed.[229]

The ideas of latent and specific heat are foundational to modern thermodynamics and physical chemistry. In fact, Black's notion of latent heat was instrumental to James Watt's understanding of the equivalence of mechanical energy to heat and the consequent redesign of a more efficient steam engine in the 1760s.[230] Though the use of Black's concepts does not require a materialist understanding of heat in modern physical science, that understanding was clearly helpful to Black in his new insights into matter theory.

2.2 David Hartley and Joseph Priestley

In order to fill out the background of Joseph Priestley's religious version of materialism, we need to glance briefly at the work of Priestley's mentor in the area of philosophy and psychology, David Hartley (1705-1757). Hartley did not contribute to the physical sciences as such, but his psychological

---

[224] H. Guerlac, 'Black, Joseph', *DSB* 2:173a-174b.
[225] Schofield, 'Joseph Priestley', p. 103.
[226] 'Experiments Upon Magnesia Alba, Quicklime...' (presented 1755, pub. 1756), based on Black's M.D. thesis at Edinburgh: 'De humore acido a cibis orto et magnesia alba' (1754); Schofield, 'Joseph Priestley', pp. 101f., 103; Leicester, *History*, pp. 133f.; Butterfield, *Origins*, p. 214; Guerlac, 'Black', p. 173b.
[227] Christie, 'Ether', pp. 102f.
[228] A.R. Hall, *Scientific Revolution*, pp. 347f.
[229] Leicester, *History*, p. 134.
[230] MacPherson, *Intellectual*, p. 103; Lord Ritchie-Calder, 'The Lunar Society of Birmingham', *Scientific American*, vol. ccxlvi (June 1982), p. 138a.

and social ideas indicate one way in which traditional notions of the creationist tradition could be carried on in a radically materialist manner.

The son of a poor Anglican clergyman in Yorkshire, Hartley at first prepared for the ministry. Though devoutly religious, he was unwilling to subscribe to the Thirty-Nine Articles after completing his degrees and was thus unable to pursue a career in the church. Consequently, he devoted himself to the practise of medicine without formal training, yet with the Christian ideal of serving rich and poor alike.[231]

Hartley deeply admired the work of Newton and pictured himself in his major work, *Observations on Man* (1749), as extending Newton's methods to the phenomena of mental and moral life.[232] For example, as a parallel to the law of cohesion for matter, Hartley developed a 'principle of association' for ideas: one idea, stimulated by sense perception, could evoke a related idea in the mind as naturally as one vibration could excite another in the brain. The moral sense in an individual then emerged as an aggregate of simple ideas relating to action and united by association. Hartley here argued against moral philosophers like Shaftesbury, Hutcheson, and Reid and moderate Newtonians like Derham who held that the moral sense and other such instincts were implanted directly by God.[233]

Even though Hartley did not attribute thought to matter itself, his emphasis on the material basis of the mind may strike the reader as being atheistic in intent. However, Hartley saw himself, like Newton before him (and Priestley after), as a champion of revealed biblical religion.[234] A comparison of his ideas with the historic creationist tradition shows that there was some merit to his claim. For one thing, Hartley argued that the world in which the human mind is formed by sensation and association is itself a system of comprehensible order and benevolence in as much as it reflects the wisdom and benevolence of God.[235] Hartley's materialistic associationism, then, was merely the means by which the traditional notions of the image of God in humanity—art, science, and morality—were made effective. It was the French materialists who first challenged the creationist basis for belief in the comprehensibility of the world.

A second way in which the radical implications of the creationist

---

[231] Jack Lindsay, 'Introduction', *Autobiography of Joseph Priestley* (Bath: Adams & Dart, 1970), p. 32; Robert M. Young, 'Priestley, Joseph', *DSB* 6:138f.

[232] *Observations on Man, His Frame, His Duties, and his Expectations* (1749). Earlier works were a pamphlet entitled *Conjecturae* (1730) and the more substantial *Progress of Happiness Deduced from Reason* (1734); Young, 'Priestley', p. 139a.

[233] Lindsay, ed., *Autobiography*, pp. 33f.

[234] Lindsay, ed., *Autobiography*, pp. 33f.

[235] J.G. McEvoy and J.E. McGuire, 'God and Nature: Priestley's Way of Rational Dissent', *Historical Studies in the Physical Sciences*, vol. vi (Russell McCormmach, ed., Princeton: Princeton Univ. Press, 1975), p. 343.

tradition were continued by Hartley was in his political philosophy. Whereas moderate British Newtonians and moral philosophers like Derham tended to treat humans as unequally gifted and to legitimate a hierarchically structured society on the basis of the biblical notions of sin and the fall of Adam, Hartely's associationistic psychology implied that all humans could strive to perfect the image within by rooting out perverse associations and replacing them with pious ones. Thus through proper education and self-examination, the effects of the Fall could be reversed and the original condition of God's creation restored.[236] However naive such views may seem to us today, they were clearly not secular or irreligious in intent. It would be instructive to investigate the possible origins of Hartley's egalitarian political views in the Christian ideals he imbibed from his medical practise. However, Hartley was far from an idealist himself. He clearly recognised the pervasiveness of selfishness among humans, particularly among professors of science who, he charged, topped all others for their 'vain-glory, self-conceit, arrogance, emulation, and envy'.[237] In many ways Hartley reminds of us of the early Christian critics of Greek science (chap. 1, sec. E).

Joseph Priestley (1733-1804) was both a devout Christian and a deterministic materialist,[238] a combination that might be thought impossible today. In fact, the combination already appeared to be an outright contradiction to Leslie Stephen (1832-1904), who lived almost exactly a century after Priestley. In his classic *History of English Thought in the Eighteenth Century* (1876), Stephen spoke approvingly of Priestley as a pioneer of free thought and political liberalism, but could not refrain from chiding him for retaining 'puerile superstitions' such as his belief in miracles and the imminent return of Christ.[239] Seeing Priestley through the late-nineteenth century eyes of Stephen gives us some idea of the change in English thought over the intervening century. It also serves as a warning to historians who would find absolute contradictions where their subjects were

---

[236] Lindsay, ed., *Autobiography*, pp. 34f.

[237] *Observations* (1749); C.A. Russell, *Science*, p. 33.

[238] Here I agree with Heimann and McGuire, who regard Priestley as a materialist ('Newtonian Forces', p. 268), rather than with Schofield, who treats him as a neomechanist on the basis of his supposed indebtedness to Boscovich (*Mechanism*, pp. 263ff., 272f.). Priestley considered himself to be a materialist, and Boscovich agreed that Priestley's views were 'materialism pure and simple'; Thackray, *Atoms*, p. 249. In a later work, Schofield refers to Priestley as a 'physicalist' like Hales but still labels Priestley's final position 'mechanist'; 'Joseph Priestley and the Physicalist Tradition in British Chemistry', pp. 103f., 108, 112. One reason for Schofield's classification is his sense of the importance of Priestley's opposition to Lavoisier's theory of elements. The latter is 'materialist' in Schofield's classification because it allows for elemental properties of matter; *Mechanism*, pp. 272f.; 'Joseph Priestley', pp. 112f.; cf. Lindsay, *Autobiography*, p. 57.

[239] Leslie Stephen, *History of English Thought in the Eighteenth Century* (1876), VIII.63f.; cf. I.64 (New York: Harcourt et al., 1962), vol. I, pp. 54, 364f.

blissfully unaware of them. As we shall see, Priestley's system of thought was highly coherent.

Having received his early education 'in all the gloom and darkness of Calvinism', Priestley spent his formative (university) years at the Dissenting Academy of Daventry. There he first read Hartley's *Observations on Man* and was encouraged by the differing views of his teachers to develop unconventional views of his own.[240] By the time he graduated from Daventry (1755), Priestley had adopted Hartley's psychological determinism and had moved from orthodox trinitarianism to Arianism (the position of Newton, though Priestley probably did not know it).

In the late 1760s, Priestley was ordained and began serving a Presbyterian church in Leeds. There he performed his first systematic experiments in chemistry,[241] studied the work of Stephen Hales,[242] continued his theological studies, and became convinced of the truth of the Socinian (Unitarian) position. He came to the conclusion that the notion of a pre-existent Logos, like that of a pre-existent soul, was a pagan intrusion into the pure teaching of Jesus.[243] At this point, Priestley was reacting most immediately to Arian teaching, according to which Christ was the first angel or a disembodied soul created by God. However, as we saw in our study of Isaac Watts, the notion of a pre-existent human soul was closely related to trinitarian thought in Dissenting academies during the eighteenth century. It appears that Priestley had this field of associations in mind when he rejected the distinctive doctrines of orthodox Christianity, in as much as the notion of the Logos he singled out for attack as pagan was that of an emanation from the mind of God rather than the Arian notion of a primordial creature.[244]

The three basic components of Priestley's thought—materialism, determinism, and Unitarianism—were all well in place in the early 1770s.[245] In 1772, Priestley gave his first paper on gases to the Royal Society of London,[246] and in 1774 he published his first major philo-

---

[240] Basil Willey, *Eighteenth Century Background*, p. 186.
[241] Priestley already performed some experiments on inflammable airs in 1766 while at Warrington Academy; R.E. Schofield, 'Priestley', *DSB* 11:143f.
[242] Schofield, 'Priestley', p. 144a; 'Joseph Priestley', p. 107.
[243] Schofield, 'Priestley', pp. 139b-140b; McEvoy and McGuire, 'God and Nature', p. 383; Erwin N. Hiebert, 'The Integration of Revealed Religion and Scientific Materialism in the Thought of Joseph Priestley', in L. Kieft and B.R. Willeford, eds., *Joseph Priestley* (Lewisburg, PA: Bucknell Univ. Press, 1980), pp. 30-33.
[244] McEvoy and McGuire, 'God and Nature', p. 333. Recall that Watts had argued for a single conscious mind in the godhead and understood the distinctive mind of the Son as the pre-existent human mind of Jesus.
[245] By 1774, according to McEvoy and McGuire, 'God and Nature', p. 329.
[246] 'Observations on Different Kinds of Air' (read before the Royal Society, March 1772); Gillispie, *Edge*, p. 208; Schofield, 'Priestley', p. 144b.

sophical work, a critique of the Scottish 'common sense' philosophy of Thomas Reid, in which he first articulated his distinctive view of Christian materialism.[247] The following year he brought out a new edition of Hartley's *Observations on Man*.[248]

Priestley was more clearly committed to materialism than Hartley was. In place of Hartley's avowed parallelism of mind and brain, he substituted (1775, 1777) a complete identity. Conscious thought was a function of bodily matter and was therefore entirely deterministic. Two conclusions followed. For one thing, there was no such thing as free will, since thoughts followed one another according to a strict law of association. Even the thoughts of the writers of Scripture were determined; they were not 'inspired' in the sense of having thoughts outside the ordinary course of their mental lives.[249] There was thus no place in Priestley for the dualism between inner freedom and outer determinism that became so popular in nineteenth-century thought.

Secondly, there was no mind or soul separate from matter; the mind or soul was just the form of the body. The soul slept when the brain ceased to function at death, and the believer had only the future restoration of the soul together with the body to look forward to.[250] Political radicalism was thus associated with mortalist ideas in Priestley much as it had been in many left-wing Puritans and sectaries of the mid-seventeenth century. The converse was also true: conservative, particularly anti-French, political views tended to occur together with support for the traditional standards of the church. The most dramatic instance of popular opposition to all Priestley stood for occurred in Birmingham in 1791 when a 'Church and King' mob, celebrating the second anniversary of the fall of the Bastille, destroyed Priestley's home as well as the Unitarian meeting house where he ministered.[251]

Among Priestley's most important contributions to science were the formulation of an inverse-square law (1767), analogous to that for gravitation, for the force surrounding an electric charge; and the 'discovery' of oxygen (1772-75). The inverse-square law for electric force was later confirmed by Henry Cavendish and Charles Augustin Coulomb (1785) and became known to generations of physics students as 'Coulomb's law', one

---

[247] *An Examination of Dr Reid's 'Inquiry into the Human Mind'* (1784); Schofield, 'Priestley', p. 143a; McEvoy and McGuire, 'God and Nature', pp. 357f. 381f.

[248] *Hartley's Theory of the Human Mind* (1775); Schofield, 'Priestley', p. 143a; McEvoy and McGuire, 'God and Nature', p. 35.

[249] Hiebert, 'Integration', pp. 32, 38, 50.

[250] *Hartley's Theory* (1775); *Disquisitions Relating to Matter and Spirit* (1777); Stephen, *History* 1:365f.; McEvoy and McGuire, 'God and Nature', p. 382; Hiebert, 'Integration', pp. 35ff., 40f.

[251] Schofield, 'Priestley', p. 141b; Hiebert, 'Integration', p. 46.

of the four fundamental laws governing electromagnetic fields.[252] Priestley's discovery of oxygen (also discovered indendently by Carl Wilhelm Scheele)[253] was important in that it helped show that common air was actually a mixture of different kinds of air and provided an experimental basis for Lavoisier's later notion of a plurality of chemical elements.

Although Priestley himself worked within the framework of Stahl's phlogiston theory of combustion (he referred to oxygen as 'dephlogisticated air'), his data made it possible for Lavoisier to substantiate an alternative (oxidation) theory of combustion based on the combination of a substance with oxygen, rather than the release of phlogiston.[254] Priestley never accepted Lavoisier's reading of the evidence, however. For him the idea of phlogiston made sense since it was an active principle in chemistry similar to the forces of electricity and gravitation that dominated physics. Priestley's sense of the unity of nature prevented him from allowing a multiplicity of elements and a combinatory mechanics like Lavoisier's that was entirely divorced from the universal principles of physics.[255]

Priestley's most important work for our purposes was his speculation (1777) concerning the relationships among the three heterodox views he held in science and theology: materialism, determinism (or philosophical necessitarianism), and Socinianism (or Unitarianism). He firmly believed that there was a mutual dependence of the three, though the source of his data was science for the first two and Scripture for the third.[256] Here we shall approach the interrelationship first from the starting point of science (to theology) and then from the angle of theology.

On the basis of Newton's concept of forces acting at a distance and the idea of some Newtonians that such forces were properties of matter itself, Priestley concluded that all matter possessed the power of sensation and thought. The attribution of mental life to humans was therefore not subjective or unscientific; it was only a statement of what was true for organised matter in general. Thus, Priestley hoped to eliminate the difficulties with Christian faith occasioned in the minds of 'philosophical unbelievers' by the traditionally held dualism of body and soul.[257]

---

[252] A.R. Hall, *Scientific Revolution*, pp. 355f.

[253] Priestley's discovery was published in the second volume (1775) of his *Experiments and Observations* (6 vols., 1774-86); Scheele's, not until 1777; Leicester, *History*, pp. 136; Schofield, 'Joseph Priestley', p. 109.

[254] Leicester, *History*, pp. 135ff.; Schofield, 'Joseph Priestley', pp. 109f.

[255] Lindsay, ed., *Autobiography*, pp. 49ff., 57f.; Schofield, 'Priestley', p. 147a; McEvoy and McGuire, 'God and Nature', pp. 397f.

[256] *Disquisitions* (1777); Heimann and McGuire, 'Newtonian Forces', p. 270; McEvoy and McGuire, 'God and Nature', pp. 328, 338; John G. McEvoy, 'Joseph Priestley: Scientist, Philosopher and Divine', *PAPS*, vol. cxxviii (Sept. 1984), pp. 193f.

[257] Heimann and McGuire, 'Newtonian Forces', pp. 270-72.

Priestley saw no inconsistency between materialism and true Christian faith, however, since all the powers of matter were derived from and continued to depend on the power of God. In fact, for Priestley, all activity in nature was really a continuous direct manifestation the activity of God. This was consistent with the biblical picture of God 'filling all in all' (1 Cor. 15:28). Priestley's God was not identical with matter; he was the inner power, life, and soul of matter—a view very close to classical Stoicism.[258]

The close association between God and matter worked two ways for Priestley. On one hand, it prevented matter from being autonomous as it seemed (to Priestley) to be in the mechanical philosophy and even in Newton. On the other hand, it made the material world eternal. Without God's power there would be no matter. Conversely (for Priestley), given God's eternal (viz. everlasting) power, there must be eternal (everlasting) matter. Of course, the possibility of an eternal creation had been considered speculatively by earlier divines like Origen and Aquinas. But Priestley's position was a form of determinism or philosophical necessitarianism quite unlike that of earlier speculative theologians. God acted always by necessity as the deterministic structure of the laws of physics seemed (to Priestley) to imply.[259]

Priestley did not equate God's eternity with everlastingness as Newton did, but he did agree with Newton in placing all the acts of God somewhere along an infinite time line.[260] However, whereas Newton could then place the 'begetting' of the Son (as an angelic creature) before creation on the line, Priestley had learned to associate the pre-existence of Christ with that of the soul and rejected both ideas as pagan notions that were inconsistent with his monistic materialism.

One can also approach Priestley's thought from the theological angle of Unitarian doctrine. Historically, it has proven difficult for Jews and Christians to attribute any personality or activity to God without positing some person or thing to which he could relate or upon which he could act. Since a commitment to the Hebrew-Christian Scriptures entails an active personal God, the further constraint that there be no multiplicity (e.g., of persons) in the godhead then implies the necessity of a creation in relation to which God can be eternally personal and active. The reality of Priestley's God was thus actualised only through the (eternal) process of creating.[261]

---

[258] Heimann and McGuire, 'Newtonian Forces', p. 281; McEvoy and McGuire, 'God and Nature', pp. 391f.; Heimann, 'Voluntarism', p. 280. Priestley rejected the monism of Spinoza because it did not distinguish God from nature; Heimann, 'Voluntarism', p. 281; McEvoy and McGuire, 'God and Nature', pp. 336, 391ff.
[259] *Letters to a Philosophical Unbeliever* (1780); Heimann, 'Voluntarism', p. 280.
[260] McEvoy and McGuire, 'God and Nature', note on p. 335.
[261] McEvoy and McGuire, 'God and Nature', p. 336.

In other words, for Priestley as for Newton, the created world took the place of the eternal Son of God in traditional Christian theology. Priestley made this substitution explicit in his *Institutes of Natural and Revealed Religion* (1782) by mimicking the Athanasian affirmation of the eternity of the Son:[262]

> ...there never was a time when this great uncaused Being did not exert his perfection, in giving life and happiness to his offspring....the creation, as it had no beginning, so neither has it any bounds....

So the material world was necessary to the full being of God, and this necessity in turn ruled out the possibility of any dualism between the two or between spirit and matter in general.

In a sense, Priestley was simply a consistent Newtonian. Newton had originally posited the existence of active principles over and above the primordial properties of matter in order to avoid the matter-spirit dualism and apparent autonomy of nature suggested by the mechanical philosophy of Descartes. On the other hand, he postulated the regulation of God's activity in accordance with invariable mathematical laws like that of universal gravitation. Finally, consistent with the need for a universal divine presence in nature, he posited an absolute space taking the traditional place of the divine Logos as the emanent effect of God (chap. 3, sec. C). Priestley merely eliminated the residual dualism in Newton: he made inert matter a function of active principles and hence of divine power[263] and arrived at a monistic system that was at once materialist, determinist, and Unitarian.

3 French Materialists from La Mettrie to Holbach

The philosophy of materialism is most dramatically represented by the radical *philosophes*, La Mettrie, Buffon, Diderot, Robinet, Holbach, and Cabanis who were active in the second half of the eighteenth century. These French materialists are the source of the familiar image of materialism as atheistic. Yet, while they emphatically rejected biblical theism, they retained many biblical insights in secularised form. Their main departure from traditional Christian doctrine was their tendency to assign divine attributes like eternity and self-determination to matter itself.

Because of their high view of the potentialities of matter, the French materialists were able to maintain their faith in the power of the human intellect (the product of organic matter) and to press that faith to unprecedented extremes, while at the same time abandoning the notion of

---

[262] *Institutes of Natural and Revealed Religion* (1782); McEvoy and McGuire, 'God and Nature', pp. 334f.
[263] Heimann, 'Voluntarism', pp. 279f.

creation in the image of God upon which that faith had traditionally rested. For example, Julien Offray de la Mettrie (1709-1751) argued that matter contained within it the potential for producing living creatures, even humans with souls, by a process of pure trial and error. As La Mettrie put it in his *System of Epicurus* (1750): 'Nature, without seeing, has made seeing eyes and, without thinking, has made a thinking machine.'[264] In other words, the Deity was not eliminated by the materialists; it (or aspects of it) were merely absorbed completely into nature.

It should be pointed out that there was nothing inherently inconsistent between believing in the transcendence of God and affirming the unlimited potentialities of matter. The two had been integrally related in the historic creationist tradition, for example, in Basil, Aquinas, and Leibniz. Since the twelfth century, however, the more prevalent view in science and theology had been that the sufficiency of matter and the direct involvement of a transcendent God were mutually exclusive alternatives by definition. At the outset of the eighteenth century, this view had been made normative by the Newtonians.

Here we shall focus on the work of Buffon, Diderot, and Holbach, whose ideas had the most direct bearing on theology and the physical sciences. All three reacted against Newton's stress on mathematical, active principles and attempted to develop more empirical theories based on the character of matter rather than on (what appeared to them to be) abstract formalisms.

3.1 Buffon

The work of Georges Louis Leclerc, Comte de Buffon (1707-1788), represents a major watershed in Western ideas about nature. Buffon's massive *Natural History* (36 vols., 1749-88) was the first comprehensive naturalistic account of the origin and evolution of the terrestrial environment.[265] Still partly under the influence of Newton,[266] Buffon was a transitional figure. Thus he did not try to account for the formation of the cosmos as a whole (as Kant was to do) or the origin of species (as Lamarck was to do). He held that most species emerged directly by spontaneous

---

[264] *Système d'Epicure* (1750); Jacques Roger, 'The Mechanistic Conception of Life', in D.C. Lindberg and R.L. Numbers, eds., *God and Nature* (Berkeley, 1986), p. 288.

[265] *Histoire naturelle, générale et particulière* (1749-88); J. Roger, 'Buffon' *DSB* 2:576b. Additional writers brought the total number of volumes up to 44 by 1803; Simon Eliot and Beverley Stern, eds., *The Age of Enlightenment* (London: Open University, 1979), 2:160.

[266] Buffon began his career as a mathematician and, prior to 1739, associated with the Newtonian party at the Academie des Sciences; he also translated Newton's *Method of Fluxions and Infinite Series* (1740) into French; Roger, 'Buffon', p.576b; John Lyon and Phillip R. Sloan, 'Introduction', *From Natural History to the History of Nature* (Notre Dame: Univ. of Notre Dame Press, 1981), pp. 18ff.; Hankins, *Science*, p. 129.

generation from organic molecules[267] as the earth gradually cooled and the prerequisite environmental conditions arose.[268] But he took a critical step beyond the modest suggestions of Whiston toward a complete natural history of the planetary system.[269] And, following the example of left-wing creationists before him (e.g., Adelard of Bath, Albert the Great, Buridan, Bacon, and Leibniz), Buffon argued that scientists should refrain from appealing to supernatural causes whenever possible (which meant, practically always).[270]

Unlike more radical materialists, however, Buffon did not deny the existence of God or the initial creation of matter out of nothing. Creation was for him still the only possible explanation for the origin, laws, and initial condition of the world. Buffon stated this traditional motif rather eloquently in his own personal hymn of creation (1764):[271]

> Minister of God's irrevocable orders, depository of his immutable decrees, Nature never separates herself from the laws which have been prescribed for her; she alters not at all the plans which have been made for her, and in all her works she presents the seal of the Eternal.

This statement is clearly a paraphrase of the idea of the relative autonomy of nature (chap. 1, sec. D) found in classic creationist texts like Ecclesiasticus 16:26-28.

But even in its original condition, all matter, for Buffon, was imbued with motion and the potential for life. As the result of its creation, it is a manifestation of the living, immense, and all-embracing power of God. Hence, nature too is 'a living power, immense, which embraces everything

---

[267] Sometime in the 1740s, Buffon rejected the notion of preformation in sperm and adopted the idea of spontaneous generation from organic matter; Roger, 'Mechanistic', p. 289. Father Needham's experiments to verify the existence of organic molecules were done with Buffon's encouragement; R.H. Westbrook, 'Needham', *DSB* 10:10a. By 1779, however, Buffon apparently concluded that the environment had become too cold for spontaneous generation; *Des époques de la nature* (1779); Roger, 'Buffon', p. 579b.

[268] *Époques*; Bréhier, *History*, 5:137. Some species like the horse and the ass (even humanity and the orangutan) could have had a common ancestor of which they were degraded forms; *De la dégénération des animaux* (1766); Herbert Butterworth, *The Origins of Modern Science, 1300-1800* (New York: Free Press, 1965), p. 238; Roger, 'Buffon', p. 580b.

[269] Buffon explained the corotation of the planets by appealing to the idea of the oblique impact of a comet, an idea he took over from Whiston; Toulmin and Goodfield, *Discovery* p. 115; Ronald L. Numbers, *Creation by Natural Law* (Seattle: Univ. of Washington Press, 1977), pp. 6f.

[270] 'In physics one must, to the best of one's ability, refrain from turning to causes outside nature'; *Théorie de la terre*; Roger, 'Buffon', p. 577b.

[271] 'Premier vue de la nature' (in 1764 ed. of *Histoire naturelle*, vol. 12); Baumer, *Modern European Thought*, p. 214.

and animates everything'.[272] Here Buffon clearly shows the influence of the pansophist ideas of Leibniz and his disciple, Christian Wolff, which had been brought to public attention in France by Madame du Châtelet in 1740.[273]

As a result of the self-sufficiency of nature, everything since the first moment of creation, even the emergence of life itself, had happened in accordance with immutable laws of nature.[274] Or, turning the problem around, human reason could now reconstruct the history of nature, going all the way back to the first moment of creation, simply on the basis of the laws and present state of the world.

In order to account for fossil remains and other geological data on strictly naturalistic assumptions, Buffon concluded that there had been a succession of seven stages, analogous to the seven 'days' of Genesis 1-2, but lasting far longer than twenty-four hours each.[275] First, the earth was formed as a hot fragment of the sun broken off by the grazing impact of a comet (after Whiston); then, as the earth cooled, rock consolidated, the mountains rose, and the organic molecules needed for life were formed in the warm seas; then huge seas temporarily covered the earth (not to be equated with the biblical flood of Noah), depositing marine fossils on otherwise dry land; when the waters receded, volcanoes became active; then land animals emerged—tropical ones at first which inhabited the entire globe, even the northern regions; later the continents separated from each other to give their present configuration; and last came the age of humanity in which the character of the earth was being transformed by technology.[276]

Thus, in contrast to later uniformitarians like Hutton, Buffon allowed for a succession of radically different eras in the history of nature. But he regarded the products of those eras (e.g., mountains and volcanoes) to have remained relatively unchanged since their initial formation.[277] In other

---

[272] 'Premier vue'; Gay, *Enlightenment* 2:153; Baumer, *Modern European Thought*, p. 210. Early vitalists like Stahl, Louis Bourquet, Albrecht von Haller, and Caspar Friedrich Wolff had differentiated between inert, inanimate matter and living matter governed by organic principles. At first, Buffon also differentiated between two kinds of matter—organic and inorganic—but by 1779 he concluded that organic molecules were themselves generated from inorganic matter under the influence of moderate heat; *Époques* (1779); Roger, 'Buffon', p. 579b.

[273] Madame du Châtelet's *Institutions de physique* (1740) was based on Wolff's *Ontologia* of 1729, French manuscripts of which were apparently sent to her by Frederick the Great in 1736; Hankins, *Science*, pp. 155f.; Lyon and Sloan, eds., *From Natural History*, pp. 20f. and note 43.

[274] Introduction to *Histoire naturelle* (1749); Randall, *Making*, p. 462.

[275] The length of Buffon's stages ranged from 3,000 to 35,000 years; Toulmin and Goodfield, *Discovery*, p. 178.

[276] *Époques*; Butterworth, *Origins*, p. 237; Glacken, *Traces*, p. 666; Roger, 'Buffon', p. 578.

[277] Baumer, *Modern European Thought*, pp. 209f.

words, Buffon dissociated the idea of unchanging laws of nature from that of a steady state of nature: immutable laws could give rise to a succession of different states; conversely, a succession of states could be understood in terms of as single set of fundamental laws.278

Buffon's estimates of how much time was required to allow all this to happen ranged from 75,000 to 3 million years.279 Though these figures seem modest by our present standards for the age of the earth (in the thousands of millions of years), they represented a profound challenge to the traditional thought of Buffon's time. In fact, the first volume of the *Natural History* was vehemently criticised by the theologians of the Sorbonne (1751), and Buffon was obliged to affirm his complete faith in the Genesis account of creation.280

Buffon's work marked a critical step in the development of modern Western thought. In retrospect, its effect can best be understood in terms of two resulting paradoxes concerning the relationship of natural science to history and the relationship of modern social optimism to cosmic pessimism, respectively. Thus, although Buffon was ostensibly less Christian than earlier representatives of the creationist tradition, his work was profoundly theological in its character and implications for modern thought.

For the first time in human history, the age of the cosmos could be estimated using scientific methods that were not based on ancient texts. This led to the first paradox: a discipline that was the outgrowth of the special history of a peculiar culture had now become so comprehensive and so well established that its view of the world appeared to be entirely independent of the discipline of history itself. Science, like history, can be understood either as a human construct or in terms of what it tells us about the world in which we live. Its credibility has become so great that we tend to identify it with the world of facts and lose sight of its dependence (for better or worse)

---

[278] Buffon contemplated two orders (views) of nature: one temporal and changing, the other eternal and immutable; 'Deux vues de la nature' (1764-65); Baumer, *Modern European Thought*, pp. 213f.

[279] Using Newton's law of cooling, Buffon estimated that 75,000-168,000 years were needed for the earth to cool from its initial incandescence; Toulmin and Goodfield, *Discovery*, pp. 175-79. The unpublished figure of 3 million years was based on his estimates of the rates of sedimentation; Roger, 'Buffon', p. 579a. Benoit de Maillet had made an even earlier estimate of the age of the earth based on observed sedimentation rates. The figure he came up with was 2 billion years. He presented these findings in the form of an imaginary dialogue between a French missionary and an Indian philosopher, entitled *Telliamed* ('de Maillet' spelled backwards; see the English translation by Albert V. Carozzi, Urbana: Univ. of Illinois Press, 1968). The published version (posthumous, 1748) toned down the original estimate to 2 million years, which was more in line with Buffon's figure; Davis A. Young, 'The Discovery of Terrestrial History', in Howard Van Till, et al., *Portraits of Creation* (Grand Rapids, Mich.: Eerdmans, 1990), p. 47.

[280] Gay, *Enlightenment*, 2:155.

on particular historical and cultural circumstances. With Buffon, we see this shift taking place before our very eyes as the whole timespan of human history becomes a mere fraction of the natural history constructed by modern science.

The other significant change was the separation of objective science from human ideals. Buffon could estimate not only the time since the creation, but also the amount of time left before the earth would become too cold to support life any more. The figure he arrived at was about 93,000 years:[281] in another 93,000 years all life would cease on earth.[282] In one sense, this was a direct challenge to biblical faith, which traditionally held that the present age would be terminated by divine intervention. But, paradoxically, Buffon's projection was an even deeper challenge to the widespread eighteenth-century faith in limitless human progress, a belief which Buffon himself shared.[283] By the time the earth became too cold, the more massive planet Jupiter would have cooled down to the point where it could support human life (Mars was already too cold). But, even though humans were extremely adaptable and could continue for thousands or even millions of years somewhere in the solar system, they would eventually become extinct like the tropical species that once inhabited the northern regions of earth.

So the elimination of the biblical projection of the future from science made it appear that the present age would last much longer than earlier generations had thought. However, it provided no hope for an age to come. In fact, Buffon's naturalistic projection ruled out any objective basis for the hopes and ideals that he and his generation had imbibed from the creationist tradition concerning the possiblity of realising human potential in a future world. Buffon had effectively separated objective scientific inquiry from his personal vision for the future of Western civilisation.[284]

The fact that the bleakness of scientifically established prospects for the future are no hindrance to continued optimism concerning the future of human life is one of the most difficult features of modern thought to account for. It may reflect the fact that Western humans have come to see themselves more in relation to their technology than to the world of nature. Buffon was acutely aware of this shift in perspective himself. Partly due to

---

[281] 93,291 years to be exact; Toulmin and Goodfield, *Discovery*, pp. 178, 181.
[282] Newton himself had speculated on the possible destruction of life on earth by the 1680 comet's falling into the sun and causing an immense explosion in about 3,000 years; 'Conduitt Memorandum'; Genuth, 'Comets', pp. 54ff. Since Newton also believed in the possibility of supernatural intervention, however, this presented no problem to his faith in universal progress.
[283] Glacken, *Traces*, pp. 663ff.
[284] Gay, *Enlightenment*, 2:154.

his work as keeper of the royal French botanical gardens,[285] he well appreciated the impact of human ordering on the environment:[286]

> The state in which we see nature today is as much our work as its; we have known how to temper and modify it, bend it to our needs and desires; we have founded, cultivated, fructified the earth: the aspect under which it presents itself is, then, very different from that of times anterior to the invention of the arts.

Buffon had understood the essential malleability of nature as taught by creationist—particularly the apocalyptic and hermetic—traditions. The biblical and patristic sanction for technology as a ministry of healing and restoration had been eliminated, however, and with it any sense of reliance on the name of Christ and the Spirit of God (chap. 1, sec. E). For Buffon, the rationalisation and transformation of nature were willed and planned by humans alone—they were the result of technology. The subject of pure science, on the other hand, was the natural state of the earth untouched by human hands and unrationalised by human intellect.[287]

The independence of modern science from history thus had as its converse the independence of social imperatives from the worldview of modern science. In other words, a dualism of objective worldview and subjective belief was emerging as the result of the fragmentation and secularisation of creationist ideals.

3.2 Diderot

Dualism certainly did not characterise the thinking of Denis Diderot (1713-84), the figure most often cited as marking the transition from static to dynamic, organismic thinking in the mid-eighteenth century.[288] The ideal of the unity of all reality was affirmed by him as it rarely has been before or since. For the mature Diderot (after 1749), nothing was fixed or exact in nature. Everything was ephemeral and in continual flux into everything else: matter, organisms, species, even the human soul.[289] The price of such

---

[285] As *intendant* of the Jardin du Roi (1739), Buffon succeeded in doubling the cultivated area; Roger, 'Buffon', p. 576b.

[286] *Époques*; Baumer, *Modern European Thought*, p. 216; cf. Glacken, *Traces*, p. 666.

[287] Hence Buffon's rejection (1744ff.) of Linnaeus's binary classification of plants based on a single characteristic for each subdivision as the imposition of human reason on nature. For Buffon, organic nature consisted of a plenum of unique individuals, an idea that went back to Leibniz and Christian Wolff, whose ideas had been popluarised in 1740 by Madame du Châtelet; Hankins, *Science*, pp. 149ff., 155f.; Lyon and Sloan, eds., *From Natural History*, pp. 20ff.; Rudwick, 'Shape', pp. 308f.

[288] E.g., Cassirer, *Philosophy*, pp. 90ff.

[289] 'Tout change; tout passe; il n'ya que le tout qui reste'; *La rève de D'Alembert* (written 1769; pub. posthumously; trans. J. Barzun, *Rameau's Nephew* [Indianapolis: Bobbs-Merrill, 1964], p. 117); Bréhier, *History*, 5:126; Randall, *Career*, 1:887f. The idea is present as early

unity, however, was the comprehensibility of the world, at least in terms of analytic reason and mathematical physics.

Diderot rejected the traditional creationist notion of the comprehensibility of the world. For one thing, he argued that nature functioned on the level of wholes, rather than just parts, and that scientific analysis was not suited to dealing with such wholes.[290] But he also abandoned the traditional underpinning of the belief in comprehensibility—the ideas that the world is created in accordance with divine reason and that humanity is created in the image of God. In contrast to contemporaries like Buffon, Diderot saw the primordial condition of the universe as one of total chaos with only momentary appearances of order like the one we presently observe.[291] From the historical-theological point of view, he was entirely consistent here. In times like Diderot's, when the tools of science seemed inadequate to the task of understanding the real world as a whole, it had always been faith in a Creator that had sustained serious scientific endeavour.

Diderot's reaction against the abstract classification and mathematisation of nature was enshrined in the multivolume *Encyclopedia of the Sciences, Arts and Trades* (1751-72; initially coedited with d'Alembert),[292] a monument to the techniques and ideals of the burgeoning small industries of Western Europe. The practical ideals embodied in the *Encyclopedia* were derived from the medieval and Renaissance tradition of lay craftsmanship: Diderot was himself the son of an artisan (a cutler) in Langres.[293] Such ideals had historically been formulated in opposition to the mathematical emphasis of Neoplatonism. They stressed practise rather than theory, an intuitive grasp of nature rather than rational analysis, and service to other humans rather than the personal prestige of the craftsman. Some earlier exponents of these ideals that we have studied are Cusa, Agricola, Paracelsus, Bacon, and Helmont.

Diderot clearly belongs in this stream of the creationist tradition even though he secularised its ideals. Like his predecessors, Diderot berated mathematical science for imposing human categories on nature and exercising the intellectual hubris of the scientist. In contrast, Diderot's ideal craftsman was dedicated to harnessing nature by intuitive understanding and respect for the nature of his material.[294] But, while Diderot believed fervently in the limitless progress of human technology and applied the

as Diderot's *Lettre sur les aveugles* (1749); Baumer, *Modern European Thought*, p. 211.
[290] Baumer, *Modern European Thought*, pp. 152, 216.
[291] Stanley L. Jaki, *The Road of Science and the Ways to God* (Chicago: Univ. of Chicago Press, 1978), p. 78 and note 74.
[292] D'Alembert backed out of the project in 1758, just after the publication of vol. VII and one year before the suspension of publication by government authorities. Publication of the plates was resumed under Diderot's sole editorship in 1762, and ten volumes of corresponding text came out in 1763; C.C. Gillispie, 'Diderot', *DSB* 4:86a-87a.
[293] Gillispie, 'Diderot', p. 86b.
[294] Gillispie, *Edge*, pp. 187-90; *idem*, 'Diderot', pp. 88f.

traditional criterion of human benefit (immortalised by Bacon), there is little indication of his preserving the biblical motive for healing or service to others.[295]

From the perspective of the historic creationist tradition the difference between Diderot and Buffon is this: Buffon secularised the idea of the comprehensibility of the world, and Diderot secularised the ideal of the ministry of healing and restoration, while both secularised the notion of the relative autonomy of nature.

3.3 Holbach

Paul Henri Thiry, Baron d'Holbach (1723-89), is best known for his *System of Nature*, first published anonymously in 1770.[296] Holbach's *System* was a manifesto for reductionism, materialism, determinism, and practical atheism.

For Holbach, motion was neither added to matter at creation (mechanical philosophy), nor sustained by active principles in accordance with divinely imposed laws (Newton). In fact, there was neither a Creator (in the biblical sense) nor a creation *ex nihilo*.[297] All such religious beliefs were merely the result of the superstitious awe in which primitive peoples held the unknown powers of nature. Now that modern science was capable of giving a rational explanation of these powers in terms of an unbroken chain of cause and effect, there was no excuse for recourse to the supernatural.[298] Holbach allowed that there might be an unknown 'Cause of causes', but only in so far as it was a necessary postulate of reason; there was no longer any room for Christian, or any other, religious faith.[299]

Instead, both motion and the forces that generated it were eternal, primary qualities of matter, as essential to its being as extension itself. The power of Holbach's argumentation rivaled that which he attributed to matter:[300]

> The idea of Nature necessarily includes that of motion. But it will be asked [by Newtonians like Voltaire], and not a little triumphantly, from whence did she derive her motion? Our reply is...that it is fair to infer, unless they can logically prove to the contrary, that it is in herself, since she is the great whole out of which nothing can exist. We say this motion is a

---

[295] Gillispie, *Edge*, pp. 173ff., 180f.; Baumer, *Modern European Thought*, p. 148.
[296] *La système de la nature, ou des lois du monde physique et du monde moral* (anon. 1770).
[297] *Système*; Randall, *Career*, pp. 914f.; Goodman, ed., *Science*, p. 281ff.
[298] *Système*; A. Vartanian, 'Holbach', *DSB* 6:469a; Goodman, ed., *Science*, pp. 286-89, 292ff.
[299] *Système*; Goodman, ed., *Science*, pp. 289ff., 293.
[300] *Système*; trans. Samuel Wilkinson, 3 vols., (London, 1820-21), 1:23 (Goodman, ed., *Science*, p. 276). Voltaire had begun his attack on materialist notions and argued that God was the source of all motion in 1765; Baumer, *Modern European Thought*, p. 212.

manner of existence that flows, necessarily, out of the nature of matter; that matter moves by its own peculiar energies; that its motion is to be attributed to the force which is inherent in itself....

In other words, once Descartes's a priori identification of matter with extension was eliminated, the source of motion could be found in matter as easily as in God.

Moreover, all reality—even human phenomena like feelings, ideas, religion, and art—was, for Holbach, just the byproduct of matter in motion:[301]

> The universe, that vast assemblage of everything that exists, presents only matter and motion: the whole offers to our contemplation nothing but an immense, an uninterrupted succession of causes and effects.

This was one of the most thorough statements of reductionism and determinism since the ancient Epicureans. The traditional creationist ideal of the unity of nature was effectively replaced by a thoroughgoing materialist view of homogeneous matter in motion.[302]

There was a profound contradiction between Holbach's deterministic, reductionistic view of the world and his personal passion for moral issues, a contradiction that has plagued modern secular thought ever since. Lacking Priestley's identification of natural self-determination with the work and will of God, Holbach had no real grounds for believing that human judgements concerning moral issues were anything other than the blind product of interacting atoms that knew nothing of human feelings or morals.[303] However, this did not prevent him from making sweeping moral judgements of his own about contemporary institutions like the church. Whereas Diderot had merely secularised the traditional Christian ideal of human benefit, Holbach turned it around and used it as a weapon against the church itself. Christianity, he charged, was responsible for most of the ills of modern society: among other things, it encouraged prejudice, depreciated nature, and discouraged self-fulfillment.[304]

---

[301] *Système*; Randall, *Making*, p. 274; Bréhier, *History* 5:128f.
[302] *Système*, Randall, *Career*, p. 913; cf. Caponigri, *History*, 3:275, 278.
[303] Holbach argued that the natural propensity of the human mind to 'endeavour to fathom' and 'strive to unravel' nature was somehow programmed into the eternal laws of nature by the Cause of causes (Goodman, ed., *Science*, p. 292), but he gave no reason for supposing the results of this propensity to be valid. La Mettrie himself denied that vice and virtue had any meaning in a deterministic view of human nature; *Discours sur le bonheur* (1748); A. Vartanian, 'La Mettrie', *DSB* 7:605a. However, the contradiction could conveniently be overlooked so long as there was no accepted mechanism for the biological evolution of human nature.
[304] Holbach's edition of Boulanger's *Christianity Unveiled* (1767; Randall, *Career*, p. 912); *Système* (1770; Bréhier, *History*, 5:130f.; F.L. Baumer, ed., *Main Currents of Western Thought* [New Haven: Yale Univ. Press, 1978], pp. 403f.).

Holbach was not unaware of the difficulty. His solution was to propose a new morality based on self-interest and the natural desire for pleasure, yet independent of any religious faith.[305] The inconsistency lay in the fact that Holbach's materialism made no room for the potentiality for valid moral judgements in the brute matter from which humans evolved. There was, therefore, no more scope for judgement in a human than in a tree. But it took a wiley pragmatist like Frederick the Great to see the dilemma:[306]

> What foolishness and what nonsense! If everything is moved by necessary [material] causes, then all counsel, all instruction, all rewards and punishments are as superfluous as inexplicable; for one might just as well preach to an oak and try to persuade it to turn into an orange tree.

A predestinarian like Calvin or a Christian materialist like Priestley could argue the compatibility of moral judgements with belief in predetermination based on the all-determining power of a moral God. Holbach tried to base morality on the study of human nature but then had only brute matter to appeal to as the basis of human nature.

3.4 Conclusion: From the Creationist Tradition to Modern Materialism

We have noted the partial continuity, fragmentation, and secularisation of creationist themes in the French materialists. There was a particularly close link between eighteenth-century materialism and seventeenth-century spiritualism and pansophism (Leibniz), with early eighteenth-century vitalism acting as the intermediary. Continuity was evident primarily in the common view of matter as imbued with relatively autonomous powers and in the lay-oriented critique of established elites on the basis of egalitarian principles. In the spiritualist tradition, however, powers were present in matter by virtue of creation and were only relatively autonomous; in materialism they became eternal and entirely autonomous. In spiritualism, moreover, the critique of established interests was based on the biblical concepts of the Spirit and the Church; in materialism it was based on the natural equality of humans by itself and was directed against the established church. The causes of this dramatic shift in theology are complex, but our survey suggests at least the following five factors.

(1) The increasingly specialised nature of natural science allowed the adoption of selected ideas from early eighteenth-century figures like Boerhaave and Stahl without reference to their Christian faith.

(2) The involvement of the French church in national politics was such as to make implacable enemies of the more radical *philosophes*, particularly

---

[305] Bréhier, *History* 5:129f.; Vartanian, 'Holbach', p, 469b; Baumer, *Modern European Thought*, p. 198.
[306] Cassirer, *Philosophy*, p. 71.

Diderot and Holbach.

(3) Newtonianism set the standard for understanding the relationship between God and nature. Newton found plenty of scope for divine activity in nature, but, in order to make room for an active God, he had to make matter itself entirely inert. In other words, Newton reinforced the underlying assumption of the mechanical philosophy that more power for matter implied less power for God.[307]

(4) Guardians of orthodoxy in France had exploited dualistic, Cartesian ideas for the defense of religion. Some *philosophes* like Voltaire had adopted the Newtonian symbiosis of God and nature. But the alternative, spiritualist, tradition had not been a vigorous movement in France as it had in England, Holland, and Germany.

(5) Underlying all these contemporary factors were several longer-range tendencies we have observed in Western thought since the twelfth century: the tendency to press the comprehensibility, unity, and relative autonomy of creation to the point of making nature entirely self-sufficient; the tendency to define the self-sufficiency of nature and the direct operation of God as mutually exclusive alternatives; and the tendency to treat the gifts of healing and social reform on secular terms in contrast to the 'spiritual' ministry of the clergy. On the basis of all these points, eighteenth-century materialism can be seen as a legitimate expression of the creationist tradition as much as the medieval Catholic emphasis on God's *potentia absoluta* or the Reformation emphasis on perpetual miracles within the sphere of God's *potentia ordinata*.

C. British and American Anti-Newtonians

There was a variety of reactions to Newton's symbiosis of science and theology in the eighteenth century. Among them were several groups that rejected Newton's ideas as encouraging monistic materialism (like that discussed in the previous section) or otherwise compromising religious principles. It is important to give these figures due consideration. Their critiques preserved some traditional features of the creationist tradition, and the alternative systems they inspired, though now mostly forgotten, were positive stimuli in their time.

There were at least three different groups of people who developed an anti-Newtonian stance:

(1) High-Church Anglican Tories like Jonathan Swift and Samuel Johnson;

---

[307] Cf. Michael J. Buckley, *At the Origins of Modern Atheism* (New Haven: Yale Univ. Press, 1987), pp. 352-55, on Malebranche (following Descartes) and Clarke (following Newton).

(2) antimaterialists like George Berkeley, Jonathan Edwards, and William Blake;

(3) the followers of John Hutchinson.

These names include some of the most prominent figures of the time. Although we cannot possibly do justice to them individually, we can gain a sense of the breadth of the interaction of humanities, science, and theology in the eighteenth century.

1 High-Church Anglican Tories

On the whole, the High-Church party viewed itself as being on the defensive against the emerging power of the Latitudinarians and Whigs in the late seventeenth and early eighteeenth centuries. The more conscientious of them (the Nonjurors) refused to declare their allegiance to William and Mary after the 'Glorious Revolution' of 1688 and were naturally bypassed in the appointments of the new regime. They briefly reasserted themselves, however, during the Tory ascendancy in the latter part of the reign of Queen Anne (1710-14). During that time, High Churchmen were responsible for two significant movements that raised up widespread concerns about the impact of the new science: one ecclesiastical—the temporary prosecution of Samuel Clarke—and the other literary—the campaign of the 'Scriblerians' against the scientific projects of the Royal Society of London.

1.1 High-Church Criticisms of Clarke

Samuel Clarke's doubts about the orthodox version of the Trinity (sec. A) were due to his subordination of revelation to reason, according to Nonjurors like George Hickes (1642-1715), the nonjuring Bishop of Thetford, and Roger North (1653-1734), a solicitor to Queen Anne. North noted what appeared to conservatives to be a certain arrogance on the part of Newtonians like Clarke: their promotion of one particular natural philosophy against all others (Cartesian, Leibnizian, etc.) was paralleled by their attempt to pass judgement on the divine essence using the tools of human reason.[308] North accepted that space was infinite but rejected Newton's notion of absolute, uncreated space as encroaching on the power of God (whereas for Newtonians it was a manifestation of divine omnipotence).[309]

It appears that the Nonjurors' sense of the transcendence of God and the majesty of his revelation was related to their high conception of the church and insistence on its autonomy in relation to the state. The dynamics of the

---

[308] E.g., in Clarke's *Demonstration of the Being and Attributes of God* (1704); L. Stewart, 'Samuel Clarke', pp. 55f.

[309] Roger North, 'Answer to Dr Clarke' (1712?); Stewart, 'Samuel Clarke', pp. 63-70.

period is strikingly similar to that of the late eleventh and early twelfth centuries, when a dispute over the relation of revelation to Aristotelian dialectic was similarly related to the struggle between papal and secular powers for control of the church (chap. 1, sec. D).

1.2 The Scriblerians and Samuel Johnson

The Scriblerians (or Scriblerus Club) included John Arbuthnot (1667-1735), Jonathan Swift (1667-1745), and Alexander Pope (1688-1744). Some of their ideas were later developed by Samuel Johnson (1709-84).

The Scriblerians were not opposed to Newtonian science as such or to the use of science for the improvement of human life[310]—Arbuthnot himself wrote an important paper for the Royal Society (1710-12) that provided the basis of modern mathematical statistics, particularly the work of Laplace.[311] But they viewed the propaganda of the Royal Society with enough detachment to see its humorous and farcical aspects. Arbuthnot and Swift were particularly concerned with the latitudinarian policies (tolerating Dissenters) of many members of the Royal Society and rejected any relaxation of conformity to the standards of the Church of England as a matter of principle. They clearly associated current science with a pragmatism that neglected the spiritual side of human nature and a self-interest that conflicted with true Christian charity.[312] The rivalry between opposing schools of scientific thought was also held to be needlessly divisive for the church.[313] To some extent their concerns paralleled the early Christian critique of Greek science (chap. 1, sec. E): according to the Scriblerians, many scientists were motivated by arrogance and pride rather than by a genuine love for God and their fellow humans.

Swift had earlier criticised the modern 'philosophers' for erecting edifices in the air in *A Tale of a Tub* (1704). The Latitudinarians' use of the phenomena of nature to infer the existence of God as its architect was parodied by his description of a religious sect that worshipped an idol in the form of a tailor because their world had the appearance of a suit of clothes.[314]

---

[310] E.g., Swift, *Gulliver's Travels* II.7 (Modern Library ed. [New York: Random House, 1958], p. 104).

[311] 'An Argument for Divine Providence, Taken from the Constant Regularity Observ'd in the Birth of Both Sexes' (1710-12), Hans Freudenthal, 'Arbuthnot', *DSB* 1:208f.

[312] Richard G. Olson, 'Tory-High Church Opposition to Science and Scientism in the Eighteenth Century', in John G. Burke, ed., *The Uses of Science in the Age of Newton* (Berkeley: Univ. of California Press, 1983), pp. 177-80 (on Arbuthnot), 181-89 (on Swift).

[313] Swift, 'On the Wisdom of this World'; Olson, 'Tory-High Church', p. 185; cf. *Gulliver's Travels* III.2 (p. 128).

[314] A.L. Rowse, *Jonathan Swift* (London: Thames & Hudson, 1975), pp. 34f.; Olson, 'Tory-High Church', pp. 186f. Olson points to Aristophanes as the model; cf. his 'Science, Scientism, and Anti-Science in Hellenic Athens', *History of Science*, vol. xvi (1978), pp. 179-99.

In *Gulliver's Travels* (1726), Swift ridiculed the Royal Society by depicting a Grand Academy of Projectors in the city of Lagado. The preposterous schemes of the members of the academy included experiments to extract sunbeams from cucumbers and plans to develop a breed of naked sheep! These 'projects' were pursued with such singlemindedness that the people's homes and lands went to ruin.[315]

Gulliver also visited the more theoretically inclined Laputians, whose obsession with the effects of a recent—and soon to return—comet clearly reflected on the theories of Newton and Whiston.[316] The Laputians thus exhibited the plight of a science-based society for which knowledge of the future was of no avail in controlling it.

Swift criticised his own culture by having Gulliver belittle the mores of the lands he visited in ways that clearly reflected back those of his home country. For instance, when the king of Brobdingnag refused to consider Gulliver's proposal to exploit the destructive potential of gunpowder, Gulliver attributed this to his 'Narrowness of Thinking' and 'unnecessary Scruple, whereof in Europe we can have no Conception'.[317] Swift's sensitivity to the abuses that could result from new technologies in the service of political interests came partly as the result of his own witnessing of the English exploitation of Ireland. Here again parallels could be drawn to earlier models, e.g., the Jewish experience of the superior technologies of colonial powers in the ancient world (chap. 1, sec. E).[318]

Though initially more sympathetic to the efforts of the 'projectors' than Swift, Samuel Johnson became increasingly critical of the apparent smugness with which Latitudinarians explained the decrees of God and the structures of nature in terms of each other. For Johnson, the will of God could not be accounted for through human deliberation,[319] and the natural

---

[315] *Gulliver's Travels* III.4f. (pp. 140-45); Marjorie Nicolson and Nora M. Mohler, 'The Scientific Background of Swift's *Voyage to Laputa*', *Annals of Science*, vol. ii (1937), pp. 299-334; G.S. Rousseau, 'Science', pp. 160f. Pat Rogers relates the reckless projects of the Lagado Academy to the South Sea Bubble mania; 'Gulliver and the Engineers', *Modern Language Review*, vol. lxx (1975), pp. 260-70. Commercial ventures were not inconsistent with the vision of the Royal Society, however, as the two were synthesised by many early eighteenth-century Newtonians.

[316] *Gulliver's Travels* III.2 (p. 129). On Newton's anticipation of the possible destruction of the world by the 1680 comet's eventual falling into the sun, see note 282 above.

[317] *Gulliver's Travels* II.7 (pp. 102f.); Olson, 'Tory-High Church', p. 189.

[318] Olson, 'Tory-High Church', p. 188.

[319] *Life of Boerhaave*; Review of Soame Jenyn's *Free Inquiry into the Nature and Origin of Evil* in the *Literary Magazine* (1750s); Olson, 'Tory-High Church', p. 195.

world was not there just to be exploited for experimental or apologetic purposes. He argued against defenders of vivisection that the insensitivity to animal pain required by physicians who conducted medical experiments on animals led directly to those physicians' callousness toward their human patients.[320] Johnson could also adopt the animal's point of view as a literary device critiquing anthropocentrism in his attack on the argument from design popularised by latitudinarian scientists and theologians: a mother vulture explains to her children that humans are endowed with a natural propensity to war so that they may have a regular supply of food.[321]

No stranger to the struggle for economic survival himself,[322] Johnson exhibited an unusual sense of moral absolutes in an age characterised by overt self-interest. In his earlier writings, he upheld the Baconian valuation of science as a means towards the amelioration of the human condition and as a vehicle for the expression of Christian charity; only when pursued as an end in itself would science become vain and destructive.[323] With the writing of *Rasselas* (1759), however, he came to the grim conclusion that the desire to control nature would lead to the madness of imagining oneself in the place of God even when exercised in the interests of Christian service.[324] At the time, Johnson was grieving over the loss of his mother—he wrote the work in an effort to pay her funeral expenses. His undoubted depression only partly accounts for his one-sided view of the current science, however: literary criticism shows the basic continuity with his earlier writings.[325]

The critique of Newtonian science by these literary figures was haphazard, and they offered no coherent alternatives. Moreover, their social and religious conservatism made them unappreciative of the legitimate needs for radical change in their society. However, their concerns about the public uses of science and technology were to find ample support from subsequent historical events. Their writings thus provided an important precedent for future generations of writers who sought to express their concerns about science and technology through the medium of fantasy and science fiction.

---

[320] *Idler*, 17 (series begun in 1758); Olson, 'Tory-High Church', p. 197; K. Thomas, *Man*, p. 178.

[321] *Idler*, 22; Olson, 'Tory-High Church', p. 196.

[322] In 1756 Samuel Richardson had to rescue Johnson from the fate of debtor's prison. In 1759 Johnson was unable to pay the expense of his mother's funeral and small debts; Warren Fleischauer, ed., *Samuel Johnson: The History of Rasselas* (Woodbury, N.Y.: Barron's, 1962), p. 2.

[323] Richard B. Schwartz, *Samuel Johnson and the New Science* (Madison: Univ. of Wisconsin Press, 1971), p. 126; Olson, 'Tory-High Church', p. 197.

[324] *Rasselas* 41-43; Olson, 'Tory-High Church', pp. 198f.

[325] Fleischauer, ed., *Samuel Johnson*, pp. 3f.

2 Anti-Materialists: Berkeley, Edwards, and Blake

George Berkeley and Jonathan Edwards were two of the most sophisticated thinkers of the eighteenth century. Far from attempting to characterise their thought as a whole, we are concerned here only with their reactions to contemporary science—the ideas of Newton in particular—and the role their theology played in those reactions.

2.1 George Berkeley

George Berkeley (1685-1753) was one of the earliest critics of Newton's natural philosophy. He aimed his attacks directly at those entities in Newton's system that acted as intermediaries between God and passive matter—absolute space and supramechanical force. Newton himself had intended by these notions to ensure an active role for God in nature, but they seemed to others to make God more remote and lead to deism and even materialism.[326]

Absolute space, Berkeley argued (1710), must either be God (which Newton denied) or else 'something beside God which is eternal, uncreated, infinite, indivisible, immutable'—either of which notions is 'pernicious and absurd'.[327] Newton's ideas of force and gravitation, on the other hand, were convenient tools for the purposes of calculation, but they had no more reality than the epicycles of medieval astronomers. For Berkeley (as for Descartes), the real causes of motion could only be spirits and therefore could not be investigated by science but were reserved for metaphysics and theology.[328]

Another arena in which issues of science and theology were warmly debated in the early eighteenth century was the treatment of diseases like smallpox. Most physicians who had been influenced by Isaac Newton advocated a program of systematic inoculation against the disease. Berkeley saw deistic tendencies in such a materialist approach. For him diseases like smallpox could not be just the result of airborn contagion, as Newtonians like James Jurin argued; they were due to the will of God. Consequently, the

---

[326] Theodore Hornberber, 'Samuel Johnson of Yale and King's College', *New England Quarterly*, vol. viii (Sept. 1935), pp. 388ff.; Marina Benjamin, 'Medicine, Morality and the Politics of Berkeley's Tar-Water', in Andrew Cunningham and Roger French, eds., *The Medical Enlightenment of the Eighteenth Century* (Cambridge: Cambridge Univ. Press, 1990), pp. 165-76.

[327] *A Treatise Concerning the Principles of Human Knowledge*, CXVII, in Everyman ed., *Berkeley: A New Theory of Vision and Other Writings* (London: Dent, 1938) pp. 173f.; Jammer, *Concepts of Space*, pp. 112, 135; F.E.L. Priestley, 'Newton', p. 331.

[328] *De motu* (1721); *Siris: A Chain of Philosophical Inquiries Concerning the Virtues of Tar-water* (1744); Jammer, *Concepts of Force*, pp. 203-7; Heimann and McGuire. 'Newtonian Forces', p. 203.

best medicine was not medical inoculation, but a God-given (seemingly occult) panacea like tar-water (a mixture of pine resin and water), which was, in effect, a drinkable manifestation of God.[329]

In effect, Berkeley challenged the Newtonian synthesis of physics and theology at its very core. Like the materialists (sec. B), he took over Newton's notion of a subtle ether and Boerhaave's idea of a cosmic fire, but he concluded that such an active substratum for empirical phenomena must be a form of mind rather than of matter.[330]

Rather than distancing God from the phenomena of nature as Newton's universal active principles appeared to do, Berkeley's idealism ran the opposite danger of emptying nature of any intrinsic lawfulness. His stated intention, however, was not to reduce the laws of nature to epiphenomena of the divine mind as much as to insist on their plurality and specificity in opposition to the universal principles of Newton.[331] His reaction to Newton was thus not unlike that of earlier Christians to Aristotle's subordination of terrestrial events to cosmic cycles. Like the church fathers and the Reformers, Berkeley saw the world as a plurality of principles which found their unity in God, rather than a hierarchy of principles subordinate to each other.

2.2 Jonathan Edwards

Like Berkeley, though independently of him, the American theologian Jonathan Edwards (1703-58) opposed current tendencies to materialism and rejected Newton's hierarchical view of God and nature. But like Newton and unlike Berkeley, Edwards followed Henry More in holding that space is a necessary manifestation (for human minds, at least) of the divine being.[332] However, he took aim at Newton's dualism between active principles or ethers and solid atomic matter.

---

[329] James Jurin, *Account of the Success of Inoculating the Small-Pox in Great Britain* (1724); Berkeley, *Siris*, 211-18; Benjamin, 'Medicine', pp. 165-66, 180-82; David Berman, *George Berkeley* (Oxford: Clarendon Press, 1994), pp. 172-76.

[330] Heimann, 'Nature', pp. 13f.; G.N. Cantor, 'The Theological Significance of Ethers', in G.N. Cantor and M.J.S. Hodge, eds., *Conceptions of Ether* (Cambridge: Cambridge Univ. Press, 1981), p. 147.

[331] Berkeley, *Principles* CVI: 'There is nothing *necessary or essential in the case*, but it depends entirely on the will of the *governing spirit*, who causes certain bodies to cleave together, or tend towards each other, according to various laws, whilst he keeps others at a fixed distance; and to some he gives a quite contrary tendency to fly asunder, just as he sees convenient.'

[332] 'Space is this necessary, eternal, infinite and omnipresent being....I have already said as much as that space is God'; 'Of Being' (*Works*, 6:203); Heimann and McGuire, 'Newtonian Forces', pp. 251, 254; Leonard R. Riforgiato, 'The Unified Thought of Jonathan Edwards', *Thought*, vol. xlvii (1972), p. 604; *Works*, 6:73ff. However, Edwards allowed that disembodied spirits might have a different view of space, 'The Mind' (*Works*, 6:99ff., 343). Also, he rejected More's notion that spirits are extended in space; *ibid.*, pp. 98, 111f.

For Newton, gravitation was a supramechanical principle imposed on matter, while solidity (hardness, impenetrability) was a primary quality of the primordial particles themselves. Whereas many Newtonians and materialists attempted to overcome the dualism by making gravitation an intrinsic property of matter, Edwards solved it by making solidity a direct manifestation of divine power like gravitation. The substratum of all qualities was God himself: space was God acting, and Newton's laws could be understood as simply describing the manner in which God acted in space and time.[333] Thus the mutual attraction[334] of atoms was like the mutual love between two minds: both were reflections of the mutuality of the eternal self-expression and love of God in the Trinity.[335]

Edwards was probably not aware of the fact that Newton was an Arian, but he did know of the cultivation of Unitarian ideas among his followers and, like Cotton Mather, he defended the traditional doctrine of the Trinity. There is no need to suppose a cause-effect relationship between the different aspects of his thought, but clearly his ability to see the activity of God equally in all the phenomena of nature was facilitated by his dynamic sense of the inner life of the Godhead.

Like Berkeley, Edwards ran the risk of emptying nature of any structure or principles of its own. The truth of divine providence was established at the expense of the doctrine of primary creation. Creation, for Edwards, was simply the first instance of the exercise of the power God has wielded ever since.[336]

Belief in the utter dependence of all things on God did not prevent Edwards from supposing that the principles of mechanics could account for the operations of nature. As in the mechanical philosophy of Descartes and Boyle, the activity of God in sustaining nature was supposed to be regular and predictable. In fact, God established the primary qualities and mechanical laws of nature at creation in such a way that all the inanimate forms of this world would be produced from a primordial chaos in accordance with those qualities and laws (as for Descartes). Only the production of plants and animals required the imposition of new structures and laws (as with Boyle

---

[333] 'Of Atoms'; 'Things to be Considered' (*Works*, 6:214ff., 234f., 241); Heimann and McGuire, 'Newtonian Forces', pp. 252ff.; Riforgiato, 'Unified Thought', pp. 600-604; *Works*, 6:49, 54, 58f., 67f.

[334] Edwards used the term 'gravitation' to include all types of mutual attraction; *Works*, 6:45, 265.

[335] Riforgiato, 'Unified Thought', pp. 604-10. The analogy between Newtonian gravitation and the love of God was also developed by William Law, based on the writings of Jacob Boehme and George Cheyne; Albert J. Kuhn, 'Nature Spiritualized: Aspects of Anti-Newtonianism', *English Literary History*, vol. xli (Fall, 1974), pp. 404ff.

[336] '...creation is the first exercise of that power in that manner'; 'Of Atoms'; cf. 'Things' (*Works*, 6:215, 241).

and Derham).³³⁷ Edwards clearly followed Descartes and Boyle rather than Newton at this point. Evidently, his sense of the utter dependence of all nature on God—primary qualities and active principles equally—made it unnecessary for him to reserve a special role for God in the formation of the solar system as Newton and Clarke had done.

Neither Berkeley nor Edwards made any significant contribution to the development of science. Nor do their critiques of Newton seem to have inspired any attempts to develop alternative systems of natural philosophy. Still, their ideas have been cited frequently in retrospect, particularly in the twentieth century, when a relational view of space has been developed not unlike Berkeley's and when matter and fields of force are being united in a single physics reminiscent of the views of Edwards.

2.3 William Blake

The visionary poet and artist William Blake (1757-1827) lived and died a full century after Newton (1642-1727). His work reflects the late eighteenth-century quest for radical reform more than the seventeenth-century quest for stability. Blake's political sympathies were with the Jacobinism and Republicanism associated with the French Revolution. He viewed Newton (together with Bacon and Locke) as a symbol of the established order rather than an original thinker in his own right. Still, Blake's fantastic symbolism captured some of the essential features of Newton's system as well as any physicist or historian could have done.

Like Berkeley, Blake took issue with two key Newtonian ideas: absolute space (and time) and supramechanical force. Both notions presupposed the complete passivity or inertness of matter. From this perspective, Newton's worldview had the same meaning as the mechanical philosophy of Descartes and Boyle. The only difference was that it substituted a Neoplatonic hierarchy of force over matter for the mechanical dualism of spirit and matter. Blake himself was more in tune with a third tradition of natural philosophy (chap. 3, sec. C), the spiritualist tradition of Paracelsus, Helmont, and Boehme, which imbued matter with living force and

---

[337] Edwards, 'Things' (*Works* 6:49, 265): 'And this being so ordered that, without doing any more, the chaoses of themselves, according to the established laws of matter, were brought into those various and excellent forms adapted to every of God's ends—omitting the more excellent works of plants and animals, which it was proper and fit God should have a more immediate hand in. So the atoms of the chaos were created, in such places, of such magnitudes and figures, that the laws of nature brought them into this form, fit in every regard for them who were to the the inhabitants.' This was written under the influence of Descartes and Boyle (*Works*, 6:47, 56) and before the maturation of Edwards's idealism.

energy.338

The problem with Newton's hierarchy for Blake was that it robbed matter of its intrinsic energy and subjected it to extraneous forces and laws. This subordination was not simply a matter of physics for Blake. It carried over into the social and political arena, where it was associated with the pragmatic (Malthusian) policies of William Pitt, the Younger, and the conventionl Augustan norms of morality and art.339

Blake's image of Newton is best known from the large colour print, dated 1795.340 Immersed in a sea of space and time, Newton bends over a scroll spread out on the sea floor and uses a compass to plot the sparse geometry of nature. His body is rooted in rock and sand and his bright reddish hair stands out against the dark expanse beyond.341 The scroll on the ground is the extension of a cloth; the cloth, in turn, is the projection of Newton's own thought.342 While heroic in his own way, Newton thus symbolises the ascendancy of calculation over imagination and of external force over the intrinsic energy of life.343 As Blake put it in *Jerusalem*

---

338 On the influence of Paracelsus, Helmont, and Boehme, see Northrop Frye, *Fearful Symmetry: A Study of William Blake* (Princeton: Princeton Univ. Press, 1947, 1969), pp. 147, 150-53; David Erdman, *Blake, Prophet Against Empire* (Princeton: Princeton Univ. Press, 1954, 1969, 1977), pp. 11-12, 144, 180 n. 11, 334 n. 14; Kathleen Raine, *Blake and Tradition* (2 vols., Princeton: Princeton Univ. Press, 1968), *passim*; idem, *William Blake* (London: Thames and Hudson, 1970), pp. 51, 56, 62, 86, 127. For a demurral, see George Mills Harper, *The Neoplatonism of William Blake* (Chapel Hill: Univ. of North Carolina Press, 1961), pp. 55-56, 299 note 26; J. Bronowski, *William Blake and the Age of Revolution* (London: RKP, 1972), p. 32, but see also p. 202 note 7. Harper locates Blake in the context of eighteenth-century Neoplatonism, particularly that of Thomas Taylor (1758-1835). Bronowski places Blake in the nonconformist artisan tradition of Boyle, Petty, and Priestley; *William Blake*, pp. 13, 140-44.

339 On the parallel between Newton's passive matter, subject to universal gravitation, and Pitt's political realism, see Erdman, *Blake*, 367-69; Bronowski, *William Blake*, pp. 129-30, 137.

340 'Newton' (1795); Raine, *William Blake*, Illus. 61 on p. 84 with note on p. 211. There are two versions of the colour print: one at the Tate Gallery in London and one in the collection of Mrs. William T. Tonner; Martin Butlin, *William Blake (1757-1827): A Catalogue of the Works of William Blake in the Tate Gallery* (London: Tate Gallery, 1957), p. 41 note 13. According to Butlin (*ibid*, p. 39), the date of 1795 was an ideal date referring to the conception of the design, not necessarily the actual execution.

341 For the symbolism of the rock of England immersed in the midst of a space-time sea, see *Jerusalem* 94:1-14; Harper, *Neoplatonism*, pp. 163-69 and cf. note 348 below.

342 Raine, *Blake and Tradition*, 2:137.

343 Newton holds the compass in his left hand. On the parallel between abstract reason, inert matter, external force, and empty space, see Frye, *Fearful Symmetry*, pp. 34, 189-90, 255; Raine, *William Blake*, p. 50; Bronowski, *William Blake*, p. 136. Cf. Blake's *Jerusalem* 91:36-37: 'Los reads the Stars of Albion! the spectre reads the Voids Between the Stars...' (ed. Erdman, p. 251). On Los's system and the 'voids', in particular, as images of Newton's cosmology, see Donald D. Ault, *Visionary Physics: Blake's Response to Newton* (Chicago: Univ. of Chicago Press, 1974), pp. 28f., 36, 39f., 117.

THE HERITAGE OF ISAAC NEWTON   329

(15:11-20, written 1804-20):[344]

> For Bacon & Newton sheathd in dismal steel, their terrors hang
> Like iron scourges over Albion, Reasonings like vast Serpents
> Infold around my limbs, bruising my minute articulations
>
> I turn my eyes to the Schools & Universities of Europe
> And there behold the Loom of Locke whose Woof rages dire
> Washd by the Water-wheels of Newton: black the cloth
> In heavy wreathes folds over every Nation; cruel Works
> Of many Wheels I view, wheel without wheel, with cogs tyrannic
> Moving by compulsion each other; not as those in Eden: which
> Wheel within Wheel in freedom revolve in harmony & peace.

The contrast of the constructions of the wheels of Newton and of those of Eden indicates the difference between a gravitational system and one based on divine energy and prophetic inspiration (cf. Ezek. 1:15-21).[345]

Blake's abhorrence of Newton's mathematical system of forces operating on inert matter carried over into a rejection of the very idea of creation as Newton and others of his time understood it. Blake regarded the God who subjected all nature to measure, number, and weight (Wisd. 11:20) as a mere demiurge who acted out of ignorance and compulsion. The most common name for this god in Blake's mythology was Urizen (pronounced like 'your reason'), best known from the colour print, 'The Ancient of Days', which served as the frontispiece of Blake's *Europe* (1794).[346] Like Newton, Urizen stoops and reaches downward toward earth, a giant compass in his left hand,[347] to create geometric structure in an otherwise dark void. The idea of the compass originally came from the text of Proverbs 8:27: 'when he drew a circle on the face of the deep'. It had also been used by John Milton in his more positive portrayal of creation by the Son of God in *Paradise Lost* (VII.224-31). In Blake's mythology, however, the imposition of mathematical form on nature signified the enslavement of the human spirit,

---

[344] *Jerusalem* 15:11-20; *Poetry and Prose of William Blake*, ed. David V. Erdmam (Berkeley: Univ. of California Press, 1982), p. 159.

[345] For the industrial, social, and cosmic significance of the wheels without wheels, see Bronowski, *William Blake*, pp. 120-24; Erdman, *Blake*, pp. 144, 339; Harper, *Neoplatonism*, pp. 179-80.

[346] 'The Ancient of Days', frontispiece to *Europe* (1794); Frye, *Fearful Symmetry*, Illus. 1, opp. p. 50, with note on p. 433; Raine, *Blake and Tradition*, Illus. 144, 2:57. There is also a colour print (also dated 1794) in the Whitworth Art Gallery of the Univ. of Manchester; Raine, *William Blake*, Illus. 54 on p. 77 with note on p. 211.

[347] The connection between Newton and Urizen is further shown by a drawing in the collection of Mrs. T. Lowinsky which shows Urizen in the pose (reversed) of Newton; Butlin, *William Blake*, p. 41 note 13. For the significance of the left hand, see Bronowski, *William Blake*, p. 135.

'Petrifying all the Human Imagination into rock & sand' (*Four Zoas* 24:6).[348]

What made Blake's critique of Newtonianism so devastating was the fact that he attacked it and the creational theology behind it, not as an agnostic, but as a believer who was thoroughly embued with biblical imagery and sensitive to biblical standards of justice. The three creationist ideas of the comprehensibility, unity, and relative autonomy of nature were no longer creative ones for this poet and prophet. They now signified abstractness, uniformity, and compulsion, and all three seemed to deny the possibility of healing and restoration that the creationist tradition had traditionally affirmed. What Blake saw in his vision was the demise of the creationist tradition in Western culture.

But Blake also saw beyond the current distortion of creationist ideas to a reconciliation of reason and imagination, external force and internal energy. According to the Introduction to Blake's *Songs of Experience* (1794), the Word of God would finally take control of the celestial mechanics of Newton ('the starry pole') and rouse the earth from its sleepy subjection to external force ('the slumberous mass').[349] Indeed, as a youthful apprentice, Jesus had already taken up the compass as one of the tools of his trade and effected a reconciliation of reason and imagination.[350]

In the words of the well-known poem in the Preface to *Milton* (1804), Jerusalem could be rebuilt in the midst of the 'dark Satanic Mills' of Newtonian physics and the nascent industrial revolution it heralded.[351] Indeed, according to Blake's *Europe* (13:1-5), only the mighty spirit of Newton himself could sound the trumpet that would signal the end of the present world order and the beginning of a new age of imagination, peace, and love.[352]

In the restored world, the representatives of abstract reason (Bacon, Newton, and Locke) would join with Blake's heroes of the imagination (Milton, Shakespeare, and Chaucer) in an antiphonal chorus of modern

---

[348] *Four Zoas* 24:6 (ed. Erdman, p. 314). The imagery of 'rock and sand' represents Ulro, the lowest level of fallen existence; Frye, *Fearful Symmetry*, pp. 48-49; Ault, *Visionary Physics*, pp. 35-43.

[349] *Songs of Experience* 30:4-15 (1794); ed. Erdman, p. 18; cf. Raine, *Blake and Tradition*, 2:26-28. On the image of sleep, cf. *Jerusalem* 15:6; and Blake's letter to Butts, 22 Nov. 1802, 87-88 (ed. Erdman, pp. 159, 722).

[350] Butlin, *William Blake*, p. 49 note 30.

[351] *Milton* 1:7-16 (1804); ed. Erdman, pp. 95-96.

[352] *Europe* 13:1-5 (ed. Erdman, p. 65). Cf. *Jerusalem* 93:21-26, where the empiricism and natural religion of Bacon, Newton, and Locke signal the dawn of a new age (ed. Erdman, p. 254).

mathematics and Renaissance art.[353] The problem with the present age was not Newton's mechanics so much as the Druid Spectre of exploitation and warfare that Blake so closely associated with it.[354] In keeping with the fourth theme of the creationist tradition (chap. 1, sec. E), Blake saw that human science and technology could not fulfill their God-given purpose unless they were directed towards the healing and restoration of the human condition.

3 The Hutchinsonians

The Hutchinsonians were a small group of thinkers inspired by (though not always agreeing with) the work of John Hutchinson (1674-1737). The most influential of them were George Horne (1730-92), president of Magdalen College (Oxford) and later bishop of Norwich, and William Jones (1726-1800), curate of Nayland. Both Horne and Jones had High-Church affiliations and were opposed to the Latitudinarianism and Deism often associated with Newton's philosophy.[355] Hutchinson's works also provided an alternative natural philosophy—more acceptable than Newton's—to conservative Anglicans like John Wesley[356] and Samuel Johnson of Yale.[357]

One basic concern of the Hutchinsonians was that the physico-theologians' interpretation of Scripture in terms of the latest scientific theories could lead to rationalism and naturalism. It was better instead to base scientific theory directly on Scripture. A second concern was that Newton's inclusion of active principles like gravitation in physical science put matter in the place of God and led straight to atheism. The

---

[353] *Jerusalem* 97:9-13 (ed. Erdman, p. 257); cf. Frye, *Fearful Symmetry*, pp. 189, 298-300; Erdman, *Blake*, p. 486. For Blake, mathematics and art had been happily married in Gothic architecture; Frye, *Fearful Symmetry*, pp. 33-34, 104, 148-50. On the significance of the 'fourfold vision', see *ibid.*, 48-50 and the references to sleep in note 349 above.

[354] On the Druid symbolism, see particularly *Jerusalem* 93:25; 94:25; 98:6; 98:48-49 (ed. Erdman, pp. 254, 257, 258); Erdman, *Blake*, pp. 212-13, 330, 463-64, 484-86.

[355] C.B. Wilde, 'Hutchinsonianism, Natural Philosophy and Religious Controversy in Eighteenth Century Britain', *History of Science*, vol. xviii (1980), p. 2

[356] Schofield, *Mechanism*, pp. 332, 337, 339. Wesley's acceptance helped propagate Hutchinson's ideas in England; Thackray, *Atoms*, note on p. 247; C.A. Russell, *Science*, p. 66. Conversely, Horne was sympathetic to the Methodists (Jones of Nayland was not) and refused to prevent Wesley from preaching in his diocese; *DNB* 9:1250b, 10:1066a.

[357] Hornberger, 'Samuel Johnson', pp. 396f.; Herbert Leventhal, *In the Shadow of the Enlightenment* (New York: New York Univ. Press, 1976), pp. 179ff. Leventhal shows how Johnson drew on writers as diverse as Berkeley (esp. *Siris*) and Hutchinson. Their common opposition to Newton was correlated with their common use of the idea of aetherial fire, from which terrestrial air derives its power, and trinitarian patterns in cosmology; *ibid.*, pp. 181-87.

fundamental constituents of the cosmos had to be entirely passive.358

So the Hutchinsonians proposed a view of nature that was more mechanistic than Newton's, but one based on a 'literal' reading of Scripture that ensured God's transcendence over nature. In contrast to those Newtonians and materialists who disliked Newton's retention of the Cartesian notion of passive matter, they insisted that matter was entirely inert and rejected Newton's notion of active principles.

Hutchinson's anti-Newtonian ideas were published under revealing titles like *Moses's Principia* (1724-27) and *Glory or Gravity* (1733-34). Scientists and theologians had been labouring, he argued, under the misimpression that Moses was not interested in the details of natural philosophy. In the original (unpointed) Hebrew language, Hutchinson argues, the word for 'heavens' (*shmym*) could also be translated as 'names' (root *shm*, assuming masculine gender). So the familiar first verse of Genesis, 'In the beginning God created the heavens [*shmym*] and the earth', also meant that God created the 'names', or fundamental constituents, of the heavens.

Genesis chapter 1 actually described three such celestial elements: fire, light, and air (or spirit), and this triad of celestial elements or 'names' was a type of the heavenly Trinity: Father, Son, and Holy Spirit.359 Like the members of the Trinity, fire, light, and air were three manifestations of a single substance—a subtle fluid that circulated throughout the cosmos. But, in contrast to God—and unlike Newton's subtle ether—this subtle fluid was completely passive and had to be set in motion by God. Moses was indeed adept in the details of natural philosophy!

Once set in motion the elemental fluid continued in perpetual circulation and became the source of all other motion in the cosmos. It was emitted from the sun in the form of fire, travelled out through the solar system in the form of light, and condensed and returned to the sun in the form of air.360 In the process of circulation, the subtle fluid initiated and sustained all of the motions we observe in nature. The pressure of outward-bound light, for example, was responsible for the cohesion and hardness of bodies; the pressure of returning air was the cause of their apparent gravitation (really their 'glory' or 'heaviness', these two words having a common root in Hebrew) towards the sun. Thus nature was in perpetual motion and did

---

358 Albert J. Kuhn, 'Glory or Gravity: Hutchinson vs. Newton', *JHI*, vol. xxii (July 1961), pp. 306, 310; C.B. Wilde, 'Matter and Spirit as Natural Symbols in Eighteenth-Century British Natural Philosophy', *BJHS*, vol. xv (July 1982), p. 106.

359 Kuhn, 'Glory', pp. 307f.; Wilde, 'Hutchinsonianism', p. 4. The analogy between the Trinity and the three primordial elements of fire, light, and air was also used by William Law in his *Appeal to All that doubt or disbelieve the Truths of the Gospel* (1740); *Works of William Law* (London, 1893), 6:118; Arthur Wormhoudt, 'Newton's Natural Philosophy in the Behmenistic Works of William Law', *JHI*, vol. x (1949), p. 421.

360 Kuhn, 'Glory', pp. 311f.

not require periodic interventions by God, as Newton thought, in order to keep it from collapsing.[361]

It is worthwhile comparing Hutchinson's anti-Newtonian system to Jonathan Edwards's. Both wished to reaffirm the traditional doctrine of the Trinity in an age of theological doubt. Both were enabled by their more dynamic view of the Godhead to view the normal operations of nature as perfect manifestations of divine power and avoid any appeal to a 'God of the gaps'. The difference was that Edwards countered Newton's dualism by making the hardness of matter an immediate expression of divine power like gravitation and viewed all such interatomic forces as analogous to the unity of the Trinity. Hutchinson, on the other hand, made gravitation and cohesion equally the result of contact forces of a mechanical fluid that more distantly reflected the dynamic of the life of the Trinity. Edwards made matter continually dependent on God, while Hutchinson emphasised God's transcendence and granted the power of perpetual motion to matter. Edwards stressed providence at the expense of primary creation; Hutchinson limited the action of God to the first moment of creation and occasional subsequent interventions[362] and left no room for regular divine providence.

For the most part, Hutchinson's physics was hopelessly inadequate to deal with the increasingly technical issues of physical science. There is one way in which his ideas may have contributed indirectly to the development of science, not in physics itself, however, but in chemistry. Newtonians held that all the phenomena of chemistry could be reduced to the varying interparticulate forces between the atoms of an homogeneous substratum of matter subject to different concentrations. A corollary was the possible transmutation of any element into any other, a phenomenon that played an essential role in Newton's world system. For example, the transmutation of cometary vapours into terrestrial water was responsible for replenishing the earth's diminishing water supply.

Hutchinson, on the other hand, took the different kinds of atoms as immutably distinct. (It is not clear how these atomic elements are related to the three 'names' discussed above.) Apparently, the incessant circulation of the subtle fluid made the transmutation of one form of matter into another appear to be superfluous.[363] At any rate, Hutchinson encouraged chemical speculation along lines radically different from those of orthodox

---

[361] Wilde, 'Hutchinsonianism', pp. 4ff. Similar ideas were promulgated by the sometime Sandemanian, Samuel Pike, *Philosophia Sacra* (1753); Cantor, 'Theological Significance', pp. 141f.

[362] Hutchinson differentiated between God's 'power mechanical', which was communicated to matter at creation, and God's 'power essential' to intervene as in the plagues of Egypt; Wilde, 'Matter', p. 106. Note the similarity to the scholastic distinction between *potentia ordinata* and *potentia absoluta*.

[363] Wilde, 'Matter', p. 107.

Newtonianism.

Several of Hutchinson's followers popularised this aspect of his thought. For example, William Jones denied the homogeneity of matter and postulated the existence of solid, immutable atoms and went so far (1781) as to infer the shapes of the atoms from the properties of the substances they constituted (earth, water, etc.).[364]

Toward the end of the century, an associate of John Dalton named George Adams (1750-95) propounded similar notions in an effort to counter the atheistic tendencies of a prominent 'modern philosopher of France' (presumably Laplace) and the materialism of Joseph Priestley, both of which he and others associated with the disruptive forces of the French Revolution.[365] It is known that Adams was influenced by Hutchinson and Jones, though it is not clear to what extent Dalton himself was also so influenced.[366] The fact that the two shared a common concern about the new wave of materialism may be sufficient to account for the similarity of their ideas.[367] But the least that can be said is that the *preparatio* for Dalton's atomic theory of chemistry was provided through the anti-Newtonianism of John Hutchinson and his followers.[368]

The eighteenth-century controversy between Newtonians and anti-Newtonians brings out the variety of roles that theology could play in the history of science. In the propaganda of the Royal Society, theology provided social legitimation to the efforts of struggling proponents of a hitherto unrecognised profession. In the writings of the Scriblerians and Samuel Johnson, it ensured that all such projects would continue to be scrutinised in terms of longer range social and spiritual goals. In the case of the Newtonians, theological notions contributed to the development of alternatives to the mechanical philosophy and granted a degree of respectability to the new science. In the case of the antimaterialists (Berkeley and Edwards) and Hutchinsonians, theology ensured the consideration of alternative paradigms once Newtonianism had become established. If there has been a positive contribution of the creationist tradition to the development of Western science, it has stemmed from its balance and flexibility in charting direction rather than from any ability to determine the contents of science.

---

[364] *Physiological Disquisitions* (1781); Thackray, *Atoms*, p. 251; Wilde, 'Matter', p. 112f.
[365] *Lectures on Natural and Experimental Philosophy* (1794); Thackray, *Atoms*, pp. 250f.; Wilde, 'Matter', p. 114.
[366] Adams was Dalton's instrument maker; Wilde, 'Matter', p. 114.
[367] On Dalton's concern about Priestley's materialistic use of Newton's ideas, see Thackray, *Atoms*, pp. 52, 248ff.
[368] Thackray, *Atoms*, p. 280.

D. Post-Newtonian Cosmogonists and Neomechanists

The final grouping of the eighteenth century we shall consider is the one that established the scope of cosmological speculation and the rigour of mathematical analysis that has characterised physical science ever since. It was thus the first group of scientists to develop a natural philosophy whose comprehensiveness could be demonstrated rather than just assumed. The nineteenth-century image of Newtonian science as an all-embracing, deterministic system is due to our present subjects, not to Newton himself. Whereas Newton understood his principles to account only for the general operation of the world system once it had been set up by God, the cosmogonists and neomechanists demonstrated the ability of those principles to ensure the permanent stability of the solar system and applied them also to the problem of the origins of all systems, both planetary and galactic.

The development of analytical mechanics was largely the work of the French mathematicians, D'Alembert, Lagrange, and Laplace. There were several others, however—English, Scots, and Germans—who first developed the ideas about cosmogony—the so-called 'nebular hypothesis'—later taken up and developed more rigorously by the French. The shift from Newtonian thought to neomechanism is most clearly seen in their work.

1 Early Cosmogonists: Wright, Ferguson, and the Early Kant

Three major innovations of the mid-eighteenth century prepared the way for the 'nebular hypothesis' of Laplace:

(1) the notion that the sun and stars were not 'fixed', but in motion under mutual gravitation (Wright);

(2) the hypothesis of a primordial condition out of which the solar system was formed by the force of gravity (Ferguson);

(3) the hypothesis of a primordial chaos out of which the entire universe was formed by the force of gravity (Kant).

Kant's contribution was partly inspired by Wright's but independent of Ferguson's.

1.1 A Dynamic Theocentric Universe: Thomas Wright

The first person known to have suggested the dynamic nature of the universe within a Newtonian framework was Thomas Wright of Durham (1711-86). Impelled by a desire to synthesise his faith with contemporary astronomy, Wright imagined (1734, 1750) the universe as an infinite expanse, with the throne of God as the source of light and power at its centre and eternal darkness (for the wicked) in the outer regions.[369] Like Thomas Digges and Giordano Bruno (whom Wright quoted) in the sixteenth century (chap. 3, sec. B), Wright argued that the unbounded power of God required an infinite universe as its proper object. At the outer fringe of the visible universe was the Milky Way, the most distant part of our own galaxy, which Wright surmised to have the form of either a ring (like one of the rings of Saturn) or a spherical shell (with the solar system located in the body of the shell). Beyond the Milky Way were an infinite number of other galaxies, all in motion, extending into the realm of outer darkness. Between light and darkness, heaven and hell, all the stars were in motion under the power of mutual gravitation.[370] Wright had penetrated the thought barrier between a static (fixed star) cosmos and a dynamic one.

The details of Wright's theory were not well worked out, but the general idea was later picked up by Kant. Eventually it led to a consistently mechanistic account of the origins of the present structure of the cosmos and became a source of concern for believers: there no longer seemed to be a role for God as Newton and others had imagined. Nor did the origin of the cosmos have any apparent relationship to the aspirations and fates of humans.

But in its inception, Wright's departure from Newtonian thought was motivated by the desire to ensure a correlation between the moral and physical orders. The sun was displaced from its central role and all the stars put in motion in order to reserve the place of unique fixed centre for the visible manifestation of the Deity.[371] The regions of the universe closest to the throne of God were designed to be the eternal home of the righteous, with different types of planetary systems suited to different degrees of reward. Beyond the Milky Way were more distant galaxies with their own stars and planetary systems: these were the future abodes of the unjust, with different types of planetary systems suited to different forms of punishment.[372] Fantastic as Wright's scheme may sound to us today, it at least prevents us from supposing any general divorce between science and theology in the mid-eighteenth century. The same is true in the case of the

---

[369] 'Elements of Existence' (written 1734; not published); *An Original Theory and New Hypothesis of the Universe* (1750). The latter made no mention of the location of the wicked. In his 'Second of Singular Thoughts' (written in the early 1770s), Wright reversed the spatial ordering of heaven and hell with the former being promoted to higher, more spacious spheres; Michael A. Hoskin, 'Wright', *DSB* 14:519b.

[370] Stanley L. Jaki, *The Milky Way* (Newton Abbot: David & Charles, 1973), pp. 183-92.

[371] '...I would next suppose, in the general Center of the whole an intelligent Principle, from whence proceeds that mystick and paternal Power, productive of all Life, Light, and the Infinity of Things', *Original Theory*; Jaki, *Milky Way*, p. 192.

[372] Jaki, *Milky Way*, pp. 184f. Newton had also speculated privately on the future abode of the Blessed in the physical heavens; Yahuda Ms. 9.2, cited in Manuel, *Religion of Isaac Newton*, pp. 101f.

second figure we are to examine, James Ferguson.

1.2 A Mechanically Formed Solar System: James Ferguson

At about the same time as Kant,[373] a Scots astronomer and instrument maker, James Ferguson (1710-76), suggested that gravitation could account for the formation of the solar system as well as its current dynamics. Isaac Watts and Jonathan Edwards had argued in a general way that the formation of all things occurred in accordance with the laws God established at the moment of creation. But neither of these men was a scientist, and neither was concerned specifically with the potential and limits of gravitation as a formative principle.

Though not proficient in current mathematics, Ferguson had a good grasp of observational astronomy and began lecturing on the new science in 1748. His *Astronomy Explained upon Sir Isaac Newton's Principles* (1756) went through thirteen editions and was translated into Swedish and German.[374] Ferguson also specialised in the production of mechanical models of the solar system which surpassed in detail and accuracy any others of the time.[375]

Ferguson assumed that the atoms that later constituted the sun and planets were originally separate and distributed (inhomogeneously) through space. They were endowed by God with the power of gravitation—each attracting all the others in accordance with Newton's inverse-square law. Given time, therefore, the atoms naturally clustered into various massive bodies, with the most massive, the stars, surrounded by a number of smaller planets. The resulting coplanarity and corotation of the planetary bodies was striking evidence of the skill of the Creator in giving the atoms the proper initial distribution and framing the laws imposed on them. But, given that initial distribution, coplanarity and corotation followed naturally without any further intervention on God's part.[376]

Ferguson's cosmogony differed from the 'nebular hypotheses' later

---

[373] Kant probably wrote his *Allgemeine Naturgeschichte* in the years prior to 1750; Ernst Cassirer, *Kant's Life and Thought* (New Haven: Yale Univ. Press, 1981), pp. 35f. and note 43. This would make the preliminary studies nearly simultaneous as well as the final editions. Cf. Pierre Estève's *Origine de l'univers* (1748), which also gave a mechanical account of the origin of the solar system: Jaki, trans., *Immanuel Kant*, pp. 23-26.

[374] *DNB* 6:1209.

[375] Laurens Laudan, 'Ferguson', *DSB* 4:565f.

[376] Ebenezer Henderson, *Life of James Ferguson* (1867, 1870), p. 483; quoted in Hector MacPherson, *The Church and Science* (London: James Clarke, 1927), pp. 140f.
> In the beginning, God brought all the particles of matter into being in those parts of open space where the Sun and planets were to be formed, and endowed each particle with an attractive power, by which these neighbouring and at first detached particles would in time come together in their respective parts of space and would form the different bodies of the Solar System.

developed by Kant and Laplace in that it did not impose the constraint of randomness in the original distribution of the atoms. Ferguson thus preserved the role for God of giving some form to the atomic matter he created. On the other hand, Ferguson did go beyond Newton in supposing that the present order of the solar system could be entirely accounted for in terms of a prior, more basic, structure.

1.3 A Cyclical Universe: Immanuel Kant

Immanuel Kant (1724-1804) is best known for his tomes of philosophy, which moved in an entirely different world of faith and thought from the relatively simple piety of Thomas Wright and James Ferguson. Kant's mature work (beginning with the *Critique of Pure Reason*, 1781) marked the clear separation in Western thought between the moral and physical orders. Kant affirmed the existence of both but grounded them in entirely different modes of human experience: the moral in the universal inner sense of ought, and the physical in sense experience and the empirical judgements of theoretical reason.[377]

Here we are concerned only with early ('precritical') work of Kant, which contributed directly to natural philosophy and still worked within a vaguely theistic framework. The point we wish to make is that the separation of moral and physical realms that characterised Kant's later thought was already implicit in this earlier work.

The aim of Kant's early natural philosophy was to reconcile the contrary views of Newton and Leibniz.[378] From the one he took the law of gravitation, the explanatory power of which had been suggested to him by Thomas Wright.[379] From the other he adopted a sense of the sufficiency of the laws of matter to account for all the phenomena of nature. But Kant went beyond both Newton and Leibniz (or behind them both to Descartes) in supposing that these principles could be applied to the previously intractable problem of the formation of the present order of the universe. The result was his *Universal Natural History and Theory of the Heavens* (1755), subtitled 'An Essay on the Constitution and Mechanical Origin of the Whole

---

[377] Dillenberger, *Protestant Thought*, p. 184. Kant already differentiated the two in his 'Über die Deutlichkeit der Grundsatze der naturlichen Theologie und Moral' (1763), reflecting the influence of Rousseau; Cassirer, *Kant's Life*, pp. 86ff., 234.

[378] Jammer, *Concepts of Space*, p. 131.

[379] Kant learned of Wright's work through an anonymous German review published in a Hamburg newspaper in 1751. The reviewer referred to Wright's ring model as having the structure of a plane (as Wright also did), thus suggesting the idea of a flat disk to Kant; Jaki, *Milky Way*, pp. 196f.

Universe Treated According to Newtonian Principles'.380

The new view of the formation of the universe Kant proposed was startlingly austere (by contemporary standards). Like Ferguson, he claimed that the entire process could be understood in terms of strictly mechanical causes.381 (Gravitation was now included in the category of the mechanical, contrary to Newton.) As the original state of the universe, however, Kant hypothesised an infinite *random* distribution of matter (not a prearranged one like Ferguson's).382 Then, on the basis of Newton's theory, the region that happened to have the highest density in this random distribution would condense most rapidly under the power of gravitation and give rise to the first galaxy of stars, each with its own planetary system. The stars would orbit around their common galactic centre just as the planets orbited around their respective suns. Gradually further regions of space would also condense so that the formation of stars and planets would continue indefinitely, extending further and further from the first galaxy (the latter being the centre of galactic formation).383

Kant still had to suppose a first moment of creation *ex nihilo* (something he considered to be a paradox in later writings).384 But that moment was now completely separated from the process of formation and stripped even of any design of the original distribution of matter with a view to the subsequent location of the sun and planets (contrary to Ferguson). The formation of the universe as it appears at present was a process governed entirely by the laws of physics acting on a random distribution of matter.

Kant not only emptied the idea of creation of most of its content, he also pushed the moment of creation back further than had ever been previously

---

380 *Allgemeine Naturgeschichte und Theorie des Himmels oder Versuch von der Verfassung und dem mechanischen Ursprunge des ganzen Weltgebäudes nach Newtonischen Grundsatzen abgehandelt*; Jaki, *Milky Way*, p. 197, note 44. Published anonymously, the original work was never distributed because the publisher went broke; Numbers, *Creation*, p. 13.

381 Kant (twice) cited Descartes's idea as follows: 'Gebet mir Materie, ich will eine Welt daraus bauen'; Vartanian, *Diderot*, note on p. 94.; W. Hastie, trans., *Allgemeine Naturgeschichte* (Ann Arbor: Univ. of Michigan Press, 1969), pp. 28f. This formulation goes back to Voltaire and Maupertius: Jaki, trans., *Immanuel Kant*, pp. 249f., note 26. According to Kant, it would be easier to understand the origins of the universe in terms of mechanical causes than the origins of a vegetable or animal; 'Preface', Hastie, *Allgemeine Naturgeschichte*, p. 29; James W. Ellington, 'Kant', *DSB* 7:232b. Here Kant clearly followed the modified mechanical philosophy of Boyle, Ray, and Derham.

382 According to Jaki, however, the initial distribution of matter was actually 'carefully contrived'; *Immanuel Kant*, pp. 32, 248, note 14.

383 Hastie, trans., *Allgemeine Naturgeschichte*, pp. 141ff.; Toulmin and Goodfield, *Discovery*, pp. 161f. Kant seems to reflect Wright's notion of the centre as a divine source of light for the cosmos as a whole; Hastie, trans., *Allgemeine Naturgeschichte*, p. 155. But he later rejects Wright's localisation of deity: Jaki, trans., *Immanuel Kant*, pp. 166f.

384 The first antinomy in the *Critique of Pure Reason* (1781); Ellington, 'Kant', p. 231b.

imagined. 'A series of millions of years and centuries' might have elapsed, he suggested, to allow the emergence of the ordered universe we see around us today.[385]

Given enough time, stars and planetary systems would not only be formed; according to Kant, they would also eventually collapse. Like Newton, Kant supposed that the orbital motion of the planets would gradually slow down so that the planets would all eventually fall into their respective suns. As a result the suns would heat up, one after the other, and explode. The incinerated remains would then provide the material for the formation of new galaxies in accordance with the same mechanical laws.[386] These cycles of formation and dissolution would continue without end. Like Hutton's theory of the earth, Kant's view of the universe was ultimately a steady-state theory, not a truly evolutionary one.

There was no conflict for Kant, however, between this austere view of the origin and destiny of the galaxies and religious faith. The immensity of an infinite universe was, as for Digges, Bruno, and Wright, a fitting expression of the infinity of God.[387] And its unlimited productivity was an indication of divine omnipotence,[388] which for Kant, as for Leibniz (against Newton), was expressed through the means of material causation, not over and above it.[389]

The prospect of the eventual destruction of our own—and every—galaxy naturally occasioned 'profound astonishment' in Kant's mind. He took consolation in the thought that the destiny of the human soul was eternal life independent of the fate of heaven and earth.[390] The soul would survive the death of its planet as well as the death of its body. In the future life, it might even visit other worlds and find a home there—temporarily.[391] But the machinery of the universe would grind on, making and unmaking

---

[385] *Allgemeine Naturgeschichte*, trans. Hastie, pp. 144f.; Toulmin and Goodfield, *Discovery*, p. 162.

[386] *Allgemeine Naturgeschichte*, trans. Hastie, pp. 153f.; Toulmin and Goodfield, *Discovery*, p. 164; Ellington, 'Kant', p. 232a.

[387] '...an idea of the work of God which is in accordance with the Infinity of the great Architect of the Universe'; Hastie, trans., *Allgemeine Naturgeschichte*, pp. 64f., 139; Toulmin and Goodfield, *Discovery*, p. 161.

[388] '...this fruitfulness of Nature...is nothing else than the exercise of Divine omnipotence'; Hastie, trans., *Allgemeine Naturgeschichte*, p. 149; Toulmin and Goodfield, *Discovery*, p. 164.

[389] *Allgemeine Naturgeschichte*, trans. Hastie, pp. 23f.: '...that its [nature's] inherent essential striving brings such a result [of order out of chaos] necessarily with it, and that this is the most splendid evidence of its dependence on that pre-existing Being who contains in Himself not only the source of these beings themselves but their primary laws of action'.

[390] Kant cited lines from a poem by Albrecht von Haller (1708-77); Hastie, trans., *Allgemeine Naturgeschichte*, pp. 154f.; Toulmin and Goodfield, *Discovery*, p. 165.

[391] 'Possibly some orbs of the planetary system are being formed to prepare new dwelling places for us under other skies,...'; Cassirer, *Kant's Life*, p. 267; Jaki, trans., *Immanuel Kant*, p. 196. Hence Kant's delight in 'the starry heaven above'.

worlds, and would never constitute a 'new heavens and a new earth in which righteousness dwells' (2 Peter 3:13).

Here the contrast between Kant's vision and that of Thomas Wright is striking. For Wright the structure and destiny of the cosmos was the vehicle of human fulfillment; for Kant it was merely a foil.[392] So the separation of the physical and moral realms that characterised Kant's later thought was not without precedent: it was already implicit in his earlier work. The concept of the universe which Kant had developed in 1755 was already that of an amoral phenomenon indifferent to the aspirations and struggles of humans. The consonance between the productivity of the universe and the power of God which Kant so highly valued had no apparent relation to the parallel correlation between moral obligation and the Supreme Good. The moral dimension of human life could not be denied, but neither could it any longer be tied to the origin and destiny of the external world.[393] Consequently, it would have to be grounded in a distinctive experience and granted a realm of discourse of its own—practical, as opposed to pure, theoretical, reason.

2 French Neomechanists: D'Alembert, Lagrange, and Laplace

The notion of classical physics as an abstract deterministic mathematical system is due to three major figures of eighteenth-century French science: d'Alembert, Lagrange, and Laplace. They were rigorously mathematical in their work and consistently agnostic on questions of ultimate causation or meaning. The lable 'neomechanist' is appropriate in that they reinterpreted Newton's dynamics by eliminating the metaphysical and theological dimension and reaffirmed the earlier (Cartesian) notion of nature as a machine.[394] In effect, the notion of what constituted a machine was broadened to include principles which Newton's generation had understood to transcend mechanics. Such a generalisation was made plausible by the fact that new types of apparatus were being developed in the eighteenth century that exploited forces like electricity and magnetism previously thought to be supramechanical.

---

[392] Kant assumed to begin with that the refinement of spirit could only be hindered by the inertia and resistance of matter; Hastie, trans., *Allgemeine Naturgeschichte*, pp. 166f.; Jaki, trans., *Immanuel Kant*, p. 167.

[393] *Pace* Cassirer, *Kant's Life*, pp. 50f., 56, 266f. Kant's consideration of the beneficial harmony of the elements on earth (Hastie, trans., *Allgemeine Naturgeschichte*, pp. 20-23) is not carried over into his discussion of the cycles of creation and destruction in the cosmos as a whole. Cassirer seems to recongise this on p. 58 in his distinction between material and formal teleology.

[394] 'Nature is a vast machine whose inner springs are hidden from us...', D'Alembert, *Mélanges*; Henry Guerlac, 'Where the Statue Stood: Divergent Loyalties to Newton in the Eighteenth Century', in Earl R. Wasserman, ed., *Aspects of the Eighteenth Century* (Baltimore: Johns Hopkins Press, 1965), p. 330.

2.1 Jean D'Alembert

Jean le Rond d'Alembert (1717-83) made a clean break with those scientists who fostered a synthesis of science and theology and laid the basis for the later positivism of Condorcet and Comte.[395] He roundly criticised Maupertius, for example, for incorporating teleological and theological notions into his science,[396] and he treated the *vis viva* controversy between Newtonians and Leibnizians as a mere battle over words with no real bearing on the content of science.[397] Scientists should get on with their work and not become embroiled in metaphysical controversies that might never be resolved. On the basis of those laws, particularly those governing organic creatures, one could indeed infer the existence and intelligence of a Being who was their author.[398] But the existence and attributes of God were of no consequence for positive science, since God could not be counted on to intervene in worldly affairs in any predictable manner.[399]

The different types of knowledge must not be confused, d'Alembert argued. The truths of revealed religion (like the Trinity or creation *ex nihilo* or transubstantiation) transcended reason.[400] They had to be supernaturally impressed on the soul by God and received through the internal sense of faith—what Pascal had referred to as the 'heart'. The truths or laws of natural science, on the other hand, could be demonstrated objectively by comparing the deductions of human reason with the observation of nature (1759).[401]

D'Alembert's emphasis on the priority of reason, as distinct from sense experience, reflected the influence of the French rationalists, Descartes and Malebranche. Like them he held that the laws of motion were logically necessary and could be derived without recourse to experiment.[402] But d'Alembert went beyond Descartes and Malebranche in holding that the laws of the world system were just those that matter *left to itself* would follow

---

[395] Hahn, 'Laplace and the Mechanistic Universe', pp. 264, 268.

[396] Ronald Grimsley, *Jean d'Alembert* (Oxford: Clarendon Press, 1963), p. 281; Odom, 'Estrangement', p. 545.

[397] Hahn, 'Laplace and the Mechanistic Universe', p. 264.

[398] 'Cosmologie' (1754); Grimsley, *Jean d'Alembert*, p. 203; Jaki, *Road*, p. 78. However, God was not transcendent for d'Alembert: 'God is nothing but matter in so far as it is intelligent'; Grimsley, *Jean d'Alembert*, pp. 204ff.

[399] Preface to *Traité de dynamique* 2nd ed. (1758); Hahn, 'Laplace and the Mechanistic Universe', p. 266.

[400] From the standpoint of human reason, 'three is not one and bread is not God'; Grimsley, *Jean d'Alembert*, p. 197.

[401] *Élémens de philosophie* (1759); Randall, *Career*, 1:890f.

[402] An idea he shared with Johann Bernouilli and Leonhard Euler; Hankins, *Science*, p. 210.

and need not have been invented and imposed by an external agent.[403] As d'Alembert's followers Condorcet and Laplace were to put it, the present state of the world could have been predicted by a rational being who knew only its initial condition and its intrinsic laws of operation.[404] Like their materialist contemporaries Diderot and Holbach, the d'Alembertians thus held that more power for matter (or mechanism in this case) meant less power for God. The tendency to define the powers of God and matter as mutually exclusive alternatives, as we have seen, dated from at least the twelfth century and had been reinforced by the mechanical philosophers and the Newtonians.

Did d'Alembert then turn his back on the creationist tradition? The underlying motifs of the creation of matter out of nothing and laws of nature as the decrees of God subject to his ratification and amendment were no longer viable ideas for d'Alembert and the neomechanists. But one must also consider the meaning of their departure from the historic tradition in context.

D'Alembert's programmatic elimination of theology from the work of the scientist did not occur in a vacuum. The context was one in which the principal proponents of French theology were incessantly in conflict with social and scientific progressives. Of course, d'Alembert particularly despised the Jesuits, who worked to suppress the *Encyclopedia*, of which he was the scientific editor (1751-58). But, to d'Alembert, all varieties of priests and theologians—Jesuits, Jansenists, Doctors of the Sorbonne, and ministers of Calvinist Geneva alike—were characterised by bigotry and intolerance. They were more concerned with the maintenance of their own authority and privileges, he felt, than with the welfare of the people. For centuries they had made open war on natural philosophy.[405] All spiritual authority, it appeared, was corrupted when exercised in a secular context.[406]

---

[403] Actually d'Alembert allowed both views as alternative possibilities, but his own view was clearly the former; *Traité*, 2nd ed. (1758); Hahn, 'Laplace and the Mechanistic Universe', p. 266.

[404] *Le Marquis de Condorcet a d'Alembert sur le système du monde* (1768); Hahn, 'Laplace and the Mechanistic Universe', p. 268:
> ...one could conceive [the universe] at every instant to be the consequence of the initial arrangement of matter in a particular order and left to its own devices. In such a case, an Intelligence knowing the state of all phenomena at a given instant, the laws to which matter is subjected, and their consequences at the end of any given time would have a perfect knowledge of the 'System of the World'.

[405] D'Alembert cited the cases of Virgil of Salzburg (erroneously stating that he was condemned by Pope Zachary) and Galileo; *Discours préliminaire;* ET, *Preliminary Discourse to the Encyclopedia of Diderot* (1751), trans. Richard N. Schwab (Chicago: Univ. of Chicago Press, 1995), p. 73; Eliot and Stern, ed., *Age*, 2:131f.

[406] *De l'abus de la critique en matière de religion*; 'Genève'; Grimsley, *Jean d'Alembert*, pp. 194ff. D'Alembert had more sympathy with persecuted groups like the French Huguenots. He also had friends among the Catholic clergy who were philosophical in spirit; Grimsley, *Jean d'Alembert*, pp. 196, 200.

If the church would not reform itself (as it did in the eleventh and sixteenth centuries), then it was up to less religious people to take the initiative: rid secular affairs (e.g., the administration of the universities) of ecclesiastical control and rid scientific research of theological presuppositions.

In retrospect, it appears that a son of the church imbued with the Christian ideal of public service could only be faithful to that aspect of his calling by attempting to eliminate the influence of the church and its theology in matters of public concern. One might say that the creationist tradition was rejected by d'Alembert. D'Alembert, on the other hand, might have argued that it was merely rescued (in a secularised form) from the power of its previous guardians.

If the principle enunciated by Jesus has any meaning for the history of science, we must take the latter possibility seriously: '...the kingdom of God will be taken away from you and given to a nation producing the fruits of it' (Matt. 21:43). If that principle applied to the priests and rabbis of Jesus' time,[407] it could have applied to those of the eighteenth century as well. It should also apply to the present guardians of the skills and tasks God has bequeathed to humans, the 'high priests' of current science and technology.

2.2 Lagrange and Laplace on the Stability of the Solar System

The science of analytical mechanics was brought to a new level of formal completion through the efforts of two of d'Alembert's protégés:[408] Joseph Louis de Lagrange (1763-1813) and Pierre Simon de Laplace (1749-1827). The two competed with each other, assisted each other, and incited each other to some of the most spectacular accomplishments of eighteenth-century physics. Among their many contributions, we shall discuss only two that occasioned attention from a theological viewpoint: Lagrange and Laplace's work on the stability of the solar system; and Laplace's cosmogonic speculations—the so-called 'nebular hypothesis'. We shall conclude with some reflections on Laplace's secular theology.

Newton, it will be recalled, had been concerned about the fact that the orbits of the planets about the sun would be perturbed when planets came close to passing comets and to each other. He concluded that the solar system was not stable, and that periodic interventions by God were required to prevent a total collapse. Leibniz ridiculed Newton's notion that God would create such an imperfect planetary system but could not show that he was wrong. Kant had followed Newton in supposing that the solar system

---

[407] According to d'Alembert, Jesus was the enemy of priests; Grimsley, *Jean d'Alembert*, p. 197.

[408] Hankins, *Science*, p. 17; Hahn, 'Laplace and the Mechanistic Universe', p. 259.

would eventually collapse but assumed with Leibniz that the system was still perfect and that God would do nothing to prevent such a collapse. The mathematical difficulty of the problems was immense, given the methods of the time, but Lagrange and Laplace approached the issues with complete confidence that they could be solved.[409] Their faith in the comprehensibility of the world can be compared to that of earlier believers like Kepler and Newton.

There were two possible types of perturbation: secular and periodic. Secular perturbations were cumulative and would lead to long-range disruption. But periodic perturbations were cyclical: a variation in one direction would subsequently be compensated for by an equal variation in the other. Lagrange and Laplace set out to demonstrate that all the perturbations among the planets were of the periodic type—even the apparently secular ones were really periodic with very long periods—and hence that the solar system was secularly stable.[410] In other words, periodic interventions by God were made superfluous by the periodic nature of the perturbations themselves. Lagrange and Laplace were clearly motivated by a belief in the perfection of the system (like Leibniz) and a desire to eliminate Newton's recourse to the supernatural.

One of the principle items of concern was the interaction of the two largest planets, Jupiter and Saturn. The orbital motion of Jupiter appeared to be gradually accelerating, while that of Saturn was slowing down. Between them, Lagrange and Laplace were able to show (1774-76) that the variations of the inclinations and eccentricities of the orbits of Jupiter and Saturn were bound by fixed limits: in other words, the values of the inclinations and eccentricities of the two planets had to oscillate about mean values within those limits. Thus the single worst threat to the stability of the solar system had been diffused by the late 1770s.[411]

In 1785-88, Laplace was able to calculate the periodicity of variations in the orbits of Jupiter and Saturn: it came out to 929 years.[412] He also showed that the total eccentricity, as well as the total inclination, of all the planetary orbits was a constant: an increase in the eccentricity or inclination of one had to be compensated for by an equivalent decrease for the others. Since the eccentricities and inclinations were all low to begin with, no planet could change its eccentricity or inclination anywhere near enough to change the overall dynamics of the system.[413] And again the alterations

---

[409] Hahn, 'Laplace and the Mechanistic Universe', p. 259.
[410] Gillispie, 'Laplace', p. 323.
[411] Bynum et al., eds., *Dictionary*, p. 57b.
[412] His first calculation gave a figure of 817 years; Hahn, 'Laplace and the Mechanistic Universe', p. 259; Gillispie, 'Laplace', p. 324b.
[413] Gillispie, 'Laplace', p. 327.

were periodic.

The foundational results of Lagrange and Laplace were synthesised at the turn of the century in Laplace's *Treatise on Celestial Mechanics* (1799-1825). Some finishing touches were that the masses of the comets were far too small to affect the stability of the planets and that the accelerating effect of solar radiation pressure was counterbalanced by the decreased gravitation of the sun due to mass loss (1805).[414] The important result was that the solar system was secularly stable as a whole, regardless of the masses of its constituents, provided that all its members orbit the sun in the same direction (which they do for the most part).[415]

The picture was somewhat complicated by the behaviour of an eighth planet, Uranus, which had been discovered by William Herschel in 1781.[416] Not only was the motion of Uranus itself found (in the 1840s) to be erratic, but two of its moons had orbits so highly inclined to the ecliptic that Herschel classified their motion as retrograde (1799).[417] Two of the asteroids discovered in the early nineteenth century also had high inclinations and so did not completely fit the assumptions of Lagrange and Laplace.[418]

No one, however, any longer resorted to Newton's God-of-the-gaps to account for the unresolved problems. For the first time in history, the level of confidence in the scientific method was high enough that the existence of gaps in the world picture could actually be used to predict new discoveries. In other words, the gaps pointed not to God, but to previously unknown constituents of the universe itself.

In the case at hand, the erratic behaviour of Uranus led to the search for new planets. Neptune was discovered in 1846, and Pluto in 1930, and the belief in the long-range stability of the solar system was largely vindicated. There are still residual perturbations in the orbits of Uranus and Neptune that have not been accounted for to this day. But the exclusion of God from scientific discourse is a *fait accompli*. It was a philosophical commitment on the part of the neomechanists that excluded God, not the result of scientific data themselves, any more than were the earlier arguments in favour of the existence of God.

---

[414] *Traité*, vol. IV (1805), Books IX and X; Gillispie, 'Laplace', pp. 354ff.

[415] Bynum et al., eds., *Dictionary*, p. 57b.

[416] William Herschel (the elder, 1738-1822) is best known for his discovery of the planet Uranus (the first new planet to be discovered since the ancients, 1781) and his demonstration (1783) that the sun was moving through space like other stars; MacPherson, *Church*, pp. 118f.

[417] Herschel gave the inclinations as 89 and 91 degrees. In 1799 he classified them both as retrograde. Laplace cited this discovery in his work but conveniently ignored the fact that the orbits of the two moons were reported to be sharply inclined to the ecliptic; Stanley L. Jaki, 'The Five Forms of Laplace's Cosmogony', *American Journal of Physics*, vol. xliv (Jan. 1976), pp. 5a and note 12, 6b.

[418] Again Laplace cited the asteroids but ignored their inclinations; Jaki, 'Five Forms', p. 7a.

2.3 The Nebular Hypothesis of Laplace

In 1796, Laplace published a semipopular *Exposition of the World System* as an introduction to his more technical work on celestial mechanics. At the very end of this work, he allowed himself to depart from his usual rigour to speculate on the problem of the origin of the solar system and the nature of the universe as a whole.[419] There is no evidence that Laplace was basing his thought on the work of Ferguson or Kant at this point. It is more likely that he was motivated by his overall ambition to reduce the areas Newton had left to divine providence.[420] As he put the question in a later (1813) edition of the *Exposition*: 'Could not this arrangement of the planets be itself a result of the laws of motion? And could not the supreme intelligence that Newton made to intervene have made it depend on a more general phenomenon [instead of intervening]?'[421]

The hypothesis Laplace developed in successive editions of the *Exposition* (1796-1824) was this: the planets were formed from the primordial solar atmosphere—a large, nearly static cloud (nebula) of gas surrounding the sun. Gravitational attraction caused the material to fall in towards the sun and hence to rotate in order to conserve angular momentum (presumably it had an initial, very small net rotation to begin with).[422] As the cloud condensed, the outer regions of material in the equatorial plane would begin to rotate fast enough to counteract the gravitational pull of the sun and would thus form stable rings around the sun. The condensing gas cloud would form a succession of rings at varying distances from the sun in this way. As these rings cooled they would break up and form smaller, planetary nebulae and eventually coalesce into the present planets with their respective moons. As a result, the planets and their moons would all rotate in the same direction and in the same plane (low inclinations) with nearly circular orbits (low eccentricities) as generally observed. The comets had highly eccentric orbits only because they were captured from material outside the solar system. The peculiar structure of the solar system was the result neither of chance, not of divine formation (as Newton held), but

---

[419] Laplace readily admitted that his speculations had absolutely no basis in observations or mathematical calculations; *Exposition du système du monde* (1796) V.6; MacPherson, *Church*, p. 143; C.G. Gillispie, 'Laplace', *DSB* 15:344a.

[420] According to Francois Russo, Laplace only came to exclude divine action in 1812; 'Theologie naturelle et secularisation de la science au XVIIIème siècle', *Recherches de science religieuse*, vol. lxvi (1978), pp. 42, 55f. However, Laplace had abandoned the idea of an interventionist God prior to his findings, not as the result of them; Hahn, 'Laplace and the Mechanistic Universe', p. 259.

[421] *Exposition*, 4th ed. (1813); Russo, 'Theologie', p. 38; cf. Jaki, 'Five Forms', pp. 8f.: 'Cet arrangement des planètes, ne peut-il pas être lui-même un effet des lois du mouvement? Et la suprême intelligence que Newton fait intervenir, ne peut-elle pas l'avoir fait dependre d'un phénomène plus générale?'

[422] Jaki, 'Five Forms', pp. 7f.

resulted from the behaviour of matter under the force of universal gravitation.[423]

Laplace had developed the nebular hypothesis out of strictly theoretical considerations—only his appeal to the universal force of gravitation was based on the empirically based research of Newton. When he became aware of William Herschel's work on stellar nebulae (early 1790s), however, Laplace believed it confirmed his ideas. Herschel, an observational astronomer, had determined that certain nebulae in space were diffuse stellar atmospheres rather than groups of unresolved stars. He concluded that the stars were in various stages of formation as the result of the gravitational collapse of the nebulae.[424] Here, then, was the same process that Laplace had postulated in the case of sun being witnessed for other stars.

Both the theoretical and observational aspects of stellar evolution were still rudimentary in the late eighteenth century, at least by today's standards. But even at this early stage, the power of cosmological speculation was appreciated. Hypotheses were not just inferred from data: they could also be conjectured on a purely speculative basis (as earlier by Descartes and Kant) and then verified (or falsified) by observations. And a new appreciation was developing for the fact that unaided human reason, while far from infallible, was also able in a remarkable way to anticipate the results of observation. Laplace referred to the agreement between his hypothesis and Herschel's results as 'a marked coincidence' and apparently did not reflect on the matter further.[425] He was, after all, still conditioned by the creationist belief in the comprehensibility of the world. The problem of this kind of coincidence would become more acute as the range of human science exceeded normal human experience while the creation faith that sustained belief in the comprehensibility of the world became increasingly marginal.

The best known aspects of Laplace's career are some of his encounters with Napoleon Bonaparte. In October of 1799, three weeks before the coup d'état that made him First Consul, Laplace formally presented to Napoleon copies of the first two volumes of the *Treatise on Celestial Mechanics*, who promised to read them if he had the time (he estimated six months to be sufficient!).[426] Appearing again before Bonaparte in 1802, Laplace is

---

[423] Ronald L. Numbers, *Creation*, pp. 8ff.; Gillispie, 'Laplace', pp. 344f.

[424] In 1791 Herschel presented a paper, 'On Nebulous Stars', before the Royal Society. In it he gave evidence that nebulae were stars with extended atmospheres rather than clusters of distinct stars. Since the nebular material was itself luminous, he reasoned, it made more sense to assume that the central star condensed out of the nebula than that the latter was emitted from the star; *Collected Scientific Papers*, 2:423; MacPherson, *Church*, pp. 144ff.

[425] *Exposition*, 5th ed., 1824; Henry H. Harte, trans., *The System of the World* (Dublin, 1830), 2:336f.; Numbers, *Creation*, p. 11. Laplace first cited Herschel's nebulae in 1812; the 4th ed. of the *Exposition* (1813) was the first to mention them; Jaki, 'Five Forms', p. 7b.

[426] Gillispie, 'Laplace', p. 346a.

reported to have defended his case—stated earlier in the *Exposition of the World System*—that natural causes could account for the formation and stability of the solar system.[427]

The best known exchange—and unfortunately the least documented—was one in which Napoleon allegedly told Laplace that he had looked through his *Exposition* to see what role God played in the formation of the solar system. According to a later source, he then remarked: 'Newton spoke of God in his book [second and third editions of the *Principia*]. I have purused yours but failed to find his name even once. How come?' And Laplace's oft-quoted reply: 'Sire, I did not need that hypothesis.'[428]

The earliest known record of this exchange dates from 1864,[429] thirty-seven years after Laplace's death and at least six decades after the purported encounter. However, the incident corresponds so well with what we do know of the exchanges between Laplace and Bonaparte that no distortion of the facts is incurred by its frequent citation. It captures the haughtiness and determination of the of the great mathematician in a single phrase: Newton's God had been retired as far as physical science was concerned.

2.4 The Secularised Theology of Laplace

As we have seen, Laplace's accomplishments were far from theologically neutral. His efforts to demonstrate the natural origin and stability of the solar system and thus the completeness of the laws of nature had a clear theological agenda. Like Leibniz, Laplace criticised Newton's appeal to supernatural intervention where the laws of physics seemed to fail as the expression of ignorance. Unlike Leibniz, however, Laplace assumed (as did other *philosophes*) that the elimination of the supernatural in scientific explanation meant that there was no longer any role for God in the natural world.

Like d'Alembert, Laplace referred to the 'Author' of nature, but worked

---

[427] See J.L.E. Dreyer, ed., *The Scientific Papers of Sir William Herschel* (2 vols., London, 1912), 1:lxii; Constance A. Lubbock, ed., *The Herschel Chronicle* (Cambridge: Cambridge Univ. Press, 1933), p. 310; Roger Hahn, 'Laplace and the Vanishing Role of God in the Physical Universe', in Harry Woolf, ed., *The Analytic Spirit* (Ithaca, N.Y.: Cornell Univ. Press, 1981), pp. 85f.
  According William Herschel's diary for 8 August 1802, Napoleon...
  held a considerable argument with him in which he differed from that eminent mathematician. The difference was occasioned by an exclamation of the First Consul's, who asked in a tone of exclamation or admiration (when we were speaking of the extent of the sidereal heavens) 'and who is the author of all this?' M. de La Place wished to shew that a chain of natural causes would account for the construction and preservation of the wonderful system; this the First Consul rather opposed.

[428] 'Sire, je n'ai pas eu besoin de cette hypothese.' Cf. Odom, 'Estrangement', p. 535.

[429] Augustus de Morgan, 'A Budget of Paradoxes', *The Athenaeum* 20 Aug. 1864; Hahn, 'Laplace and the Vanishing Role', note on p. 85.

to eliminate the remaining grounds for any reference to God in discussions of nature. Whereas Newton and others were motivated to heroic lengths by their desire to demonstrate the activity of God in nature, Laplace was motivated by the desire to eliminate whatever grounds remained for such reference to the divine. In effect, his science was his religion.[430]

The secular theology of Laplace is most clearly seen in his statements of the ideal of physical determinism. As he first put it in 1773:[431]

> The present state of the system of nature is evidently a result of what it was in the preceding instant, and if we conceive of an Intelligence who, for a given moment, embraces all the relations of being in this Universe, it will also be able to determine for any instant of the past or future their respective positions, motions, and generally all their affections....

On the more practical level of celestial mechanics, for instance:

> ...in order to determine the state of the system of these large bodies in past or future centuries, it is enough for the mathemetician that observation provide him with their positions and speeds at any given instant.

Earlier statements of this ideal had been made by d'Alembert (1758) and Condorcet (1768),[432] but it was Laplace that gave it its first precise formulation and whose words have been quoted ever since as typifying the determinism of classical physics. Given the positions and velocities (and masses) of all of the bodies in the universe, both past and future were entirely determined (in principle) by Newton's laws.

The 'Intelligence' Laplace hypothesised (following Condorcet) stood for the set of laws governing the universe. It was a device for affirming the existence of such a set of laws as an ideal towards which human science aspired. The fact that the universe had such laws and the fact that human

---

[430] Jean Pelseneer, 'La religion de Laplace', *Isis*, vol. xxxvi (1946), pp. 159f.

[431] In 'Recherches 1. sur l'integration des equations differentielles...', a paper read on 10 Feb. 1773 and published in 1776 in *Memoires de mathématique et de physique...année 1773*; *Oeuvres complètes de Laplace* (14 vols., Paris, 1878-1912), 8:144f.; Hahn, 'Laplace and the Mechanistic Universe', pp. 268f.; Gillispie, 'Laplace', p. 285. Compare the more frequently quoted statement from the introductory essay to Laplace's 1812 *Théorie analytique des probabilités; A Philosophical Essay on Probabilities*, trans. Frederick Wilson Truscott and Frederick Lincion Emory (New York: Dover, 1951), p. 4; cf. the Andrew I. Dole, trans. (New York: Springer-Verlag, 1995), p. 2:
> Given for one instant an intelligence which could comprehend all the forces by which nature is animated and the respective situation of the beings who compose it—an intelligence sufficiently vast to submit these data to analysis...for it, nothing would be uncertain and the future, as the past, would be present to its eyes.

Like Kepler's faith in the comprehensibility of the world, Laplace's belief in determinism was the motivation for his research, not its result.

[432] D'Alembert, Preface to the 2nd ed. (1758) of the *Traité de dynamique*; Condorcet, *Sur le système du monde* (1768); Hahn, 'Laplace and the Mechanistic Universe', pp. 266ff.

minds were able to grasp them, however imperfectly, were taken on faith. The comprehensibility, unity, and autonomy of the universe could thus be affirmed without reference to God.

The secular ideals of d'Alembert and Laplace are important because they have become the motivating ideals of many physical scientists of the nineteenth and twentieth centuries. We are suggesting that they are the product of a particular history and theology, not a necessary concomitant of progressive science. Those who single-mindedly pursue scientific research will generally be motivated by ideals of some sort. The ideals in question must be ones that appear to vindicate faith in the comprehensibility, unity, and relative autonomy of the world. As long as Christian theology was clearly associated with these beliefs, it was conducive to good scientific work and constructive criticism of science, as we have seen. But, in the seventeenth and eighteenth centuries, a long-range tendency to view the active role of God and the innate properties of matter as alternative modes of explanation gained credence to the extent that the ideals needed for the furtherance of science could not so readily be sustained by positive Christian commitment.

From an historical viewpoint, this development resulted from the tendency of theological parties to define themselves by excluding others at the expense of the integrity of the creationist tradition as a whole. This is perhaps most clearly seen in the naturalist-conservative split of the twelfth century (chap. 1, sec. D) and the latitudinarian-spiritualist controversy in seventeenth-century England (chap. 3, sec. C). There is no historical reason to believe that the development was inevitable or that it could not have followed a different course. Therefore, there is no reason to suppose that the future development of science might not find its motivation once again in the creationist tradition, particularly if its present secular orientation is found to fail.

CHAPTER FIVE

THE CREATIONIST TRADITION AND THE EMERGENCE OF
POST-NEWTONIAN MECHANICS
(Nineteenth and Early Twentieth Centuries)

A. THE NINETEENTH-CENTURY CONTEXT

An operational faith in God as creator was a vital factor in the development of all branches of science until the late eighteenth century. It constituted a tradition—the 'creationist tradition'—which provided the matrix of faith for the professional endeavours of Western European natural philosophers Catholic, Protestant, and Nonconformist from Bede and Adelard of Bath to Boscovich, Hutton, and Dalton.

The situation in the nineteenth century was radically different for both theology and science. On the theological side, the problem was that few parish ministers or theologians were adequately informed or even concerned about current developments in the physical sciences.[1] The dramatic developments of nineteenth-century geology and evolutionary theory attracted attention, but the equally revolutionary changes in basic physics[2] and chemistry passed largely unnoticed or misunderstood. As most people perceived them, developments in physical science were far removed from the practical exigencies of parish life and religious experience. This, paradoxically enough, at a time when basic discoveries were being

---

[1] George Peacock reorganised the Cambridge Observatory and introduced symbolic algebra at the university; after 1839 he served as Dean of Ely Cathedral but continued to be active in university reform and on Royal Society committees; Walter (Susan) F. Cannon, 'Scientists and Broad Churchmen', *Journal of British Studies*, vol. iv (Nov. 1964), pp. 69f. Thomas Chalmers (1780-1847), founder of the Free Church of Scotland, was one of the last theologians to discuss the physical sciences, but he was not conversant with major developments in physics and astronomy after Laplace and Cuvier; Cannon, 'The Problem of Miracles in the 1830s', *Victorian Studies*, vol. iv (Sept. 1960), p. 16; Numbers, *Creation*, p. 21. Chalmers's pupil, James McCosh, later president of Princeton College, was mostly concerned with the theory of evolution; Gary Scott Smith, *The Seeds of Secularization* (Grand Rapids, Mich.: Eerdmans, 1985), pp. 97f., 101f. German critic of scientific materialism, Otto Zöckler, was likewise concerned with the issue of evolution; Frederick Gregory, *Nature Lost: Natural Science and the German Theological Traditions of the Nineteenth Century* (Cambridge Mass.: Harvard Univ. Press, 1992), pp. 112-59.

[2] In the nineteenth century, the term 'physics' was redefined to include electricity, magnetism, optics, thermodynamics, and mechanics; David Knight, *The Age of Science* (Oxford: Basil Blackwell, 1986), pp. 160f., 166. Medicine was no longer described as 'physic'.

made—for example, the laws of electricity and magnetism—that would transform the technological structures of everyday life more than any of the scientific advances of previous centuries.

There were at least three reasons for this paradox of apparent irrelevance in the face of revolutionary change. One was the fact that nineteenth-century physics had become highly mathematical and often dealt with entities that were not directly observable: this made it less accessible to the average lay person. This movement away from common-sense reality and toward greater abstraction became progressively more pronounced with the development of field theory and statistical mechanics in the late nineteenth century, and again with the development of relativity theory and quantum menchanics in the early twentieth century.

A second reason was that, as the industrial revolution progressed, the public gradually developed a truncated image of physical science. In the popular mind, science was generally equated with a fictitious method of objectivity and induction, on the one hand, and with impressive new technologies and the latest consumer goods, on the other.[3] There was very little public awareness of the basic science that lay behind the new technologies, much less the history of their development or the complex creative processes and theological motifs that were involved in that history. Personal ideals and objective methods and artifacts were generally relegated to two different spheres of existence.[4]

This leads us to a third factor. The more creative theologians of the nineteenth century had to defend the place of religion in a world for which this truncated view of science had become a major norm for evaluating truth claims. As a consequence, they tended to redefine the sphere of theology and religion in terms of the personal, experiential, and moral side of life (e.g., Friedrich Schleiermacher, Matthew Arnold, Albrecht Ritschl). The validity of rational, natural theology was increasingly questioned, and the idea of divine creation was reduced to a useful fiction (Kant) or a statement of the human experience of absolute dependence (Schleiermacher).

On the scientists' side of the issue, there was a corresponding tendency toward a segregation of personal religious convictions from the professional discipline of research and publication. In France, a positivist style of science had come into vogue already in the late eighteenth century with the work of d'Alembert and Laplace (chap. 4, sec. D). In Germany, England, and America, a similar progression occurred in the second half of the nineteenth century—it was seen as an effective way of guarding the autonomy of the

---

[3] James Burke, *The Day the Universe Changed* (Boston: Little, Brown and Co., 1985), pp. 279, 284f., 289, 291.

[4] Significantly, Pacey ends his history of the role of ideas and ideals in the development of technology (*Maze of Ingenuity*) in the mid-nineteenth century.

various scientific communities in the face of political and ecclesiastical interests.[5] Toward the end of the century, some apologists for science even argued that organised religion had always been hostile to the development of their discipline.[6] Whatever the merits of this particular view of history (the evidence has been reviewed and found wanting in earlier chapters), it was itself an indication of an increasingly intentional separation of the sciences from religious faith, rather than an ongoing conflict in the nineteenth century itself.

This broad historical shift in European thought calls for a corresponding shift in our approach to the historical creationist tradition. We are not able to review here the development of the entire breadth of the physical sciences as we have attempted to do in earlier chapters. Nor are we concerned here with the ideas of 'science' in the inductivist tradition of John Stuart Mill or in the idealist tradition of post-Kantian philosophy and theology—neither idea being closely related to the actual practise of science in the nineteenth century, as we shall argue.

Instead, we shall focus on one continuing tradition of scientific research, the Anglo-Scottish tradition, which persisted in appealing to the idea of creation through most of the nineteenth century—long enough to bring us within view of the revolutionary developments of twentieth-century physics (even though those developments were primarily due to German and Danish scientists, rather than English or Scottish ones).

The main disadvantage of this procedure is that it bypasses many scientific and philosophical developments in continental Europe that influenced the British scientists we shall discuss. We will only be able to mention some of the relevant French and German contributors to nineteenth-century physical science in passing. But, since our theme is the creationist tradition, we must restrict our present coverage to a group of physicists and chemists for whom the idea of creation was still a consciously held idea and an instrumental factor in the development of science.

In the Middle Ages, the Renaissance, and the seventeenth century, the creationist tradition had been much broader than the scope of natural philosophy itself, and we had to cover a variety of peripheral movements of thought in order to give a coherent historical account. Even in the eighteenth century, we considered divergent movements in relation to the creationist tradition, since they either reinterpreted it (materialists) or reacted

---

[5] Frank M. Turner, 'The Victorian Conflict between Science and Religion: A Professional Dimension', *Isis*, vol. lxix (Sept. 1978), pp. 356-76; Stephen Brush, *The Temperature of History* (New York: Burt Franklin, 1978), pp. 26f.

[6] Owen Chadwick, *The Secularization of the European Mind in the Nineteenth Century* (Cambridge: Cambridge Univ. Press, 1975), chap. 7.

against it (some neomechanists). But, in the nineteenth century, for the first time in Western history, the creationist tradition became narrower than the physical sciences and survived only in a few local traditions. The history of science and the history of theology became two separate tracks, with only a modest degree of overlap.

The idea of creation was no longer viable as a vigorous international tradition, nor even as a public reality in Great Britain itself. However, one remnant survived long enough to guide the chief architects of post-Newtonian mechanics into areas far removed from the experiences of everyday physical reality, areas for which guidelines were not available from the existing stock of scientific knowledge.

And here we shall stop. Tracing the more subtle ways in which creationist ideas have persisted in twentieth-century physics itself would require another book and an entirely different methodology—one oriented more to individual perspectives than to an ongoing tradition of shared ideas.

Why the persistence of an Anglo-Scottish stream of the creationist tradition through the nineteenth century? The roots of this phenomenon go back to the eighteenth century. The heritage of Isaac Newton and his English and Scottish followers was still a potent force in Britain. Conservative nationalistic sentiments were partly responsible, as evidenced by the English reaction to the radical philosophy and politics of revolutionary France and the efforts of English scientists to defend the priority of their countrymen's contributions to scientific discoveries ranging from the calculus in the eighteenth century to the conservation of energy in the nineteenth.[7]

We recall also that the stance of the established church toward science was perceived to be more moderate in the England of the Latitudinarians than it was in the France of the Jesuits and *philosophes*. But there had also been a major revival of evangelical Christianity in the British Isles that provided its own stimulus in the relations of science and theology, particularly in the cases of Michael Faraday and James Clerk Maxwell, as we shall see below.

The major issue in nineteenth-century physics and chemistry was still the meaning and validity of the mechanical philosophy inherited from the seventeenth and eighteenth centuries. Unfortunately for the inquiring student, present-day historians are seriously divided on the handling of this issue: some see the beginnings of a reaction against the mechanical model with Faraday and Maxwell around the mid-nineteenth century;[8] others treat

---

[7] Peter M. Harman, *Energy, Force, and Matter* (Cambridge: Cambridge Univ. Press, 1982), pp. 63, 66.

[8] Barbara Giusti Doran, 'Origins and Consolidation of Field Theory in Nineteenth-Century Britain', HSPS, vol. vi (1975), pp. 133-260.

the entire development through Maxwell (and his immediate successors) as a series of variations within the mechanical tradition.[9] The problem is complicated by the fact that the very meaning of terms involved in the science of mechanics changed with time. Phenomena like energy, forces, and fields, which were at first thought of as supramechanical, gradually became assimilated to the mechanical worldview as they were harnessed by increasingly sophisticated types of machinery. If energy and force fields were not mechanical in the original Cartesian sense, at least they were believed to have their 'mechanical equivalents'. The increasing interdependence between science and technology with the progress of the industrial revolution meant that the idea of mechanism could easily be generalised to include new phenomena almost as fast as they could be discovered. In fact, the revolutionary developments of twentieth-century physics have followed much the same pattern.

It seems best to allow for both continuity and variation within the mechanical tradition throughout the nineteenth century. It will be convenient, therefore, to divide our period into three successive stages: a movement away from the existing mechanical model in the early nineteenth century (Oersted, Davy, and Faraday); a strong reaffirmation and formalisation of the mechanical philosophy in the middle of the century (Whewell, Joule, and Kelvin); and an extensive generalisation of the mechanical worldview by James Clerk Maxwell in the latter half of the century, which pointed the way toward the theory of relativity and quantum mechanics in the early twentieth century. To a degree, there is an analogy here to our treatment of the seventeenth century (chap. 3, sec. C), with James Clerk Maxwell cast in the role of a second Isaac Newton.

---

[9] Daniel M. Siegel, 'Thomson, Maxwell, and the Universal Ether in Victorian Physics', in G.N. Cantor and M.J.S. Hodge, eds., *Conceptions of Ether* (Cambridge: Cambridge Univ. Press, 1981), pp. 263f.; *idem*, 'Mechanical Image and Reality in Maxwell's Electromagnetic Theory', in P.M. Harman, ed., *Wranglers and Physicists* (Manchester: Manchester Univ. Press, 1985), 199f.; Harman, *Energy*, pp. 1f., 5f., 44, 58, 148; Introduction to *idem*, ed., *Wranglers*, pp. 2f. Harman locates the departure from the mechanical view of nature along with the positivist critique of Mach (1883), the energetics of Ostwald (1891), and electromagnetic theory of Lorentz (1895); *Energy*, 118f., 151ff. Brush identifies a 'neo-romantic' revolution against mechanism in the last decades of the century, after Maxwell, with Ostwald, Poincaré, Zermelo, and Mach (*Temperature*, pp. 12f., 24, 67f., 75ff., 85, 94f.); the twentieth century then opens with the 'neo-realism' of Planck and Einstein (*ibid.*, pp. 126ff.). D.B. Wilson identifies a late Victorian positivist reaction against the earlier theologically based realism beginning with Karl Pearson (1892), H.H. Poynting (1899), and Horace Lamb (1904); 'Concepts of Physical Nature: John Herschel to Karl Pearson', in U.C. Knoepflmacher and G.B. Tennyson, eds., *Nature and the Victorian Imagination* (Berkeley: Univ. of California Press, 1977), pp. 201-215.

B. THE MECHANICAL PHILOSOPHY CHALLENGED
(OERSTED, DAVY, AND FARADAY)

Our principal interest in this section is the case of Michael Faraday (1791-1867), whose work on magnetism established the basis of modern field theory. First, however, we must look briefly at the two major antecedents of Faraday's ideas, the Dane, Hans Christian Oersted (1777-1851), and the Cornishman, Humphry Davy (1778-1829).

Oersted and Davy exemplify the complex interaction of intellectual influences in nineteenth-century science. It has often been argued that they were influenced by the idealism of Immanuel Kant and the *Naturphilosophie* ('philosophy of nature') of Friedrich Schelling. We have discussed the mechanical philosophy of the early (precritical) Kant, but the complex metaphysics of the mature Kant and that of Schelling lie outside the scope of a work focusing on the relations of natural science and creational theology. The principal elements of relevance in their philosophies of nature are the emphasis on the unity of all forces in nature and the conversion (or conservation) of one force into another, particularly under conditions of severe constraint or confinement.[10] In addition to their immediate German sources, both ideas have deep roots in motifs of the creationist tradition like the unity of nature and the conservation of force.

1 Hans Christian Oersted

Oersted's experimental discovery of electromagnetism (the connection of electricity and magnetism) certainly reflected some of Kant's and Schelling's ideas[11]—he had defended Kant's philosophy in his doctoral thesis at the University of Copenhagen (1799) and studied the work of Schelling and other philosophers of nature while travelling in Germany in the early 1800s.[12] But Oersted also departed from the German idealist tradition in

---

[10] Leslie Pearce Williams traces these ideas to Kant's *Metaphysische Anfangsgründe der Naturwissenschaften* (1786), which postulated two fundamental forces (*Gründkrafte*)—one attractive and the other repulsive—on the basis of which he postulated a relationship between electricity, magnetism, heat, and light; 'Kant, *Naturphilosophie* and Scientific Method', in R.N. Giere and R.S. Westfall, eds., *Foundations of Scientific Method* (Bloomington: Univ. of Indiana Press, 1973), p. 15; idem, 'Oersted', *DSB* 10:183b; cf. Harman, 'Concepts', pp. 129f.

[11] Schelling's influence is stressed by Mason, 'Scientific Revolution II', p. 174; R.C. Stauffer, 'Speculation and Experiment in the Background of Oersted's Discovery of Electromagnetism', *Isis*, vol. xlviii (1957), pp. 33-50; T.S. Kuhn, 'Energy Conservation', in M. Clagett, ed., *Critical Problems*, p. 338; Williams, 'Kant', pp. 8, 16. More recently, Williams has shifted the emphasis to Kant; 'Oersted', pp. 182a-184b.

[12] Stauffer, 'Speculation', pp. 34f.; Williams, 'Oersted', pp. 182b-184a; H.A.M. Snelders, 'Oersted's Discovery of Electromagnetism', in Andrew Cunningham and Nicholas Jardine, eds., *Romanticism and the Sciences* (Cambridge: Cambridge Univ. Press, 1990), pp. 231f.

significant ways that reflect his Danish family background (his father was an apothecary) and practical training in pharmacy.[13]

As early as 1803, Oersted stated that the governing principles of heat, electricity, and light were the same as those of magnetism, and that all could be treated in 'one united physics'.[14] By 1812 he had developed an experimental arrangement that would demonstrate the inner connection between electricity and magnetism. If an electric current were passed through a resistant wire, he said, it would give off heat, light, and even magnetism, depending on the narrowness of the gauge of the wire.[15] Eight years later, in the course of one of his classroom demonstrations, Oersted found what he had predicted. When he held a magnetic compass under (or over) a wire carrying an electric current, he found that the compass needle pointed across the wire rather than in a northerly direction. Certainly, this meant that electric currents had magnetic effects. But it also showed that, contrary to Newtonian thought, all forces in nature were not central forces—that is, all forces were not necessarily directed towards or away from the centre of their source. In the case of Oersted's demonstration, the magnetic force was directed at right angles to both the direction of the electric current and an imaginary line drawn from the nearest point on the wire to the compass needle. So the magnetic force was not directed along the wire or toward it; rather, it circled around the wire, with the direction of the circling depending on the direction of the electric current.[16]

Before considering the philosophical ideas that lay behind Oersted's discovery, we should say a word about the technical advances that made it possible. In 1795, the Italian, Alessandro Volta, had produced the first electric battery consisting of a series of rods of two different metals (like silver and zinc) connected by a conducting medium. Announced in London in 1800, the invention of the 'voltaic pile' meant that, for the first time in history, electric currents could be produced and controlled at will.[17] Oersted had developed his own version of the voltaic pile in 1801 and used it in his experiments thereafter.[18] The demonstration of the relationship between

---

[13] Stauffer, 'Speculation', p. 39; Williams, 'Oersted', pp. 182a, 183a.

[14] *Materialen zu einer Chemie des neunzehnten Jahrhunderts* (1803), cited in Snelders, 'Oersted's Discovery', p. 235.

[15] Oersted, *Ansicht der chemischen Naturgesetze durch die neueren Entdeckungen gewonnen* (Berlin, 1812); Stauffer, 'Speculation', p. 38; Williams, 'Kant', p. 16; *idem*, 'Oersted', pp. 184a-185a; Snelders, 'Oersted's Discovery', p. 235.

[16] *Experimenta circa effectum conflictus electrici in acum magneticam* (1820); Mason, 'Scientific Revolution II', p. 174; Williams, 'Oersted', p. 185.

[17] Trevor H. Levere, *Poetry Realized in Nature: Samuel Taylor Coleridge and Nineteenth-Century Science* (Cambridge: Cambridge Univ. Press, 1981), p. 33.

[18] Williams, 'Oersted', p. 182b.

electricity and magnetism would have been prohibitively difficult without the steady supply of electric current that the voltaic battery made possible.

The previously mentioned influence of Schelling's *Naturphilosophie* on Oersted has become a commonplace of historical literature. But it has also been questioned. The idea of the unity and mutual interchange of the forces of nature had also had a long history. The immediate source of the idea for Oersted may well have been the natural philosophies of Kant and Schelling.[19] But the broader background was the eighteenth-century Newtonian tradition of Boerhaave and Hutton that envisaged an active fiery substance (originally coming from the sun) which circulated through the cosmos, appearing variously as heat, light, and electricity (chap. 4, sec. A).

In Schelling's thought, the laws of nature were identical to the laws of reason as understood by the human mind. In other words, it was theoretically possible to infer the laws of nature through pure reason without scientific experiments at all.[20] In his early years, Oersted had accepted some of the speculative notions of German *Naturphilosophie*. Exponents of the French tradition of Laplace and Lavoisier convinced him to be more critical.[21] Oersted did not give up his philosophic ideals. But he had his training in the experimental tradition of pharmacy to fall back on, and this enabled him to develop the apparatus and experience that made his discovery of electromagnetism possible.[22]

The importance of Kant's influence on Oersted also has to be questioned. The ideas of the intelligibility and unity of nature so prominent in Oersted's thought had been major themes of the creationist tradition, particularly of the alchemical tradition, since the Middle Ages. Quite unlike Kant, Oersted attributed the correspondence between human reason and the divine plan to the fact that humans had been created in the image of God. Hence, human reason could, if guided by experiment, begin to comprehend creation; the laws of science were not simply useful fictions, imposed by humans, as they were for Kant.[23]

---

[19] Williams, 'Oersted', p. 183b; *idem*, 'Faraday', *DSB* 4:530.

[20] Williams, 'Kant', p. 8. The antiempirical bias of Schelling's system is stressed by Frederick Gregory, 'Kant's Influence on Natural Scientists in the German Romantic Period', in R.P.W. Visser et al., eds., *New Trends in the History of Science* (Amsterdam: Rodopi, 1989), pp. 59-62.

[21] Williams, 'Oersted', pp. 183a, 184a.

[22] Williams, 'Kant', pp. 9f.; Hans Eichner, 'The Rise of Modern Science and the Genesis of Romanticism', *Publications of the Modern Language Association*, vol. xcvii (1982), p. 24a.

[23] Oersted, *The Soul in Nature* (London: Dawsons, 1966), pp. 384, 450f.; selections cited in Williams, 'Kant', pp. 8f.; *idem*, 'Oersted', p. 183; Snelders, 'Oersted's Discovery', p. 238; cf. Immanuel Kant, *Critique of Pure Reason*, A678, B706. Williams notes the difference but attributes it to the fact that Oersted, like many of his contemporaries, misinterpreted Kant. But, if Oersted was as well versed in Kant's writings as Williams claims, then he must have been conscious of this serious departure and his own theological reasons for it.

2 Sir Humphry Davy

If Oersted's main contribution was relating electricity to magnetism, Davy's was relating electricity to chemistry and the realisation that electricity was a fundamental property of all matter.

Again, the discovery was overtly experimental. Like Oersted, Davy experimented with voltaic piles, working in the early 1800s at the Royal Institution of Great Britain (London). In 1806 he announced his conclusion that electrical force was responsible for the molecular structure of matter: the force that held different elements together in a compound (e.g., hydrogen and oxygen in water) had to be electric, he reasoned, since these elements could be isolated from their compounds in voltaic piles (again the importance of techniques of producing and controlling electric currents).[24]

Davy's roots, not unlike Oersted's, went back to the alchemical tradition of experimentation as a means of realising unity with nature. His experiments with electric discharges, for instance, were viewed as microcosmic (we would say 'laboratory') reproductions of lightning in the atmosphere.[25] Davy even compared one of the elements he discovered (potassium) to one of the fundamental substances imagined by the alchemists (Paracelsus's principle of salt, chap. 3, sec. A).[26] Like Oersted, too, he was influenced by the Boerhaavian concept of the fundamental unity and interconvertibility of active principles or forces of nature.[27] In fact, Boerhaave had already attempted to analyse chemical affinities and reactions in terms of the active substance of fire, and Gowin Knight and James Hutton had related the latter to electricity (although all three thought in terms of imponderable fluids rather than forces in the early Newtonian sense).[28]

Davy also believed in the the existence of a single material substance underlying all the chemical elements—hence his negative reaction to Dalton's hypothesis of irreducible atoms corresponding to each chemical element (chap. 4, sec. C). Again, the roots of Davy's thought go back to the eighteenth century: alongside the expansive fiery substance of

---

[24] 'On Some Chemical Agents of Electricity' (Bakerian Lecture, 1806; pub. 1807); A.R. Hall, *Scientific Revolution*, pp. 363ff.; P.M. Heimann (Harman), 'Conversion of Forces and the Conservation of Energy', *Centaurus*, vol. xviii (1974), p. 152; Levere, *Poetry*, pp. 32f.

[25] T.H. Levere, 'The Rich Economy of Nature', in Knoepflmacher and Tennyson, eds., *Nature and the Victorian Imagination*, p. 191.

[26] Bakerian Lecture of 1807; D.M. Knight, 'Davy', *DSB* 3:602a.

[27] 'Essay on Heat, Light and the Combinations of Light' (1799); Harman, 'Conversion', p. 152.

[28] Harman draws attention to Davy's affinities with Hutton; 'Conversion', p. 152. Levere points out the correspondence with Knight's ideas; 'Faraday, Matter and Natural Theology', *BJHS*, vol. iv (Dec. 1968), p. 97. But neither mentions Boerhaave's work in spite of the fact that it was widely known and more relevant to practical chemistry.

Boerhaave, Knight and Hutton had postulated ordinary gravitational matter as a second primary substance. The most important influence on Davy was apparently British Newtonianism rather than German *Naturphilosophie*.[29]

Davy's insistence on the fundamental unity of matter was all the more remarkable in view of the fact that he discovered more than his share of new elements himself (e.g., sodium, potassium, calcium, and magnesium). His willingness to tolerate increasing diversity in the theory of nature was apparently facilitated by the conviction that, underlying all the so-called elements, there was a fundamental unity of substance and that this unity would eventually manifest itself in experiments.[30]

Even more clearly than in the case of Oersted, we know that Davy's belief in the simplicity of nature was sustained by his faith in its Creator. His memoirs record the conviction that the unified law governing matter was due to his belief in 'an energy of mutation impressed by the will of the Deity'. On this basis, he expected to 'discover simplicity and unity of design' and 'an extensive field for sublime investigation'.[31]

In other words, Davy assumed as a matter of principle some of the basic themes of the historic creationist tradition.[32] In fact, Davy's opposition to Priestley's materialism was based on his belief that matter would be inert if God had not imposed active powers on it. In support of this belief, Davy showed that charcoal and diamond were chemically the same in spite of their outward differences. In this respect Davy was not so much a Romantic as a loyal Newtonian.[33]

---

[29] Levere, 'Faraday', pp. 96f.; *idem*, *Poetry*, p. 35. According to Levere, even Coleridge associated Davy's ideas with British natural philosophy; *Poetry*, p. 23; *pace* L. Pearce Williams who had previously argued for the influence of Kant and Boscovich; *Michael Faraday* (London: Chapman & Hall, 1964), pp. 66, 78.

[30] Thackray, 'Matter', pp. 51, 53; Levere, 'Rich Economy', p. 197; Harman, *Energy*, p. 124.

[31] John Davy (Humphry's brother), *Memoirs of...H. Davy* (2 vols., London, 1836), 1:76; cited in Levere, 'Faraday', pp. 97f. Similarly, the different arrangements of nature 'appear as sounds of one voice, impulses of one eternal intelligence'; undated lecture note quoted in T.H. Levere, *Affinity and Matter* (Oxford: Clarendon Press, 1971), p. 67.

[32] In addition to the creational themes of the unity ('unity of design') and relative autonomy of nature ('energy of mutation impressed by the will of the Deity'), we find the ideal of healing and restoration. For example, Davy explicitly distinguished 'true alchemical philosophers' from charlatans on the basis of their constant goals of ameliorating the condition of humanity and supporting the interests of religion. In the British context, these twin goals are clearly Baconian motifs (chap. 3, sec. B); see T.H. Levere, 'Humphry Davy, "The Sons of Genius", and the Idea of Glory', in Sophie Forgan, ed., *Science and the Sons of Genius* (London: Science Reviews, 1980), pp. 37, 49.

[33] On Davy's Newtonianism, see Levere, *Affinity and Matter*, pp. 6, 24-29, 34, 61, 64f.; David Knight, 'Romanticism and the Sciences', in Cunningham and Jardine, eds., *Romanticism and the Sciences*, p. 21. I take issue here with Colin Russell, who has portrayed Davy as more of a Romantic than an orthodox Anglican; *Cross-Currents*, p. 181. Davy may not have been any more orthodox than Newton was, but he clearly deviated from the Neoplatonism of Romantic friends like Coleridge in his Lockean empiricism; Levere, *Affinity*

The same sort of faith in the ultimate oneness of matter was responsible for the famous hypothesis of Davy's younger contemporary, William Prout (1785-1850), which anticipated discoveries of late nineteenth and twentieth centuries. Prout argued in 1815-16 that hydrogen, the lightest element, was the universal matter out of which all other elements were composed. Therefore, the weights of all other atoms and molecules should be exact multiples of the weight of hydrogen.[34] Though Prout's hypothesis failed to account for nonintegral atomic weights of elements like chlorine and magnesium (which normally consist of a mixture of isotopes), it was not forgotten. J. J. Thomson resurrected the idea in 1897 when he postulated a second universal type of matter, later known as the electron,[35] and it was further developed in Ernest Rutherford's 1911 model of the atom, which consisted of hydrogen nuclei (protons) and electrons, which, in turn, was the basis of Niels Bohr's early quantum theory of the atom.[36]

3 Michael Faraday

Michael Faraday's principal contributions were his discovery of electromagnetic induction (1831)—a means of generating an electric current using magnets (the converse of Oersted's effect)—and his subsequent conception of the idea of a magnetic field and its possible connection with the propagation of light.

Faraday served as Davy's lab assistant from 1813 until 1827, when he became Davy's successor at the Royal Institution.[37] Davy's ideas about the identity of chemical affinity with electrical force provided the basis of Faraday's early experiments.[38] Significantly, Davy had also developed some tentative analogies between chemical reactions and the nature of light: the two ends of the light spectrum, he suggested, were analogous to the positive and negative poles of the voltaic battery.[39] Together with the eighteenth-century theology of nature they mediated, these ideas provided what Peter Harman has termed the 'conceptual framework' for Faraday's ongoing research programme.[40]

and Matter, p. 33; *idem*, 'Humphry Davy', pp. 40-44.

[34] 'On the Relation between the Specific Gravities of Bodies in their Gaseous State and the Weights of their Atoms' (anon., 1815); James Kendall, 'The Adventures of an Hypothesis', *Proceedings of the Royal Society of Edinburgh*, vol. lxiii A (1949), pp. 1ff.; John L. Speller, 'William Prout and His Hypothesis', in *Technology Studies Resource Center Working Papers*, vol. i (1984), pp. 46ff.

[35] G.P. Thomson, *J.J. Thomson* (Garden City, N.Y.: Doubleday/Anchor, 1966), p. 59.

[36] Kendall, 'Adventures', p. 7.

[37] Williams, *Michael Faraday*, pp. 84ff.; Knight, 'Davy', p. 603a; William Gould, ed., *Lives of the Georgian Age* (New York: Barnes & Noble, 1978), pp. 127, 138.

[38] Harman, *Energy*, p. 34.

[39] Williams, *Michael Faraday*, p. 84. A passage in Davy's 1799 'Lecture', omitted in the final, printed edition, stated that electricity might be condensed light; Knight, 'Davy', p. 599b.

[40] Davy, 'Relations of Chemical Affinity, Electricity, Heat, Magnetism and other powers of Matter' (1834); Harman, 'Conversion', pp. 149, 154f.

Faraday supplemented Davy's idea of the unity of forces with the idea that, since force was a divinely created entity, it could be neither augmented nor diminished by natural causes.[41] Like Newton, Faraday saw force as transcending the properties of ordinary (ponderable) matter. But whereas force for Newton was an ephemeral principle, likely to decay without divine sustenance and replenishment, for Faraday it was as stable and permanent as matter itself. The philosophic difference here partly reflects the improvements in technology—by Faraday's time it was possible efficiently to generate forces like electricity and to store them for later use. Consequently, divine activity was even more closely confined to the established laws of nature for Faraday than for Newton. The only exception to the laws of nature that was relevant to science was the event of creation itself. No other supernatural event or act of God was called for, at least, in matters of natural philosophy.

Since force could neither be augmented nor diminished, Faraday reasoned, no force could act at a distance—that is, no force emanating from one body could act on another without being transmitted through the intervening space in some manner. Otherwise, if it was observed that the force grew stronger as the bodies approached each other, such an intensification would require the creation of force out of nothing (the creation or the disappearance of potential energy in the terms of later physics).[42]

If force was really transmitted from one body to the other, there must be some way of representing this process visually. Observing the pattern of magnetic filings suspended over a magnet, Faraday developed the concept of magnetic 'lines of force', and, in 1845, he introduced the term 'magnetic field' to describe the overall structure of the pattern.[43] At first, lines of force were perhaps just a convenient model of representation for Faraday, but in the early 1850s he clearly stated that they were not just symbols but fundamental entities in nature.[44] In Faraday's view, space appeared to be full of forces and powers, and the balance of reality was not so much in isolated bodies of matter as in the space between (and within) them.[45] This was a significant departure from the more dualistic, Newtonian stance of Davy

---

[41] *Seventeenth Series of Experimental Researches* (1840); Harman, 'Conversion', p. 154; idem, *Energy*, p. 35.

[42] 'On the Conservation of Forces' (1857); Harman, *Energy*, 60f.

[43] Harman, *Energy*, pp. 6, 72, 75. The idea of the magnetic field, however, was first introduced in Faraday's 'On Some New Electro-Magnetical Motions, and on the Theory of Magnetism' (1821); Williams, 'Faraday', p. 533b.

[44] 'On the Physical Character of the Lines of Force' (1852); Williams, *Michael Faraday*, pp. 439f., 444-50; Harman, *Energy*, pp. 78, 80f.

[45] 'On Static Electrical Inductive Action' (1843); 'A Speculation touching Electric Conduction and the Nature of Matter' (1844); Williams, *Michael Faraday*, pp. 374ff.; Harman, *Energy*, 76f. According to Levere, Faraday already had the idea in the 1830s, though he did not state it until 1844; 'Faraday', pp. 98ff.

(active forces and inert matter) and was more akin to the idea of the primacy of force in Boscovich and Priestley (chap 4, secs. A, B).[46] And like Boscovich and Priestley (and Newton), Faraday was encouraged to think in terms of the primacy of force by his association of active powers in the universe with the activity of God.[47]

In 1846, Faraday articulated a further implication of the idea of lines of force—that vibrations of the lines could account for the propagation of visible light.[48] The notion that light and magnetism were manifestations of the same underlying type of matter had been stated by Gowin Knight among others. Oersted had proven a deep connection between electricity and magnetism, and Faraday had provided further evidence in his work on electromagnetic induction. But Faraday's explanation in terms of lines of force gave physicists a way of defining these connections more exactly.

We have already noted some of the ways in which creational theology played a role in Faraday's investigations. Much of this could be accounted for in terms of the general theological framework shared by other scientists discussed in this chapter. However, Faraday belonged not to the Church of England, but to a dissenting group known as the Sandemanians. A good deal of attention has been devoted to the religious beliefs of this group since the pioneering biography of Faraday written by L. Pearce Williams appeared in 1964.[49]

The Sandemanians had their origins in two evangelical movements that began in England and Scotland in the 1730s. In spite of their disestablishmentarian and revivalist roots, the sect's theology developed along decidedly cognitive lines in the following decades, under the leadership of Robert Sandeman.[50] Sandeman emphasised the ideas of nature as God's creation and the Bible as God's revealed word. Humans could only come to know God through his self-revalation in Scripture, but on the basis of that

---

[46] The influence of Boscovich is argued by Williams, though he admits that Faraday did not cite Boscovich until his 1844 'Speculation'; *Michael Faraday*, pp. 78f., 89, 376ff.; 'Faraday', pp. 529f. *Pace* Williams, Harman argues for the influence of Priestley, though he admits there is no concrete evidence and that Faraday never cited Priestley's name, possibly due to its association with unorthodox (Unitarian) theology (to say nothing of radical politics); Heimann and McGuire, 'Newtonian Forces', 305; Heimann, 'Faraday's Theories of Matter and Electricity', *BJHS*, vol. v (June 1971), pp. 235ff., 247-53; Harman, *Energy*, p. 77.

[47] Williams, *Michael Faraday*, p. 103; Levere, 'Faraday', pp. 101f.; David Gooding, 'Metaphysics versus Measurement: The Conversion and Conservation of Force in Faraday's Physics', *Annals of Science*, vol. xxxvii (Jan. 1980), p. 28.

[48] 'Thoughts on Ray-Vibrations' (1846); Williams, *Michael Faraday*, p. 380.

[49] Williams, *Michael Faraday*. pp. 2-6, 102-6; cf. R.E.D. Clark, 'Michael Faraday on Science & Religion', *Hibbert Journal*, vol. lxv (Summer 1965), pp. 144-47; Levere, 'Faraday', pp. 103ff.; G.N. Cantor, 'Reading the Book of Nature: The Relation Between Faraday's Religion and His Science', in D. Gooding and F. James, eds., *Faraday Rediscovered* (Basingstoke: Macmillan, 1985), pp. 71ff.

[50] Levere, 'Faraday', p. 103.

faith they could also view nature as a book of signs manifesting the Creator's eternal power and godhead (Rom. 1:20).[51]

According to Williams and others, the startling boldness of Faraday's theorising about the spatial ordering of magnetic fields was facilitated by his deep belief that nature's laws were themselves the product of a rational mind. The biblical idea of creation implied that there was an underlying order and unity to the phenomena of nature—by virtue of the forces impressed on matter by the Creator—and that the human mind, fallible as it was, could, when guided by experiments, formulate ideas that reflected those laws.[52]

In other words, a revival of biblical thought had occasioned a reaffirmation of some of the basic tenets of the historic creationist tradition: the comprehensibility, unity, and relative autonomy of nature (chap. 1, secs. B-D). Whereas much of the popular natural theology of the seventeenth, eighteenth, and nineteenth centuries exploited the apparent congruities of nature as evidences of the existence of God, Faraday, reflecting his more biblicist roots, relied on his faith in God as the motivation for seeking analogies in nature.[53] As he wrote during the years he was developing his idea of magnetic lines of force and their relation to the propagation of light:[54]

> I am struggling to exert my poetical ideas just now for the discovery of analogies and remote figures...for I think that is the true way (corrected by judgement) to work out a discovery.

It should be pointed out, however, that Faraday's theological perspective may have restricted his vision of science, even if it had a positive effect on the progress of physics as a whole. Around the middle of the century, a new generation of physicists (considered in the next section) developed a more precise formulation of the law of the conservation of energy. Faraday had already postulated the constancy of force but resisted making his idea quantitative and did not accept the more precise definitions which distinguished force from energy—the latter being the equivalent of force applied over an interval of distance ('work'). It is possible that Faraday's view of force as the medium of God's activity in the world led him to resist

---

[51] Sandeman, *Law of Nature Defended by Scripture* (1760); Cantor, 'Reading', pp. 71f.

[52] Williams, *Michael Faraday*, p. 83; idem, 'Faraday', p. 527b. Faraday's reasoning is found in his memo on 'Matter' (1844; Levere, 'Faraday', pp. 103, 107); 'Speculation' (1844; Gooding, 'Metaphysics', p. 28); and *Lectures on Physico-Chemical Philosophy* (1847; Gooding, 'Metaphysics', p. 29).

[53] Williams, *Michael Faraday*, pp. 103f.; Levere, 'Faraday', p. 103; Gooding, 'Metaphysics', note on p. 3.

[54] Written in 1845; Williams, *Michael Faraday*, p. 443.

precise definition and mathematical formalism that might suggest a reduction to mechanistic principles like energy and work.[55]

C. The Mechanical Philosophy Restated and Formalised (Whewell, Joule, and Kelvin)

Diverse as they were in other respects, the three figures we consider in this section—Whewell, Joule, and Kelvin—had two things in common. First, in comparison with the relatively qualitative contributions of Oersted, Davy, and Faraday, they all placed greater emphasis on the precise, quantitative aspects of physics. In fact, Whewell and Kelvin made significant contributions to the progress of mathematical physics.

Secondly, whereas Faraday viewed the operations of forces as the manifestation of God's activity in nature, the figures considered here all viewed nature in strictly mechanical terms. Whewell and Kelvin sought the activity of God at the limits of mechanical explanation. But Joule viewed the perfection of the mechanical system of the world as itself a manifestation of divine wisdom and power.

A comparison with Newton's thought gives another way of differentiating Whewell, Joule, and Kelvin from Faraday. We recall that Newton's worldview was one of active principles (forces) operating on inert matter. In these terms, both Faraday and the figures considered here were 'Newtonian', but in different senses: Faraday emphasised the one side of Newton's thought, the idea of active principles, and the figures considered here stressed the other side, the inertness of matter. So, if Faraday was Newtonian in his distinctive emphasis on the supramechanical character of the forces of nature, Whewell, Joule, and Kelvin were more Newtonian in their appeal to divine activity in the origin of the mechanical systems of nature and (except for Joule) in their periodic restoration.

1 William Whewell

William Whewell (1794-1866, an almost exact contemporary of Michael Faraday) was a first-generation member of what has been called the 'Cambridge school' or 'Cambridge network', a group of scholars who either studied or taught at Cambridge and who were responsible for introducing the new techniques of calculus developed in the late eighteenth century by the

[55] Gooding, 'Metaphysics', pp. 2f.

French and using them to formalise mathematical physics.[56] Whewell himself wrote influential textbooks on mechanics and geology and was influential in promoting 'mixed mathematics', which included mechanics, hydrodynamics, and astronomy, as well as pure mathematics, into the curriculum.[57] More than anyone, he was responsible for reestablishing the mechanical philosophy as the framework for investigating the phenomena of nature.[58]

Whewell is the only figure considered in this chapter who was an ordained clergyman as well as a scientist. There were others among his contemporaries, like George Peacock, who combined the two callings,[59] but theirs was the last generation for which it was still common for members of the Royal Society of London to be members of the clergy.[60] In fact, Whewell's coining of the terms 'scientist' and 'physicist' was a harbinger of the emerging professionalism that would fundamentally alter the common perception of the relation between science and theology.[61]

Whewell's advocacy of the mechanical philosophy did not prevent him from viewing God as deeply involved in the history of nature. The two ideas were rendered compatible by his stress on the limits of the strictly mechanical account of things. One historian has described his position as a rational version of Catastrophism. In fact, it grew out of Whewell's criticism of the views of the geologist Charles Lyell, for which he himself had coined the label 'Uniformitarian'.[62]

Whewell made a clear distinction between mechanical and historical accounts of nature, the latter being particularly important in historical sciences like cosmogony and geology.[63] There were three basic steps in the development of any science. The first step was the accumulation of data about what was observed to have happened in the history of nature and the formation of inductive generalisations to describe the data. The second step was the determination of historical sequences of cause and effect in which all

---

[56] Walter (Susan) F. Cannon uses the phrase, 'Cambridge network', to include 'scientists, historians, dons, and other scholars', beginning with John Herschel, Charles Babbage, and George Peacock; 'Scientists and Broad Churchmen', pp. 65-88. P.M. Harman (following Edmund Whittaker) uses the phrase, 'Cambridge school', in the more restrictive sense of mathematically trained scientists; Harman, ed., *Wranglers*, p. 1.

[57] Cannon, 'Scientists', pp. 68, 71; Harman, *Wranglers*, p. 2.

[58] Harman, *Wranglers*, p. 3.

[59] Cannon, 'Scientists', p. 70.

[60] Turner, 'Victorian Conflict', pp. 360, 366ff. Turner documents a dramatic shift in the membership and leadership of the Royal Society between 1860 and 1880.

[61] The term 'scientist' was coined in 1834; R.E. Butts, 'Whewell', *DSB* 14:292b; Turner, 'Victorian Conflict', p. 360; C.A. Russell, *Cross-Currents*, p. 191.

[62] Cannon, 'Problem', pp. 7f., 26f.

[63] Whewell, *Philosophy of the Inductive Sciences founded upon their History* (2 vols., 1840); Cannon, 'Problem', p. 26.

the phenomena in a series could be traced back to the first cause in that sequence. Only in the third step were theoretical ideas like those of mechanics applied to the data thus gathered and organised.[64] Here Whewell distinguished empirical data from the *a priori* ideas or laws of thought.[65] The fact that God had created the world, together with the divine attributes of goodness and wisdom, guaranteed that the laws governing nature must be rational and simple enough for the human mind to comprehend.[66] Most important of these laws was the Law of Continuity, which stated that every event must have a sufficient cause (whether natural or supernatural) in order that there be no gaps in the overall account of things. The laws of mechanics could account for all the phenomena after the first cause in each sequence. But, in accordance with the Law of Continuity, the first cause of each sequence of phenomena, which was inexplicable in strictly mechanical terms, had to be referred back to a transcendent, supernatural cause.[67]

According to Whewell, there were several cases of such causal sequences for which the Principle of Continuity required the direct action of God as the initial event in the series. The most obvious case was that of the creation of the cosmos: even if the formation of the solar system could be accounted for in terms of Laplace's nebular hypothesis (chap. 4, sec. D), it ultimately had to be traced back to a moment of creation and a transcendent First Cause which could not itself be mechanical.[68] So natural science itself could teach us nothing positive about the beginning of things: the most it could (and should) do was to indicate that there must be such a beginning.[69]

Of course, in order to support his contention that divine creation was historically continuous with natural history, Whewell had to rule out the possiblity of an eternal, self-existent world. In this connection, he appealed to the recent work of Johann Encke, who had computed the periodic return of a short-period comet (now known as Encke's comet) and then found that its period was shorter than predicted and was continually decreasing, presumably due to the resistance of the interplanetary medium. According to Whewell, this deterioration of motion proved that the solar system—and, by

---

[64] Whewell, *Philosophy*; Crosbie Smith, 'Geologists and Mathematicians', in Harman, ed., *Wranglers*, pp. 67-72. Smith notes Whewell's indebtedness to John Herschel for his ideas on induction (*ibid.*, p. 67).

[65] The Kantian influence is brought out by P.M. Harman, 'Edinburgh Philosophy and Cambridge Physics', in Harman, ed., *Wranglers*, p. 222.

[66] Whewell, *Astronomy and General Physics considered with reference to Natural Theology* (first Bridgewater Treatise, 1834); Cannon, 'Problem', p. 10.

[67] *Philosophy*; Cannon, 'Problem', p. 27.

[68] *Astronomy*; Cannon, 'Problem', pp. 12f.; C. Smith, 'Natural Philosophy and Thermodynamics', *BJHS*, vol. ix (1976), p. 304.

[69] *Philosophy*; P.M. Heimann, 'The Unseen Universe', *BJHS*, vol. vi (1972), pp. 74f.

inference, the cosmos—could not have been eternal. Like a watch that was still ticking, it must have been wound up a finite time ago.[70]

Other examples of causal sequences which pointed to divine initiative were the formation of new strata in the geological record, the emergence of new biological species (as evidenced in the fossil record), and the origin of human intelligence. None of these beginnings could be accounted for in terms of the nebular hypothesis or in terms of any other mechanical account of nature.[71]

Thus, in contrast to the Uniformitarianism of Charles Lyell, Whewell described the history of the cosmos as a series of cycles, each beginning with the emergence of qualitatively new forms of matter and life and initiated by a fresh exertion of divine power.[72] The fossil record showed that the material world was inert and dead in itself. Millions of years had passed, for instance, before intelligence had emerged on earth. Without divine intervention, there would have been no intelligence, and probably not even any life at all.[73]

Whewell's version of mechanical philosophy incorporated traditional creationist ideas like the comprehensibility of the world and the relative autonomy of nature. But it was severely limited by the traditional mechanical idea of matter in motion and did not do justice to the concepts of energy, forces, and fields that were beginning to dominate mathematical physics. Still, Whewell's views established the consistency of the mechanical model with Christian faith and made it acceptable to a new generation of physicists for whom it could be broadened to include the new concepts.

2 James Prescott Joule

Joule is best known for his experimental proof of the exact equivalence of the heat produced and the work done (force times distance) in the process of producing it, whether mechanical, electrical, or chemical (combustion). As developed in the writings of William Thomson (Lord Kelvin) and Hermann von Helmholtz, this equivalence provided the basis of the law of the

---

[70] *Astronomy*; Cannon, 'Problem', p. 13.

[71] *Astronomy* (Cannon, 'Problem', p. 12); *Philosophy* (Heimann, 'Unseen Universe', pp. 74f.).

[72] *Philosophy*; Cannon, 'Problem', p. 28. In *Of the Plurality of Worlds* (anon., 1853), Whewell argued on a second front against the idea of a universal law of development advocated by Robert Chambers's *Vestiges of the Natural History of Creation* (1844); John Hedley Brooke, 'Natural Theology and the Plurality of Worlds', *Annals of Science*, vol. xxxiv (May 1977), pp. 230, 233.

[73] By analogy, it was not likely that there was intelligent life in the vast reaches of spaces beyond the earth; Brooke, 'Natural Theology', p. 234.

conservation of energy, one of the most powerful principles in modern physics.

In support of his notion of the mechanical equivalence of heat, Joule appealed to the once and for all character of divine creation. Once God had established the world and all the forces of nature, no additional act of sustenance or repair was ever needed.[74] 'The grand agents of nature are, by the Creator's fiat, indestructible', so the forces God had created remained constant in their operation for all time.[75] Therefore, none of the force expended in a physical process like the generation of heat could be lost to nature. The force expended and the heat produced were simply two different forms of the same thing; therefore, there must be a precise quantitative equivalence between them. The idea of creation thus provided what Peter Harman has called a framework for the interpretation of Joule's experimental discoveries.[76]

Similar ideas about the ceaseless energy of nature had been a recurring theme in the creationist tradition since the second century BC (chap. 1, sec. D) and had already been articulated in modern scientific terms by Leibniz and Faraday. But Joule translated this general idea into an exact mathematical formula: every single foot-pound (or erg or joule) of force expended over distance must be transformed into its exact equivalent of heat.[77] Another difference was the fact that Joule viewed electromagnetic phenomena strictly in terms of the mechanical force expended over distance rather than in terms of independent fields of force, the way that Faraday did. In fact, using Faraday's principle of electromagnetic induction, Joule could show how mechanical force was converted into electricity and the electricity was converted, in turn, into heat.[78] Heat, chemistry, and electromagnetism were all just manifestations of mechanical power.

The establishment of exact equivalency between mechanical force and heat raised the question of the relationship between the macroscopic and atomic levels of matter. At the macroscopic level, energy could appear in two basic forms: the energy of motion (Leibniz's *vis viva*) and the attraction between bodies due to forces like gravitation (the terms 'kinetic energy' and

---

[74] Heimann, 'Conversion', pp. 148f.; Harman, *Energy*, p. 40.

[75] 'On the Califoric Effects of Magneto-Electricity and on the Mechanical Value of Heat', postscript added in press (1843); Heimann, 'Conversion', pp. 156f.; Henry John Steffens, *James Prescott Joule and the Concept of Energy* (Dawson: Science History Publications, 1979), p. 48; cf. Joule's 'On Matter, Living Force, and Heat' (1847); Steffens, *James Prescott Joule*, p. 133.

[76] Heimann, 'Conversion', pp. 149, 156.

[77] According to his first published determination (1843), the expenditure of force needed to increase the temperature of a pound of water by one degree Farenheit was 770 foot-pounds, close to the presently accepted value of 778 foot-pounds; Steffens, *James Prescott Joule*, pp. 36, 47f.

[78] Harman, *Energy*, p. 40.

'potential energy' were introduced by Kelvin and Tait in 1867).[79] Some physicists (at this time, chiefly French) held that there was a third macroscopic form: subtle fluids corresponding to the phenomena of heat ('caloric') and electricity. But Davy and Faraday had rejected the idea of subtle fluids and treated heat, like chemical affinity, in terms of interparticulate forces at the atomic level.[80] Joule also rejected subtle fluids, adding the theological argument that the conservation of the heat fluid by itself required the creation of mechanical force out of nothing in heat engines and was therefore impossible.[81]

What, then, happened to mechanical force when it disappeared at the macroscopic level and was converted into heat? What did it look like at the atomic level? Joule at first suggested an electromechanical model of the atom in order to define its temperature. Each atom had a tiny atmosphere of electricity, and the heat was stored in the rotation of these atmospheres, with the rate of rotation being determined by the temperature. In a later, more narrowly mechanical, model, he even imagined tiny weights being attached to the rotating atoms by ropes: mechanical force could then be transferred to the weights by lifting them small distances.[82]

The attempt to model the dynamics of atoms, which began with Joule in the 1840s, was to be a major problem in physics in the latter half of the nineteenth and the early twentieth century. The question of the nature of atoms was an ancient one, but it took on a new sense of urgency partly as the result of the rejection of subtle fluids (Davy and Faraday) and partly due to the precise definition of the equivalence between mechanical work and heat (Joule). Hitherto, macroscopic phenomena had been treated in their own terms—in terms of Newton's laws of motion, gravitation, and the newly discovered laws of electricity and magnetism—without too much concern about their relationship to the microscopic properties of matter. Most scientists, Joule among them, followed Isaac Newton in postulating what was known as the 'analogy of nature'—the supposition that the same fundamental properties and laws that applied to macroscopic objects also were applicable at the atomic level.[83] Joule was one of those who acknowledged that the real properties of atoms might forever be beyond the reach of human science. Nonetheless, he insisted on the importance of

---

[79] William Thomson and Peter Tait, *Treatise on Natural Philosophy* (1867). The terms 'dynamical energy' and 'statical energy' had already been used by Kelvin in 1851; Harman, *Energy*, pp. 5f., 58, 59.
[80] Davy, 'Essay on Heat...' (1799); Williams, *Michael Faraday*, pp. 67, 84; Levere, *Poetry*, p. 22; Harman, *Energy*, p. 19.
[81] Steffens, *James Prescott Joule*, p. 48.
[82] Harman, *Energy*, p. 40.
[83] Newton's 'Third Rule of Reasoning' (*Principia*, ed. F. Cajori, p. 398).

visualisable models of atomic phenomena in accordance with the Newtonian analogy of nature.[84]

But now that macroscopic mechanical work was believed to have an atomic-level equivalent in heat, some account had to be given of how that equivalent was to be represented at the atomic level. Any complete account of physics, therefore, had to specify a model of the atom in terms that could be related to the properties and laws of macroscopic phenomena. Ultimately, these efforts would lead to the revolutionary new ideas of quantum mechanics and the field theory of particles. In the nineteenth century, however, natural philosophers could only speculate about the nature of the atoms by constructing hypothetical models like those of Joule and, later on, Kelvin and Maxwell.

Joule's vision of the universe was that of a well-oiled machine that would never run down. The winds still blew as strongly, and the currents of water flowed with the same force as they ever did, at least since time of the Deluge, in spite of the continual conversion of mechanical force into molecular heat by friction.[85] In one of his more poetic moments, Joule even compared the mechanics of the universe to Ezekiel's vision of the chariot of God:[86]

> Thus it is that order is maintained in the universe—nothing is deranged, nothing ever lost, but the entire machinery, complicated as it is, works smoothly and harmoniously. And though, as in the awful vision of Ezekiel, 'wheel may be in the middle of wheel', and everything may appear complicated and involved in the apparent confusion and intricacy of an almost endless variety of causes, effects, conversions, and arrangements, yet is the most perfect regularity preserved—the whole being governed by the sovereign will of God.

Joule thus gave the mechanical philosophy its strongest possible statement, and, contrary to Laplace (chap. 4, sec. D), God was still very much a part of the overall picture! Whewell, like Newton, had seen the motion of the universe as an ephemeral thing—always tending to decrease. But Joule followed a path more like that of Leibniz in seeing the apparent diminishment of motion merely as its becoming temporarily invisible—being converted into other forms at an atomic level and capable of being restored to the macroscopic level at some future time. Both were versions of the mechanical philosophy, but whereas the Catastrophist, Whewell, saw the

---

[84] Steffens, *James Prescott Joule*, p. 22.

[85] 'On Matter...' (1847); Steffens, *James Prescott Joule*, p. 133. In 1844, Kelvin had argued on thermodynamic grounds that, in the earth's remote past, the dissipation of heat must have been more rapid and hence the winds more vigorous; Jed Z. Buchwald, 'Thomson', *DSB* 13:382b.

[86] Steffens, *James Prescott Joule*, p. 134.

activity of God at the boundaries of mechanical causation, Joule saw it in the design of the machine itself—a view more like that of the Uniformitarians, Hutton and Lyell, whom Whewell had opposed.[87]

3 William Thomson (Lord Kelvin)

Born in Belfast and educated at the universities of Glasgow and Cambridge, William Thomson was knighted by Queen Victoria in 1866 for the contribution of his calculations to the success of the trans-Atlantic telegraph cable completed that year.[88] In spite of the anachronism, we shall refer to him by his title of Lord Kelvin in order to avoid confusion with several other nineteenth-century scientists with the name Thomson. We shall review Kelvin's scientific and theological ideas in two areas: (1) the theory of the relationship of heat and mechanical work—the science to which Kelvin himself gave the name of 'thermodynamics'[89]—based on the work of Joule; (2) the theory of electricity and magnetism, based on the work of Faraday. As we shall see, the mechanical philosophy of Joule and the field theory of Faraday represent the two poles between which much of Kelvin's thought moved.

3.1 Kelvin on Thermodynamics

Kelvin first came into contact with Joule in 1847 when the latter presented his results on the mechanical equivalency of heat at a meeting of the British Association for the Advancement of Science. In fact, Kelvin was the first to recognise the immense significance of Joule's work and to synthesise it with the findings of other physicists. In particular, he was perplexed by the apparent contradiction between Joule's experiments, in which mechanical work and heat were shown to be mutually convertible, and Joseph Fourier's work on the conduction of heat through solids, where heat was dissipated and work was irreversibly lost to the system.[90] In the pivotal years of 1848-52, Kelvin resolved the paradox by giving precise definition to the concept of 'energy' as the basic quantity that was conserved, whether in the mechanical form of matter in motion or work done against a force, or in the

---

[87] Heimann points out the similarity to Hutton's views; 'Conversion', pp. 149f.

[88] The cable was laid between Ireland and Newfoundland. The first two attempts in 1857 and 1858 failed. Kelvin argued that the latter failure was due to the use of too high voltages, the consequent statical charging of the insulating material, and the lowering of the rate of signal transmission. He had developed the instruments needed to measure such low voltages as were required; Buchwald, 'Thomson,' pp. 386f.

[89] Kelvin coined the term, 'thermodynamics', in 1854; Harman, *Energy*, p. 45. Kelvin had used the phrase,' thermo-dynamic engine', already in his 1849 'Account of Carnot's theory'; C. Smith, 'Natural Philosophy', p. 310.

[90] Jean Baptiste Joseph Fourier, *Théorie analytique de la chaleur* (1822). Fourier's first papers on the subject were in 1807 and 1811; Harman, *Energy*, pp. 27ff.

molecular form of chemical bonding or heat.[91] Based on this definition, the first law of thermodynamics (independently formulated by Rudolf Clausius in 1850)[92] could be stated quite simply in terms of the conservation of energy: the net heat input had to balance the mechanical energy extracted from any system going through a complete thermodynamic cycle (i.e., returning to its initial thermodynamic state).[93] Thus far Kelvin was in complete agreement with Joule. He even gave a theological justification for the principle of conservation similar to that offered by Joule and undoubtedly influenced by him: since only God could create or destroy anything, a fundamental quantity like energy must be conserved in all natural processes.[94]

However, a second principle was also at work in thermodynamic systems, as pointed out by Clausius in 1850: according to this second law of thermodynamics, heat always flowed from hotter bodies to colder ones, never the other way around. Unless it was efficiently converted into mechanical work by a suitable apparatus, heat continually spread itself out, tending towards a state of uniformity or dissipation.[95] Thus a new paradox arose: whereas the first law of thermodynamics ensured absolute permanence for energy, the second law required universal dissipation. Kelvin resolved the problem by clearly distinguishing between dissipation and destruction. The energy dissipated in accordance with the second law of thermodynamics was not destroyed—that would have contradicted the first law; it simply became unavailable for useful work. Conservation and dissipation then were thus twin principles at the basis of thermodynamics.[96]

The dissipation of energy implied that there was a definite direction to thermodynamic processes. Nature changes in ways that tend to dissipate heat rather than concentrate it. Like Whewell's (and Newton's) arguments about the degradation of motion in the solar system, then, Kelvin's version of the second law of thermodynamics implied that the universe could not be

---

[91] Paper of 1848-49; Draft of 'On the Dynamical Theory of Heat' (1851); C. Smith, 'Natural Philosophy', pp. 308, 310f.; Harman, *Energy*, pp. 49ff.

[92] Clausius, 'Über die bewegende Kraft der Wärme' (1850); Harman, *Energy*, pp. 52ff. The basic ideas had already been articulated by Sadi Carnot (1824) and Emile Clapeyron (1834); *ibid.*, pp. 45-49.

[93] Draft of 'Dynamical Theory' (1851); Harman, *Energy*, p. 56.

[94] 1848 letter to Joule (C. Smith, 'Natural Philosophy', p. 308); Draft of 'Dynamical Theory'; cf. 'On a Universal Tendency in Nature to the Dissipation of Mechanical Energy' (1852; C. Smith, 'Natural Philosophy', p. 314; Steffans, *James Prescott Joule*, p. 139; Harman, *Energy*, pp. 57f.).

[95] Draft of 'Dynamical Theory'; C. Smith, 'Natural Philosophy', p. 311.

[96] 'Dynamical Theory' (1851; C. Smith, 'Natural Philosophy', p. 313; Harman, *Energy*, p. 56); 'Universal Tendency' (1852; C. Smith, 'Natural Philosophy', p. 314).

infinitely old.[97] It must have been created in a such a way that energy was stored in concentrated sources because, after the moment of creation, it would automatically tend toward states of ever increasing dissipation. This was entirely different from the notion of the permanence of force in the cosmos as taught by Joule and Uniformitarians like Lyell.[98] Nonetheless, according to Kelvin, it had its own theological correlate in the biblical idea that history had a direction and moved towards an end.[99]

The end toward which the universe moved on its own, however, was one of complete dissipation, or, to use a term coined by Rudolf Clausius in 1865, a state of maximum 'entropy'.[100] Kelvin realised this ominous cosmic implication of the second law of thermodynamics already in 1851: in an unpublished draft of a paper written that year, he even cited Psalm 102:26 ('all of them shall wax old like a garment', AV) for biblical support.[101] The following year, he published a paper, 'On a Universal Tendency in Nature to the Dissipation of Mechanical Energy', in which he concluded that, barring some supernatural intervention, the earth would eventually become unfit for human habitation.[102] But, ten years later, he wrote more optimistically that civilisation could progress indefinitely if the universe was infinite, since there would then be an inexhaustible supply of free energy that could be tapped.[103] Clearly there was an ongoing struggle in Kelvin's mind between Leibniz's and Joule's notion of the perfection and permanence of the cosmic

---

[97] The 1851 draft of 'Dynamical Theory' stated that the motions of the planets were gradually losing more *vis viva* than could probably be compensated for by the energy from the sun. William Meikleham, who was briefly Kelvin's natural philosophy professor at Glasgow (*c.* 1837-40; David B. Wilson, 'Kelvin's Scientific Realism', in *The Philosophical Journal*, vol. xi [1974], pp. 42, 44), recommended that his students read Whewell's *Astronomy*, in which the discussion of the degradation of comet Encke's motion was discussed; C. Smith, 'Natural Philosophy', pp. 303f., 311f.

[98] 'The Doctrine of Uniformity in Geology Briefly Refuted' (1865); Loren Eisley, *Darwin's Century* (Garden City, N.Y.: Doubleday/Anchor, 1961), pp. 239f.; Harman, *Energy*, p. 67. Kelvin had already used thermodynamic arguments (based on Fourier's theory of heat diffusion) against Lyell's geology in 1844 and 1862. The theoretical upper limit he established on the age of the sun and earth was 400 million years; Eisley, *Darwin's Century*, p. 238; Buchwald, 'Thomson', pp. 382f.

[99] Draft of 'Dynamical Theory'; Harman, *Energy*, pp. 57, 66.

[100] 'Die Entropie der Welt strebt einem Maximum zu'; Clausius, 'Über verschiedene für die Anwendung bequeme Formen der Hauptgleichungen der mechanischen Wärmetheorie' (1865); Justin Wintle, *Makers of Nineteenth Century Culture* (London: RKP, 1982), p. 122a; Harman, *Energy*, pp. 65, 67.

[101] Draft of 'Dynamical Theory'; C. Smith, 'Natural Philosophy', p. 312.

[102] Steffans, *James Prescott Joule*, p. 141; C. Smith, 'Natural Philosophy', p. 314.

[103] 'On the Age of the Sun's Heat' (1862); Jaki, *Science*, p. 294 and notes; cf. Eisley, *Darwin's Century*, p. 238. In 1875, Kelvin's associate, Peter Guthrie Tait, and Balfour Stewart published *The Unseen Universe*, a theological defense of mechanistic physics in which they argued, as the title suggests, for the existence of eternal, inexhaustible, invisible sources of energy in the universe; Heimann, 'Unseen Universe', pp. 77f.; Harman, *Energy*, p. 68.

machinery, on the one hand, and Newton's and Whewell's sense of its instability and transience, on the other.

Kelvin's major contribution to thermodynamics was thus his formalisation and synthesis of the findings of other scientists. He had a remarkable ability to identify the apparent contradictions among these results and then to rethink the concepts involved and to make precise definitions in such a way as to resolve those contradictions. Since the willingness and ability to contemplate paradoxes was to become one of the major themes of late nineteenth- and early twentieth-century physics, is worthwhile noting the theological framework in which this penchant occurred in the work of Kelvin.

As we have noted, Kelvin found a theological parallel to the conservation of energy in the biblical teaching of the stability of God's creation. He also found a parallel to the gradual dissipation of free energy in the theological notion of the directionality and finality of the biblical concept of history. Even if these parallels are questionable in themselves, the very ability to recognise apparent paradox in science as well as in theology is significant. In both cases, Kelvin took over ideas that were available from others and provided a creative synthesis.

Other than the bare tolerance of paradox, the most important theological aspect of Kelvin's synthesis was his understanding of the attributes of God. According to the traditional theology Kelvin had imbibed at Glasgow, God was eternal and his counsels stood for all time. It followed that, with the exception of miracles, the physical world must operate at all times and in all places in accordance with the same basic laws (the moral realm was another matter).[104] Both the conservation of energy and the universal dissipation of heat (free energy) were, therefore, evidences of divine design. Kelvin's world was far more mechanical than Faraday's. Hence, his God, like that of naturalists and mechanists since the twelfth century, was somewhat more remote than Faraday's, but the tokens of his existence were equally in evidence for both.[105]

3.2 Kelvin on Electricity and Magnetism

A similar struggle with paradox characterised Kelvin's work in the area of electricity and magnetism. Here Kelvin adopted Faraday's idea of fields of

---

[104] C. Smith, 'Natural Philosophy', pp. 299, 308, 313. For Kelvin's differentiation between uniform determinism, with the exception of miracles, in the physical world and freedom in the moral, see his Draft of 'Dynamical Theory'; Wilson, 'Kelvin's Scientific Realism', p. 58.

[105] The remoteness of God's presence and activity can also be seen by comparing Tait and Stewart's 'Unseen Universe' with Faraday's fields of force. For Faraday, God's activity was present at the empirical level. For Tait and Stewart, however, it was displaced to the other end of a hierarchy of invisible ethers; Cantor, 'Theological Significance', p. 140.

force, but he interpreted it within a mechanical framework much closer to Joule's ideas than to Faraday's own outlook. Kelvin was willing to adopt Faraday's notion of the magnetic field's independence of matter in terms of the mathematical formalism he helped to develop.[106] But, whereas Faraday had seen the lines of force as an alternative to the idea of an ethereal medium, Kelvin repeatedly tried to explain the transmission of electromagnetic force through space in terms of a mechanical ether and even proposed models of rotating molecular atmospheres or vortices, ideas similar to those proposed by Joule and which had since been refined by Rankine and Helmholtz.[107]

Partly out of deference to Faraday (as indicated by their correspondence), Kelvin carefully qualified his ethereal mechanisms as mere models, analogies, or illustrations: they were not to be taken literally as physical realities.[108] But he persisted in speculating all the same. Kelvin's long-range commitment to the importance of mechanical models was particularly remarkable in view of the difficulties he encountered in constructing one that was stable—by the late 1880s, he had virtually abandoned the hope of success.[109]

Still, all through his long life, Kelvin insisted, as Joule had done before him, that the demonstration of the intelligibility of electromagnetic phenomena depended on the eventual development of a viable mechanical model of the ether, however hypothetical it might be.[110] The result was a

---

[106] 'A Mathematical Theory of Magnetism' (1851); Harman, *Energy*, p. 82.

[107] 'On a Mechanical Representation of Electric, Magnetic, and Galvanic Forces' (1847) represented propagation of these forces in terms of the linear and rotational strain of an elastic, solid medium. A paper of 1856 explained the magneto-optic rotation discovered by Faraday in terms of Rankine's theory of molecular atmospheres in a fluid ether. An unpublished ms. of 1858 represented molecules in terms of minute eddies in a fluid Universal Plenum. 'On Vortex Atoms' (1867) represented the propagation of action along Helmholtz's vortex filaments in a perfect fluid, analogous to smoke rings. Kelvin's *Baltimore Lectures* (delivered, 1884; pub., 1904) abandoned the vortex theory but still disallowed a strictly electromagnetic theory of light and insisted on an elastic, gyroscopic ether; Robert H. Silliman, 'William Thomson: Smoke Rings and Nineteenth-Century Atomism', *Isis*, vol. liv (1963), pp. 461-74; Siegel, 'Thomson', pp. 242, 244ff.; Harman, *Energy*, pp. 79-84, 99ff. One of the reasons for Kelvin's commitment to the ether concept was the success of Fresnel's wave theory of light; D.B. Wilson, 'Kinetic Atom', *AJP*, vol. xlix (March 1981), p. 218b.

[108] Letter to Faraday, June 1847; Harman, *Energy*, p. 82.

[109] D.B. Wilson, 'Kinetic Atom', p. 220a. The stability of the vortex-atom model became a serious problem for Kelvin in 1887; Silliman, 'William Thomson', p. 472; Siegel, 'Thomson', p. 262.

[110] *Elements of Natural Philosophy* (coauthored with W.P.G. Tait, 1867); 'Nineteenth-Century Clouds over the Dynamical Theory of Heat and Light (lecture, 1900; pub. as an appendix to the *Baltimore Lectures*, 1904); Silliman, 'William Thomson', pp. 464f.; Harman, *Energy*, p. 149. D.B. Wilson presents Kelvin's alternatives as those posed by two schools of French physics: the 'Newtonian' or 'astronomical view' of nature of Laplace and Poisson (also Coulomb, Biot, and Navier); and the formalist or positivist school (which also claimed Newton as its patron) of Fourier (also Ampère); 'Kinetic Atom', p. 217; cf. Robert

sharp differentiation in Kelvin's thought between mathematical formalism and the physical understanding of reality. This was one paradox he never quite resolved.[111] At the end of his career, Kelvin saw his efforts to understand the relationship between fields and matter at the atomic level as a complete failure.[112]

The reason for Kelvin's recurring attempts to construct mechanical models for electromagnetic phenomena was partly theological. Kelvin's training in natural philosophy at the universities of Glasgow and Cambridge had been in the natural theology tradition of Samuel Clarke, William Derham, Thomas Reid, and William Paley (chap. 4, sec. A).[113] According to this tradition, one should expect to find evidences of design in creation particularly in the unforseen analogies between entirely different natural phenomena.[114] From the very beginning of his career, Kelvin had been able to identify and develop some of these analogies, particularly that between the equations of electrostatics and those for the flow of heat through a conducting body.[115] Amazingly, the exact same equations applied in the two cases.

The formal, mathematical analogy was enough by itself to provide the fondly sought evidence of design in nature, but it also suggested that some deeper connection was probably involved between the two phenomena. If Joule's dynamical theory of heat was correct in positing an atomic basis for the phenomenon of heat, mustn't there also be an atomic basis for electricity and magnetism?[116] The same could be said on the basis of the analogy between the propagation of light and waves of a fluid like water, which had been firmly established by A.J. Fresnel in 1821. Could the analogy be purely formal, or did it indicate a real similarity between the two

Fox, 'The Rise and Fall of Laplacian Physics', HSPS, vol. iv (1974), pp. 89-136.

[111]Harman, *Energy*, pp. 79-83; Buchwald, 'Thomson', pp. 375f., 377a.

[112]Address on the Jubilee of Kelvin's appointment as professor of natural philosophy at the Univ. of Glasgow (1896); D.B. Wilson, 'Kinetic Atom', p. 220a, note 35.

[113]C. Smith, 'Natural Philosophy', pp. 298ff.; D.B. Wilson, 'The Thought of Late Victorian Physicists', *Victorian Studies*, vol. xv (Sept. 1971), pp. 34f.; *idem*, 'Kelvin's Scientific Realism', pp. 42-48.

[114]Royal Institute Lecture (1860); D.B. Wilson, 'Concepts', p. 206. In 1871, Kelvin criticised Darwin's theory of natural selection for its apparent denial of divine design; 'Presidential Address to the British Association' (1871; D.B. Wilson, 'Kelvin's Scientific Realism', p. 50; *idem*, 'Concepts', p. 206). On belief in analogies between different phenomena of nature in the Renaissance Pythagorean tradition, see D. Koenigsberger, *Renaissance Man*, pp. 1ff. (on Leonardo), 7ff. (on Alberti), 90ff.

[115]Kelvin developed the analogy in 1842. Maxwell developed the idea of a 'flux' of lines of force; Harman, *Energy*, pp. 29f., 79, 87; *idem*, ed., *Wranglers*, pp. 7f. Richard Olson argues that the critical investigation of analogies was characteristic of Scottish common sense realism; *Scottish Philosophy*, pp. 48-53 (on Reid), 114-21 (on Stewart), 146-49 (on William Hamilton), 226-28 (on J.D. Forbes), though he does not develop the point for Kelvin's background as he does for that of Maxwell (*ibid.*, pp. 288-93). Harman, on the other hand, views Maxwell's emphasis on physical analogies as due to the influence of George Gabriel Stokes and Lord Kelvin; 'Edinburgh Philosophy', p. 203.

[116]Siegel, 'Thomson', pp. 240f.; Buchwald, 'Thomson', pp. 383f.

phenomena? If the latter, mustn't there be a medium through which light waves propagated, a medium with a molecular structure analogous to water's? Thus Kelvin's belief in the divine design of nature sustained his belief in the existence of an ether with mechanical properties, even though his ingenious mechanical models failed to verify the idea.

Kelvin's theological conviction of the deep inner connections of nature was one aspect of what we have called the creationist tradition. However, his insistence that any acceptable mechanical description of phenomena be given in clearly visualisable terms was another matter. Like Newton, Kelvin insisted that only qualities that were absolutely universal to physical bodies (qualities that were could not be absent or even present in varying degrees) could be accepted as irreducible and primary in mechanical theory. According to Newton, the only attributes that could be regarded as primary were extension, incompressibility, and inertia.[117] Other properties of matter, such as elasticity, had to be explained in terms of these more fundamental ones. Otherwise, they were not admissible.[118] In fact, Kelvin even hoped to account for inertia in terms of molecular vortices.[119] In these strictly Newtonian terms, the mechanical approach to the physics of atoms was eventually found to fail. The future of atomic physics lay not with the mechanical philosophy of Kelvin, but with a more flexible approach that would allow the mechanical philosophy to be generalised. The beginnings of such an approach were due to the efforts of James Clerk Maxwell.

D. THE MECHANICAL PHILOSOPHY GENERALISED (MAXWELL)

James Clerk Maxwell (1831-79) was a Scot who studied at the universities of Edinburgh (1847-50) and Cambridge (1850-56). Like Kelvin, therefore, he was trained in Scottish philosophical theology[120] and the

---

[117] Newton's 'Third Rule of Reasoning' allowed as 'universal qualities of all bodies whatsoever' only those 'which admit neither intensification nor remission of degrees, and which are found to belong to all bodies within the reach of our experiments' (*Principia*, ed. F. Cajori, p. 398). Whewell had regarded it as 'a mere rule of prudence' but accepted it all the same; D.B. Wilson, 'Herschel and Whewell's Version of Newtonianism', *JHI*, vol. xxxv (Jan. 1974), pp. 90f.

[118] Thus Kelvin was critical of Maxwell's allowance of unreduced elasticity; Buchwald, 'Thomson', pp. 384b, 385f.

[119] Silliman, 'William Thomson', pp. 469f.

[120] According to Richard Olson, the principal Scottish influence was from Thomas Reid and Dugald Stewart, transmitted by J.D. Forbes (a protégé of William Whewell, however) and William Hamilton of Edinburgh. Characteristic of this 'common sense realism' was an emphasis on the guidance of mathematical formalism by physical insight and analogy (*Scottish Philosophy*, pp. 225-26, 232-35, 290-91) and a sense of the artificiality of all scientific theories as abstractions or partial views of nature (*ibid.*, 293f.). Harman criticises

tradition of mixed mathematics and natural theology established by Whewell and others at Cambridge.[121] He frequently cited the basic creationist ideas of laws impressed on nature (at one point citing Wisd. 11:20: 'thou hast arranged all things by measure and number and weight') and the divine image in humanity, resulting in the comprehensibility, unity, and relative autonomy of the world.[122] The God who had revealed himself to Abraham, Isaac, and Jacob was for Maxwell the same as the God of nature (cf. Ps. 146:5f.).[123] The act of divine creation was a presupposition of all science, and, though it was not itself open to scientific explanation, it was clearly evidenced in seemingly inexplicable features of the cosmos such as the identical properties of molecules in all parts of the universe, near and far.[124] The fact that Maxwell limited scientific investigation to the cosmos as it exists today and excluded any consideration of its origin reflects the basically Newtonian (pre-evolutionary) framework within which he and other British physicists of his time worked.

Maxwell had affinities with both of the categories of nineteenth-century physical scientists we have reviewed. Like Faraday, he approached natural philosophy in an intuitive way and regarded the electromagnetic field as an independent entity, not reducible to molecular mechanics. Yet, like Kelvin, he also speculated on mechanical models for the ether and carried on the process of formalising classical dynamics.

Maxwell belongs in a category of his own, however, due to the skill with which he synthesised and generalised the ideas of others, particularly in two areas. First, there was his pioneering work in statistical mechanics. Maxwell attempted to bridge the gap between the laws of thermodynamics and the dynamics of gaseous molecules, while recognising the relative

Olson for exaggerating the Scottish influence, that of Hamilton, in particular; 'Edinburgh Philosophy', pp. 202ff., 221. However, Harman does allow Hamilton's influence in Maxwell's early writings; *ibid.*, pp. 213-14.

[121]Cannon, 'Scientists', pp. 72f. Paul Theerman argues that the evangelical piety of the Rev. C.B. Tayler and the Coleridgean influence of F.D. Maurice (via the Apostles' Club) moved Maxwell away from the rationalist approach of Paley, which had previously been dominant at Cambridge; 'James Clerk Maxwell and Religion', *AJP*, vol. liv (April 1986), pp. 312f. Consequently, Maxwell based his convictions about nature on his creational beliefs, not the other way around. The same emphasis is found in Faraday and Kelvin, however, and need not be derived directly from evangelical or Coleridgean influence.

[122]'Molecules' (BAAS Lecture, 1873; W.D. Niven, ed., *The Scientific Papers of James Clerk Maxwell*, 2 vols. [Cambridge: Cambridge Univ. Press, 1890], 2:377); Theerman, 'Maxwell', p. 316a.; cf. Wilson, 'Thought', pp. 35f.

[123]Letter to Lewis Campbell, March, 1852; Ivan Tolstoy, *James Clerk Maxwell* (Edinburgh: Canongate, 1981), p. 58.

[124]Presidential Address to the Mathematical and Physical Sections of the BAAS (1870); 'Molecules' (1873); 'Atom' (Encyclopedia Britannica Article [*c.* 1875], in which Maxwell takes account of Prout's hypothesis; *Scientific Papers*, 2:224f., 375ff., 478-84); Heimann, 'Unseen Universe', p. 75. Maxwell referred to Thomas Chalmers's idea (published in his *Natural Theology*) that divine design could be seen in the original 'collocation' of matter; 'Molecules', p. 377; 'Atom', p. 483; P.M. Heimann, 'Molecular Forces, Statistical Representation and Maxwell's Demon', *SHPS*, vol. i (1970), pp. 202f. and note 102.

independence of those two fields and the essential role of statistics in the overall worldview of natural philosophy. Second, Maxwell developed a generalised version of the mechanical philosophy that incorporated Faraday's concept of independent fields and so opened the way for developments of twentieth-century physics. As in the case of Kelvin, we shall treat these two areas in turn, beginning with statistical mechanics.

1 Maxwell on Statistical Mechanics

Maxwell was the first to recognise that there was no straightforward relationship between the macroscopic laws of thermodynamics and the dynamics of molecules as understood in Newtonian terms. In particular, he departed from the strictly mechanical interpretation of thermodynamic functions introduced by Rudolf Clausius (1862) and Ludwig Boltzmann (1866).[125]

Clausius had interpreted the thermodynamic properties of heat-energy and entropy as straightforward measures of the kinetic energy and spatial configuration of the molecules.[126] In the 1860s, however, Maxwell developed a radically new approach to the concepts of thermodynamics,[127] using the ideas and methods of the newly developed science of statistics.[128] Concepts like temperature and entropy only applied to large aggregates, he concluded, and were meaningless when applied to individual molecules.[129] The temperature of a system was still determined by the average of the kinetic energies of the constituent molecules, but the energy af any given molecule was largely random given the macroscopic, thermodynamic parameters.

Energy was still strictly conserved, in accordance with the first law of thermodynamics, and was time invariant. However, the dissipation of energy required by the second law of thermodynamics was a very different matter. Not only did it imply a direction in time, as Kelvin had pointed out, but it

---

[125]Maxwell ignored Clausius's work in the first edition of his *Theory of Heat* (1871; C.W.F. Everitt, 'Maxwell', *DSB* 9:227a, gives 1870 as the date) but referred to it, after becoming aware of Clausius's criticisms, in the corrected edition of 1872 (the one which is still in print); Harman, *Energy*, p. 66.

[126]Harman, *Energy*, pp. 65f., 141.

[127]Maxwell first worked out the distribution function for the number of molecules within a given velocity range in 'Illustrations of the Dynamical Theory of Gases' (1860); Harman, *Energy*, pp. 129f.; Everitt, 'Maxwell', p. 218.

[128]Maxwell's interest in statistics was aroused by J.D. Forbes at Edinburgh in 1848. He later studied the works of Laplace, Boole, Herschel's 1856 review of Adolphe Quetelet's *Theory of Probability*, and Henry Buckle's *History of Civilization in England* (1857); Everitt, 'Maxwell', pp. 218f.; Ian Hacking, 'Nineteenth Century Cracks' in the Concept of Determinism', *JHI*, vol. xliv (July 1983), p. 471; cf. Theodore M. Porter, *The Rise of Statistical Thinking, 1820-1900* (Princeton: Princeton Univ. Press, 1986).

[129]Heimann, 'Molecular Forces', pp. 189f.; idem, *Energy*, pp. 66, 131.

was not mechanically determined: it was only the most probable course of the system's temporal development.[130] In other words, we cannot be absolutely certain that energy will dissipate (i.e., that entropy will increase) in any given time interval. Thermodynamic systems are not deterministic in the same sense that mechanical systems operating in accordance with Newton's laws of motion are.[131] But though we cannot be absolutely certain, we can be morally certain that, for all practical purposes, energy will dissipate over extended periods of time.[132]

The distinction Maxwell made here between absolute (mechanical) and moral certainty can only be understood with reference to the parallel issue of free will and determinism—an issue that he addressed in 1856 as a university student at Cambridge and again in a paper of 1873.[133] Maxwell argued that all descriptions, including scientific theories, are at best abstractions or partial views of the complex reality of nature.[134] For example, according to one description, we experience freedom in our ability to act as moral agents, while, according to another, we observe strict determinism in our scientific analysis of the motions of bodies.

Kant had resolved the paradox of free will and determinism by making freedom an unanalysable precondition for moral action while, at the same time, requiring a completely deterministic account of the external, objective world. Maxwell's approach was rather different.[135] For him, the laws of

---

[130] Harman, *Energy*, pp. 139f.

[131] The fact that the dynamics of the individual molecules was not determined in the laws of thermodynamics was brought out by Maxwell's use of the 'kinetic [rather than dynamical] theory of gases' after 1870-71; Heimann, 'Molecular Forces', pp. 199ff. In 1873, Maxwell argued that even the equations of Newtonian dynamics were not deterministic in special cases (pointed out earlier by Saint-Venant and Boussinesq): they had no well-defined (finite) solutions near singularities; 'Does the Progress of Physical Science Tend to Give Any Advantage to the Opinion of Necessity (or Determinism) Over that of the Contingency of Events and the Freedom of the Will?' (talk to the Eranus Club, consisting of former members of the Apostles' Club, 1873; printed in Lewis Campbell and William Garnett, *The Life of James Clerk Maxwell* [London, 1882], pp. 483-89); cf. 'Paradoxical Philosophy', (*c.* 1878; *Scientific Papers*, p. 760); Stephen G. Brush, *Statistical Physics and the Atomic Theory of Matter* (Princeton: Princeton Univ. Press, 1983), p. 89; Hacking, 'Nineteenth Century Cracks', pp. 461, 464f.

[132] 'Molecules'; Harman, *Energy*, p. 131.

[133] 'Analogies' (lecture to the Apostles' Club, 1856; in Campbell and Garnett, eds., *Life*, pp. 235-44).

[134] Cf. William M. Rankine's idea of 'abstractive theories', based on the ideas of Thomas Reid and John Gregory; Olson, *Scottish Philosophy*, p. 45. For Maxwell, even number was an abstraction, for nature tends to be both more unified than our distinctions and more diverse than our generalisations; 'Analogies' (Campbell and Garnett, eds., *Life*, pp. 236f.).

[135] Maxwell had read Kant's *Critique of Pure Reason* in German at Edinburgh in 1850; Tolstoy, *James Clerk Maxwell*, p. 40. Kantian ideas would also have been picked up from Whewell, who had posed some of the questions Maxwell addressed in 'Analogies', according to Harman, 'Edinburgh Philosophy', p. 213.

mind and the laws of nature were both aspects of reality created by God,[136] but only partial aspects. One could, therefore, arrive at a view of reality entailing either freedom or determinism depending on how one focussed the instruments of observation and analysis on the events involved:[137]

> The dimmed outlines of phenomenal things all merge into one another unless we put on the focussing glass of theory and screw it up sometimes to one pitch of definition, and sometimes to another, so as to see down into the different depths through the great millstone of the world.

The paradox of moral freedom and mechanical determinism, then, provided the paradigm for Maxwell's later analysis of thermodynamics, where he found an analogous paradox of statistical laws and directionality, on the one hand, and mechanical determinism and time-invariance, on the other.[138]

Theoretically, there could be massive random fluctuations in the dynamics of the molecules that would reverse the dissipation of energy, just as, theoretically speaking, it would be possible for millions of pieces of glass to converge and form a flawless window. Maxwell employed the idea of an imaginary demon (dubbed Maxwell's 'sorting demon' by Kelvin) in order to show that such fluctuations could conceivably be engineered in such a way as to contradict the uniform experience of what we observe in nature, but without violating the laws of Newtonian mechanics.[139] In other words, we may accept the validity of the second law of thermodymamics, even though it is not absolutely, mechanically certain, just as we trust in the sovereignty of God in all matters and ignore the possible activity of mischievous demons. The second law of thermodynamics could not be reduced to the principles of Newtonian mechanics:[140] it involved an additional supplementary principle which was never clearly defined, but which amounted, in Maxwell's context, to trust in the sovereignty and good will of God.

The student of twentieth-century physics can not fail to see here a remarkable preview of one of Einstein's philosophical ideas. According to Einstein, the laws of nature must reflect the fact that, though God is subtle, he is not deceptive or devious.[141] Einstein stated his own view in the

---

[136] Innaugural Address at Aberdeen (1856); Theerman, 'Maxwell', p. 314b.

[137] Later Maxwell speaks of 'plac[ing] the telescope of theory in proper adjustment, to see not the physical events which form the subordinate foci of the disturbance propagated [as the consequence of our actions] through the universe, but the moral foci where the true image of the original act is reproduced'; 'Analogies' (Campbell and Garnett, eds., *Life*, pp. 237, 241f.).

[138] 'Molecules' (*Scientific Papers*, 2:374); Heimann, 'Molecular Forces', pp. 189, 202.

[139] Letter to Tait, 1867; *Theory of Heat*; Harman, *Energy*, pp. 139f.; Everitt, 'Maxwell', p. 227a; Jaki, *Science*, p. 297.

[140] Letter of 1870; Harman, *Energy*, p. 142.

[141] 'Raffiniert ist der Herrgott aber boshaft ist er nicht'; cited by John Stachel, *Science* vol. 218 (3 Dec. 1982), p. 989c.

context of the later debate concerning the completeness of quantum mechanics and used it to argue against the legitimacy of indeterminism in science. However, the underlying faith in the reliability of natural law, grounded in the trustworthiness of God, was characteristic of the tradition he inherited from nineteenth-century physicists like Maxwell.

One can also find a preview of Niels Bohr's principle of complementarity in the ideas of Maxwell just discussed. Like Maxwell, Bohr viewed nature as a complex reality which could not be captured in any one mode of description. Complementary descriptions of nature were required as a consequence of the differing 'possibilities of definition' in human science.[142] Like Maxwell, Bohr cited both the paradox of moral freedom and physical determinism and the statistical nature of the laws of thermodynamics as examples of these complementary modes of definition.[143] The nature of the historical relationship between Maxwell, Einstein, and Bohr will be explored further in the concluding section of this chapter as a way of showing the long-range influence of creationist ideas on twentieth-century physics.

2 Maxwell on Electricity and Magnetism

We have seen how persistent Kelvin was in the search for mechanical models of the ether sustaining electric and magnetic fields. Maxwell was more ambiguous. As in the case of Kelvin, we shall sketch some of the shifts in his thinking on the subject and then reflect on the theological aspects.

In spite of his many reservations, Maxwell still viewed the programme of the physical sciences as one of mechanical explanation as it was originally projected by Descartes, Charleton, and Boyle in the seventeenth century (chap. 3, sec. C).[144] Like Kelvin, he made persistent efforts to develop a mechanical model for electromagnetic phenomena, even using Kelvin's idea of ether vortices, not as a literal physical description of reality, but as an analogy or heuristic illustration to aid in comprehension and to demonstrate that mechanical explanation of the propagation of electromag-

---

[142]Bohr, *Atomic Theory and the Description of Nature* (London: Cambridge Univ. Press, 1934), pp. 52ff.; idem, 'Causality and Complementarity', *Philosophy of Science*, vol. iv (1937), pp. 295f.; *Essays 1958-1962* (New York: Wiley/Interscience, 1963), p. 5.

[143]Bohr, *Atomic Theory*, p. 24; idem, *Atomic Physics and Human Knowledge* (New York: Wiley, 1958), pp. 11, 78; 'Chemistry and the Quantum Theory of Atomic Constitution' (Faraday Lecture, 1930), *Journal of the Chemical Society* (1932), part 1, pp. 349-84.

[144]'Physical Sciences', Encyclopedia Britannica, 9th ed. (1875); Harman, *Energy*, p. 1. Maxwell's early commitment to the mechanical philosophy is stressed by Harman, ed., *Wranglers*, pp. 8f.

netic force was possible, at least, in principle.[145] Maxwell temporarily abandoned the effort to construct a model of the electromagnetic field itself in the mid 1860s[146] but continued to the end of his life to seek molecular models of the ether that supposedly underlay the field.[147] In his magnum opus, the *Treatise on Electricity and Magnetism* (1873), he made a point of formulating the equations governing the phenomena in terms of abstract parameters that made no explicit reference to the dynamics of bodies of any sort. For example, the conservation of energy was made a fundamental principle, but not expressed in terms of the dynamics of ether particles.[148]

Thus Maxwell followed Faraday in treating the field as an independent reality in its own right: the electric and magnetic fields could even be viewed as reservoirs of their own form of energy, a form comparable to that of the kinetic energy, potential energy, and heat associated with material bodies.[149] But, unlike Faraday, Maxwell mathematised the field concept along the lines of Lagrange's and Kelvin's treatment of Newtonian dynamics.[150] He also developed a physical model in which the energy stored in the field was just the sum of the kinetic energy and potential energy of the ether particles, and electromagnetic force was transmitted from particle to particle through the ether, again drawing on Kelvin's concept of molecular vortices.[151]

As heuristic illustrations, these mechanical models did more than just illustrate the mathematical equations for the fields as they were already known. Even though an endless variety of models could be constructed that were consistent with the equations,[152] the one that Maxwell developed in most detail turned out to have implications which suggested important changes in the equations themselves.[153] Maxwell found that, in order to devise a model in which the ether vortices could be interlocked without

---

[145]'On Physical Lines of Force' (1861-62); 1867 letter to Tait (comparison to an orrery); Harman, *Energy*, pp. 6, 89, 92.

[146]'A Dynamical Theory of the Electromagnetic Field' (1865); Harman, *Energy*, pp. 6, 72. Maxwell began his 'tactical retreat' from mechanical models in 1864 according to Siegel, 'Thomson', pp. 259f.

[147]Siegel, 'Mechanical Image', pp. 199f.

[148]*Treatise on Electricity and Magnetism*, 2 vols. (1st ed., 1873); Harman, *Energy*, pp. 70, 95.

[149]'Dynamical Theory'; Harman, *Energy*, p. 94; Everitt, 'Maxwell', p. 210a.

[150]*Treatise* (3rd ed., 1891), 2:176f. On Maxwell's use of Thomson and Tait's *Treatise*, see Harman, *Energy*, pp. 70, 95, 97.

[151]Maxwell, 'Dynamical Theory'; *Treatise* (1873); Harman, *Energy*, pp. 84, 95; William Berkson, *Fields of Force* (New York: Halsted, 1974), pp. 174f. Berkson describes Kelvin's and Maxwell's interest in mechanical models as a 'Cartesian metaphysic'; *Fields*, pp. 126, 136ff., 149, 172, 181.

[152]Harman, *Energy*, p. 94.

[153]'Physical Lines' (1861-62). However, the altered equations could be, and were, presented without reference to the ether model used in deriving them; 'Note' (1868); Harman, *Energy*, p. 109.

cancelling each other out, he had to postulate an imaginary electric current passing between rows of the vortices. It turned out that this imaginary current (later called the 'displacement current') made a contribution of its own to the magnetic field in spaces where there was no real current but only a time-varying electric field created by the buildup of electric charge nearby. The inclusion of the term representing this contribution in the equations revealed an unforseen symmetry between the electric and magnetic field equations and led to a mathematical solution, according to which the fields propagated through space as waves. Calculation then showed that the electromagnetic waves would travel through space at approximately the speed of light (300,000 kilometers per second).[154] Light could, therefore, be understood as a special form of electricity and magnetism.

One result of Maxwell's tinkering with mechanical models was thus the unification of the science of optics with that of electricity and magnetism. A further result was the discovery of radio waves (by Heinrich Hertz in 1888) and the eventual development of the technologies on which modern communications are based.

In many ways, Maxwell's vascillations on the relation of fields to molecular mechanics are similar to Kelvin's. However, they differed in two important ways. First, in the late 1860s and 1870s, Maxwell broke with the Newtonian 'analogy of nature'—the requirement that any description of atoms or molecules be made strictly in terms of the primary qualities that apply to macroscopic bodies.[155] First and foremost of the primary qualities of gross matter was extension. Since the time of Descartes, extension had normally been viewed as the very essence of matter.[156] It was axiomatic, therefore, that no two bodies could occupy the same portion of space. Maxwell wondered: why couldn't the properties and laws of submicroscopic, atomic objects be radically different from those of macroscopic bodies with which we are familiar in everyday experience?[157] Perhaps two atomic

---

[154]'Physical Lines' (*Scientific Papers*, 1:500); Harman, *Energy*, pp. 92f.; Everitt, 'Maxwell', p. 209a. Maxwell later derived the wave equation for light without recourse to the mechanical model in 'A Dynamical Theory of the Electromagnetic Field' (1864-65); Berkson, *Fields*, pp. 177f.

[155]Heimann, 'Maxwell', pp. 211ff. In his review of the 2nd (1879) ed. of Thomson and Tait's *Principles*, Maxwell cited Berkeley's critique of Newton's idea of a material substratum for activity in nature; *Scientific Papers*, 2:781; Harman, 'Edinburgh Philosophy', p. 223.

[156]An exception was Boscovich's infinitesimal point model of matter; cf. chapter 4, section A above.

[157]Ms. entitled, 'Dimensions of Physical Quantities' (early 1870s); 'Does the Progress of Physical Science' (1873; Campbell and Garnett, eds., *Life*, p. 439); Heimann, 'Maxwell', pp. 211f.; idem, 'Molecular Forces', pp. 198f.

objects could even coincide in their location in space.[158] The unity of nature and analogies among its various levels did not require complete uniformity!

Thus, in spite of the immense strides of physical science in his own lifetime and contrary to the views of some of his contemporaries, Maxwell did not see the work of physics as anywhere near completion. In fact, he concluded one of his last papers with the statement that it was time for physicists to adopt an attitude of 'thoroughly conscious ignorance that is the prelude to every real advance in knowledge'.[159] The advances of nineteenth-century physics had raised unforseen problems like those associated with the discovery of atomic spectra. In Maxwell's view, therefore, new phenomena and entire new fields of research were waiting to be discovered which would require the development of entirely new forms of thought.[160]

This open, adventurous outlook was not just based on personal experience or the history of physics. For Maxwell, it was based on his belief in 'the unsearchable riches of creation' (cf. Job 5:9; Eph. 3:8) and 'the untried fertility of those fresh minds into which these riches will continue to be poured'.[161] In other words, it was based on the twin aspects of the doctrine of creation: the cosmos as the work of an infinite and all-powerful God, and the human mind as created in the image of God and hence capable of comprehending the mysteries of nature.[162] The courage and open-mindedness that Maxwell modelled for his twentieth-century followers was deeply rooted in the creationist tradition.

The second difference from Kelvin was that Maxwell was able to step back from the problem of mechanical models far enough to reflect on its similarity to the paradoxes he had encountered in his work on thermodynamics. Maxwell then resolved the problem by affirming that both field and particle pictures were valid in their own way: it was good to have two ways of looking at a subject and not confine oneself to a single perspective.[163] In

---

[158]'Dynamical Theory of Gases' (1867; *Scientific Papers*, 2:33); 'Atom' (*c*. 1875; *Scientific Papers*, 2:448); P.M. Heimann, 'Maxwell and the Modes of Consistent Representation', *AHES*, vol. vi (1970), p. 212; *idem*, 'Molecular Forces', p. 198.

[159]*Nature* 20 (1877), p. 242; Everitt, 'Maxwell', p. 223a.

[160]Presidential Address (1870; *Scientific Papers*, 2:227).

[161]'Introductory Lecture on Experimental Physics'(1871?; *Scientific Papers*, 2:244); Martin Goldman, *The Demon and the Aether* (Edinburgh: Paul Harris, 1983), pp. 186f.

[162]'...the laws of matter and the laws of mind are derived from the same source, the source of all wisdom and truth', Innaugural Address at Aberdeen (1856); '[we] elicit the living truth, not from the facts [alone], but either from the utterer of facts or the giver of Reason, which two are one, or reason would never decipher facts', Letter to Campbell, 7 Aug. 1857 (Campbell and Garnett, eds., *Life*, p. 273); D.B. Wilson, 'Kelvin's Scientific Realism', p. 52; Theerman, 'Maxwell', p. 314b.

[163]'On Faraday's Lines of Force' (1856); Olson, *Scottish Philosophy*, p. 296. According to Olson, it is therefore possible to view the varying attitudes towards mechanical models in Maxwell's successive writings as different ways of focusing his lens of analysis on the

his major treatise on the subject (1873), he expressed a personal preference for the more holistic field theory of Faraday. But he backed away from Faraday's notion that fields were the primary reality and made them dependent on the presence of material bodies as their sources and sinks.[164] Likewise, Maxwell shied away from the pure formalism of field equations[165] and argued for the use of concrete mechanical models as a means of coordinating the mathematical formalism with physical reality.[166] Then again, in 1870, he stated that both approaches were human abstractions which appealed to physicists accustomed to different styles of thought.[167]

There was an important ambiguity here that has had repercussions in the twentieth-century debates about field theory. Granted Maxwell's general idea that reality is a plural unity or a unified plurality: are we to view the formal equations of the field approach as exhibiting the true wholeness of things and view the molecular models as abstractions? Or are we to identify the composite of field and particle pictures—the formal and the mechanical—as representing the whole of which both field and particle pictures are abstractions? Albert Einstein and Niels Bohr can both be viewed as disciples of Maxwell in as much as each of them adopted one of these approaches in the quest for harmony and wholeness.[168]

E. CONCLUSION: THE CONTRIBUTION OF THE CREATIONIST TRADITION TO TWENTIETH-CENTURY PHYSICS (EINSTEIN AND BOHR)

We have traced the history of the creationist tradition in relation to the physical sciences from the second century BC to through the nineteenth century AD. Major contributors to the sciences during those twenty-one centuries were frequently inspired by the belief that God had created all things in accordance with laws of his own devising, laws which made the

---

subject at hand; *Scottish Philosophy*, pp. 295ff.; *pace* Harman, *Wranglers*, pp. 8f.

[164] *Treatise* (3rd ed., 1891), 2:176f.; Heimann, 'Maxwell', p. 211; Berkson, *Fields*, p. 175.; Olson, *Scottish Philosophy*, pp. 294, 297.

[165] Hence Maxwell's emphasis on 'embodied mathematics'; Olson, *Scottish Philosophy*, pp. 302f.

[166] Harman, *Wranglers*, pp. 9f.

[167] Presidential Address (1870; *Scientific Papers*, 2:219f.); Heimann, 'Molecular Forces', p. 206; Robert Kargon, 'Model and Analogy in Victorian Science: Maxwell's Critique of the French Physicists', *JHI*, vol. xxx (July 1969), p. 435; Olson, *Scottish Philosophy*, pp. 297f. A comparison with 'On Action at a Distance' and 'Faraday' (*Scientific Papers*, 2:318, 359f.) shows that Maxwell had the French mathematical physicists (e.g., Poisson and Ampère) and Faraday in mind as respective examples of the formal and geometrical styles.

[168] Thus Thomas F. Torrance argues that Maxwell's theory was plagued by a residual dualism of fields and particles, and that this dualism was later eliminated by Einstein; *Christian Theology and Scientific Culture* (New York: Oxford Univ. Press, 1981), pp. 54-55; idem. *Transformation and Convergence in the Frame of Knowledge* (Grand Rapids, Mich.: Eerdmans, 1984), p. 234.

world comprehensible to humans and gave the world a degree of unity and relative autonomy, and that God had sent his Son and poured out his Spirit to initiate a worldwide ministry of healing and restoration.

We have also found that the creationist tradition began to unravel in the twelfth century with the polarisation between theologies emphasising the workings of nature and the truths of reason, on the one hand, and the supernatural and the suprarational mysteries of revelation, on the other. This was not a conflict between science and religion. Both sides of the issue were rooted in the creationist tradition and both made significant contributions to the development of science. At various junctures, attempts were made to synthesise nature and supernature in a recovery of biblical thought, often in conjunction with extrabiblical philosophies like those of Aristotle, Neoplatonism, and hermeticism. Nonetheless, the process of fragmentation and secularisation continued to reassert itself until the decline of the creationist outlook as an international, public tradition in the late eighteenth and nineteenth centuries.

It is beyond dispute that the the creationist tradition made significant contributions to the rise and development of both medieval and classical (seventeenth- to nineteenth-century) physics. The major breakthroughs in astronomy, medicine, mechanics, chemistry, thermodynamics, and electricity and magnetism were all associated with theological ideas related to God and creation. It would appear, however, that the triumph in the nineteenth century of individualism in religion and professionalism in the sciences had severely reduced the likelihood that scientific developments of the twentieth century would be embedded in a similar theological matrix.

Certainly physicists of the twentieth century are more diverse in their religious beliefs. Many would be reluctant to identify themselves with any theological tradition at all. And even where particular beliefs may be held privately, they are not as likely to play a dynamic role in the choice of science as a profession or in the quest for insight into nature as they were in medieval and early modern times. In any case, scientists no longer include prayers in their professional writings, as Kepler did, or draw attention to the existence of God, as Maxwell still did as late as the 1870s. The most that can be said is that a few scientists have allowed the possibility of God's existence in their more popular writings.

In this final section, however, we shall argue that remnants of the creationist tradition played a key role in the foundations of twentieth-century physics in the work of its two principal founders, Albert Einstein (1879-1955) and Niels Bohr (1885-1962). We cannot hope to do justice to the philosophies of either Einstein or Bohr in their own right—we shall not even treat them separately. But we must ask what contribution, if any, the creationist tradition has made in the case of these two figures that form a bridge from the nineteenth to the twentieth century. If ideas and beliefs have a momentum of their own, we might expect to find traces in their work of the same beliefs that inspired their predecessors, particularly those in the tradition of Michael Faraday and James Clerk Maxwell, whom Einstein and Bohr so admired and emulated.[169]

It is well known that Einstein and Bohr differed strongly on many issues, particularly those concerning the adequacy of the quantum-mechanical formalism, the development of which they both did so much to further. Einstein's 1905 paper on the photoelectric effect first showed that light was quantised in units (later called 'photons') whose momentum and energy were directly related to the wavelength and frequency of the light waves. It is to Einstein also that we owe the mathematical formula for the probability of the radiation of light from an atom (1916).[170] Subsequently, the ideas of discontinuity and statistical explanation became basic ingredients of quantum mechanics. In spite of Einstein's pioneering work in these areas, he himself insisted on continuity and completeness of dynamical description (not to be equated with 'determinism' in the classical sense) and saw this as required by the field theory of Faraday and Maxwell.[171]

Bohr's 1913 theory of the hydrogen atom provided the first working model of the new mechanics describing the interaction of atoms and light. Bohr also provided the most influential interpretation of the fully developed quantum mechanics of the late 1920s with his principles of 'correspondence' and 'complementarity'. Unlike Einstein, however, he judged these developments to be consonant with overall principles of natural philosophy and argued for their being foundational, if not final, in the progess of modern physics.[172] Whereas Einstein pointed to Maxwell's field theory as the precedent for his own work, Bohr looked back to the beginnings of

---

[169]Einstein first learned of Maxwell's theories (though probably not Maxwell's theology) through German accounts like those of August Foppl and Ludwig Boltzmann; Gerald Holton, *Thematic Origins of Scientific Thought* (Cambridge, Mass.: Harvard Univ. Press, 1973), pp. 201-2, 205-7; Abraham Pais, *'Subtle Is the Lord...': The Science and the Life of Albert Einstein* (Oxford: Clarendon Press, 1982), pp. 66-67. He saw his special theory of relativity as the natural completion of the work of Faraday, Maxwell, and Lorentz; Pais, *Subtle*, p. 30. Bohr's contact with Maxwell's ideas came through J.J. Thomson and Ernest Rutherford, with whom he worked in 1911-12 at Cambridge and Manchester.

[170]Martin J. Klein, 'Einstein', *DSB* 4:317.

[171]Maxwell's representation of physical reality by continuous fields was, for Einstein, the most profound and fruitful innovation in physics since the time of Newton; 'Clerk Maxwell's Influence on the Evolution of the Idea of Physical Reality' (1931), *The World As I See It* (London: John Lane the Bodely Head, 1935), pp. 157, 159-60; Pais, *Subtle*, p. 319.

[172]'On closer consideration, the present formulation of quantum mechanics, in spite of its great fruitfulness, would seem to be no more than a first step in the necessary generalization of the classical mode of description...', 'Causality and Complementarity', p. 294.

atomic theory under Maxwell and his successors, J.J. Thomson and Ernest Rutherford.[173]

Einstein and Bohr were both products of late nineteenth-century European culture. Ethnically a German Jew, Einstein was steeped in the literature of nineteenth-century German philosophy; Bohr was raised in cosmopolitan Copenhagen (his mother was also Jewish), where the primary influences mediated were those of nineteenth-century England and Germany. He further came 'under the spell of Cambridge and the inspiration of the great English physicists' (Thomson, Jeans, Larmor, and Rutherford) during his postgraduate studies in 1911-12.[174]

Though formal religion was not taken seriously in either of their families, both Einstein and Bohr struggled with religious questions in their youths.[175] Both expressed appreciation for the religious sense as that was understood in the liberal, romantic vein of the nineteenth century. Einstein spoke of a 'cosmic religious feeling' that was common to creative scientists

---

[173]'Maxwell and Modern Theoretical Physics', *Nature*, vol. 128 (24 Oct. 1931), pp. 691-92. Thomson had studied under Maxwell (1876ff.), succeeded to the chair of physics at the Cavendish Laboratory at Cambridge (1884), which Maxwell had helped to found, edited the 3rd ed. of Maxwell's *Treatise on Electricity and Magnetism* (1891), adopted Faraday's and Maxwell's view of electromagnetic fields as independent of the molecular structure of matter (1893), and carried on Maxwell's quest for a mechanically stable model of the atom (1904, 1910, the latter of which may already have influenced Bohr's doctoral thesis of 1911); G.P. Thomson, *J.J. Thomson*, pp. 26-29, 85, 131-32; P.M. Heimann, 'Maxwell and the Modes', pp. 207-8; J.L. Heilbron, 'Rutherford-Bohr Atom', *AJP*, vol. xlix (March 1981), pp. 224-26, 230; J.L. Heilbron and T.S. Kuhn, 'The Genesis of the Bohr Atom', *HSPS*, vol. i (1969), pp. 216, 224, 226-27, 243. Rutherford (a New Zealander of Scots ancestry) did postgraduate studies under Thomson at the Cavendish (1895ff.), collaborated with him on an important paper on X-rays (1896), developed an alternative, 'nuclear model' of the atom (1911), and succeeded to the chair of physics at the Cavendish (1919); E.N. da C. Andrade, *Rutherford and the Nature of the Atom* (Garden City, N.Y.: Doubleday/Anchor, 1964), pp. 25-27, 35-38, 156-57; Heilbron, 'Rutherford-Bohr Atom', pp. 224-25.

[174]Bohr, 'Maxwell', p. 691a. Cf. Bohr's reference to 'the early and intimate contacts which it was my good fortune to entertain with the great school of English physicists', 'Newton's Principles and Modern Atomic Mechanics', in *Newton Tercentenary Celebrations*, The Royal Society of Great Britain (Cambridge: Cambridge Univ. Press, 1947), p. 61. According to Heilbron and Kuhn ('Genesis', p. 223), Niels's decision to do postdoctoral work at Cambridge was partly due to the fact that his father, Christian Bohr, had instilled in him a deep love for all things English. As he said in a later interview, 'I considered first of all Cambridge as the center of physics.' Cf. Bohr's reference to 'the unbroken tradition upheld here in Cambridge connecting his [Maxwell's] life and his work with our time', 'Maxwell', p. 691a. Note also the self-revealing comment in Bohr's correspondence: 'Sometimes, e.g. today at the big lectures (by Jeans and Larmor), I feel that there is so much, entire worlds, that one is permitted to peep into, and that I am so small and incompetent, more incompetent than you and others suspect...'; Niels Blaedel, *Harmony and Unity: The Life of Niels Bohr* (Madison, Wis.: Science Tech, 1988), pp. 39-40.

[175]There were no distinctively Jewish precepts or rites in Einstein's immediate family, and, though he was later instructed in the elements by a distant relative and at the Luitpold Gymnasium, he never became a *bar mitzvah* or mastered Hebrew; Pais, *Subtle*, pp. 35-36, 38. On the critical attitude towards religion in Bohr's home, see Léon Rosenfeld, 'Bohr', *DSB* 2:240a.

and religious mystics alike.[176] Bohr referred to a 'universal religious feeling' that exists in every age, particularly among poets, and which is in intimate harmony with insight into nature.[177] Both Einstein and Bohr recognised the great religious and philosophical traditions of other cultures, though it should be noted that they knew Indian and Chinese thought mostly through the German adaptations of Schopenhauer and Schiller.[178] Einstein developed his own version of certain fundamental Jewish truths he once identified as 'Mosaic'.[179] Bohr was well versed in the German poets, particularly Schiller and Goethe.[180] He was also fond of Kierkegaard's *Stages on Life's Way*, though he did not agree with the thought of Kierkegaard as a whole.[181]

The reason neither Einstein nor Bohr was willing to adopt a positive theological stance was apparently that they both associated religious teachings with narrow-mindedness. After a brief period of religious devotion in his youth,[182] Einstein rejected what he called the 'anthropomorphic character' of the 'God of Providence' as portrayed in the Hebrew Bible[183] and eschewed any suggestion of personality in God[184] or of the miraculous in God's dealings with humans.[185] Einstein's unfavourable references to the

---

[176]'Religion and Science' (1930), in *World*, pp. 25-28.

[177]Blaedel, *Harmony*, pp. 158-59; cf. John Honner, *The Description of Nature: Niels Bohr and the Philosophy of Quantum Physics* (Oxford: Clarendon Press, 1987), pp. 182-87.

[178]Einstein knew Buddhism through Schopenhauer; 'Religion and Science', in *World*, p. 26. Bohr knew the sayings of Confucius through Schiller's translation; Henry J. Folse, *The Philosophy of Niels Bohr* (Amsterdam: Elsevier/North Holland, 1985), p. 54.

[179]When required to state his religious affiliation, Einstien wrote, 'Mosaisch'; Pais, *Subtle*, pp. 192-93.

[180]Blaedel, *Harmony*, pp. 20, 156-60.

[181]Folse cautions against drawing direct lines of influence between Kierkegaard and Bohr, *Philosophy*, p. 47; cf. Holton, *Thematic Origins*, pp. 144-47. Dugald Murdoch, however, is more positive on this score, *Niels Bohr's Philosophy of Physics* (Cambridge: Cambridge Univ. Press, 1987), p. 228.

[182]At the age of eleven, Einstein composed songs in honor of God which he sang on his way to school; Pais, *Subtle*, p. 38.

[183]'Religion and Science', in *World*, pp. 24-26; 'Science and Religion II' (1941), in *Ideas and Opinions* (New York: Alvin Redman, 1954), p. 46.

[184]Einstein spoke of the individual existence associated with personal ambitions as a 'sort of prison' and 'the shackles of personal hopes and desires'; 'Religion and Science', in *World*, p. 25; 'The Religiousness of Science', *ibid.*, p. 28; 'Science and Religion II', in *Ideas*, pp. 48-49. He characterised his own personal development, in contrast, as an attempt to liberate himself from the 'chains of the merely-personal' (*aus den Fesseln des Nur-Persönlichen*); 'Autobiographical Notes' in P.A. Schilpp, ed., *Albert Einstein: Philosopher-Scientist* (Evanston, Ill.: Library of Living Philosophers, 1949), p. 4; cf. Pais, *Subtle*, pp. 38, 39. Einstein attributed the conflict between science and religion to the concept of a personal God and called on religious leaders to give up the doctrine; 'Science and Religion II', in *Ideas*, pp. 47-48.

[185]'The man who is thoroughly convinced of the universal operation of the law of causation cannot for a moment entertain the idea of a being who interferes with the course of events...', 'Religion and Science', in *World*, p. 26; cf. 'Science and Religion II', in *Ideas*, p. 48. Similarly, Einstein described the process of maturing in thought as 'a continuous flight away from the miraculous' (*eine beständige Flucht aus dem 'Wunder'*); 'Autobiographical

'moral religion' and the 'social or moral conception of God' in this connection suggest that he associated the idea of a 'personal' God with the moralism that characterised much German religion, both Jewish and Christian, in his early years.[186]

Bohr was opposed, in principle, to any formal system or dogma that claimed to be the whole truth. Even with respect to his own attempt at a universal synthesis, the principle of complementarity, he disavowed any overall system or doctrine of ready-made precepts, and he never attempted to give a formal definition.[187] Accordingly, he thought of religion primarily in terms of a 'universal feeling' and rejected any attempt to 'freeze it' in terms of the concepts of any given period of human history.[188] Bohr referred to the anthropomorphic notion of a supernatural power with whom people could bargain for favours as a figment of primitive imaginations[189] and did not take the possibility of historical revelation seriously.

It is likely that both Einstein and Bohr were influenced in their views by the evolutionary theory of religion developed by Herbert Spencer, Edward Tylor, and Andrew Lang.[190] Einstein, in particular, described an evolution of religion from a primitive stage, in which humans conceived of God in their own image, through the higher religions of social and moral value to the vision, held by a few, of a cosmic God.[191]

The fact that for Einstein and Bohr the biblical teachings of the synagogue and church had little to do with the serious issues of science and society serves to confirm our observations concerning the decline of the creationist tradition in Western culture. Since the twelfth century, miracle had become increasingly viewed as the antithesis of natural law, and faith in a personal God had been gradually isolated from its moorings in the history of nature and culture. The positive faith of Einstein and Bohr, however, points to another important fact, seemingly at variance with the first: the

---

Notes', p. 8; cf. Pais, *Subtle*, p. 37.

[186]'Religion and Science', in *World*, pp. 24-26. In his early writings, Einstein carefully separated the cosmic religious feeling he espoused from the social demand for a moral code which he regarded as a 'purely human affair'; *World*, pp. 26-27; 'Religiousness', *ibid.*, p. 28. In later writings, however, he defined religion in the classical Ritschlian terms as concerning matters of value rather than matters of fact ('Science and Religion II', in *Ideas*, p. 45), and he called for an 'ethical culture' to counteract the 'stifling of human considerations by a "matter-of-fact" habit of thought' and described religion purified of superstition as the spring of all moral action ('The Need for Ethical Culture' [1951], in *Ideas*, pp. 53-54).

[187]Leon Rosenfeld, *Selected Papers*, ed. R.S. Cohen and J.J. Stachel (Dordrecht: Reidel, 1979), p. 532; quoted in Blaedel, *Harmony*, p. 196.

[188]Blaedel, *Harmony*, pp. 158-59.

[189]Blaedel, *Harmony*, p. 28. Bohr's comment corresponds closely to Einstein's descriptions of the primitive 'religion of fear' and 'religion of the naive man'; *World*, pp. 24, 28; *Ideas* p. 46.

[190]Eric J. Sharpe, *Comparative Religion* (New York: Scribner's, 1975), chaps. 3-4.

[191]Einstein used the phrase 'mankind's spiritual evolution' in 'Science and Religion II', in *Ideas*, pp. 46, 49.

survival of creationist themes in the absence of the tradition that originally mediated and sustained them.

If we were to characterise the primary object of the respective faiths of Einstein and Bohr with a single word, that word would be 'harmony'. Both Einstein and Bohr spoke of harmony in metaphysical, and even reverential, terms that would traditionally have been reserved for God.

For Einstein, the physical world was an incarnation of reason which, though manifest in various laws and principles, was inaccessible to the human mind in its profoundest depths.[192] Thus physics itself was a quest of religious proportions. The true scientist was enraptured by 'the harmony of natural law, which reveals an intelligence of such superiority that, compared with it, all the systematic thinking and acting of human beings is an utterly insignificant reflection'.[193]

The enterprise of physics, as Einstein understood it, was based on the conviction that the entire cosmos was governed by what Leibniz had called the 'pre-established harmony' of the parts. For instance, when Einstein described the work of Max Planck—discoverer of the quantum of action (1900)—he used words, as Abraham Pais has pointed out, that described his own conviction and experience as well as Planck's:[194]

> The longing to behold...pre-established harmony is the source of the inexhaustible persistence and patience with which we see Planck devoting himself to the most general problems of our science without letting himself be deflected by goals which are more profitable and easier to achieve....The emotional state which enables such achievements is similar to that of the religious person or the person in love; the daily pursuit does not originate from a design or a program [of one's own choice or invention] but from a direct need.

This statement may readily be compared to the teachings of church fathers like Irenaeus and Basil or the writings of Christian natural philosophers like Paracelsus and Bacon, or Kepler and Newton. It shares with them its ideal of selfless service as well as its belief in the unity and harmony of the world. It

---

[192] 'Science and Religion II', in *Ideas*, p. 49.

[193] 'Religiousness', in *World*, p. 28. 'I believe in Spinoza's God, who reveals himself in the harmony of all being [or 'among all people'], not in a God who concerns himself with the fate and actions of men', Letter to R. Herbert S. Goldstein, published in the New York *Times*, 25 April 1929, quoted by Arnold Sommerfeld in Arne Naess, 'Einstein, Spinoza, and God', in A. van der Merwe, ed., *Old and New Questions in Physics, Cosmology, Philosophy, and Theoretical Biology* (New York: Plenum, 1983), p. 684; for another translation, see Antonina Valentin, *The Drama of Albert Einstein* (New York: Doubleday, 1954), p. 102; cf. Schilpp, ed., *Albert Einstein*, pp. 659-60.

[194] 'Principles of Scientific Research', address delivered at the Physical Society in Berlin on the occasion of Planck's sixtieth birthday, 1918; trans. Pais, *Subtle*, pp. 26-27; cf. Einstien, *World*, p. 126.

also indicates that, however much he reacted against the current understanding of the 'personality' of God, Einstein's experience of the divine presence was not entirely an impersonal one like that generally associated with Spinoza's philosophy, with which Einstein's theological views are often compared.[195] The quest of the scientist, for Einstein, is comparable to the religious affections and the passion of lovers.

Einstein once stated that he often read the Hebrew Bible (in German translation),[196] and he particularly admired the cosmic sense of the Psalms and some of the prophets.[197] At age eighteen, the young Einstein described strenuous labour and the contemplation of God's nature as 'the angels which, reconciling, fortifying, and yet mercilessly severe, will guide me through the tumult of life'.[198] Undoubtedly, this reference to angels was a figure of speech for Einstein, but the sense of personal calling and guidance was every real, and it never left him. Einstein, like Newton, viewed himself and his work as an instrument in the hands of the Lord.[199]

The pre-established harmony Einstein believed in manifested itself not only in the unity and harmony of the diverse phenomena of nature. It was also the basis of the physicist's intellectual self-confidence—the confidence that the human mind, finite and fallible though it might be, was capable of discerning the basic outline of the order of the cosmos. The aim of science, according to Einstein, was to comprehend human experience in as much breadth as possible while, at the same time, describing it with a simplicity and economy of assumptions. These two objectives, comprehensiveness and simplicity, are, of course, potentially in conflict with each other. Yet the belief that they could exist side by side and be accomplished together was indispensible for scientific progress.[200] Einstein spoke of his own experience of the 'sublimity and marvellous order which reveal themselves both in nature and in the world of thought' and praised the 'deep conviction

---

[195]On the impersonal character of Spinoza's God (in contrast to that of the Romantics), see, e.g., Stuart Hampshire, *Spinoza* (Harmondsworth: Penguin, 1951), pp. 39-40. T.F. Torrance rightly distances Einstein from Spinoza at this point, *Christian Theology and Scientific Culture* (New York: Oxford Univ. Press, 1981), 59-60. Arne Naess defends Einstein's citations of Spinoza by arguing that, even for Spinoza, God was revealed in or 'expressed through' nature (*natura naturata*) but was not identical with it. In other words, according to Naess, both Spinoza and Einstein were panentheists; 'Einstein', pp. 683-85. But for Spinoza, finite beings are 'in God' only as 'modes' or modifications of substance are in their substance (*Ethics* I.xv). Moreover, the difference between Spinoza's strictly deductive rationalism and Einstein's hypothetico-deductive epistemology must surely reflect a significant difference in their respective views of God.

[196]So he wrote to his former religion instructor in 1929; Pais, *Subtle*, p. 38.

[197]'Religion and Science', in *World*, pp. 25-26.

[198]Letter of 1897; Pais, *Subtle*, p. 41.

[199]Pais, *Subtle*, p. 30.

[200]'Message to the Italian Society for the Advancement of Science' (1950), in *Ideas*, p. 357.

of the rationality of the universe' that enabled early scientists like Kepler and Newton to persist in their efforts in spite of social isolation and personal hardship.[201] Here we have the clearest possible evidence of the influence of the historic creationist tradition at the root of twentieth-century physics.[202]

Einstein's intellectual self-confidence has sometimes been taken as a sign of arrogance. He felt that he could judge the adequacy of scientific ideas and their compatibility with the divine mind. He was completely certain of the truth of his general theory of relativity, for instance, and felt that any failure of experiments to verify it could only reflect unfavourably on the consistency of God.[203] He was equally certain that notions like chance were incompatible with the divine mind and therefore could not play a fundamental role in physics.[204]

Niels Bohr worked with a similar confidence. He allowed for a greater degree of plurality and apparent paradox in physical science than did Einstein. But he accepted, even accentuated, these paradoxes in the belief that they would ultimately be resolved in a broader unity and harmony of the phenomena. For instance, in describing his semiclassical model of the hydrogen atom (1913), Bohr stated that he wanted to emphasise the conflict between the classical framework and the new quantum postulate in order 'that it may also be possible in the course of time to discover a certain coherence in the new ideas'.[205]

The goal of science for Bohr, as for Einstein (and for Maxwell before), was to articulate principles that would exhibit the greatest possible harmony while covering the widest possible range of phenomena, however conflicting

---

[201] 'Religion and Science', in *World*, pp. 25, 27; cf. 'Johannes Kepler', *ibid.*, pp. 141-42; 'Science and Religion, II': 'faith in the possibility that the regulations valid for the world of existence are rational, that is, comprehensible to reason', in *Ideas*, p. 46; 'Religion and Science: Irreconcilable?' (1948), *ibid.*, p. 52.

[202] Frederick Ferré has variously described this as the 'heuristic' and 'supportive' contribution of religion to science; 'Einstein on Religion and Science', *American Journal of Theology and Philosophy*, vol. i (Jan. 1980), pp. 21-22.

[203] 'Da köennt' mir halt der liebe Gott leid tun. Die Theorie stimmt doch.' ('I would have had to pity our dear God. The theory is correct all the same.'). A remark made to Ilse Rosenthal-Schneider at Berlin University in 1919, just after news was received of Eddington's experimental verification of the bending of starlight predicted by general relativity; I. Rosenthal-Schneider, *Reality and Scientific Truth: Conversations with Einstein, von Laue, and Planck* (Detroit: Wayne State Univ. Press, 1980), p. 74; cf. Pais, *Subtle*, p. 30.

[204] 'It seems hard to look in God's cards. But I cannot for a moment believe that he plays dice and makes use of "telepathic" means [action at a distance] (as the current quantum theory alleges he does)', Letter of 1942; Pais, *Subtle*, p. 440. Cf. Bohr, *Atomic Physics and Human Knowledge* (New York: Wiley, 1958), p. 47; *The Born-Einstein Letters* (New York: Walker and Co., 1971), pp. 149, 199.

[205] 'On the Spectrum of Hydrogen' (1913), in *The Theory of Spectra and Atomic Constitution* (Cambridge: Cambridge Univ. Press, 1922), p. 19.

the phenomena might appear to be in everyday experience. As he put it in his 1954 essay on 'Unity of Knowledge':[206]

> This attitude may be summarized by the endeavour to achieve a harmonious comprehension of ever wider aspects of our situation, recognizing that no experience is definable without a logical frame and that any apparent disharmony can be removed only by an appropriate widening of the conceptual framework.

In fact, Bohr cited (1947) Einstein's special and general theories of relativity as models in this respect: the special theory reconciled the separate ideas of space and time in a four-dimensional manifold, and the general theory further combined the space-time manifold with universal gravitation. Thus, for Bohr, through the work of Einstein, 'our whole world picture achieved a higher degree of unity and harmony than ever before'.[207] Niels Blaedel has aptly summarised this ideal by giving his biography of Bohr the title, *Harmony and Unity*.[208]

Bohr accepted the adequacy of classical physics as the foundation of the new developments in quantum theory in spite of the fact that the new developments called into question many of its basic teachings. As he put it in a lecture on the centenary of Maxwell's birth (1931):[209]

> We must, in fact, realise that the unambiguous interpretation of any measurement must be essentially framed in terms of the classical physical theories, and we may say that in this sense the language of Newton and Maxwell will remain the language of physicists for all time.

Bohr, like Einstein, viewed his work as rooted in that of his predecessors, particularly those of the British tradition of natural philosophy from Newton to Maxwell.

Bohr's insistence on the retention of classical terminology has sometimes been taken as an indication of Kantian idealism. He took the subject matter of physics to be human measurement and human language almost as much as physical reality itself. Recent studies like those of Henry Folse and Dugald Murdoch have shown, however, that Bohr was concerned

---

[206]'Unity of Knowledge' (1954), in *Atomic Physics,* p. 82. Other texts and references on Bohr's ideal of harmony are conveniently summarised by Folse, *Philosophy*, pp. 13-14.

[207]'Newton's Principles', p. 57. The way in which Bohr hit upon the phrase 'unity and harmony' while preparing this lecture (Newton tercentary, 1942) is humorously described by Abraham Pais, *apud* Blaedel, *Harmony*, p. 187.

[208]The original Danish title is *Harmoni og Enhed* (Copenhagen, 1985).

[209]'Maxwell', pp. 691, 692.

about the problems involved in the objective description of nature and was neither an idealist nor a Kantian in the proper sense.[210]

Both Einstein and Bohr must be viewed against the background of the ideals of nineteenth-century physics into which they were indoctrinated. The nineteenth century was an era of great confidence: confidence in the rationality of nature and confidence in the power of the human in tellect—convictions which were based on the creationist ideas of the divine law in the universe and the divine image in humanity.

This confidence had always played a role in the physical sciences, as we have seen in previous chapters of this book. Paradoxically, its importance has actually been heightened in the twentieth century due to the general acceptance of Darwin's theory of evolution. According to Darwin, the basic characteristics of the human species, including our mental capabilities, have been determined by the conditions necessary for the survival of the race over the millions of years of human evolution. The absence of any role in this theory for the divine image in humanity might well have called into question the possibility of humans comprehending areas that were completely beyond the experience of fossil hominids and unrelated to their survival. Thus, in going beyond the relatively commonsensical teachings of classical physics and delving into the mysteries of relativity and quantum theory, modern physics has exercised faith in one of the central teachings of the creationist tradition at a time when that tradition has been generally regarded as irrelevant to science.

The philosophies of Einstein and Bohr are more like the Platonic and Pythagorean ideas that the church transmitted to the modern West than like the Judeo-Christian idea of creation itself. But they could still be said to be within the creationist tradition conceived in the broad sense of its ancient Near Eastern roots and the parallel Jewish and Greek developments that the church later synthesised. The idea of creation is no longer assumed by scientists in their work, but the values the idea embodied and transmitted live on in the lives and minds of many physical scientists independently of, or even in the absence of, personal religious faith.

Therefore, Jews and Christians who are aware of their theological tradition can afford to be appreciative of today's natural philosophers, as much as the early Jews and Christians were of the natural philosophers of their own time. They can also afford to be as critical. Where the comprehen-

---

[210]Folse, *Philosophy*, pp. 49-51, 53, 216, 217-220; Murdoch, *Bohr's Philosophy*, pp. 102, 224-33. Both Folse and Murdoch associate Bohr with James's pragmatism rather than with Kant or even neo-Kantianism. James's ideas were probably mediated by Harold Høffding, Bohr's philosophy professor and family friend, though neither Høffding nor Bohr was as strongly instrumentalist as James was. According to Murdoch, 'Høffding maintains that reality itself is to some extent intelligible—to a certain extent we can penetrate into reality itself. This is exactly the attitude that Bohr takes' (*Bohr's Philosophy*, p. 227).

sibility of the world is a deeply held conviction, we can recognise that as a legitimate manifestation of faith. Where sacrifices are made to alleviate the suffering of humanity through medicine and technology, we can appreciate that as part of the Spirit's work in our world. On the other hand, where the power and privileges awarded scientists in modern society are enjoyed as ends in themselves or applied in unjust ways, we must recall the ultimate source of human science and the ultimate goals for which the gifts of creativity and healing were poured out on us.

For the time being, science and technology are immensely successful and profitable enterprises. People with talent and ambition may be drawn to them out of self-interest as well as for humanitarian ideals. But the future is uncertain. The time may come when it takes self-sacrifice and courage to develop the ideas and invent the techniques needed for further progress. The time may even come when an operational faith, supported by a religious community and its creeds, will provide insight and inspiration for the pioneers of new scientific developments as it did in Western Europe from the Middle Ages to the nineteenth century. If that time does come, other theological traditions beside the Judeo-Christian may well play an important role. Only then will we be in a position to assess the full import of theology as a whole, and of the creationist tradition in particular, to the history of science.

RETROSPECT AND PROSPECT

A. The Creationist Tradition Summarised

History can never be constrained to a single theme, and this book is no exception. But my primary interest has been to trace the history of the creationist tradition, a historical tradition based on belief in the creation of the world by a wise and powerful God who intends good for it.[1] For centuries, the creationist tradition provided a theological framework for the development of Western science, medicine, and technology as well as other secular disciplines.[2] I have identified four basic themes in the tradition: the comprehensibility of the natural world; the unity of heaven and earth; the relative autonomy of nature; and the ministry of healing and restoration (chap. 1).

The creationist tradition had its roots in ancient Near Eastern mythology and Old Testament texts that relate God, humanity, and nature. It was first consciously articulated by the sages of Second Temple Judaism and the fathers of the early Christian church, partly as a way of addressing the challenges of the Hellenistic era, when Platonic, Pythagorean, and Stoic ideas were influential. In the Middle Ages, the disciplines of physics, cosmology, alchemy, and medicine developed in Western Europe in response to the impact of Aristotelian and Arabic natural philosophy. A polarisation between naturalist and conservative wings of the tradition began to surface as various attempts were made to synthesise the new Aristotelian science with creational theology. While we usually speak of the creationist tradition being 'Judeo-Christian' for simplicity, it would really be more accurate to use a more inclusive label like 'Hellenistic-Jewish-Arabic-Christian'.

The basic themes of the creationist tradition were revived in the late Middle Ages, the Renaissance, and the Protestant Reformation. These ideas provided inspiration, guidance, and assurance for the work of the early founders of modern science.[3] In the eighteenth century, however, an effort

---

[1] The historic creationist tradition has nothing to do with creation science. See 'Introduction', section A.

[2] Daniel J. Boorstin traces the history of Western art in terms of its relation to the biblical idea of creation; *The Creators: A History of Heroes of the Imagination* (New York: Random House, 1992). Other trajectories can be discerned in areas like management, government, and city planning.

[3] John Hedley Brooke discusses a variety of ways in which religious beliefs and rational science have interacted historically. In various cases, religious beliefs have served as presuppositions, provided sanction, supplied motivation, imposed regulations, provided criteria of selection, and even played a constitutive role for scientific work; *Science and Religion*, pp. 19-33.

was made to dissociate natural philosophy from the hardened dogmas of the churches, particularly in France, and widespread opposition between scientific work and creational theology occurred for the first time. In the nineteenth century, the basic themes continued to be influential in British natural philosophy, but, due to the secularisation of science and the privatisation of theology, it is no longer possible to speak of a creationist tradition being shared by scientists as a profession.

Since the creationist tradition provides a bridge between physical science and biblical theology, recognition of its importance could have an impact on our understanding of both science and theology. I draw this work to a conclusion by pointing out some of the differences such recognition could make.

B. Implications for Our View of Science and Technology

In the light of the creationist tradition, we would do well to reexamine the complex problem of autonomy and relationality in the sciences. As stated in the introduction to this book, the idea of the epistemological autonomy of science makes it difficult to understand how we are capable of probing phenomena so remote from the environmental conditions under which our brains first evolved. Similarly, the idea of the autonomy of nature makes it nearly impossible for us to integrate our spiritual and ethical selves with our science-based view of the world. And the idea of the ethical autonomy of science and technology makes it difficult to examine moral choices within the scientific disciplines themselves.

Searching the creationist tradition for insights will not resolve all aspects of the problem of autonomy and relationality, but it can provide a framework in which the three issues mentioned can be rationally addressed. I shall briefly sketch the implications of such a framework for each of them in turn.

A creational perspective means, first of all, that scientific work is not epistemologically autonomous; rather, it depends on the grace of a God who created us in the divine image. Johannes Kepler, perhaps the most famous exponent of this belief, argued that the universe is governed by divinely ordained laws that we can recognise and grasp conceptually and, further, that God invites humans to share in his thoughts in this way.[4] As Kepler saw it, professional training and persistent effort were certainly required. But, even if our received notions do not match the phenomena of nature, we have the ability to revise them appropriately.

---

[4] Kepler's letter to Herwart von Hohenburg, 9 April 1599. See chapter 3, section B1 above.

The creative prodecures that scientists rely on are, therefore, indirectly related to those by which the world itself was created. We may never have (or be able to afford) the technology to collect all the data we need to resolve our questions about the universe. And we will probably never be able to adapt the structure of our brains to meet all the demands of scientific research. Still, we may rationally believe that we can design experiments and interpret the data in those areas of nature that we can reach. The world is comprehensible to the extent that we have access to it.

Second, in a creationist perspective nature is not entirely autonomous but functions in response to the word of God. The patron saint of this perspective, Basil of Caesarea, put it this way: the command of God is the foundation of the universe and its effects are seen in the continuous processes of nature throughout time.[5] Therefore, our common experience of life energy and psychic motivation is somehow related (indirectly) to the energy and momentum of the cosmos in which we live. Our inner subjectivity and the external world around us are both responses to the same divine Word. It follows that the autonomy of nature does not exclude the healing or reformation of human life. Of course, any medical operation may have complications, and any new technology will have unintended, often harmful, consequences. Both medicine and technology are fallible. Still, we believe that steps can be made toward wholeness. The healing and restoration of humans are possible in the long run even though they are not determined or guaranteed by the initial course of nature. In this sense, the autonomy of nature is a relative (or relational) autonomy.[6]

---

[5] Basil, *Hexaemeron* V.1, 10. See chapter 1, section D3, 'Basil of Caesarea'.

[6] The paradoxical semantics of the phrase 'relative autonomy' is based on Ian Ramsey's analysis of models and qualifiers; *Religious Language*, chapter 2 (see the discussion in chap. 1, sec. D). Howard J. Van Till uses the phrases, 'functional integrity' and 'gapless economy', which avoid the paradoxical element; 'Can the Creationist Tradition Be Recovered?' *Perspectives on Science & Christian Faith*, vol. xliv (Sept. 1992), p. 180. The phrase, 'functional integrity', captures the dynamic sense of 'relative autonomy' very nicely.

The idea of a 'gapless economy' of the created world is helpful, but also problematic. To be sure, it is possible to view physical processes as being gapless, provided that what counts as a gap is defined appropriately in terms of the laws governing those processes. So, for example, we believe that quantum phenomena have no gaps when they are viewed in terms of the principles of quantum physics (Schrodinger's equation and the complementarity principle), even though quantum phenomena do have gaps when analysed in terms of classical physics (precise momentum and location). Probably biological processes can be viewed in gapless terms as well. However, the relationship between such levels of organisation is more difficult to formulate. While each level may be gapless in its own terms, there may may be gaps between the levels, especially when the phenomena of one level cannot be reduced to the laws of another. Confirming the existence of a gapless economy, in this case, would require finding a more comprehensive framework that unifies the laws of physics and biology in the same way that quantum physics unifies physics and chemistry, i.e., without reducing one to the other.

In spite of the possibility of healing and restoration, the scope of application for healing and restoration may be more limited than that of the relative autonomy of nature. In the creational perspective, the 'nature' that is to be healed and restored is primarily that of human beings. The prophets were called on to heal children and even slaves (Luke 7:3), but never animals. On the other hand, the stipulations of the Old Testament covenant included domesticated animals as well as children and servants (Exod. 20:10, 17).

So the extension of the ministry of healing to captive animals, cultivated land, and even the natural environment has a certain plausibility, particularly where the welfare of these creatures has an impact on human well-being (Exod. 15:25; 2 Kings. 2:19-22; 2 Chron. 7:14). However, it would be wise also to recognise that there are limits to the power of humanity, particularly in those portions of nature that lie beyond the scope of human cultivation. Humans may be entitled to cultivate and domesticate as much of God's creation as they can care for (Gen. 1:26-28; Ps. 8:6-8). But many aspects of the natural world will probably always defy human management (Job 38-41; Ps. 104). At the end of the day, some things have to be left to the God to whom all creatures must turn for sustenance and healing.[7]

Finally, from a creational perspective, science and technology are not ethically autonomous disciplines either. They are callings, like the traditional callings of medicine and ministry, that are publically accountable in a shared social context.[8] In the words of Francis Bacon, undoubtedly the greatest proponent of this norm, all scientific work should be done 'for the glory of the Creator and the relief of man's estate'; it is not just a 'shop, for profit or sale'.[9]

Theologians and ethicists, therefore, should not be called in on difficult cases as though they were experts with some sort of monopoly on social and moral wisdom. The professions have a common calling, and they should have a common social vision (which, however, does not guarantee a common social agenda). In all of these ways, a recovery of the values of the creationist tradition can make a significant contribution to our view of the sciences.

---

[7] Pss. 104:21, 27-30; 145:15-16; cf. Job 38:41; Pss. 65:9-13; 104:10-16; 147:9. For further discussion, see my article, 'The Integrity of Creation and the Social Nature of God', *Scottish Journal of Theology*, vol. il (1996), pp. 261-90.

[8] For an excellent historical analysis of the differences between the traditional professions and the twentieth-century, technologically based ones, see William M. Sullivan, *Work and Integrity: The Crisis and Promise of Professionalism in America* (New York: HarperCollins, 1995), chaps. 2-4.

[9] Francis Bacon, *Works*, 6:33-4, 94, 134. See above, chapter 3, section B4, 'Ministry of Healing and Restoration'.

C. Implications for Creational Theology

An understanding of the creationist tradition also has implications for our understanding of theology. To begin with, we may suggest a new way of outlining the implications of the biblical doctrine of creation. In collecting the historical material for this book, I have used the simple four-part outline noted above (see also chapter 1). But several distinctions and connections have arisen that suggest a way of reorganising the material.

As noted in chapter one, the idea of the comprehensibility of the world can be understood as combining two premises: the world is intrinsically comprehensible, based on the belief that it is created by (and therefore comprehensible to) God; and the world is also open to human comprehension, based on the belief that humanity is created in the image of God. In other words, the investment of creation with its own logic (not necessarily the same as human logic) gives it epistemic depth, but it does not place it completely beyond human understanding.

So the idea of comprehensibility has two aspects: one cosmological and the other anthropological. One is about the intrinsic logic of creation; the other tells us something about the human ability to grasp that logic. The two aspects taken together could be described under the heading of the 'open intelligibility of creation'. Belief in the open intelligibility (or the comprehensibility) of the world is the epistemological basis of all natural science.

It so happens that the three other themes of the creationist tradition can be reformulated along similar lines. The ideas of the unity and relative autonomy of nature (chapter 1, sections C and D) can easily be synthesised. They are really the space-like and time-like dimensions of what we may call the 'integrity of creation'. In other words, creation has two kinds of integrity: structural and functional.[10] God has invested it with both unity and regularity, or, in the language of modern field theory, with both symmetry and invariance.[11]

On the other hand, the investment of nature with its own unity and regularity does not place it completely beyond human intervention. The

---

[10] As noted above, Howard Van Till refers to the relative autonomy of nature as 'functional integrity'. Along the same lines, we might refer to the unity of heaven and earth as the 'structural integrity' of the created world.

[11] A basic principle of field theory is that for every physical symmetry there is a principle of invariance or conservation. Spatial unity is essentially symmetry with respect to linear translation (and rotation). The relative autonomy of nature is exemplified in the principle of the conservation of momentum (and angular momentum). So the pairing of spatial unity and relative autonomy is a simple case of the field principle of symmetry and invariance. It is important to note that the principle of symmetry and invariance applies perfectly well even in the indeterministic microworld of quantum mechanics. Chaotic phenomena are also governed by what are called 'strange attractors', in spite of the fact that they appear to be irregular.

course of nature is open to human alteration, particularly for the purposes of healing and restoration (chap. 1, sec. E). So the idea of the integrity of creation can be paired with the possibility of healing and restoration, much as the idea of the intrinsic logic of creation is paired with the possibility of human comprehension.

As with the heading of the open intelligibility of creation, we have two aspects to deal with here: one cosmological, and one anthropological. One is about the inherent integrity of creation; the other tells us something about the human ability to govern it. The two aspects can then be referred to together under the heading of the 'flexible integrity of creation'. Belief in the flexible integrity of nature (the unity and relative autonomy of nature plus the possiblity of healing) is presupposed in all medicine and technology.

I suggest, therefore, that the implications of the theology of creation can then be formulated under two major headings—one epistemological (open intelligibility) and the other practical (flexible integrity)—where both headings have cosmological and anthropological aspects.

This way of presenting the theology of creation appears to be rather different from traditional formulations. In retrospect, however, we may recognise in these two major headings the basic themes of the biblical traditions from which they originated. The idea of open intelligibility is derived largely from the wisdom tradition in the Old Testament and in Jewish apocalyptic literature. In the early church it was championed by the Apologists, particularly those of the Alexandrian tradition (chap. 1, sec. B). Its main focus was the human need for wisdom and knowledge in the midst of bewildering complexity.

The idea of flexible integrity is derived from prophetic traditions like Isaiah's healing of Hezekiah (2 Kings. 20:1-11; Isa. 38:1-8) and Jeremiah's words of consolation for Israel (Jer. 33:1-9, 19-26). In the early church it was championed by the monastic tradition, particularly by Basil of Caesarea, who modelled his own ministry to the poor on the examples of Moses, Elijah, and Jesus (chap. 1, sec. E). Its main focus is the human need for lifegiving power in the midst of suffering and death.

So, in spite of its applicability to the history of science, our reformulation of creational theology is more archaic than modern. It is deeply rooted in two of the major traditions of the Hebrew Bible. Yet there is one other biblical tradition that was covered in our historical survey that we have not yet included in our reformulation.

The twin themes of wisdom and lifegiving power were synthesised by Paul in his teaching about the Cross. According to Paul, both are rooted in Jesus Christ, who as the crucified one is 'the power of God and the wisdom of God' (1 Cor. 1:23-24). This text serves to remind us of the ultimate grounding of the cosmological and anthropological aspects in the Word of

God, through whom all things were created (John 1:1-4; 1 Cor. 8:6; Col. 1:15-17). It also reminds us of the ethical aspect of the ministry of healing—the criterion of human welfare and social benefit. The gifts of wisdom and power can both all too easily be exploited for personal gain in the absence of the model of the Son of Man who 'came not to be served, but to serve and to give his life a ransom for many' (Mark 10:44).[12]

As we think of the future of science and technology and the civilisation we have built on them, we need to be reminded not only of the wisdom and prophetic traditions, but also of the tradition of the martyrs—all those who were willing to sacrifice their personal desires for the welfare of others. The idea of personal sacrifice is not popular in most political circles today. In fact, it is not even popular in our churches and synagogues. But it will need to be raised if we are to recover the historic, Jewish-Christian creationist tradition as described in this work.

D. Implications for Dichotomies in Theology

An understanding of the creationist tradition can affect our understanding of theology in more general ways as well. Western thought is infested with dichotomies like those of reason and faith, God and the world, natural law and divine activity, and medicine and prayer. We also need to reexamine these in light of the creationist tradition.

First, from a creational perspective, analytic reason is not strictly an alternative to religious faith. Traditionally, the human ability to grasp hidden structures was viewed as a gift from God (Job 32:8; Wisd. 7:15-22; Ecclus. 1:1-10; 17:1-11; John 1:9). We rely on this ability in our thinking, and we often take it for granted, just as we rely on light to see where we are going. Therefore, the sustained use of reason is an exercise in faith. For scientists, in fact, the use of reason can be something like an experience of the presence of God.[13]

Second, the transcendence of God need not imply that God is removed from creation as is often claimed (by both supporters and critics of the idea

---

[12] I deal with the implications of the prophetic, wisdom (or royal), and martyrological traditions in more detail in 'The Incarnation and the Trinity: Two Doctrines Rooted in the Offices of Christ', in Iain R. Torrance, ed., *Theological Dialogue Between Orthodox and Reformed Churches*, vol. 3 (Edinburgh: Scottish Academic Press), in press.

[13] For example, Eugene Wigner states: 'The miracle of the appropriateness of the language of mathematics for the formulation of the laws of physics is a wonderful gift which we neither understand nor deserve. We should be grateful for it and hope that it will remain valid in future research and that it will extend...to wide branches of learning'; Wigner, *Symmetries and Reflections*, p. 237. The wonder, gratitude, and hope expressed here is a deeply religious sentiment. Compare Einstein's observation about the work of Max Planck, quoted in chapter 5, section E.

of transcendence).[14] In biblical and patristic theology, God transcends all things by filling and embracing them (Jer. 23:24; Wisd. 1:7-8; Ecclus. 43:26; Acts 17:28; Rom. 8:36; Eph. 4:6). All things are in God even though all things do not contain God (chap. 1, sec. C). Prior to the spatialisation of the God-world relation in medieval Aristotelianism (chap. 2, sec. B), there was never systemic conflict between God's transcendence and God's immanence.

Similarly, the lawful operation of nature need not be seen as an alternative to the activity of God. Nor does the action of God necessarily violate the laws of nature. Instead, the laws and energies of nature should be viewed as direct expressions of the word and act of God and continuously responsive to the divine will (Ps. 148:1-8; Jer. 31:35-36; Wis. 1:14; Ecclus. 16:26-28; 43:9-10).

Finally, the spiritual and material aspects of healing should not be separated from each other. All healing comes from God. Physical healing and restoration are sometimes granted in response to prayer (Gen. 20:17; Num. 12:13-15; Pss. 30:2-3; 107:19-20; Acts 4:24-30). But they also depend on the energies and medicinal properties invested in creation, and they usually require the skill of trained practitioners (Ecclus. 38:1-8). Normally both prayer and medicine are called for (2 Kings. 20:1-7).

In short, Western theology needs to be healed from the dichotomies that have governed it for centuries. The great systems of the Western world have thrived on clearly defined lines of demarcation and the logic of mutual exclusion. But dichotomous thinking has taken its toll as well. We need to recover the nonduality of matter and spirit, God and nature, as taught in Scripture and celebrated by our spiritual ancestors as recently as a few centuries ago.

---

[14] E.g., in Sallie McFague, *Metaphorical Theology: Models of God in Religious Language* (London: SCM Press, 1982), pp. 187, 189, a book that is otherwise very helpful on issues of language about God.

BIBLIOGRAPHY

Agricola, George Bauer, *De Re Metallica*. Translated by Herbert Clark Hoover and Lou Henry Hoover, New York, 1950.
Agrippa, Henry Cornelius, *Three Books of Occult Philosophy*. Edited by Willis F. Whitehead, London, 1651, 1879; New York, 1971.
Albright, William F., 'Neglected Factors in the Greek Intellectual Revolution', *PAPS*, cxvi (1972): 225-42.
Alexander, H.G., ed., *The Leibniz-Clarke Correspondence Together with Extracts from Newton's 'Principia' and 'Optics'*, Manchester, 1956.
Alic, Margaret, *Hypatia's Heritage: A History of Women in Science from Antiquity through the Nineteenth Century*, Boston, 1986.
Althaus, Paul, *The Theology of Martin Luther* (*Die Theologie Martin Luthers*, Gütersloh, 1963). Translated by Robert C. Schultz, Philadelphia, 1966.
Amundsen, Darrel W., 'Medical Deontology and Pestilential Disease in the Late Middle Ages', *JHMAS*, xxxii (1977): 403-21.
——, 'Medicine and Faith in Early Christianity', *BHM*, lvi (1982): 326-50.
—— and Ferngren, Gary B., 'Medicine and Religion: Early Christianity through the Middle Ages', in M.E. Marty and K.L. Vaux, eds., *Health, Medicine, and the Faith Traditions*, Philadelphia, 1982.
Anderson, Bernhard W., ed., *Creation in the Old Testament*, Issues in Religion and Theology, vol. vi, Philadelphia, 1984.
Andrade, E.N., *Rutherford and the Nature of the Atom*, Garden City, N.Y., 1964.
Aquinas, Thomas, *The Division and Methods of the Sciences (Questions V and IV of his Commentary on the 'De Trinitate' of Boethius)*. Translated by Armand A. Maurer, Toronto, 1963.
——, *Summa Contra Gentiles*, 5 vols. Translated by Anton C. Pegis, Garden City, N.Y., 1955-57; Notre Dame, Ind., 1975.
Armstrong, A.H., ed., *The Cambridge History of Later Greek and Early Medieval Philosophy*, Cambridge, 1967.
Arnold, Clinton E., *Ephesians: Power and Magic: The Concept of Power in Ephesians in Light of its Historical Setting*, Cambridge, 1989.
Arthur, Richard T.W., 'Newton's Fluxions and Equably Flowing Time', *SHPMP*, xxvi (1995): 323-51.
Asimov, Isaac, *Asimov's Biographical Encyclopedia of Science and Technology*, 2nd ed., Garden City, N.Y., 1982.
Ault, Donald, *Visionary Physics: Blake's Response to Newton*, Chicago, 1974.
Austin, William H., 'More', *DSB*, 9:509-10.
Babcock, William S., 'A Changing of the Christian God: The Doctrine of the Trinity in the Seventeenth Century', *Interpretation*, xlv (1991): 33-46.
Bacon, Francis, *The New Organon and Related Writings*. Edited by Fulton H. Anderson, Indianapolis, 1960.
——, *The Works of Francis Bacon*, 14 vols. Edited by James Spedding, Robert Leslie Ellis, and Douglas Denon Heath, London, 1857-74; 1887-97.
Bacon, Roger, 'The Errors of the Doctors according to Friar Roger Bacon of the Minor Order'. Translated by Mary Catherine Welborn, *Isis*, xviii (1932): 26-62.
——, *Letter Concerning the Marvelous Power of Art and of Nature and Concerning the Nullity of Magic*. Translated by Tenney L. Davis, Easton, Penn., 1923.
——, *The Opus Majus* 2 vols. Translated by Robert Belle Burke, 1st ed., 1928; reprinted New York, 1962.
Baker, Derek, ed., *Sanctity and Secularity: The Church and the World*, Studies in Church History, vol. x, Oxford and New York, 1973.
Barnouw, Jeffrey, 'The Separation of Reason and Faith in Bacon and Hobbes, and Leibnitz's Theodicy', *JHI*, xlii (1981): 607-28.

Baron, Hans, 'Calvinist Republicanism and Its Historical Roots', *CH*, viii (1939): 30-42.
Barrow, Isaac, *Theological Works of Isaac Barrow,* 9 vols. Edited by Alexander Napier, Cambridge, 1859.
Basil of Caesarea, *Ascetical Works*, Writings of Saint Basil, vol. i. Translated by M. Monica Wagner, New York, 1950.
——, *Homélies sur l'hexaéméron*. Edited by Stanislas Giet, Paris, 1968.
Battenhouse, Roy W., ed., *A Companion to the Study of St. Augustine*, New York, 1955; Grand Rapids, Mich., 1979.
Baumer, Franklin L., *Modern European Thought: Continuity and Change in Ideas, 1600-1950*, New York, 1977.
——, ed., *Main Currents of Western Thought: Readings in Western European Intellectual History from the Middle Ages to the Present*, 4th ed., New Haven, Conn., 1978.
Baumstark, Anton, *Geschichte der syrischen Literatur mit Ausschluss der christlich-Palästinenischen Texte*, Bonn, 1922; Berlin, 1968.
Beall, Otho T., Jr., and Shyrock, Richard H., *Cotton Mather: First Significant Figure in American Medicine*, Baltimore, 1954.
Becco, Anne, 'Leibniz et Francois-Mercure van Helmont: Bagatelle pour des Monades', *Studia Leibnitiana*, vii (1978): 119-41.
Bede, The Venerable, *A History of the English Church and People.* Translated by Leo Sherley-Price, Harmondsworth, 1955, 1968.
——, *Opera de Temporibus*. Edited by Charles W. Jones, Cambridge, Mass., 1943.
Beer, Arthur and Peter, eds., *Kepler: Four Hundred Years*, Vistas in Astronomy, vol. xviii, Oxford, 1975.
Benjamin, Marina, 'Medicine, Mortality, and the Politics of Berkeley's Tar-Water', in A. Cunningham and R. French, eds., *Medical Enlightenment*, Cambridge, 1990.
Benson, Robert L, and Constable, Giles, eds., *Renaissance and Renewal in the Twelfth Century,* Cambridge, Mass., 1982.
Berger, Peter, Berger, Brigitte, and Kellner, Hansfried, *The Homeless Mind: Modernization and Consciousness*, Harmondsworth, 1973.
Berkel, Klaas van, *Isaac Beeckman (1588-1637) en de Mechanisering van het Wereldbeeld*, Amsterdam, 1983.
Berkson, William, *Fields of Force: The Development of a World View from Faraday to Einstein,* New York, 1974.
Berman, David, *George Berkeley: Idealism and the Man,* Oxford, 1994.
Bettoni, Efrem, *Saint Bonaventure (S. Bonaventura*, Brescia, n.d.). Translated by Angelus Gambatese, Notre Dame, Ind., 1964.
Bietenholz, Peter G., *Basle and France in the Sixteenth Century: The Basle Humanists and Printers in their Contacts with Francophone Culture,* Geneva, 1971.
Birkenmajer, Alexandre, 'Le Commentaire inédit d'Erasme Reinhold sur le *De revolutionibus* de Nicolas Copernic', in *La science au siezième siècle,* Paris, 1960.
Bizer, Ernst, 'Reformed Orthodoxy and Cartesianism', in Robert W. Funk, ed., *Translating Theology into the Modern Age,* Journal for Theology and the Church, vol. ii, Tübingen, 1965.
Blaedel, Niels, *Harmony and Unity: The Life of Niels Bohr (Harmoni og Enhed,* Copenhagen, 1985). Translated by Geoffrey French, Madison, Wis., 1988.
Bloch, Marc, *Land and Work in Medieval Europe: Selected Papers (Mélanges historiques,* Paris, 1966). Translated by J.E. Anderson, London, 1967; New York, 1969.
Bodde, Derk, *Chinese Thought, Society, and Science: The Intellectual and Social Background of Science in Pre-modern China,* Honolulu, 1991.
Bohr, Niels, *Atomic Physics and Human Knowledge,* New York, 1958.
——, *Atomic Theory and the Description of Nature,* London, 1934.
——, 'Causality and Complementarity', *Philosophy of Science,* iv (1937): 289-98.
——, 'Chemistry and the Quantum Theory of Atomic Constitution', *Journal of the Chemical Society* (1932, pt.1): 349-84.
——, *Essays 1958-1962 on Atomic Physics and Human Knowledge,* New York, 1963.
——, 'Maxwell and Modern Theoretical Physics', *Nature*, cxxviii (1931): 691-92.
——, 'Newton's Principles and Modern Atomic Mechanics', in The Royal Society of Great

Britain, *Newton Tercentenary Celebrations,* Cambridge, 1947.
——, *The Theory of Spectra and Atomic Constitution: Three Essays,* Cambridge, 1922.
Bolman, C. Gordon, et al., *The English Presbyterians from Elizabethan Puritanism to Modern Unitarianism,* London, 1968.
Bonaventure, *The Mind's Road to God.* Translated by George Boas, Indianapolis, 1953.
——, *The Soul's Journey into God,* Classics of Western Spirituality. Translated by Ewert H. Cousins, New York, 1978.
Bornkamm, Heinrich, *Luther's World of Thought* (*Luthers geistige Welt,* Gütersloh, n.d). Translated by Martin H. Bertram, Saint Louis, 1958.
Bougerol, J. Guy, *Introduction to the Works of Bonaventure.* Translated by José de Vinck, Paterson, N.J., 1964.
Bourke, Vernon J., 'St Augustine and the Cosmic Soul', *Giornale di metafisica*, ix (1954): 431-40.
Bouwsma, William J., 'Calvin and the Renaissance Crisis of Knowing', *Calvin Theological Journal,* xvii (1982): 190-211.
Boyle, Robert, *The Works of the Honourable Robert Boyle,* 6 vols. Edited by Thomas Birch, London, 1772.
Box, G. H., ed., *The Apocalypse of Abraham,* London, 1919.
Brandt, William, *The Shape of Medieval History,* New Haven, Conn., 1960.
Brann, Noel L., 'The Shift from Mystical to Magical Theology in the Abbot Trithemius (1462-1516)', in J.R. Sommerfeldt and T.H. Seiler, eds., *Studies in Medieval Culture,* vol.ii, Kalamazoo, Mich., 1977.
Bréhier, Émile, 'La création continuée chez Descartes', *Sophia,* v (1937): 3-10.
——, *The History of Philosophy,* 7 vols. (*Histoire de la philosophie,* Paris, 1930-38). Translated by Wade Baskin, Chicago, 1963-69.
Broad, C.D., *Leibniz: An Introduction.* Edited by C. Lewy, Cambridge, 1975.
Bronowski, J., *William Blake and the Age of Revolution,* London, 1972.
Brooke, John Hedley, 'Natural Theology and the Plurality of Worlds: Observations on the Brewster-Whewell Debate', *Annals of Science,* xxxiv (1977): 221-86.
——, *Science and Religion: Some Historical Perspectives,* Cambridge, 1991.
Brown, Colin, *Miracles and the Critical Mind,* Grand Rapids, Mich., 1984.
Brown, Peter, *Society and the Holy in Late Antiquity,* Berkeley, Calif., 1982.
Brundell, Barry, 'Gassendi between Religion and Science', *Quadricentenaire de la naissance de Pierre Gassendi, 1592-1992,* vol. i, Digne-les-Bains, 1994.
——, *Pierre Gassendi: From Aristotelianism to a New Natural Philosophy,* Dordrecht, 1987.
Brush, Stephen G., *Statistical Physics and the Atomic Theory of Matter from Boyle and Newton to Landau and Onsager,* Princeton, N.J., 1983.
——, *The Temperature of History: Phases of Science and Culture in the Nineteenth Century,* New York, 1978.
Buchwald, Jed Z., 'Thomas, William', *DSB,* xiii: 374-88.
Buckley, Michael J., *At the Origins of Modern Atheism,* New Haven, Conn., 1987.
Bullen, A. H., *Elizabethans,* New York, 1924, 1962.
Bullough, Geoffrey, 'Bacon and the Defence of Learning', in B. Vickers, ed., *Essential Articles for the Study of Francis Bacon,* Hamden, Conn., 1968.
Bullough, Vern L., 'Guy de Chauliac', *DSB,* iii: 218-19.
Burke, James, *The Day the Universe Changed,* Boston, 1985.
Burke, John G., 'Hermeticism as a Renaissance Worldview', in Robert S. Kinsman, ed., *The Darker Vision of the Renaissance,* Berkeley, Calif., 1974.
——, *The Uses of Science in the Age of Newton,* Berkeley, Calif., 1983.
Burnett, Charles, 'Scientific Speculations', in Peter Dronke, ed., *History of Twelfth-Century Western Philosophy,* Cambridge, 1988.
Burtt, Edwin Arthur, *The Metaphysical Foundations of Modern Physical Science: A Historical and Critical Essay,* London, 1925, 1932; New York, 1952.
Busson, Henri, *Le rationalisme dans la littérature française de la Renaissance (1533-1601),* Paris, 1957.
Butler, Dom Cuthbert, *Benedictine Monachism: Studies in Benedictine Life and Rule,* London, 1919, 1924.

Butterfield, Herbert, *The Origins of Modern Science, 1300-1800*, New York, 1965.
Butts, Robert E., 'Whewell', *DSB* xiv: 292-95.
—— and Davis, John W., eds., *The Methodological Heritage of Newton*, Toronto, 1970.
Calinger, Ronald S., 'The Newtonian-Wolffian Controversy (1740-1759)', *JHI*, xxx (1989): 319-30.
Callus, D.A., ed., *Robert Grosseteste, Scholar and Bishop: Essays in Commemoration of the Seventh Centenary of his Death*, Oxford, 1955.
Calvin, John, *Calvin's New Testament Commentaries (CNTC)*, 12 vols. Edited by David W. and Thomas F. Torrance, Edinburgh, 1959-73.
——, *Commentaires de Jehan Calvin sur le Livre des Psaumes*, 2 vols., Paris, 1859.
——, *Concerning Scandals*. Translated by John W. Fraser, Grand Rapids, Mich., 1978.
——, *Institutes of the Christian Religion*, Library of Christian Classics vols. xx, xxi. Translated by Ford Lewis Battles. Edited by John T. McNeill, London, 1960.
——, *Sermons on the Ten Commandments*. Translated by Benjamin W. Farley, Grand Rapids, Mich., 1980.
——, *Treatises Against the Anabaptists and Against the Libertines*. Translated by Benjamin W. Farley, Grand Rapids, Mich., 1982.
——, 'A Warning Against Judiciary Astrology and Other Prevalent Curiosities'. Translated by Mary Potter, *Calvin Theological Journal*, xvii (1983): 157-89.
Campbell, Sheila, Hall, Bert, and Klausner, David, eds., *Health, Disease, and Healing in Medieval Culture*, New York, 1992.
Cannon, Susan (Walter) F., 'Buckland', *DSB*, ii: 566-72.
——, 'The Problem of Miracles in the 1830s', *Victorian Studies*, iv (1960): 5-32.
——, 'Scientists and Broad Churchmen: An Early Victorian Intellectual Network', *Journal of British Studies*, iv (1964): 65-88.
Cantor, Geoffrey N., 'Reading the Book of Nature: The Relation Between Faraday's Religion and His Science', in D. Gooding and F. James, eds., *Faraday Rediscovered*, Basingstoke, 1985.
——, 'The Theological Significance of Ethers', in G. N. Cantor and M.J.S. Hodge, eds., *Conceptions of Ether*, Cambridge, 1981.
—— and M.J.S. Hodge, eds., *Conceptions of Ether: Studies in the History of Ether Theories, 1740-1900*, Cambridge, 1981.
Cantor, Norman F., *Medieval History: The Life and Death of a Civilization*, New York, 1963, 1969.
Caponigri, A. Robert, *A History of Western Philosophy*, vol. III, *Philosophy from the Renaissance to the Romantic Age*, Notre Dame, Ind., 1963.
Carey, John, 'Ireland and the Antipodes: The Heterodoxy of Virgil of Salzburg', *Speculum*, lxiv (1989): 1-10.
Casini, Paolo, *Filosofia e fisica da Newton a Kant*, Torino, 1978.
——, *L'universo macchina: Origini della filosofia newtoniana*, Bari, 1969.
Caspar, Max, *Kepler*. Translated by C. Doris Hellman, London, 1959.
Cassiodorus, Senator, *An Introduction to Divine and Human Readings*. Translated by Leslie Webber Jones, New York, 1946, 1969.
Cassirer, Ernst, *The Individual and the Cosmos in Renaissance Philosophy* (*Individuum und Kosmos in der Philosophie der Renaissance*, Leipzig, 1927). Translated by Mario Domandi, New York, 1963; Philadelphia, 1972.
——, *Kant's Life and Thought* (*Kants Leben und Lehre*, 1918). Translated by James Haden, New Haven, Conn., 1985.
——, *The Philosophy of the Enlightenment* (*Die Philosophie der Aufklärung*, Tübingen, 1932). Translated by Fritz C. Koelln and James P. Pettegrove, Princeton, 1951; Boston, 1955.
——, Kristeller, Paul Oskar, and Kandall, John Herman, Jr., eds., *The Renaissance Philosophy of Man*, Chicago, 1948.
Centre International de Synthèse, *Copernic: La représentation de l'univers et ses conséquences épistémologiques*, Paris, 1975.
Chadwick, Henry, *Boethius: The Consolations of Music, Logic, Theology, and Philosophy*, Oxford, 1981.

——, *Early Christian Thought and the Classical Tradition: Studies in Justin, Clement, and Origen*, New York, 1966.
Chadwick, Owen, *The Secularization of the European Mind in the Nineteenth Century*, Cambridge, 1975.
Chapek, Milich, 'The Conflict between the Absolutist and the Relational Theory of Time before Newton', *JHI*, xlviii (1987): 595-608.
Charlesworth, James H., ed., *The Odes of Solomon: The Syriac Texts*, Oxford, 1973.
Chenu, M.-D., *Nature, Man, and Society in the Twelfth Century: Essays on New Theological Perspectives in the Latin West* (*La théologie au douzième siècle*, Paris, 1957). Edited and translated by Jerome Taylor and Lester K. Little, Chicago, 1968.
——, *La Théologie comme science au XIIIe siècle*; 3rd ed., Paris, 1957.
Choisy, Eugène, *Calvin et la science*, Geneva, 1931.
Christie, J.R.R., 'Ether and the Science of Chemistry: 1740-1790', in G.N. Cantor and M.J.S. Hodge, eds., *Conceptions of Ether*, Cambridge, 1981.
Clagett, Marshall, *Greek Science in Antiquity*, rev. ed., New York, 1963.
——, *The Science of Mechanics in the Middle Ages*, Madison, Wis., 1959.
——, ed., *Critical Problems in the History of Science*, Madison, Wis., 1959.
Clark, George Norman, *Science and Social Welfare in the Age of Newton*, Oxford, 1937, 1949.
——, *The Seventeenth Century*, Oxford, 1929, 1947.
Clark, Robert E.D., 'Michael Faraday on Science and Religion', *Hibbert Journal*, lxv (1967): 144-47.
Clericuzio, Antonio, 'Carneades and the Chemists: A Study of *The Sceptical Chymist* and Its Impact on Seventeenth-Century Chemistry', in M. Hunter, ed., *Robert Boyle Reconsidered*, Cambridge, 1994.
——, 'From Van Helmont to Boyle: A Study of the Transmission of Helmontian Chemical and Medical Theories in Seventeenth-Century England', *BJHS*, xxvi (1993): 303-34.
——, 'A Redefinition of Boyle's Chemistry and Corpuscular Philosophy', *Annals of Science*, xlvii (1990): 561-89.
——, 'Robert Boyle and the English Helmontians', in Z.R.W.M. von Martels, ed., *Alchemy Revisited*, Leiden, 1990.
Clerval, A., *Les écoles de Chartres au moyen âge*, Chartres, 1895.
Clucas, Stephen, 'The Atomism of the Cavendish Circle: A Reappraisal', *Seventeenth Century*, ix (1994): 247-73.
Clulee, Nicholas H., *John Dee's Natural Philosophy, Between Science and Religion*, London, 1988.
Coleman, William, 'Providence, Capitalism, and Environmental Degradation', *JHI*, xxxvii (1976): 27-44.
Colie, Rosalie L., *Light and Enlightenment: A Study of the Cambridge Platonists and the Dutch Arminians*, Cambridge, 1957.
Colligan, J. Hay, *The Arian Movement in England*, Manchester, 1913.
Collingwood, R.G., *The Idea of Nature*, London, 1945.
Collins, James, *Descartes' Philosophy of Nature*, Oxford, 1971.
Copenhaver, Brian P., 'Jewish Theologies of Space in the Scientific Revolution', *Annals of Science*, xxxvii (1980): 489-548.
——, 'Natural Magic, Hermeticism, and Occultism in Early Modern Science', in D.C. Lindberg and R.S. Westman, eds., *Reappraisals*, Cambridge, 1990.
Copernicus, Nicholas, *On the Revolutions*. Edited by Jerzy Dobrycki; translated by Edward Rosen, Baltimore, 1978.
——, *On the Revolutions of the Heavenly Spheres*. Translated by A.M. Duncan, Newton Abbot, 1976.
Copleston, Frederick C., *A History of Philosophy*, 9 vols., Westminster, Md., 1946-75; New York, 1962-94.
Coudert, Alison, 'A Cambridge Platonist's Kabbalist Nightmare', *JHI*, xxxvi (1975): 633-52.
——, 'Henry More, the Kabbalah, and the Quakers', in R. Kroll et al., ed., *Philosophy, Science, and Religion*, Cambridge, 1992.
——, *Leibniz and the Kabbalah*, Dordrecht, 1995.

——, 'A Quaker-Kabbalist Controversy: George Fox's Reaction to Francis Mercury Van Helmont', *Journal of the Warburg and Courtauld Institutes*, xxxix (1976): 171-89.
Courtonne, Yves, *Saint Basile et l'Hellénisme: Étude sur la rencontre de la pensée chrétienne avec la sagesse antique dans l'Hexaméron de Basile le Grand*, Paris, 1934.
Couturier, Guy, 'La Vision du Conseil divin: étude d'une forme commune au prophétisme et à l'apocalyptique', *Science et esprit*, xxxvi (1984): 5-43.
Crocker, Robert, 'Mysticism and Enthusiasm in Henry More', in S. Hutton, ed., *Henry More*, Dordrecht, 1990.
Crombie, A.C., *Augustine to Galileo: The History of Science, AD 400-1650*, 2 vols., London, 1952; Oxford, 1957, 1961; Harmondsworth, 1969 (American ed. entitled *Medieval and Early Modern Science*, New York, 1959).
——, 'Descartes', *DSB*, 4:51-55.
——, 'Grosseteste's Position in the History of Science', in D.A. Callus, ed., *Robert Grosseteste, Scholar and Bishop*, Oxford, 1955.
——, 'Mersenne', *DSB*, 9:316-22.
——, 'Quantification in Medieval Physics', *Isis*, lii (1961): 143-60.
——, *Robert Grosseteste and the Origins of Experimental Science, 1100-1700*, Oxford, 1953.
——, 'The Significance of Medieval Discussions of Scientific Method for the Scientific Revolution', in Marshall Clagett, ed., *Critical Problems in the History of Science*, London, 1989.
——, ed., *Scientific Change: Historical Studies in the Intellectual, Social, and Technical Conditions for Scientific Discovery and Technical Invention, From Antiquity to the Present*, London, 1963.
Cunningham, Andrew, 'Getting the Game Right: Some Plain Words on the Identity and Invention of Science', *SHPS*, xix (1988): 365-89.
——, 'How the *Principia* Got Its Name; or, Taking Natural Philosophy Seriously', *History of Science* xxix (1991): 377-92.
—— and French, Roger, eds., *The Medical Enlightenment of the Eighteenth Century*, Cambridge, 1990.
—— and Jardine, Nicholas, eds., *Romanticism and the Sciences*, Cambridge, 1990.
Curry, Patrick, ed., *Astrology, Science, and Society: Historical Essays*, Woodbridge, Suffolk, 1987.
Dahm, John J., 'Science and Apologetics in the Early Boyle Lectures', *CH*, xxxix (1970): 172-86.
Daisomont, M., *Le Clergé catholique devant l'astronomie suivi d'un Essai sur le conflit entre l'astronomie et l'Église Romaine au xviiᵉsiècle*, Bruges, 1950.
d'Alembert, Jean Le Rond, *Preliminary Discourse to the Encyclopedia of Diderot*. Translated by Richard N. Schwab, Chicago, 1995.
Dales, Richard C., 'The De-Animation of the Heavens in the Middle Ages', *JHI*, xli (1980): 531-50.
——, 'Discussions of the Eternity of the World during the First Half of the Twelfth Century', *Speculum*, lvii (1982): 495-508.
——, *The Intellectual Life of Western Europe in the Middle Ages*, Washington, D.C., 1980.
——, *The Scientific Achievement of the Middle Ages*, Philadelphia, 1973.
——, 'A Twelfth-Century Concept of the Natural Order', *Viator*, ix (1978): 179-92.
Dan, Joseph, ed., *The Early Kabbalah*, New York, 1986.
Davidson, Herbert A., 'John Philoponus as a Source of Medieval Islamic and Jewish Proofs of Creation', *Journal of the American Oriental Society*, lxxxix (1969): 357-91 (reworked in *Proofs for Eternity*, New York, 1987 [listed below]).
——, *Proofs for Eternity, Creation and the Existence of God in Medieval Islamic and Jewish Philosophy*, New York, 1987.
Davies, W.D., and Finkelstein, Louis, eds., *The Cambridge History of Judaism*, vol. i, Cambridge, 1984.
Davis, Arthur P., *Isaac Watts: His Life and Works*, New York, 1943.
Davis, Edward B., 'Newton's Rejection of the "Newtonian World View": The Role of Divine Will in Newton's Natural Philosophy', *Fides et Historia*, xxii (1990): 6-20; reprinted in *Science and Christian Belief*, iii (1991): 103-117.

Dawson, Christopher, *Mediaeval Religion (Forwood Lectures, 1934) and Other Essays*, New York, 1934.
Dawtry, Anne F., 'The *Modus Medendi* and the Benedictine Order in Anglo-Norman England', in W.J. Sheils, ed., *The Church and Healing*, Oxford, 1982.
Dear, Peter, *Mersenne and the Learning of the Schools*, Ithaca, New York, 1988.
Deason, Gary Bruce, 'The Philosophy of a Lord Chancellor: Religion, Science, and Social Stability in theWork of Francis Bacon', unpub. diss., Princeton Theological Seminary, 1977.
Debus, Allen G., 'Becher', *DSB*, i:548-51.
——, 'The Chemical Philosophers: Chemical Medicine from Paracelsus to Van Helmont', *History of Science*, xii (1974): 235-59.
——, *The Chemical Philosophy: Paracelsian Science and Medicine in the Sixteenth and Seventeenth Centuries*, 2 vols., New York, 1977.
——, *Chemistry, Alchemy, and the New Philosophy, 1550-1700: Studies in the History of Science and Medicine*, London, 1987.
——, *The English Paracelsians*, London, 1965.
——, *Man and Nature in the Renaissance*, Cambridge, 1978.
——, 'Mathematics and Nature in the Chemical Texts of the Renaissance', *Ambix*, xv (1968): 1-28.
——, 'The Paracelsian Compromise in Elizabethan England', *Ambix*, viii (1960): 71-97.
——, 'Science vs. Pseudo-Science: The Persistent Debate', in *Chemistry*, London, 1987.
——, ed., *Science, Medicine, and Society in the Renaissance: Essays to Honor Walter Pagel*, 2 vols., New York, 1972.
Décarreaux, Jean, *Monks and Civilization from the Barbarian Invasions to the Reign of Charlemagne* (*Les Moines et la Civilisation*, Paris, 1962). Translated by Charlotte Haldane, London, 1964.
DeKosky, Robert K., *Knowledge and Cosmos: Development and Decline of the Medieval Perspective*, Washington, D.C., 1979.
de Santillana, Giorgio, *The Crime of Galileo*, Chicago, 1955.
Descartes, René, *Le Monde, ou Traité de la lumière*. Translated by Michael Sean Mahoney, New York, 1979.
——, *Oeuvres de Descartes*, 12 vols. plus supplement. Edited by Charles Adam and Paul Tannery, Paris, 1897-1913; reprinted 1957-58.
——, *The Philosophical Works of Descartes*, 2 vols. Translated by Elizabeth S. Haldane and G.R.T. Ross, Cambridge, 1931.
Dijksterhuis, E.J., *The Mechanization of the World Picture* (*De Mechanisering van het Wereldbeeld*, Amsterdam, 1950). Translated by C. Dikshoorn, Oxford, 1961; Princeton, N.J., 1986.
Dillenberger, John, *Protestant Thought and Natural Science: A Historical Interpretation*, Garden City, N.Y., 1960; London, 1961.
Dillon, John, *The Middle Platonists*, Ithaca, N.Y., 1977.
Dobbs, Betty Jo Teeter, *Alchemical Death and Resurrection: The Significance of Alchemy in the Age of Newton*, Washington, D.C., 1990.
——, *The Foundations of Newton's Alchemy, or 'The Hunting of the Green Lion'*, Cambridge, 1975.
——, 'Newton as Final Cause and First Mover', *Isis*, lxxxv (1994): 633-43.
——, 'Newton's Alchemy and His "Active Principle" of Gravitation', in P.B. Scheurer and G. Debrock, eds., *Newton's Scientific and Philosophical Legacy*, Dordrecht, 1988.
——, 'Newton's Alchemy and His Theory of Matter', *Isis*, lxxiii (1982): 511-28.
——, 'Newton's *Commentary* on the *Emerald Tablet* of Hermes Trismegistus: Its Scientific and Theological Significance', in I. Merkel and A.G. Debus, eds., *Hermeticism*, Washington, D.C., 1988.
—— and Jacob, Margaret, *Newton and the Culture of Newtonianism*, Atlantic Highlands, N.J., 1995.
Dodd, C.H., *The Bible and the Greeks*, London, 1935.
Dolan, John Patrick, ed., *Unity and Reform: Selected Writings of Nicholas of Cusa*, Notre Dame, Ind., 1962.

—— and Adams-Smith, William N., *Health and Society: A Documentary History of Medicine*, New York: Seabury Press, 1978.
Dols, Michael W., 'The Origins of the Islamic Hospital: Myth and Reality', *BHM*, lxi (1987): 367-90.
Doran, Barbara Giusti, 'Origins and Consolidation of Field Theory in Nineteenth Century Britain: From the Mechanical to the Electromagnetic View of Nature', *HSPS*, vi (1975): 133-260.
Dorn, Harold, *The Geography of Science*, Baltimore, 1991.
Drake, Stillman, *Galileo Studies: Personality, Tradition, and Revolution*, Ann Arbor, 1970.
—— and Drabkin, I.E., eds., *Mechanics in Sixteenth-Century Italy: Selections from Tartaglia, Benedetti, Guido Ubaldo, and Galileo*, Madison, Wis., 1969.
Dreyer, J.L.E., *A History of Astronomy from Thales to Kepler*, Cambridge, 1906: 2nd. ed. rev. by W.H. Stahl, New York, 1953 (orig. ed. entitled *History of the Planetary Systems from Thales to Kepler*).
Dronke, Peter, ed., *A History of Twelfth-Century Western Philosophy*, Cambridge, 1988.
Duffy, Eamon, '"Whiston's Affair": The Trials of a Primitive Christian, 1709-14', *JEH*, xxvii (1976): 129-50.
Duhem, Pierre, 'Physics, History of', *Catholic Encyclopedia*, vol. ix, New York, 1911.
——, *To Save the Phenomena: An Essay on the Idea of Physical Theory from Plato to Galileo* (*Sozein ta Phainomena: Essai sur la notion de théorie physique de Platon à Galilée*, Paris, 1908). Translated by Edmund Doland and Chaninah Maschler, Chicago, 1969.
——, *Le système du monde: Histoire des doctrines cosmologiques de Platon à Copernic*, 10 vols., Paris, 1913-1959.
Easton, Stewart C., *Roger Bacon and His Search for a Universal Science: A Reconsideration of the Life and Work of Roger Bacon in the Light of His Own Stated Purposes*, Oxford, 1952.
Eckenrode, Thomas R., 'The Growth of a Scientific Mind: Bede's Early and Late Scientific Writings', *Downside Review*, xciv (1976): 197-212.
——, 'Venerable Bede as a Scientist', *American Benedictine Review*, xxii (1971): 486-507.
Edelstein, Ludwig, *Ancient Medicine: Selected Papers of Ludwig Edelstein*, Baltimore, 1967.
——, 'Motives and Incentives for Science in Antiquity', in A.C. Crombie, ed., *Scientific Change*, London, 1952.
——, 'Recent Trends in the Interpretation of Ancient Science', *JHI*, xiii (1952): 573-604.
Edwards, Jonathan, *Freedom of the Will*, Works of Jonathan Edwards, vol. i. Edited by Paul Ramsey, New Haven, Conn., 1957, 1985.
——, *Jonathan Edwards: Scientific and Philosophical Writings*, Works of Jonathan Edwards, vol. vi. Edited by Wallace E. Anderson, New Haven, Conn., 1980.
Eichner, Hans, 'The Rise of Modern Science and the Genesis of Romanticism', *Publications of the Modern Language Association*, xcvii (1982): 8-31.
Einstein, Albert, 'Autobiographical Notes', in P.A. Schilpp, ed., *Albert Einstein*, Evanston, Ill., 1949.
——, *Ideas and Opinions*, London, 1954.
——, *The World As I See It*, (*Mein Weltbild*, Amsterdam, 1934). Translated by Alan Harris, London, 1935.
—— and Born, Max and Hedwig, *The Born-Einstein Letters*. Translated by Irene Born, New York, 1971.
Eisley, Loren, *Darwin's Century: Evolution and the Men Who Discovered It*, Garden City, N.Y., 1958, 1961.
Eliot, Simon, and Stern, Beverley, eds., *The Age of Enlightenment: An Anthology of Eighteenth-Century Texts*, 2 vols., London, 1979.
Elmer, Peter, 'Medicine, Religion, and the Puritan Revolution', in R. French and A. Wear, eds., *Medical Revolution*, Cambridge, 1989.
——, 'Medicine, Science, and the Quakers: The "Puritanism-Science" Debate Reconsidered', *Journal of the Friends Historical Society*, liv (1981): 265-86.
Epstein, Perle, *Kabbalah: The Way of the Jewish Mystic*, Garden City, N.Y., 1978.
Erdman, David V., *Blake: Prophet Against Empire*, Princeton, N.J., 1954, 1969, 1977.
Eusebius of Caesarea, *Preparation for the Gospel*, 2 vols. Translated by Edwin Hamilton

Gifford, Oxford, 1903; Grand Rapids, Mich., 1981.
Evans, Joan, *Monastic Life at Cluny*, Oxford, 1931; Hamden, Conn., 1968.
Everitt, C.W.F., 'Maxwell', *DSB*, ix:198-230.
Farrington, Benjamin, *Francis Bacon: Philosopher of Industrial Science*, New York, 1949.
——, *The Philosophy of Francis Bacon: An Essay on Its Development from 1603 to 1609 with New Translations of Fundamental Texts*, Liverpool, 1964; Chicago, 1966.
Fauvel, John, et al., eds., *Let Newton Be!*, Oxford, 1988.
Fedwick, Paul Jonathan, ed., *Basil of Caesarea, Christian, Humanist, Ascetic,* 2 vols., Toronto, 1981.
Ferguson, James P., *An Eighteenth-Century Heretic: Dr. Samuel Clarke*, Kineton, 1976.
——, *The Philosophy of Dr. Samuel Clarke and Its Critics*, New York, 1974.
Ferré, Frederick, 'Einstein on Religion and Science', *American Journal of Theology and Philosophy*, i (1980): 21-28.
Festugière, André-Jean, *Hermétisme et Mystique Païenne*, Paris, 1967.
Ficino, Marsilio, *Three Books on Life*. Edited and translated by Carol V. Kaske and John R. Clark, Medieval and Renaissance Texts and Studies 57, Binghamton, 1989.
Field, J.V., *Kepler's Geometrical Cosmology*, Chicago, 1988.
——, 'A Lutheran Astrologer: Johannes Kepler', *Archive for History of Exact Sciences*, xxxi (1984): 189-272
Fierz, Markus, *Naturwissenschaft und Geschichte: Vorträge und Aufsätze*, Basel, 1988.
Figala, Karin, 'Die exacte Alchemie von Isaac Newton', *Verhandlungen der Naturforschers-Gesellschaft Basel*, xciv (1984): 157-228.
Fisch, Harold, 'The Scientist as Priest: A Note on Robert Boyle's Natural Theology', *Isis*, xliv (1953): 252-65.
Fix, Andrew, *Prophecy and Reason: The Dutch Collegiants in the Early Enlightenment*, Princeton, N.J., 1991.
Flint, Valerie J., 'The Early Medieval "Medicus", the Saint—and the Enchanter', *Social History of Medicine*, ii (1989): 127-45.
——, *The Rise of Magic in Early Medieval Europe*, Princeton, N.J., 1991.
Folse, Henry J., *The Philosophy of Niels Bohr: The Framework of Complementarity*, Amsterdam, 1985.
Force, James E., *William Whiston: Honest Newtonian*, Cambridge, 1985.
Forgan, Sophie, ed., *Science and the Sons of Genius: Studies on Humphry Davy*, London, 1980.
Fox, Robert, 'The Rise and Fall of Laplacian Physics', *HSPS*, iv (1974): 89-136.
Franklin, Allan, *The Principle of Inertia in the Middle Ages*, Boulder, Colo., 1976.
Franklin, Carmela Vircillo, et al., trans., *Early Monastic Rules: The Rules of the Fathers and the Regula Orientalis*, Collegeville, Minn., 1982.
French, Peter J., *John Dee: The World of an Elizabethan Magus*, London, 1972.
French, Roger, and Wear, Andrew, eds., *The Medical Revolution of the Seventeenth Century*, Cambridge, 1989
Frye, Northrop, *Fearful Symmetry: A Study of William Blake*, Princeton, N.J., 1947, 1969.
Funkenstein, Amos, *Theology and the Scientific Imagination from the Middle Ages to the Seventeenth Century*, Princeton, N.J., 1986.
Gabbey, Alan, 'Henry More and the Limits of Mechanism', in S. Hutton, ed., *Henry More*, Dordrecht, 1990.
——, 'Philosophia Cartesiana Triumphata: Henry More (1646-1671)', in Thomas M. Lennon et al., eds., *Problems of Cartesianism*, Kingston, Ont., 1982.
Gadol, Joan, *Leon Battista Alberti: Universal Man of the Early Renaissance*, Chicago, 1969.
Gale, George, Jr., 'Leibniz' Dynamical Metaphysics and the Origins of the Vis Viva Controversy', *Systematics*, xi (1973): 184-207.
Galileo, *Galilei Opere,* 20 vols. Edited by A. Favaro, Florence, 1890-1909.
Gammie, John G. et al., eds., *Israelite Wisdom: Theological and Literary Essays in Honor of Samuel Terrien*, Missoula, Mont., 1978.
Garin, Eugenio, *Science and the Civic Life in the Italian Renaissance (Scienza e vita civile nel Rinascimento italiano*, 1965). Translated by Peter Munz, Garden City, N.J., 1969.
Gascoigne, John, *Cambridge in the Age of the Enlightenment: Science, Religion, and Politics*

from the Restoration to the French Revolution, Cambridge, 1989.
Gask, George E., and Todd, John, 'The Origin of Hospitals', in E. Ashworth Underwood, ed., Science, Medicine, and History, vol. i, London, 1953.
Gay, Peter, *The Enlightenment, An Interpretation*, vol. ii: *The Science of Freedom*, London, 1969, 1973.
Gelbart, Nina Rattner, 'The Intellectual Development of Walter Charleton', *Ambix*, xviii (1971): 149-68.
Genuth, Sara Schechner, 'Comets, Teleology, and the Relationship of Chemistry to Cosmology in Newton's Thought', *Annali dell'Instituto e Museo di Storia della Scienza di Firenze*, x (1985): 31-65.
——, 'Newton and the Ongoing Teleological Role of Comets', in N.J.W. Thrower, *Standing on the Shoulders*, Berkeley, 1990.
Georgi, Dieter, *The Opponents of Paul in Second Corinthians (Die Gegner des Paulus im 2. Korintherbrief*, Neukirchen-Vluyn, 1964), Philadelphia, 1986.
Gibson, Margaret, ed., *Boethius: His Life, Thought and Influence*, Oxford, 1981.
Giet, Stanislas, *Les idées et l'action sociales de Saint Basil*, Paris, 1941.
Gillispie, Charles Coulston, 'Diderot', *DSB*, iv:84-90.
——, 'Laplace', *DSB*, xv:273-403.
Gilson, Étienne, *History of Christian Philosophy in the Middle Ages*, New York, 1955.
——, *The Philosophy of St Bonaventure*. Translated by Dom Illtyd Trethowan and Frank J. Sheed, New York, 1938; Patterson, N.J. 1965.
——, *The Spirit of Mediaeval Philosophy* (Gifford Lectures, 1931-1932). Translated by A.H.C. Downes, London and New York, 1936.
Gingerich, Owen, ed., *The Nature of Scientific Discovery: A Symposium Commemorating the 500th Anniversary of the Birth of Nicolaus Copernicus*, Washington, D.C., 1975.
Ginsburg, Christian D., *The Essenes: Their History and Doctrines; The Kabbalah: Its Doctrines, Development and Literature*, London, 1955.
Glacken, Clarence J., *Traces on the Rhodian Shore: Nature and Culture in Western Thought from Ancient Times to the End of the Eighteenth Century*, Berkeley, 1967.
Goldman, Martin, *The Demon and the Aether: The Story of James Clerk Maxwell the Father of Modern Science*, Edinburgh, 1983.
Goldstein, Thomas, *Dawn of Modern Science: From the Arabs to Leonardo da Vinci*, Boston, 1980.
González, Justo L., *The Story of Christianity*, 2 vols., San Francisco, 1984.
Goodenough, Erwin R., *By Light, Light: The Mystic Gospel of Hellenistic Judaism*, New Haven, Conn., 1935.
Gooding, David C., 'Metaphysics versus Measurement: The Conversion and Conservation of Force in Faraday's Physics', *Annals of Science*, xxxvii (1980): 1-29.
——, and Adams, Frank A.J.L., eds., *Faraday Rediscovered: Essays on the Life and Work of Michael Faraday, 1791-1867*, Basingstoke, 1985; New York, 1989.
Goodman, D.C., ed., *Science and Religious Belief, 1600-1900: A Selection of Primary Sources*, Dorchester, 1973.
Gorman, Peter, *Pythagoras, A Life*, London, 1979.
Gosselin, Edward A., 'Bruno's "French Connection": A Historiographical Debate', in I. Merkel and A.G. Debus, eds., *Hermeticism*, Washington D.C., 1988.
Gould, Stephen Jay, 'Hutton's Purposeful View', *Natural History*, xci (May 1982): 6-12.
Gould, William, ed., *Lives of the Georgian Age, 1714-1837*, New York, 1978.
Grant, Edward, 'Comment' on the Church and Academic Freedom in the Middle Ages, a session of the American Society of Church History, Holland, Mich., 21 April 1983.
——, 'The Condemnation of 1277, God's Absolute Power, and Physical Thought in the Late Middle Ages', *Viator*, x (1979): 211-44.
——, 'Cosmology', in D. Lindberg, ed., *Science in the Middle Ages*, Chicago, 1978.
——, 'Late Medieval Thought, Copernicus, and the Scientific Revolution', *JHI*, xxiii (1962): 197-220.
——, 'Medieval and Seventeenth-Century Conceptions of an Infinite Void Space beyond the Cosmos', *Isis*, lx (1969): 39-60.
——, 'Motion in the Void and the Principle of Inertia in the Middle Ages', *Isis*, lv (1964):

265-92.
———, *Much Ado About Nothing: Theories of Space and Vacuum from the Middle Ages to the Scientific Revolution*, Cambridge, 1981.
———, *Physical Science in the Middle Ages*, New York, 1971; Cambridge, 1977.
———, 'Place and Space in Medieval Physical Thought', in Peter K. Machamer and Robert G. Turnbull, eds., *Motion and Time, Space, and Matter*, Columbus, Ohio, 1976.
———, *Planets, Stars, and Orbs: The Medieval Cosmos, 1200-1687*, Cambridge, 1994.
Grant, R., 'Hutton's Theory of the Earth', in L.J. Jordanova and R.S. Porter, eds., *Images of the Earth*, Chalfont St Giles, 1978.
Grant, Robert M., *Miracle and Natural Law in Graeco-Roman and Early Christian Thought*, Amsterdam, 1952.
Green, Peter, *Alexander to Actium: The Historical Evolution of the Hellenistic Age*, Berkeley, 1990.
Greene, Robert A., 'Henry More and Robert Boyle on the Spirit of Nature', *JHI*, xxiii (1962): 451-74.
Gregory, Frederick, 'Kant's Influence on Natural Scientists in the German Romantic Period', in R.P.W. Visser, et al., eds., *New Trends*, Amsterdam, 1989.
———, *Nature Lost: Natural Science and the German Theological Traditions of the Nineteenth Century*, Cambridge, Mass., 1992.
Grimsley, Ronald, *Jean d'Alembert (1717-83)*, Oxford, 1963.
Gross, Charlotte, 'Twelfth-Century Concepts of Time: Three Reinterpretations of Augustine's Doctrine of Creation *Simul*', *JHP*, xxiii (1985): 325-38.
Grosseteste, Robert, *On Light (De Luce)*. Translated by Clare C. Riedl, Milwaukee, Wis., 1978.
Gruenwald, Ithamar, *Apocalyptic and Merkavah Mysticism*, Leiden, 1980.
———, *From Apocalypticism to Gnosticism: Studies in Apocalypticism, Merkavah Mysticism and Gnosticism*, Frankfurt am Main, 1988.
———, 'Knowledge and Vision: Towards a Clarification of Two "Gnostic" Concepts in the Light of Their Alleged Origins', *Israel Oriental Studies*, iii (1973): 63-107; reprinted in *From Apocalypticism to Gnosticism*, Frankfurt am Main, 1988.
———, 'Some Critical Notes on the First Part of *Sefer* Yezira', *Révue des études Juives*, cxxxii (1973): 475-512
Guerlac, Henry, 'Black, Joseph', *DSB*, ii:173-83.
———, *Newton on the Continent*, Ithaca, N.Y., 1981.
———, 'Where the Statue Stood: Divergent Loyalties to Newton in the Eighteenth Century', in Earl R. Wasserman, ed., *Aspects of the Eighteenth Century*, Baltimore, 1965.
Gunther, John J., *St Paul's Opponents and Their Background: A Study of Apocalyptic and Jewish Sectarian Teachings*, Leiden, 1973.
Haase, Rudolf, 'Kepler's Harmonies, between Pansophia and Mathesis Universalis', in A. and P. Beer, *Kepler*, Oxford, 1975.
Hacking, Ian, 'Nineteenth Century Cracks in the Concept of Determinism', *JHI*, xliv (1983): 455-75.
Hadas, Moses, *Hellenistic Culture: Fusion and Diffusion*, New York, 1959.
Haddad, Rachid, 'Hunayn ibn Ishaq, Apologiste Chrétien', in Gérard Troupeau, ed., *Hunayn Ibn Ishaq*, Leiden, 1975.
Hahn, Roger, 'Laplace and the Mechanistic Universe', in D.C. Lindberg and R.L. Numbers, eds., *God and Nature*, Berkeley, 1985.
———, 'Laplace and the Vanishing Role of God in the Physical Universe', in H. Woolf, ed., *The Analytic Spirit*, Ithaca, N.Y., 1981.
Hall, A. Rupert, 'Desauguliers', *DSB*, iv:43-46.
———, *From Galileo to Newton, 1630-1720*, New York, 1963; Dover, 1981.
———, 's'Gravesande', *DSB*, v:509-511.
———, *The Scientific Revolution, 1500-1800: The Formation of the Modern Scientific Attitude*, London, 1954, 1962; Boston, 1966; rev. ed., *The Revoltuion in Science 1500-1750*, London, 1983.
Hall, Marie Boas, 'Boyle', *DSB* 2:377-82.
———, 'Digby', *DSB* 4:95-96.

——, 'An Early Version of Boyle's Sceptical Chymist', *Isis*, xlv (1954): 153-68.
Hall, Thomas S., *History of General Physiology, 600 BC to AD 1900*, 2 vols., Chicago, 1975 (orig. ed. entitled *Ideas of Life and Matter*, Chicago, 1969).
Haller, William, *Elizabeth I and the Puritans*, Folger Booklets on Tudor and Stuart Civilization, Ithaca, N.Y., 1964.
Hammer, William, 'Melanchthon, Inspirer of the Study of Astronomy, With a Translation of His Oration in Praise of Astronomy (*De Orione*, 1553)', *Popular Astronomy*, lix (1951): 308-19.
Hampshire, Susan, *Spinoza*, Harmondsworth, 1951.
Hands, A.R., *Charities and Social Aid in Greece and Rome*, Ithaca, N.Y., 1968.
Hankins, Thomas L., *Science and the Enlightenment*, Cambridge, 1985.
Hannaway, Owen, *The Chemists and the Word: The Didactic Origins of Chemistry*, Baltimore, 1975.
Harbison, E. Harris, *The Christian Scholar in the Age of the Reformation*, New York, 1956.
Harman (Heimann), Peter M., 'Concepts of Inertia: Newton to Kant', in M.J. Osler and P.L. Farber, eds., *Religion, Science, and Worldview*, Cambridge, 1985.
——, 'Conversion of Forces and the Conservation of Energy', *Centaurus*, xviii (1974) 147-61.
——, 'Edinburgh Philosopohy and Cambridge Physics: The Natural Philosophy of James Clerk Maxwell', in P.M. Harman, ed., *Wranglers and Physicists*, Manchester, 1985.
——, *Energy, Force, and Matter: The Conceptual Development of Nineteenth-Century Physics*, Cambridge, 1982.
——, 'Faraday's Theories of Matter and Electricity', *BJHS*, v (1971): 235-57.
——, 'Maxwell and the Modes of Consistent Representation', *AHES*, vi (1970): 171-213.
——, 'Molecular Forces, Statistical Representation and Maxwell's Demon', *SHPS*, i (1970): 189-211.
——, '"Nature Is a Perpetual Worker": Newton's Aether and Eighteenth-Century Natural Philosophy', *Ambix*, xx (1973): 1-25.
——, 'The Unseen Universe: Physics and the Philosophy of Nature in Victorian Britain', *BJHS*, vi (1972): 73-79.
——, 'Voluntarism and Immanence: Conceptions of Nature in Eighteenth-Century Thought', *JHI*, xxxix (1978): 271-83.
——, ed., *Wranglers and Physicists: Studies on Cambridge Mathematical Physics in the Nineteenth Century*, Manchester, 1985.
—— and McGuire, J.E., 'Newtonian Forces and Lockean Powers: Concepts of Matter in Eighteenth-Century Thought', *HSPS*, Philadelphia, 1971.
Harries, Karsten, 'The Infinite Sphere: Comments on the History of a Metaphor', *JHP*, xiii (1975): 5-15.
Haskins, Charles Homer, *The Renaissance of the Twelfth Century*, Cleveland and New York, 1957.
——, *Studies in the History of Medieval Science*, Cambridge, Mass., 1927.
Hatfield, Gary C., 'Force (God) in Descartes' Physics', *SHPS*, x (1979): 113-40.
Heal, Felicity, and O'Day, Rosemary, eds., *Church and Society in England: Henry VIII to James I*, London, 1977.
Heilbron, John L., *Elements of Early Modern Physics*, Berkeley, 1982.
——, 'Rutherford-Bohr Atom', *AJP*, xlix (1981): 223-31.
—— and Kuhn, Thomas S., 'The Genesis of the Bohr Atom', *HSPS*, i (1969).
Heimann, Peter M., see Harman, Peter M.
Helleman, Wendy E., ed., *Christianity and the Classics. The Acceptance of a Heritage*, Lanham, N.Y., 1990.
Hellman, C. Doris, and Swerdlow, Noel M., 'Peurbach', *DSB*, Suppl. I: 473-79.
Helmont, Jean Baptiste van, *Oriatrike, or Physicke Refined* (*De Lithiasi*; from *Opuscula medica inaudita*, 2nd ed., Amsterdam, 1648). Translated by John Chandler, London, 1662.
Hengel, Martin, *Judaism and Hellenism: Studies in Their Encounter in Palestine during the Early Hellenistic Period*, 2 vols., (*Judentum und Hellenismus*, Tübingen, 1973). Translated by John Bowden, London, 1974.

——, *Property and Riches in the Early Church: Aspects of a Social History of Early Christianity* (*Eigentum und Reichtum in der frühen Kirche*, Stuttgart, 1973). Translated by John Bowden, London, 1974.
Heninger, S.K., Jr., *The Cosmographical Glass: Renaissance Diagrams of the Universe*, San Marino, 1977.
Henry, John, 'Atomism and Eschatology: Catholicism and Natural Philosophy in the Interregnum', *BJHS*, xv (1982): 211-39.
——, 'Boyle and Cosmical Qualities', in M. Hunter, ed., *Robert Boyle Reconsidered*, Cambridge, 1994.
——, 'Henry More versus Robert Boyle: The Spirit of Nature and the Nature of Providence', in S. Hutton, ed., *Henry More*, Dordrecht, 1990.
——, 'Occult Qualities and the Experimental Philosophy: Active Principles in Pre-Newtonian Matter Theory', *History of Science*, xxiv (1986): 335-81.
——, '"Pray Do Not Ascribe That Notion to Me": God and Newton's Gravity', in J.E. Force and R.H. Popkin, eds., *The Books of Nature and Scripture*, Dordrecht, 1994.
Heyd, Michael, 'Descartes—An Enthusiast *malgré lui*?' in D.S. Katz and J.I. Israel, *Sceptics*, Leiden, 1990.
——, 'Orthodoxy, Non-Conformity and Modern Science: The Case of Geneva', in Myriam Yardeni, ed., *Modernité et non-conformisme en France à travers les âges*, Leiden, 1983.
Hiebert, Erwin N., 'The Integration of Revealed Religion and Scientific Materialism in the Thought of Joseph Priestley', in L. Kieft and B.R. Willeford, Jr., eds., *Joseph Priestley*, Lewisburg, Penn., 1980.
Hill, Christopher, *The World Turned Upside Down: Radical Ideas During the English Revolution*, New York, 1972.
Hillgarth, J.N., *Ramon Lull and Lullism in Fourteenth-Century France*, Oxford, 1971.
Hine, William L., 'Mersenne and Alchemy' in Z.R.W.M. von Martels, ed., *Alchemy Revisited*, Leiden, 1990.
Hobhouse, Stephen, *Selected Mystical Writings of William Law*, London, 1948.
Holstun, James, *A Rational Millennium: Puritan Utopias of Seventeenth-Century England and America*, New York, 1987.
Holton, Gerald, *Thematic Origins of Scientific Thought: Kepler to Einstein*, Cambridge, Mass., 1973.
Home, R.W., 'Force, Electricity, and the Powers of Living Matter in Newton's Mature Philosophy of Nature', in M.J. Osler and P.L. Farber, eds., *Religion, Science, and Worldview*, Cambridge, 1985.
Honner, John, *The Description of Nature: Niels Bohr and the Philosophy of Quantum Physics*, Oxford, 1987.
Hooper, John, *Early Writings of John Hooper, D.D. (Parker Society Works of the Reformed English Church*, vol. xx). Edited by Samuel Carr, Cambridge, 1843.
Hooykaas, R., 'Beeckman', *DSB*, i:566-68.
——, 'Humanisme, science et réforme, 1515-72', *Free University Quarterly*, v (1958): 167-294.
——, *Natural Law and Divine Miracle: A Historical Critical Study of the Principle of Uniformity in Geology, Biology and Theology*, Leiden, 1959.
——, *Religion and the Rise of Modern Science*, Edinburgh, 1972.
Hornberger, Theodore, 'Samuel Johnson of Yale and King's College: A Note on the Relation of Science and Religion in Provincial America', *New England Quarterly*, viii (1935): 378-97.
Hoskin, Michael A., *Stellar Astronomy: Historical Studies*, 1982.
Howe, Herbert M., 'A Root of Helmont's Tree', *Isis*, lvi (1965): 408-19.
Huffman, William H., *Robert Fludd and the End of the Renaissance*, London, 1988.
Hughes, Philip Edgcumbe, *LeFèvre, Pioneer of Ecclesiastical Renewal in France*, Grand Rapids, Mich., 1984.
——, *Theology of the English Reformers*, Grand Rapids, Mich., 1980.
Hunter, Michael, 'Alchemy, Magic and Moralism in the Thought of Robert Boyle', *BJHS*, xxiii (1990): 387-410.

——, 'Science and Heterodoxy: An Early Modern Problem Reconsidered', in D.C. Lindberg and R.S. Westman, eds., *Reappraisals*, Cambridge, 1990.
——, *Science and the Shape of Orthodoxy: Intellectual Change in Late Seventeenth-Century Britain*, Woodbridge, 1995.
——, ed., *Robert Boyle Reconsidered*, Cambridge, 1994.
Hunter, William B., Jr., 'The Seventeenth Century Doctrine of Plastic Nature', *HTR*, xliii (1950): 197-213.
Hutchison, Keith, 'Supernaturalism and the Mechanical Philosophy', *History of Science*, xxi (1983): 297-333.
——, 'Towards a Political Iconology of the Copernican Revolution', in P. Curry, ed., *Astrology, Science and Society*, Woodbridge, 1987.
Hutton, Sarah, ed., *Henry More (1614-1687): Tercentenary Studies*, Dordrecht, 1989.
Hyma, Albert, *The Christian Renaissance: A History of the 'Devotio Moderna'*, New York, 1925.
Hyman, Arthur, and Walsh, James J., eds., *Philosophy in the Middle Ages: The Christian, Islamic, and Jewish Traditions*, Indianapolis, Ind.,1973.
Idel, Moshe, *Golem: Jewish Magical and Mystical Traditions on the Artificial Anthropoid*, Albany, N.Y., 1990.
——, 'Hermeticism and Judaism', in I. Merkel and A.G. Debus, eds., *Hermeticism and the Renaissance*, Washington, 1988.
——, *Kabbalah: New Perspectives*, New Haven, Conn., 1988.
——, *The Mystical Experience in Abraham Abulafia*. Translated from the Hebrew by Jonathan Chipman, Albany, N.Y., 1988.
Iliffe, Rob, '"Making a Shrew": Apocalyptic Hermeneutics and the Sociology of Christian Idolatry in the Work of Isaac Newton and Henry More', in J. E. Force and R.H. Popkin, eds., *The Books of Nature and Scripture*, Dordrecht, 1994.
Iltis, Carolyn, 'Leibniz and the *Vis Viva* Controversy', *Isis*, lxii (1971) 21-35.
——, 'The Leibnizian-Newtonian Debates: Natural Philosophy and the *Vis Viva* Controversy', *BJHS*, vi (1973): 343-77.
Irmscher, J., 'The Pseudo-Clementines', in Wilhelm Schneemelcher and R. McL. Wilson, eds., *New Testament Apocrypha, Vol. II*, London, 1965.
Jacob, James R., 'Boyle's Atomism and the Restoration: Assault on Pagan Naturalism', *Social Studies of Science*, viii (1978): 211-33.
——, 'Boyle's Circle in the Protectorate: Revelation, Politics, and the Millennium', *JHI*, xxxviii (1977): 131-40.
——, 'The Ideological Origins of Robert Boyle's Natural Philosophy', *Journal of European Studies*, ii (1972): 1-21.
——, *Robert Boyle and the English Revolution: A Study in Social and Intellectual Change*, New York, 1977.
Jacob, Margaret C., 'John Toland and the Newtonian Ideology', *Journal of the Warburg and Courtauld Inst.*, xxxii (1969): 307-31.
——, 'Newtonianism and the Origins of the Enlightenment: A Reassessment', *ECS*, xi (1977): 1-25.
——, *The Newtonians and the English Revolution, 1689-1720*, Ithaca, N.Y., 1976.
——, *The Radical Enlightenment: Pantheists, Freemasons, and Republicans*, London, 1982.
——, 'Science and Social Passion: The Case of Seventeenth-Century England', *JHI*, vol. xliii (1982): 331-39.
Jacquart, Danielle, 'The Introduction of Arabic Medicine into the West: The Question of Etiology', in S. Campbell et al., eds., *Health, Disease and Healing*, New York, 1992.
Jaki, Stanley L., *he Road of Science and the Ways to God* (Gifford Lectures, 1974-75 and 1975-76), Chicago, 1978.
——, *Science and Creation: From Eternal Cycles to an Oscillating Universe*, Edinburgh, 1974.
Jammer, Max,, *Concepts of Force: A Study in the Foundations of Dynamics*, Cambridge, Mass., 1957; New York, 1962.
——, *Concepts of Mass in Classical and Modern Physics*, Cambridge, Mass., 1961.
——, *Concepts of Space: The History of Theories of Space in Physics* 2nd ed., Cambridge,

Mass., 1969.
——, *Einstein und die Religion*, Konstanz, 1995.
Jarzombek, Mark, *On Leon Baptista Alberti, His Literary and Aesthetic Theories*, Cambridge, Mass., 1989.
Jeske, Jeffrey, 'Cotton Mather: Physico-Theologian', *JHI*, xlvii (1986): 583-94.
Johnson, Francis R., *Astronomical Thought in Renaissance England: A Study of the English Scientific Writings from 1500 to 1645*, Baltimore, 1937.
——, 'Thomas Hood's Inaugural Address as Mathematical Lecturer of the City of London (1588)', *JHI*, iii (1942): 94-106.
Johnson, Harold J., ed., *The Medieval Tradition of Natural Law*, Kalamazoo, Mich., 1987.
Jolivet, Jean, 'The Arabic Inheritance', in Peter Dronke, ed., *The History of Twelfth-Century Western Philosophy*, Cambridge, 1988.
Jones, Rufus M., *Spiritual Reformers in the 16th and 17th Centuries*, London, 1914; Boston, 1959.
Jordan, W.K., *Philanthropy in England, 1480-1660: A Study of the Changing Pattern of English Social Aspirations*, London, 1959.
Josephus, Flavius, *The Works of Flavius Josephus*, 4 vols. Translated by William Whiston, Philadelphia, 1833; reprinted, Grand Rapids, Mich., 1974.
Kadushin, Max, *The Rabbinic Mind*, 3rd ed., New York, 1972.
Kahn, Charles H., *Anaximander and the Origins of Greek Cosmology*, New York, 1960.
Kaiser, Christopher B., 'Calvin, Copernicus, and Castellio', *Calvin Theological Journal*, xxi (1986): 5-31.
——, 'Calvin's Understanding of Aristotelian Natural Philosophy: Its Extent and Possible Origins', in Robert V. Schnucker, ed., *Calviniana: Ideas and Influence of Jean Calvin*, Sixteenth Century Essays and Studies 10, Kirksville,1988.
——, *Creation and the History of Science*, London, 1991.
——, 'The Creationist Tradition in the History of Science', *Perspectives on Science and Christian Faith*, xlv (June 1993): 80-89.
——, *The Doctrine of God: An Historical Survey*, London, 1982.
——, 'The Early Christian Belief in Creation: Background for the Origins and Assessment of Modern Western Science', *Horizons in Biblical Theology*, ix (Dec. 1987): 1-30.
——, 'The Early Christian Critique of Greek Science', *Patristic and Byzantine Review*, i (1982): 211-16.
——, 'Faith and Science and the W.C.C.', *Reformed World*, xxxv (1979): 330-36.
——, 'From Biblical Secularity to Modern Secularism: Historical Aspects and Stages', in S. Marianne Postiglione and Robert Brungs, eds., *Secularism versus Biblical Secularity*, St. Louis, 1994.
——, 'The Integrity of Creation and the Social Nature of God', *Scottish Journal of Theology*, il (1996): 261-90.
——, 'The Laws of Nature and the Nature of God', in Jitse Van der Meer, ed., *Facets of Faith and Science*, vol. 4, Lanham, Md., 1996.
——, 'Scientific Work and Its Theological Dimensions: Toward a Theology of Natural Science', in Jitse Van der Meer, ed., *Facets of Faith and Science*, vol. i, Lanham, Md., 1996.
Kant, Immanuel, *Universal Natural History and Theory of the Heavens*, trans. W. Hastie, Ann Arbor, 1969.
——, *Universal Natural History and Theory of the Heavens*, trans. Stanley L. Jaki, Edinburgh, 1981.
Kargon, Robert H., *Atomism in England from Hariot to Newton*, Oxford, 1966.
——, 'Atomism in the Seventeenth Century', *DHI*, i:132-41.
——, 'Model and Analogy in Victorian Science: Maxwell's Critique of the French Physicists', *JHI*, 30 (1969): 423-36.
——, 'Walter Charleton, Robert Boyle, and the Acceptance of Epicurean Atomism in England', *Isis*, lv (1964): 184-92.
Kearney, Hugh, *Science and Change, 1500-1700*, New York, 1971.
Kee, Howard Clark, *Medicine, Miracle and Magic in New Testament Times*, Cambridge, 1986.

Kendall, James, 'The Adventures of an Hypothesis', in *Proceedings of the Royal Society of Edinburgh*, lxiii (1949-50): part A, 1-17.
Kepler, Johannes, *Gesammelte Werke*, 20 vols. to 1988. Edited by Max Caspar, W. Von Dyck, et al., Munich, 1938-88.
Kieckhefer, Richard, *Magic in the Middle Ages*, Cambridge, 1990.
Kieft, Lester, and Willeford, Bennett R., Jr., eds., *Joseph Priestley: Scientist, Theologian, and Metaphysician*, Lewisburg, 1980.
Kiernan, Colm, *The Enlightenment and Science in Eighteenth-Century France*, Banbury, 1973.
King, Lester S., 'Stahl', *DSB*, lxii: 599-606.
Kinsman, Robert S., ed., *The Darker Vision of the Renaissance: Beyond the Fields of Reason*, Berkeley, 1974.
Klaaren, Eugene M., *Religious Origins of Modern Science: Belief in Creation in Seventeenth-Century Thought*, Grand Rapids, Mich.,1977.
Knight, David M., *The Age of Science: The Scientific World-view in the Nineteenth Century*, Oxford, 1986.
——, 'Romanticism and the Sciences', in A. Cunningham and N. Jardine, eds., *Romanticism and the Sciences*, Cambridge, 1990.
Knoepflmacher, U.C., and Tennyson, G.B., eds., *Nature and the Victorian Imagination*, Berkeley, 1977.
Knoll, Paul W., 'The Arts Faculty at the University of Cracow at the End of the Fifteenth Century', in R.S. Westman, ed., *The Copernican Achievement*, Berkeley, 1975.
Knowles, David, *Christian Monasticism*, New York, 1969.
——, *The Evolution of Medieval Thought*, London, 1962.
Kocher, Paul H., 'Paracelsian Medicine in England: The First Thirty Years (ca. 1570-1600)', *JHMAS*, ii (1947): 451-80.
——, *Science and Religion in Elizabethan England*, New York, 1969.
Koenigsberger, Dorothy, *Renaissance Man and Creative Thinking: A History of Concepts of Harmony, 1400-1700*, Atlantic Highlands, N.J., 1979.
Koester, Helmut, 'NOMOS PHUSEOS: The Concept of Natural Law in Greek Thought', in J. Neusner, ed., *Religions in Antiquity*, Leiden, 1968.
Koestler, Arthur, *The Sleepwalkers: A History of Man's Changing Vision of the Universe*, New York, 1959.
Koryé, Alexandre, *The Astronomical Revolution (La révolution astronomique*, Paris, 1961), London, 1973.
——, *From the Closed World to the Infinite Universe*, Baltimore, 1957.
Kovach, Francis J., and Shahan, Robert W., eds., *Albert the Great: Commemorative Essays*, Norman, Okla., 1980.
Kozhamthadam, Job, *The Discovery of Kepler's Laws*, Notre Dame, 1994.
Koziol, Geoffrey, 'Lord's Law and Natural Law', in H.J. Johnson, ed., *Medieval Tradition*, Kalamazoo, Mich.,1987.
Krafft, Fritz, 'Guericke', *DSB*, v:574-76.
Kretzmann, Norman, Kenny, Anthony, and Pinborg, Jan, eds., *The Cambridge History of Later Medieval Philosophy: From the Rediscovery of Aristotle to the Disintegration of Scholasticism, 1100-1600*, Cambridge, 1982.
Kristeller, Paul Oskar, *Eight Philosophers of the Italian Renaissance*, Stanford, 1964.
——, 'Ficino and Pomponazzi on the Place of Man in the Universe', *JHI*, v (1944): 220-42.
Kubrin, David, 'Keill, John', *DSB*, 7: 275ff.
——, 'Newton and the Cyclical Cosmos: Providence and the Mechanical Philosophy', *JHI*, xxviii (1967): 325-46.
——, 'Newton's Inside Out! Magic, Class Struggle, and the Rise of Mechanism in the West', in H. Woolf, ed., *The Analytic Spirit*, Ithaca, N.Y., 1981.
Kuhn, Albert J., 'Glory or Gravity: Hutchinson vs. Newton', *JHI*, xxii (1961): 303-332.
——, 'Nature Spiritualized: Aspects of Anti-Newtonianism', *English Literary History*, xli (1974): 400-412.
Kuhn, Thomas S., *The Copernican Revolution: Planetary Astronomy in the Development of Western Thought*, Cambridge, Mass., 1957.

———, 'Energy Conservation as an Example of Simultaneous Discovery', in M. Marshall Clagett, ed., *Critical Problems*, Madison, Wis., 1969.
Kumar, Deepak, ed., *Science and Empire: Essays in Indian Context (1700-1947)*, Delhi, 1991.
Kung, Joan R., 'Review Essay on Magic, Reason and Experience', *Nature and System*, iv (1982): 101-7.
Landels, J.G., *Engineering in the Ancient World*, Berkeley, 1981.
Langford, Jerome J., *Galileo, Science and the Church*, rev.ed., Ann Arbor, 1971.
Laplace, Pierre Simon de, *A Philosophical Essay on Probabilities*. Translated by Frederick Wilson Truscott and Frederick Lincoln Emory, New York, 1951.
———, *Philosophical Essay on Probabilities*. Translated by Andrew I. Dole, New York, 1995.
Lattimer, Hugh, *Sermons*, Parker Society Works of the Reformed English Church, vol. xxviii. Edited by George Elwes Corrie, Cambridge, 1844.
Laudan, Larry, *Science and Hypothesis: Historical Essays on Scientific Methodology*, Dordrecht, 1981.
Lawn, Brian, *The Salernitan Questions: An Introduction to the History of Medieval and Renaissance Problem Literature*, Oxford, 1963.
Lawrence, Christopher, 'The Power and the Glory: Humphry Davy and Romanticism', in A. Cunningham and N. Jardine, eds., *Romanticism and the Sciences*, Cambridge, 1990.
Leclerc, Lucien, *Histoire de la médecine arabe, exposé complet des traductions du Grec*, vol. i, Paris, 1876; New York, 1971.
Le Déaut, Roger, 'Philanthropia dans la littérature greque jusqu'au Nouveau Testament', in *Mélanges Eugène Tisserant*, vol. i, Vatican City, 1964.
Leff, Gordon, *Bradwardine and the Pelagians: A Study of His 'De Causa Dei' and Its Opponents*, Cambridge Studies in Medieval Life and Thought, vol. v, Cambridge, 1957.
———, *Medieval Thought: St Augustine to Ockham*, Harmondsworth, 1958.
———, *Paris and Oxford Universities in the Thirteenth and Fourteenth Centuries: An Institutional and Intellectual History*, New York, 1968.
Le Goff, Jacques, *Time, Work, and Culture in the Middle Ages* (*Pour un autre Moyen Age: Temps, travail et culture en Occident*, 1977). Translated by Arthur Goldhammer, Chicago, 1980.
Leicester, Henry M., *The Historical Background of Chemistry*, New York, 1956, 1971.
LeMahieu, D.L., *The Mind of William Paley: A Philosopher and His Age*, Lincoln, Neb., 1976.
Lemay, Helen Rodnite, 'Science and Theology at Chartres: The Case of the Supracelestial Waters', *BJHS*, vol. x (1977): 226-36.
Lemay, Richard, *Abu Ma'shar and Latin Aristotelianism in the Twelfth Century: The Recovery of Aristotle's Natural Philosophy Through Arabic Astrology*, Beirut, 1962.
Leventhal, Herbert, *In the Shadow of the Enlightenment: Occultism and Renaissance Science in Eighteenth-Century America*, New York, 1976.
Levere, Trevor H., *Affinity and Matter: Elements of Chemical Philosophy, 1800-1865*, Oxford, 1971.
———, 'Faraday, Matter and Natural Theology', *BJHS*, iv (1968): 95-107.
———, 'Humphry Davy, "The Sons of Genius", and the Idea of Glory', in Sophie Forgan, ed., *Science and the Sons of Genius*, London, 1980.
———, *Poetry Realized in Nature: Samuel Coleridge and Nineteenth-Century Science*, Cambridge, 1981.
———, 'The Rich Economy of Nature: Chemistry in the Nineteenth Century', in U.C. Knoepflmacher and G.B. Tennyson, *Nature and the Victorian Imagination*, Berkeley, 1977.
Levine, Joseph M., 'Latitudinarians, Neoplatonists, and the Ancient Wisdom', in R. Kroll, et al., eds., *Philosophy, Science, and Religion*, Cambridge, 1992.
Lindberg, David C., *The Beginnings of Western Science: The Europoean Scientific Tradition in Philosophical, Religious and Institutional Context*, Chicago, 1992.
———, 'Science and the Early Christian Church', *Isis*, lxxiv (1983): 509-30. Reprinted in D.C. Lindberg and R.L. Numbers, eds., *God and Nature*, Berkeley, 1986.
———, 'The Transmission of Greek and Arabic Learning to the West', in D.C Lindberg, ed.,

Science in the Middle Ages, Chicago, 1978.
——, ed., *Science in the Middle Ages*, Chicago, 1978.
—— and Numbers, Ronald L., eds., *God and Nature: Historical Essays on the Encounter Between Christianity and Science*, Berkeley, 1986.
—— and Westman, Robert S., eds., *Reappraisals of the Scientific Revolution*, Cambridge, 1990.
Linden, Stanton J., 'Francis Bacon and Alchemy: The Reformation of Vulcan', *JHI*, xxxv (1974): 547-60.
Lindsay, Jack, ed., *Autobiography of Joseph Priestley*, Bath, 1970.
Lloyd, G.E.R., *The Revolutions of Wisdom: Studies in the Claims and Practice of Ancient Greek Science*, Berkeley, 1987.
Loemker, Leroy E., 'Boyle and Leibniz', *JHI*, xvi (1955): 22-43
——, *Struggle for Synthesis: The Seventeenth Century Background of Leibniz's Synthesis of Order and Freedom*, Cambridge, Mass., 1972.
Lovelace, Richard F., *The American Pietism of Cotton Mather: Origins of American Evangelicalism*, Grand Rapids, Mich.,1979.
Lubbock, Constance A., *The Herschel Chronicle*, Cambridge, 1933.
Ludwig, Günter, *Cassiodor: Über den Ursprung der abendländischen Schule*, Frankfurt am Main, 1967.
Lull, Raymond, *Selected Works of Ramon Llul (1232-1316)*, 2 vols. Edited by Anthony Bonner, Princeton, 1985.
Luther, Martin, *D. Martin Luthers Werke, Weimarer Ausgabe*, 90 vols., Weimar, 1883-1968; Graz, 1966-70.
——, *Luther's Works: American Edition*, 55 vols. Edited by Jaroslav Pelikan and Helmut T. Lehmann, St Louis and Philadelphia, 1955-76.
Lyon, John, and Sloan, Phillip R., eds., *From Natural History to the History of Nature: Readings from Buffon and His Critics*, Notre Dame, 1981.
McClaughlin, Trevor, 'Censorship and Defenders of the Cartesian Faith in Mid-Seventeenth Century France', *JHI*, xl (1979): 563-81.
McCluskey, Stephen C., 'Gregory of Tours, Monastic Timekeeping, and Early Christian Attitudes to Astronomy', *Isis*, lxxxi (1990): 9-22.
McCullough, W. Stewart, *A Short History of Syriac Christianity to the Rise of Islam*, Chico, Calif., 1982.
Macdonald, A.J., *Authority and Reason in the Early Middle Ages, 1931-1932*, London, 1933.
McEvoy, John G., 'Joseph Priestley: Scientist, Philosopher and Divine', *PAPS*, cxxviii (1984): 193-99.
—— and McGuire, J.E., 'God and Nature: Priestley's Way of Rational Dissent', *HSPS*, vol. vi (1975): 325-404.
McGuire, J.E., 'Boyle's Conception of Nature', *JHI*, xxxiii (1972): 523-42.
——, 'Existence, Actuality and Necessity: Newton on Space and Time', *Annals of Science*, xxxv (1978): 463-508.
——, 'Force, Active Principles and Newton's Invisible Realm', *Ambix*, xv (1968): 154-208.
——, 'Neoplatonism and Active Principles: Newton and the *Corpus Hermeticum*', in R.S. Westman and J.E. McGuire, eds., *Hermeticism and the Scientific Revolution*, Los Angeles, 1977.
——, 'Newton on Place, Time, and God: An Unpublished Source', *BJHS*, xi (1978): 114-29.
——, 'Transmutation and Immutability: Newton's Doctrine of Physical Qualities', *Ambix*, xiv (1967): 69-95.
—— and Rattansi, P.M., 'Newton and the Pipes of Pan', *Notes and Records of the Royal Society of London*, xxi (1966): 108-43.
Machamer, Peter K., and Turnbull, Robert G., eds., *Motion and Time, Space, and Matter: Interrelations in the History and Philosophy of Science*, Columbus, Ohio, 1976.
McInerny, Ralph M., *A History of Western Philosophy, Vol. II: Philosophy from St Augustine to Ockham*, Notre Dame, Ind., 1970.
——, *St Thomas Aquinas*, Notre Dame, Ind., 1982.
MacKinney, Loren C., *Early Medieval Medicine: With Special Reference to France and Chartres*, Baltimore, 1937.

——, 'Medical Ethics and Etiquette in the Early Middle Ages: The Persistence of the Hippocratic Ideals', *BHM*, xxvi (1952): 1-31.
McLaughlin, Mary Martin, *Intellectual Freedom and Its Limitations in the University of Paris in the Thirteenth and Fourteenth Centuries*, New York, 1977.
McMullin, Ernan, 'Augustine of Hippo', *DSB*, i: 333-38.
——, 'Medieval and Modern Science: Continuity or Discontinuity?', *International Philosophical Quarterly*, v (1965): 103-29.
——, ed., *The Concept of Matter in Greek and Medieval Philosophy*, Notre Dame, Ind., 1963.
——, ed., *Evolution and Creation*, Notre Dame, Ind., 1985.
McNutt, Paula M., *The Forging of Israel: Iron Technology, Symbolism, and Tradition in Ancient Society*, Sheffield, 1990.
MacPherson, Hector, *A Century of Intellectual Development*, Edinburgh, 1907.
——, *The Church and Science: A Study of the Inter-Relation of Theological and Scientific Thought*, London, 1927.
——, *The Intellectual Development of Scotland*, London, 1911.
Macrobius, *Commentary on the Dream of Scipio*. Translated by William Harris Stahl, New York, 1952.
McVaugh, Michael, 'Arnald of Villanova', *DSB*, i: 289-91.
Mahoney, Edward P., ed., *Philosophy and Humanism: Essays in Honor of Paul Oskar Kristeller*, Leiden, 1976.
Maier, Anneliese, *An der Grenze von Scholastik und Naturwissenschaft*, 2nd ed., Rome, 1952.
——, *Metaphysische Hintergründe der spätscholastischen Naturphilosophie*, Rome, 1955.
——, *On the Threshold of Exact Science: Selected Writings on Late Medieval Natural Philosophy*. Edited by Steven D. Sargent, Philadelphia, 1982.
——, *Die Vorläufer Galileis im 14. Jahrhundert: Studien zur Naturphilosophie der Spätscholastik*, Rome, 1949.
——, *Zwei Grundprobleme der scholastischen Naturphilosophie: Das Problem der intensiven Grösse, die Impetus theorie*, 2nd ed., Rome, 1951.
——, *Zwischen Philosophie und Mechanik*, Rome, 1958.
Maimonides, Moses, *Treatise on Resurrection*. Translated by Fred Rosner, New York, 1982.
Manschreck, Clyde Leonard, *Melanchthon the Quiet Reformer*, Nashville, 1958; Westport, Conn., 1975.
Manuel, Frank E., *The Religion of Isaac Newton*, Oxford, 1974.
—— and Manuel, Fritzie P., *Utopian Thought in the Western World*, Cambridge, Mass., 1979.
Marcel, Pierre, 'Calvin et Copernic, la légende ou les faits? La science et l'astronomie chez Calvin', *La revue réformée*, xxxi (1980): 1-210.
Marmorstein, A., *The Old Rabbinic Doctrine of God, Vol. I: The Names and Attributes of God*, New York, 1927, 1968.
Martels, Z.R.W.M. von, ed., *Alchemy Revisited*, Leiden, 1990.
Martin, Hubert, Jr., 'The Concept of *Philanthropia* in Plutarch's Lives', *American Journal of Philology*, lxxxii (1961): 164-75.
Martin, Julian, *Francis Bacon, the State and the Reform of Natural Philosophy*, Cambridge, 1992.
——, 'Natural Philosophy and Its Public Concerns', in S. Pumfrey et al., ed., *Science, Culture and Popular Belief*, Manchester, 1991.
Marty, Martin E., and Vaux, Kenneth L., eds., *Health, Medicine and the Faith Traditions: An Inquiry into Religion and Medicine*, Philadelphia, 1982.
Mason, Stephen F., *A History of the Sciences: Main Currents of Scientific Thought*, London, 1953; New York, 1962.
——, 'The Scientific Revolution and the Protestant Reformation, II: Lutheranism in Relation to Iatrochemistry and the German Nature-Philosophy', *Annals of Science*, ix (1953): 154-75.
Mather, Cotton, *The Christian Philosopher: A Collection of the Best Discoveries in Nature with Religious Improvements (1721)*, Gainesville, Fla., 1968.
——, *The Christian Philosopher*. Edited by Winten U. Solberg, Urbana, Ill., 1994.
Mathias, Peter, ed., *Science and Society, 1600-1900*, Cambridge, 1972.

Maurer, Armand A., ed., *St Thomas Aquinas, 1274-1974: Commemorative Essays*, Toronto, 1974.
Maxwell, James Clerk, *The Life of James Clerk Maxwell with a Selection from His Correspondence and Occasional Writings and a Sketch of His Contributions to Science*. Edited by Lewis Campbell and William Garnett, London, 1882; New York, 1969.
——, *The Scientific Papers of James Clerk Maxwell*, 2 vols. Edited by W.P. Niven, Cambridge, 1890; New York, 1965.
May, Gerhard, *Creatio Ex Nihilo: The Doctrine of 'Creation Out of Nothing' in Early Christian Thought (Schöpfung aus dem Nichts*, Berlin, 1978). Translated by A.S. Worrall, Edinburgh, 1994.
Mebane, John S., *Renaissance Magic and the Return of the Golden Age*, Lincoln, Neb., 1989.
Melanchthon, Philip, *Corpus Reformatorum: Melanchthonis Opera quae supersunt omnia*, 28 vols. Edited by Carolus Gottlieb Bretschneider, Saxony, 1834-60.
——, *On Christian Doctrine: Loci Communes, 1555*. Translated by Clyde Leonard Manschreck, New York, 1965.
Mendelsohn, J. Andrew, 'Alchemy and Politics in England, 1649-1665', *Past and Present*, cxxxv (1992): 30-78.
Merchant, Carolyn, *The Death of Nature: Women, Ecology, and the Scientifc Revolution*, San Francisco, 1980.
——, 'The Vitalism of Francis Mercury Van Helmont: His Influence on Leibniz', *Ambix*, xxvi (1979): 170-83.
Merkel, Ingrid, and Debus, Allen G., eds., *Hermeticism and the Renaissance: Intellectual History and the Occult in Early Modern Europe*, Washington, D.C., 1988.
Methuen, Charlotte, 'The Role of the Heavens in the Thought of Philip Melanchthon', *JHI*, lvii (1996): 385-404.
Meyer, R.W., *Leibnitz and the Seventeenth-Century Revolution (Leibniz und die europäische Ordnungskrise*, Hamburg, 1948). Translated by J.P. Stern, Cambridge, 1952.
Meyerhof, Max, 'New Light on Hunain Ibn Ishaq and his Period', *Isis*, viii (1926): 685-724.
Miller, Timothy S., *The Birth of the Hospital in the Byzantine Empire*, Supplements to *BHM*, New Series, No. 10, Baltimore, 1985.
——, 'Byzantine Hospitals', *Dumbarton Oaks Papers*, xxxviii (1984): 53-64.
——, 'The Knights of Saint John and the Hospitals of the Latin West', *Speculum*, liii (1978): 709-33.
Milton, John R., 'The Origin and Development of the Concept of the "Laws of Nature"', *Archive for European Sociology*, xxii (1981): 173-95.
Mingana, A., ed., *Job of Edessa: Encyclopaedia of Philosophical and Natural Sciences as Taught in Baghdad about AD 817, or Book of Treasures*, Woodbrooke Scientific Publications, vol. i, Cambridge, 1935.
Mintz, Samuel I., 'Hobbes', *DSB*, vi:444-51.
Mittelstrass, Jürgen, and Aiton, Eric J., 'Leibniz: Physics, Logic, Metaphysics', *DSB*, viii:150-60.
Monheit, Michael L., '"The ambition for an illustrious name": Humanism, Patronage, and Calvin's Doctrine of the Calling', *Sixteenth Century Journal*, xxiii (1992): 267-87.
Montgomery, John Warwick, *Cross and Crucible: Johann Valentin Andreae (1586-1654), Phoenix of the Theologians*, 2 vols., The Hague, 1973.
Moody, Ernest A., 'Galileo and Avempace: The Dynamics of the Leaning Tower Experiment', *JHI*, xii (1955): 163-93, 375-422. Reprinted in *Studies*, Berkeley, 1975.
——, *Studies in Medieval Philosophy, Science and Logic: Collected Papers, 1933-1969*, Berkeley, 1975.
Moore, Marian A., trans., 'A Letter of Philip Melanchthon to the Reader', *Isis*, i (1959): 145-50.
Moran, Bruce Thomas, 'The Universe of Philip Melanchthon: Criticism and Use of the Copernican Theory', *Comitatus*, iv (1973): 1-23.
Morison, E.F., *St Basil and His Rule: A Study in Early Monasticism*, London, 1912.
Mourelatos, Alexander P., ed., *The Pre-Socratics: A Collection of Critical Essays*, Garden City, N.Y., 1974.
Mulligan, Lotte, '"Reason", "Right Reason", and "Revelation" in Mid-Seventeenth Century

England', in Brian Vickers, ed., *Occult and Scientific Mentalities*, Cambridge, 1984.
Murdoch, Dugald R., *Niels Bohr's Philosophy of Physics*, Cambridge, 1987.
Murdoch, John E., *Album of Science: Antiquity and Middle Ages*, New York, 1984.
—— and Sylla, Edith Dudley, eds., *The Cultural Context of Medieval Learning*, Dordrecht, 1975.
Nadler, Steven M., 'Arnauld, Descartes, and Transubstantiation: Reconciling Cartesian Metaphysics and Real Presence', *JHI*, xlix (1988): 229-46
Nasr, Seyyed Hossein, *An Introduction to Islamic Cosmological Doctrines*, Cambridge, Mass., 1964; rev. ed., Boulder, Colo., 1978.
——, 'Islamic Conception of Intellectual Life', *DHI*, ii:638-52.
——, *Islamic Science: An Illustrated Study*, London, 1976.
Nebelsick, Harold P., *Circles of God: Theology and Science from the Greeks to Copernicus*, Edinburgh, 1985.
——, *Renaissance and Reformation and the Rise of Science*, Edinburgh, 1992.
Needham, Joseph, *The Grand Titration: Science and Society in East and West*, London, 1969.
Neugebauer, O., 'On the Computus Paschalis of "Cassiodorus"', *Centaurus*, xxv (1982): 292-302.
Neusner, Jacob, ed., *Religions in Antiquity: Essays in Memory of Erwin Ramsdell Goodenough*, Leiden, 1968.
Newman, Louis Israel, *Jewish Influence on Christian Reform Movements*, New York, 1925.
Newton, Isaac, *The Correspondence of Isaac Newton*, 7 vols. Edited by H.W. Turnbull, et al., Cambridge, 1959-77.
——, *Isaac Newton's Papers and Letters on Natural Philosophy*. Edited by I. Bernard Cohen, Cambridge, Mass., 1958.
——, *Opticks, or A Treatise of the Reflections, Refractions, Inflections & Colours of Light*, based on the 4th ed., London, 1931; New York, 1952.
——, *Sir Isaac Newton's Mathematical Principles of Natural Philosophy and His System of the World translated into English by Andrew Motte in 1729*, 2 vols. Edited by Florian Cajori, Berkeley, 1934, 1962.
——, *Theological Manuscripts*. Edited by Herbert McLachlan, Liverpool, 1950.
——, *Unpublished Scientific Papers of Isaac Newton*. Edited by A. Rupert Hall and Mary Hall Boas, Cambridge, 1962.
Nicolson, Marjorie Hope, and Mohler, Nora M., 'The Scientific Background of Swift's *Voyage to Laputa*', *Annals of Science*, ii (1937): 299-334.
Numbers, Ronald L., *Creation by Natural Law: Laplace's Nebular Hypothesis in American Thought*, Seattle, 1977.
—— and Sawyer, Ronald C., 'Medicine and Christianity in the Modern World', in M.E. Marty and K.L. Vaux, eds., *Health, Medicine and the Faith Traditions*, Philadelphia, 1982.
Oakley, Francis, 'Christian Theology and the Newtonian Science: The Rise of the Concept of the Laws of Nature', *CH*, xxx (1961): 433-57.
——, *Omnipotence, Covenant, and Order: An Excursion in the History of Ideas from Abelard to Leibniz*, Ithaca, N.Y., 1984.
——, *The Western Church in the Later Middle Ages*, Ithaca, N.Y., 1979.
Oberman, Heiko A., *Archbishop Thomas Bradwardine, A Fourteenth Century Augustinian: A Study of His Theology in Its Historical Context*, Utrecht, 1957.
——, *Masters of the Reformation: The Emergence of a New Intellectual Climate in Europe* (*Werden und Wertung der Reformation*, Tübingen, 1977). Translated by Dennis Martin, Cambridge, 1981.
——, 'Reformation and Revolution: Copernicus' Discovery in an Era of Change', in Owen Gingerich, ed., *The Nature of Scientific Discovery*, 1975.
Ockham, William of, *Philosophical Writings*. Translated by Philotheus Boehner, Edinburgh, 1957; Indianapolis, 1964.
O'Connor, Daniel, and Oakley, Francis, eds., *Creation: The Impact of an Idea*, New York, 1969.
Odom, Herbert H., 'The Estrangement of Celestial Mechanics and Religion', *JHI*, xxvii (1966): 533-48.

O'Donnell, James J., *Cassiodorus*, Berkeley, 1979.
Oersted, Hans Christian, *The Soul in Nature*. Translated from German by Leonora and Joanna B. Horner, London, 1852, reprinted, 1966.
Olby, R.C., et al., eds., *Companion to the History of Modern Science*, London, 1990.
O'Leary, De Lacy, *How Greek Science Passed to the Arabs*, London, 1949.
Olson, Richard G., *Science Deified and Science Defied: The Historical Significance of Science in Western Culture*, 3 vols., Berkeley, 1982.
——, *Scottish Philosophy and British Physics, 1750-1880: A Study in the Foundations of the Victorian Scientific Style*, Princeton, N.J., 1975.
——, 'Tory-High Church Opposition to Science and Scientism in the Eighteenth Century: The Works of John Arbuthnot, Jonathan Swift, and Samuel Johnson', in John G. Burke, ed., *The Uses of Science in the Age of Newton*, Berkeley, 1983.
O'Meara, Dominic J., *Pythagoras Revived: Mathematics and Philosophy in Late Antiquity*, Oxford, 1989.
——, ed., *Neoplatonism and Christianity*, Studies in Neoplatonism, vol. iii, Albany, N.Y., 1982.
Oresme, Nicole, *Le Livre du Ciel et du Monde*. Translated from Latin by Albert D. Menut, Madison, Wis., 1968.
Osler, Margaret J., 'Baptizing Epicurean Atomism: Pierre Garsendi on the Immortality of the Soul', in M. J. Osler and P.L. Farber, eds., *Religion, Science and Worldview*, Cambridge, 1985.
—— and Farber, Paul Lawrence, eds., *Religion, Science, and Worldview: Essays in Honor of Richard W. Westfall*, Cambridge, 1985.
Otten, Willemien, 'Nature and Scripture: Demise of a Medieval Analogy', *HTR*, lxxxviii (1995): 257-84.
Otzen, Benedikt, 'Heavenly Visions in Early Judaism: Origin and Function', in *In the Shelter of Elyon: Essays on Ancient Palestinian Life and Literature in Honor of G.W. Ahlström*, *Journal for the Study of the Old Testament*, Supplement Series, vol. xxxi (1984):199-215.
Ovitt, George, Jr., *The Restoration of Perfection: Labor and Technology in Medieval Culture*, New Brunswick, N.J., 1987.
Ozment, Steven E., *Mysticism and Dissent: Religious Ideology and Social Protest in the Sixteenth Century*, New Haven, Conn., 1973.
Pacey, Arnold, *The Maze of Ingenuity: Ideas and Idealism in the Development of Technology*, Harmondsworth, 1974; Cambridge, Mass., 1976.
Pagel, Walter, *Joan Baptista Van Helmont, Reformer of Science and Medicine*, Cambridge, 1982.
——, 'Paracelsus', *DSB*, x:304-13.
——, *Paracelsus: Introduction to Philosophical Medicine in the Era of the Renaissance*, Basel, 1958.
——, 'The Religious and Philosophical Aspects of van Helmont's Science and Medicine', Supplements to the *BHM*, No. 2, Baltimore, 1994.
——, 'Religious Motives in the Medical Biology of the XVIIth Century', *BHM*, iii (1935):97-128, 213-31, 265-312.
Pais, Abraham, *'Subtle is the Lord . . .' The Science and the Life of Albert Einstein*, Oxford, 1982.
Pannekoek, Anton, *A History of Astronomy* (*De Groei van ons Wereldbeeld*, Amsterdam, 1951), London, 1961.
Pannenberg, Wolfhart, *Theology and the Philosophy of Science (Wissenschaftstheorie und Theologie*, Frankfort am Main, 1973). Translated by Francis McDonagh, London, 1976.
Paracelsus, Theophrastus, *Four Treatises of Theophrastus von Hohenheim Called Paracelsus*. Edited by Henry E. Sigerist, Baltimore, 1941.; New York, 1979.
——, *Selected Writings* (*Theophrastus Paracelsus: Lebendiges Erbe*, Zurich, 1942). Translated by Norbert Guterman, New York, 1951; Princeton, N.J., 1958.
——, *Volumen Medicinae Paramirum*. Translated by Kurt F. Leidecker, Supplements to the *BHM*, No. 11, Baltimore, 1949.
Parent, J.M., *La doctrine de la création dans l'ecole de Chartres*, Paris, 1938.
Park, Katherine, 'Medicine and Society in Medieval Europe, 500-1500', in A. Wear, ed.,

Medicine in Society, Cambridge, 1992.
Pearson, Birger A., 'Jewish Elements in *Corpus Hermeticum* I (Poimandres)', in R. van den Broek and M.J. Vermaseren, eds., *Studies in Gnosticism and Hellenistic Religions*, Leiden, 1981.
Pedersen, Olaf, *The Book of Nature*, Vatican City, 1992.
———, 'The Development of Natural Philosophy, 1250-1350', *Classica et Mediaevalia*, xiv (1953): 86-155.
Pegis, Anton Charles, ed., *The Wisdom of Catholicism*, New York, 1949.
Pelseneer, Jean, 'La Réforme du XVIe siècle à l'origine de la science moderne', *La science au siezième siècle*, Paris, 1960.
———, 'La religion de Laplace', *Isis*, xxxvi (1946): 158-60.
Perdue, Leo G., *Wisdom and Cult: A Critical Analysis of the Views of Cult in the Wisdom Literatures of Israel and the Ancient Near East*, Missoula, Mont., 1977.
Perkins, William, *The Works of William Perkins*, Courtenay Library of Reformation Classics, vol. iii. Edited by Ian Breward, Appleford, 1970.
Peters, F.E., *Aristotle and the Arabs: The Aristotelian Tradition in Islam*, New York, 1968.
Pfizenmaier, Thomas C., 'Was Newton an Arian?', *JHI*, lviii (1997): 57-80.
Phillips, E.D., *Aspects of Greek Medicine*, London, 1973.
Pines, Shlomo, 'Un précurseur bagdadien de la théorie de l'impetus', *Isis*, xliv (1953): 247-51.
———, *Studies in Abu'l-Barakat al-Baghdadi: Physics and Metaphysics*, Collected Works, vol. i, Jerusalem, 1979.
Placher, William, *Readings in the History of Christian Theology*, 2 vols., Philadelphia, 1988.
Plattard, Jean, 'Le système de Copernic dans la littérature francaise au XVIe siècle', *Revue du seizième siècle*, i (1913): 220-37.
Poole, Reginald Lane, *Illustrations of the History of Medieval Thought and Learning*, London, 1888, 2nd ed., 1920.
Popkin, Richard H., 'The Third Force in Seventeenth-Century Philosophy: Scepticism, Science and Biblical Prophecy', *Nouvelles de la République des Lettres*, i (1983): 35-63.
——— and Force, James E., eds., *The Books of Nature and Scripture: Recent Essays on Natural Philosophy, Theology, and Biblical Criticism in the Netherlands of Spinoza's Time and the British Isles of Newton's Time*, Dordrecht, 1994.
Porter, Theodore M., *The Rise of Statistical Thinking, 1820-1900*, Princeton, N.J., 1986.
Power, J.E., 'Henry More and Isaac Newton on Absolute Space', *JHI*, xxxi (1970): 289-96.
Priestley, F.E.L., 'The Clarke-Leibniz Controversy', in R.E. Butts and J.W. Davis, eds., *The Methodological Heritage of Newton*, Toronto, 1970.
———, 'Newton and the Romantic Concept of Nature', *University of Toronto Quarterly*, xvii (1948): 323-36.
Principe, Lawrence M., 'Boyle's Alchemical Pursuits', in M. Hunter, ed., *Robert Boyle Reconsidered*, Cambridge, 1994.
———, 'Robert Boyle's *Dialogue on Transmutation*', paper presented at the annual meeting of the History of Science Society, Madison, Wis., 3 November 1991.
Pumfrey, Stephen, et al., eds., *Science, Culture and Popular Belief in Renaissance Europe*, Manchester, 1991.
Rahman, Fazlur, *Health and Medicine in the Islamic Tradition: Change and Identity*, New York, 1989.
Raine, Kathleen, *Blake and Tradition*, 2 vols., Princeton, N.J., 1968.
———, *William Blake*, London, 1970.
Ramsey, Ian T., *Religious Language: An Empirical Placing of Theological Phrases*, London, 1957.
Rand, Edward Kennard, *Founders of the Middle Ages*, Cambridge, Mass., 1928; New York, 1957.
Randall, John Herman, Jr., *Aristotle*, New York, 1960.
———, *The Career of Philosophy, Vol. I: From the Middle Ages to the Enlightenment*, New York, 1962.
———, *The Making of the Modern Mind: A Survey of the Intellectual Background of the Present Age*, New York, 1976.

Rattansi, Piyo, 'Newton and the Wisdom of the Ancients', in J. Fauvel, ed., *Let Newton Be!*, Oxford, 1988.
Rattansi, Pyarali, M., 'Alchemy in Raleigh', *Ambix*, xii (1965): 127.
———, 'The Helmontian-Galenist Controversy in Restoration England', *Ambix*, xii (1964): 1-23.
———, 'The Intellectual Origins of the Royal Society', *Notes and Records of the Royal Society of London* xxiii (1968): 129-43.
———, 'Paracelsus and the Puritan Revolution', *Ambix*, xi (1963): 24-32.
———, 'The Social Interpretation of Science in the Seventeenth Century', in Peter Mathias, ed., *Science and Society, 1600-1900*, Cambridge, 1972.
Redondi, Pietro, *Galileo Heretic* (*Galileo Eretico*, Rome, 1983). Translated by Monique Aymard, Princeton, N.J., 1987.
Redwood, John, *Reason, Ridicule and Religion: The Age of Enlightenment in England*, London, 1976.
Rees, Graham, 'Francis Bacon's Semi-Paracelsian Cosmology', *Ambix*, xxii (1975): 81-101.
Reese, James M., *Hellenistic Influence on the Book of Wisdom and Its Consequences*, Analecta Biblica, vol. xli, Rome, 1970.
Resnick, Irven M., 'Peter Damian on the Restoration of Virginity: A Problem for Medieval Theology', *Journal of Theological Studies*, xxxix (1988): 125-34.
Reuchlin, Johann, *La Kabbale (De arte cabalistica)*. Translated by Francois Secret, Paris, 1973.
———, *On the Art of the Kabbalah*. Translated by Martin and Sarah Goodman, New York, 1983.
Rice, Eugene F., 'The *De Magia Naturali* of Jacques Lefèvre d'Étaples', in E.P. Mahoney, ed., *Philosophy and Humanism*, Leiden, 1976.
Riddell, Edwin, ed., *Lives of the Stuart Age, 1603-1714*, New York, 1976.
Riforgiato, Leonard R., 'The Unified Thought of Jonathan Edwards', *Thought*, xlvii (1972): 599-610.
Righini, M.L. Bonelli, and Shea, William R., eds., *Reason, Experiment, and Mysticism in the Scientific Revolution*, New York, 1975.
Ritchie-Calder, Lord, 'The Lunar Society of Birmingham', *Scentific American*, ccxlvi (June 1982): 136-45.
Robbins, Frank Egleston, *The Hexaemeral Literature: A Study of the Greek and Latin Commentaries on Genesis*, Chicago, 1912.
Robinson, H. Wheeler, *Inspiration and Revelation in the Old Testament*, Oxford, 1946.
Rochot, Bernard, 'Gassendi', *DSB*, v:284-90.
Rogal, Samuel J., 'Pills for the Poor: John Wesley's *Primitive Physick*', *Yale Journal of Biology and Medicine*, li (1978): 81-90.
Roger, Jacques, 'Buffon', *DSB*, ii:576-82.
———, 'The Mechanistic Conception of Life', in D.C. Lindberg and R.L. Numbers, eds., *God and Nature*, Berkeley, 1986.
———, 'Whiston', *DSB*, xiv:295f.
———, ed., *Sciences de la Renaissance*, Paris, 1973.
Rogers, Pat, 'Gulliver and the Engineers', *Modern Language Review*, lxx (1975): 260-70.
Rosen, Edward, 'Regiomontanus', *DSB*, xi:348-52.
———, 'Was Copernicus a Neoplatonist', *JHI*, xliv (1983): 667-69.
———, trans., *Three Copernican Treatises*, New York, 1971.
Rosenfeld, Léon, 'Bohr', *DSB*, ii:239-55.
Rosenthal, Franz, *The Classical Heritage in Islam* (*Das Fortleben der Antike im Islam*, Zürich, 1965). Translated by Emile and Jenny Marmorstein, Berkeley, 1975.
Rosenthal-Schneider, Ilse, *Reality and Scientific Truth: Discussions with Einstein, von Laue, and Planck*, Detroit, 1980.
Ross, David, *Aristotle* 5th ed., London, 1949.
Rossi, Paolo, *Francis Bacon: From Magic to Science* (*Francesco Bacone: Dalla magia alla scienza*, Bari, 1957). Translated by Sacha Rabinovitch, London, 1968.
———, 'Hermeticism, Rationality and the Scientific Revolution', in M.L. Righini Bonelli and W.R. Shea, eds., *Reason, Experiment, and Mysticism*, New York, 1975.

——, 'The Legacy of Ramon Lull in Sixteenth-Century Thought', *Mediaeval and Renaissance Studies*, v (1961): 182-213.
——, *Philosophy, Technology, and the Arts in the Early Modern Era* (*I Filosofi e le macchine*, Milan, 1962). Translated by Salvator Attanasio, New York, 1970.
Rothkrug, Lionel, *Opposition to Louis XIV: The Political and Social Origins of the French Enlightenment*, Princeton, N.J., 1965.
Rousseau, George S., 'Science', in Pat Rogers, ed., *The Context of English Literature, The Eighteenth Century*, London, 1978.
Rowen, Herbert H., ed., *From Absolutism to Revolution, 1648-1848*, New York, 1968.
Rubin, Stanley, *Medieval English Medicine*, London, 1974.
Ruby, Jane E., 'The Origins of Scientific "Law"', *JHI*, xlvii (1986): 341-59.
Rudwick, Martin J.S., 'The Shape and Meaning of Earth History', in D.C. Lindberg and R.L. Numbers, eds., *God and Nature*, Berkeley, 1985.
Ruestow, Edward G., *Physics at Seventeenth and Eighteenth-Century Leiden: Philosophy and the New Science in the University*, The Hague, 1973.
Russell, Colin A., *Cross-Currents: Interactions Between Science and Faith*, Leicester, 1985.
——, *Science and Social Change in Britain and Europe, 1700-1900*, London and New York, 1983.
Russo, François, 'Role respectif du Catholicisme et du Protestantisme dans le développement des sciences au XVIe et XVIIe siècles', *Cahiers d'histoire mondiale*, iii (1957): 854-80.
——, 'Théologie naturelle et sécularisation de la science au XVIIIème siècle', *Recherches de science religieuse*, lxvi (1978): 27-62.
Sailor, Danton B., 'Newton's Debt to Cudworth', *JHI*, xlix (1988): 511-18.
Sambursky, Shmuel, *The Physical World of Late Antiquity*, London, 1962.
——, *The Physical World of the Greeks*. Translated from Hebrew by Merton Dagut, London, 1956.
Sanders, Jack T., *Ben Sira and Demotic Wisdom*, Chico, Calif., 1983.
Sangwan, Satpal, 'Why Did the Scientific Revolution Not Take Place in India?', in D. Kumar, ed., *Science and Empire*, Delhi, 1991.
Sarton, George, *Six Wings: Men of Science in the Renaissance*, Bloomington, Ind., 1957.
Saveson, J.E., 'Descartes' Influence on John Smith, Cambridge Platonist', *JHI*, xx (1959): 255-63.
——, 'Differing Reactions to Descartes Among the Cambridge Platonists', *JHI*, xxi (1960): 560-67.
Sawyer, Deborah F., ed., *Midrash Aleph Beth*, Atlanta, 1993.
Scarborough, John, *Roman Medicine*, Ithaca, N.Y., 1969.
Schaya, Leo, *The Universal Meaning of the Kabbalah* (*L'homme et l'absolu selon la Kabale*, Paris, 1958). Translated by Nancy Pearson, London, 1971.,
Scheurer, P.B., and Debrock, G., eds., *Newton's Scientific and Philosophical Legacy*, Dordrecht, 1988.
Schilpp, Paul Arthur, ed., *Albert Einstein: Philosopher-Scientist*, Evanston, Ill., 1949.
Schipperges, Heinrich, *Die Benediktiner in der Medizin des frühen Mittelalters*. Edited by Erich Kleineidam and Heinz Schürmann, *Erfurter Theologische Schriften*, vol. vii, Leipzig, 1964.
Schmid, H.H., 'Creation, Righteousness, and Salvation: "Creation Theology" as the Broad Horizon of Biblical Theology', ('Schöpfung, Gerechtigkeit und Heil', *Zeitschrift für Theologie und Kirche*, vol. lxx [1973]: 1-19). Translated by B.W. Anderson and D.G. Johnson in B.W. Anderson, ed., *Creation*, Philadelphia, 1984.
Schneemelcher, Wilhelm, ed., *New Testament Apocrypha* 2 vols. (*Neutestamentliche Apokryphen*, Tübingen, 1959-64). Translated by Ernest Best et al., edited by R. McL. Wilson, Philadelphia, 1963-65.
Schneer, Cecil J., 'The Renaissance Background to Crystallography', *American Scientist*, lxxi (1983): 254-63.
Schoedel, William R., '"Topological" Theology and Some Monistic Tendencies in Gnosticism', in Martin Krause, ed., *Nag Hammadi Studies*, vol. iii, Leiden, 1972.
Schofield, Robert E., 'Joseph Priestley and the Physicalist Tradition in British Chemistry', in L. Kieft and B.R. Willeford, Jr., eds., *Joseph Priestley*, Lewisburg, Penn., 1980.

——, *Mechanism and Materialism: British Natural Philosophy in an Age of Reason*, Princeton, N.J., 1970.
——, 'Priestley, Joseph', *DSB*, xi:139-47.
Scholem, Gershom, *Kabbalah*, Jerusalem, 1974.
——, *Origins of the Kaballah* (*Ursprung und Anfänge der Kabbala*, Berlin, 1962). Translated by Allan Arkush, edited by R.J. Zwi Werblowsky, Philadelphia, 1987.
Schreiner, Susan E., *The Theater of His Glory: Nature and the Natural Order in the Thought of John Calvin*, Durham, 1991.
Schröder, Gerard, 'Crollius, Oswald', *DSB*, iii:471-2.
Schuler, Robert M., 'Some Spiritual Alchemies in Seventeenth-Century England', *JHI*, xli (1980): 293-318.
Schwartz, Richard B., *Samuel Johnson and the New Science*, Madison, Wis., 1971.
Schweizer, Eduard, *The Good News According to Matthew*. Translated from German by David E. Green, Atlanta, 1975.
Seçret, Francois, *Les Kabbalistes chrétiens de la Renaissance*, Paris, 1964.
Segal, Alan F., 'Hellenistic Magic: Some Questions of Definition', in R. van den Broek and M.J. Vermaseren, eds., *Studies in Gnosticism and Hellenistic Religions*, Leiden, 1981.
Segal, J.B., *Edessa 'The Blessed City'*, Oxford, 1970.
Sevenster, J.N., *The Roots of Pagan Anti-Semitism in the Ancient World*, Leiden, 1975.
Shackelford, Jole, 'Early Reception of Paracelsian Theory: Severinus and Erastus', *Sixteenth Century Journal*, xxvi (1995): 123-35.
Shapin, Steven, 'Of Gods and Kings: Natural Philosophy and Politics: in the Leibniz-Clarke Disputes', *Isis*, lxxii (1981): 187-214.
—— and Schaffer, Simon, *Leviathan and the Air-Pump: Hobbes, Boyle, and the Experimental Life*, Princeton, N.J., 1985.
Shapiro, Barbara J., 'Latitudinarianism and Science in Seventeenth-Century England', *Past and Present*, xl (1968): 16-41.
Sharpe, Eric J., *Comparative Religion: A History*, New York, 1975.
Shea, William R., *The Magic of Numbers and Motion: The Scientific Career of René Descartes*, Canton, Mass., 1991.
Sheils, W.J., ed., *The Church and Healing*, Oxford, 1982.
Sherrard, Philip, 'The Desanctification of Nature', in D. Baker, ed., *Sanctity and Secularity*, Oxford, 1973.
Sherwin, Byron L., 'In Partnership with God: Health, Healing, and Jewish Tradition', unpub. ms., 1986.
Shumaker, Wayne, *The Occult Sciences in the Renaissance: A Study in Intellectual Patterns*, Berkeley, 1972.
Siegel, Daniel M., 'Mechanical Image and Reality in Maxwell's Electromagnetic Theory', in P.M. Harman, ed., *Wranglers and Physicists*, Manchester, 1985.
——, 'Thomson, Maxwell, and the Universal Ether in Victorian Physics', in G.N. Cantor and M.J.S. Hodge, eds., *Conceptions of Ether*, Cambridge, 1981.
Silliman, Robert H., 'William Thomson: Smoke Rings and Nineteenth-Century Atomism', *Isis*, liv (1963): 461-74.
Silverman, Kenneth, *The Life and Times of Cotton Mather*, New York, 1985.
Simon, Elliott M., 'Bacon's *New Atlantis*: The Kingdom of God and Man', *Christianity and Literature*, xxxviii (1988): 43-61.
Simon, Gérard, 'Kepler's Astrology: The Direction of a Reform', in A. and P. Beer, *Kepler*, Oxford, 1975.
Simpson, James Y., *Landmarks in the Struggle Between Science and Religion*, London, 1925.
Singer, Dorothea Waley, *Giordano Bruno: His Life and Thought with Annotated Translation of His Work on the Infinite Universe and Worlds*, New York, 1950.
Sinnema, Donald, 'Aristotle and Early Reformed Orthodoxy: Moments of Accommodation and Antithesis', in W.E. Helleman, ed., *Christianity and the Classics*, Lanham, N.Y., 1990.
Smith, Crosbie, 'Geologists and Mathematicians: The Rise of Physical Geology', in P.M. Harman, ed., *Wranglers and Physicists*, Manchester, 1985.
——, 'Natural Philosophy and Thermodynamics: William Thomson and the "Dynamical

Theory of Heat'", *BJHS*, ix (1976): 293-319.
Smith, Gary Scott, *The Seeds of Secularization: Calvinism, Culture, and Pluralism in America, 1870-1915*, Grand Rapids, Mich., 1985.
Snelders, H.A.M., 'Oersted's Discovery of Electromagnetism', in A. Cunningham and N. Jardine, eds., *Romanticism and the Sciences*, Cambridge, 1990.
Snow, Robert E., 'How Did We Get Here? A Brief Sketch of the Historical Background of the Science-Theology Tension', in H.J. Van Till, et al., *Portraits of Creation*, Grand Rapids, Mich., 1990.
Sorabji, Richard, *Time, Creation and the Continuum: Theories in Antiquity and the Early Middle Ages*, London and Ithaca, N.Y., 1983.
——, ed., *Philoponus and the Rejection of Aristotelian Science*, London and Ithaca, N.Y., 1987.
Sorrell, Roger D., *St. Francis of Assisi and Nature: Tradition and Innovation in Western Christian Attitudes toward the Environment*, New York, 1988.
Southern, Richard W., *Robert Grosseteste: The Growth of an English Mind in Medieval Europe*, Oxford, 1986.
Southgate, B.C., '"Forgotten and Lost": Some Reactions to Autonomous Science in the Seventeenth Century', *JHI*, l (1989): 249-68.
Speller, John L., 'William Prout (1785-1850) and His Hypothesis: The Religious Dimension in Grand Unified Theories', in David Schenck, ed., *Science, Philosophy and Religion*, Bethlehem, Penn., 1984.
Spitz, Lewis W., *The Religious Renaissance of the German Humanists*, Cambridge, Mass., 1963.
Stahl, William Harris, *Roman Science: Origins, Development, and Influence on the Later Middle Ages*, Madison, Wis., 1962.
Staudenbauer, C.A., 'Platonism, Theosophy, and Immaterialism: Recent Views of the Cambridge Platonists', *JHI*, xxxv (1974): 157-69.
Stauffer, Richard, 'L'attitude des réformateurs à l'égard de Copernic', in *Copernic: La représentation de l'univers et ses conséquences épistémologiques*, Centre International de Synthèse, Paris, 1975,
Stauffer, Robert C., 'Speculation and Experiment in the Background of Oersted's Discovery of Electromagnetism', *Isis*, xlviii (1957): 33-50.
Steffens, Henry John, *James Prescott Joule and the Concept of Energy*, Dawson, 1979.
Stein, Ludwig, *Leibniz und Spinoza: Ein Beitrag zur Entwicklungsgeschichte der Leibnizischen Philosophie*, Berlin, 1890.
Stek, John H., 'What Says the Scripture?', in H.J. Van Till, et al., *Portraits of Creation*, Grand Rapids, Mich., 1990.
Steneck, Nicholas H., 'Medieval Science Viewed from Afar', paper delivered at the Sixteenth International Congress on Medieval Studies, Kalamazoo, Mich., 9 May 1981.
——, *Science and Creation in the Middle Ages: Henry of Langenstein (d. 1397) on Genesis*, Notre Dame, Ind., 1976.
Stephen, Leslie, *History of English Thought in the Eighteenth Century*, New York, 1962.
Sternagel, Peter, *Die Artes mechanicae im Mittelalter*, Kallmung über Regensburg, 1966.
Stewart, Larry, *The Rise of Public Science: Rhetoric, Technology, and Natural Philosophy in Newtonian Britain, 1660-1750*, Cambridge, 1992.
——, 'Samuel Clarke, Newtonianism, and the Factions of Post-Revolutionary England', *JHI*, xlii (1981): 53-72.
Sticker, Bernhard, *Erfahrung und Erkenntnis: Vorträge und Aufsätze zur Geschichte der naturwissenschaftlichen Denkweisen*, Hildesheim, 1976.
Stiefel, Tina, 'The Heresy of Science: A Twelfth-Century Conceptual Revolution', *Isis*, lxviii (1977): 347-62.
——, '"Impious Men": Twelfth-Century Attempts to Apply Dialectic to the World of Nature', in Pamela O. Long, *Science and Technology in Medieval Society*, New York, 1985.
——, *The Intellectual Revolution in Twelfth-Century Europe*, New York, 1985.
——, 'Science, Reason and Faith in the Twelfth Century: The Cosmologists' Attack on Tradition', *Journal of European Studies*, vi (1976): 1-16.

Stillman, John Maxson, *The Story of Alchemy and Early Chemistry*, New York, 1960 (orig. ed. entitled, *Story of Early Chemistry*, 1924).
Stock, Brian, 'Experience, Praxis, Work, and Planning in Bernard of Clairvaux's Observations on the *Sermones in Cantica*', in J.E. Murdoch and E.D. Sylla, eds., *Cultural Context*, Dordrecht, 1975.
Stokes, Francis Griffin, trans., *Epistolae obscurorum vivorum: On the Eve of the Reformation, Letters of Obscure Men*, London, 1909; repr. w/o Latin text, New York, 1964; Philadelphia, 1972.
Stone, Michael E., *Scriptures, Sects, and Visions: A Profile of Judaism from Ezra to the Jewish Revolts*, Philadelphia, 1980.
Stromberg, Roland N., *Religious Liberalism in Eighteenth-Century England*, Oxford, 1954.
Strong, Edward W., 'Barrow and Newton', *JHP*, viii (1970): 155-72.
Strunz, Franz, *Geschichte der Naturwissenschaften im Mittelalter*, Stuttgart, 1910.
Stupperich, Robert, *Melanchthon* (German ed., Berlin, 1960). Translated by Robert H. Fischer, London, 1966.
Sullivan, William M., *Work and Integrity: The Crisis and Promise of Professionalism in America*, New York, 1995.
Tabor, James D., *Things Unutterable: Paul's Ascent to Paradise in Its Greco-Roman, Judaic and Early Christian Contexts*, Lanham, Md., 1986.
Tcherikover, Victor, *Hellenistic Civilization and the Jews*. Translated by S. Applebaum, Philadelphia, 1959.
Telle, Joachim, ed., *Analectica Paracelsica: Studien zum Nachleben Theophrast von Hohenheims im deutschen Kulturgebiet der frühen Neuzeit*, Stuttgart, 1994.
Temkin, Oswei, *Hippocrates in a World of Pagans and Christians*, Baltimore, Md., 1991.
——, 'Science and Society in the Age of Copernicus', in O. Gingerich, ed., *The Nature of Scientific Discovery*, Washington, D.C., 1975.
Thackray, Arnold, *Atoms and Powers: An Essay on Newtonian Matter-Theory and the Development of Chemistry*, Cambridge, Mass., 1970.
——, 'Matter in a Nutshell: Newton's *Opticks* and Eighteenth-Century Chemistry', *Ambix*, xv (1968): 29-53.
Theerman, Paul, 'James Clerk Maxwell and Religion', *AJP*, liv (1986): 312-17.
Thomas, Keith, *Man and the Natural World: A History of the Modern Sensibility*, New York, 1983.
——, *Religion and the Decline of Magic*, New York, 1971.
Thomas, Roger, 'Presbyterians in Transition', in C.G. Bolman et al., eds., *The English Presbyterians*, London, 1968.
Thompson, Thomas L., *Early History of the Israelite People from the Written and Archeological Sources*, Leiden, 1992.
Thomson, George Paget, *J.J. Thomson, Discoverer of the Electron*, Garden City, N.Y., 1966.
Thorndike, Lynn, *University Records and Life in the Middle Ages*, New York, 1944, 1971.
Tiede, David Lenz, *The Charismatic Figure as Miracle Worker*, Missoula, Mont., 1972.
Todd, Margo, *Christian Humanism and the Puritan Social Order*, Cambridge, 1987.
Tolan, John, *Petrus Alfonsi and His Medieval Readers*, Gainesville, Fla., n.d.
Tolstoy, Ivan, *James Clerk Maxwell: A Biography*, Edinburgh, 1981.
Tomlinson, Gary, *Music in Renaissance Magic: Toward a Historiography of Others*, Chicago, 1993.
Toon, Peter, *The Emergence of Hyper-Calvinism in English Nonconformity, 1689-1765*, London, 1967.
Torrance, Thomas F., *Calvin's Doctrine of Man*, London, 1949; Grand Rapids, Mich., 1957.
——, *Space, Time and Incarnation*, London, 1969.
——, *Transformation and Convergence in the Frame of Knowledge: Explorations in the Interrelations of Scientific and Theological Enterprise*, Grand Rapids, Mich., and Belfast, 1984.
Toulmin, Stephen, and Goodfield, June, *The Architecture of Matter*, Harmondsworth, 1965.
——, *The Discovery of Time*, Harmondsworth, 1967.
——, *The Fabric of the Heavens: The Development of Astronomy and Dynamics*, New York, 1961.

Trevor-Roper, Hugh R., *Religion, the Reformation and Social Change and Other Essays*, London, 1967.
Trinterud, Leonard J., *Elizabethan Puritanism*, New York, 1971.
Troupeau, Gérard, ed., *Hunayn Ibn Ishaq: Collection d'articles publiée à l'occasion du onzième centenaire de sa mort, Arabica*, xxi, fasc. 3, Leiden, 1975.
Turner, Frank M., 'The Victorian Conflict between Science and Religion: A Professional Dimension', *Isis*, lxix (1978): 356-76.
Underwood, E. Ashworth, ed., *Science, Medicine and History: Essays on the Evolution of Scientific Thought and Medical Practice written in honour of Charles Singer*, 2 vols., London, 1953.
Urbach, Ephraim E., *The Sages—Their Concepts and Beliefs*, 2 vols. Translated from Hebrew by Israel Abrahams, Jerusalem, 1975.
Vailate, Ezio, 'Leibniz and Clarke on Miracles', *JHP*, xxxiii (1995): 563-91.
Valentin, Antonina, *The Drama of Albert Einstein*, New York, 1954.
Van den Broek, R., and M.J. Vermaseren, eds., *Studies in Gnosticism and Hellenistic Religions presented to Gilles Quispel on the Occasion of his 65th Birthday*, Leiden, 1981.
Van der Merwe, Alwyn, ed., *Old and New Questions in Physics, Cosmology, Philosophy, and Theoretical Biology*, New York, 1983.
Van de Wetering, John E., 'God, Science and the Puritan Dilemma', *New England Quarterly*, xxxviii (1965): 494-507.
Van Engen, John, 'Theophilus Presbyter and Rupert of Deutz: The Manual Arts and Benedictine Theology in the Early Twelfth Century', *Viator*, xi (1980): 147-63.
Van Gelder, H.A. Enno, *The Two Reformations in the 16th Century: A Study of the Religious Aspects and Consequences of Renaissance and Humanism*, The Hague, 1961.
Van Helmont, *see* Helmont
Van Steenberghen, Fernand, *Thomas Aquinas and Radical Aristotelianism*, Washington, D.C., 1980.
Van Till, Howard J., 'Can the Creationist Tradition Be Recovered? Reflections on *Creation and the History of Science*', *Perspectives on Science and Christian Faith*, xliv (1992): 178-85.
——, et al., *Portraits of Creation: Biblical and Scientific Perspectives on the World's Formation*, Grand Rapids, Mich., 1990.
Vartanian, Aram, *Diderot and Descartes: A Study of Scientific Naturalism in the Enlightenment*, Princeton, N.J., 1953.
——, 'Holbach', *DSB*, xi: 468-69.
——, 'La Mettrie', *DSB*, vii: 605-7.
Vartanian, Pershing, 'Cotton Mather and the Puritan Transition into the Enlightenment', *Early American Literature*, vii (1973): 213-24.
Verbeke, Gerard 'Some Later Neoplatonic Views on Divine Creation and the Eternity of the World', paper presented at the Third International Conference of the International Society for Neoplatonic Studies, Washington, D.C., 12 October 1978.
Vesalius, Andreas, *The Epitome of Andreas Vesalius*, Yale Medical Library Publication, No. 21. Translated by L.R. Lind, New York 1949.
Vickers, Brian, 'Analogy vs. Identity: The Rejection of Occult Symbolism, 1580-1680', in B. Vickers, ed., *Occult and Scientific Mentalities*, Cambridge, 1984.
——, ed., *Essential Articles for the Study of Francis Bacon*, Hamden, Conn., 1968.
——, ed., *Occult and Scientific Mentalities in the Renaissance*, Cambridge, 1984.
Viner, Jacob, *The Role of Providence in the Social Order: An Essay in Intellectual History*, Princeon, N.J., 1972.
Visser, R.P.W., et al., eds., *New Trends in the History of Science*, Amsterdam, 1989.
Von Rad, Gerhard, *Old Testament Theology*, 2 vols. (*Theologie des Alten Testaments*, Munich, 1957-60). Translated by D.M.G. Stalker, London, 1962-65.
Walker, D.P., *The Ancient Theology: Studies in Christian Platonism from the Fifteenth to the Eighteenth Century*, London, 1972.
——, *Spiritual and Demonic Magic from Ficino to Campanella*, London, 1958.
Wallace, William A., 'Experimental Science and Mechanics in the Middle Ages', *DHI*,

ii:196-205.
——, 'The Philosophical Setting of Medieval Science', in D.C. Lindberg, ed., *Science in the Middle Ages*, Chicago, 1978.
——, *Prelude to Galileo: Essays on Medieval and Sixteenth-Century Sources of Galileo's Thought*, Boston Studies in the Philosophy of Science, vol. lxii, Dordrecht, 1981.
——, 'Science (Scientia)', *New Catholic Encyclopedia*, vol. xii, New York, 1967.
Walsh, James J., *The Popes and Science: The History of the Papal Relations to Science during the Middle Ages and Down to Our Own Time*, New York, 1908.
Ward, Benedicta, *Miracles and the Medieval Mind: Theory, Record and Event, 1000-1215*, Philadelphia, 1982.
Warner, Margaret Humphreys, 'Vindicating the Minister's Medical Role: Cotton Mather's Concept of the *Nishmath-Chajim* and the Spiritualization of Medicine', *JHMAS*, xxxvi (1981): 278-95.
Wasserman, Earl R., *Aspects of the Eighteenth Century*, Baltimore, 1965.
Wear, Andrew, ed., *Medicine in Society: Historical Essays*, Cambridge, 1992.
Weatherhead, Leslie D., *Psychology, Religion and Healing: A Critical Study of All the Non-physical Methods of Healing*, London, 1951.
Webster, Charles, 'English Medical Reformers of the Puritan Revolution: A Background to the "Society of Chemical Physitians"', *Ambix*, xiv (1967): 16-41.
——, *From Paracelsus to Newton: Magic and the Making of Modern Science*, November 1980, Cambridge, 1982.
——, *The Great Instauration: Science, Medicine and Reform, 1626-1660*, London, 1975; New York, 1976.
——, 'Henry More and Descartes: Some New Sources', *BJHS*, iv (1969): 359-64.
——, 'Turner, William', *DSB*, xiii:501f.
——, ed., *Samuel Hartlib and the Advancement of Learning*, London, 1970.
Wedel, Theodore Otto, *The Medieval Attitude Toward Astrology Particularly in England*, New Haven, Conn., 1920; Hamden, Conn., 1968.
Weisheipl, James A., 'The Celestial Movers in Medieval Physics', in J.A. Weisheipl, ed., *The Dignity of Science*, 1961.
——, 'The Commentary of St Thomas on the *De Caelo* of Aristotle', *Sapientia*, xxix (1974): 11-34.
——, 'The Concept of Matter in Fourteenth Century Science', in E. McMullin, ed., *The Concept of Matter*, Notre Dame, Ind., 1963.
——, *The Development of Physical Theory in the Middle Ages*, Ann Arbor, 1959, 1971.
——, 'Motion in a Void: Aquinas and Averroes', in A.A. Maurer, ed., *St Thomas Aquinas*, Toronto, 1974.
——, *Nature and Motion in the Middle Ages*, Studies in Philosophy and the History of Philosophy, vol. xi, Baltimore, 1985.
——, 'The Nature, Scope, and Classifications of the Sciences', in D.C. Lindberg, ed., *Science in the Middle Ages*, Chicago, 1978.
——, ed., *Albertus Magnus and the Sciences: Commemorative Essays 1980*, Toronto, 1980.
——, ed., *The Dignity of Science: Studies in the Philosophy of Science Presented to William Humbert Kane*, Washington, D.C., 1961.
Wendel, François, *Calvin: The Origins and Development of His Religious Thought* (*Calvin*, Paris, 1950). Translated by Philip Mairet, London, 1963.
Westbrook, R.H., 'Needham', *DSB*, x:9-11.
Westfall, Richard S., 'The Career of Isaac Newton: A Scientific Life in the Seventeenth Century', *American Scholar*, i (1981): 341-53.
——, *The Construction of Modern Science: Mechanisms and Mechanics*, New York, 1971; Cambridge, 1977.
——, 'The Influence of Alchemy on Newton', in J. Chance and R.O. Wells, eds., *Mapping the Cosmos*, Houston, 1985.
——, *Never at Rest: A Biography of Isaac Newton*, Cambridge, 1975, 1980.
——, 'Newton and Alehemy', in B. Vickers, ed., *Occult and Scientific Mentalities*, Cambridge, 1984.
——, 'The Rise of Science and the Decline of Orthodox Christianity: A Study of Kepler,

Descartes, and Newton', in D.C. Lindberg and R.L. Numbers, *God and Nature*, Berkeley, 1985.
——, 'The Role of Alchemy in Newton's Career', in M.L. Righini Borelli and William R. Shea, eds., *Reason, Experiment, and Mysticism*, New York, 1975.
——, *Science and Religion in Seventeenth-Century England*, New Haven, Conn., 1958; Ann Arbor, Mich., 1973.
Westman, Robert S., 'Nature, Art, and Psyche: Jung, Pauli, and the Kepler-Fludd Polemic', in B. Vickers, ed., *Occult and Scientific Mentalities*, Cambridge, 1984.
——, ed., *The Copernican Achievement*, UCLA Center for Medieval and Renaissance Studies Contributions, vol. vii, Berkeley, 1975.
——, and McGuire, J.E., *Hermeticism and the Scientific Revolution*, Los Angeles, 1977.
Wetherbee, Winthrop, *The 'Cosmographia' of Bernardus Silvestris*, New York, 1973.
——, 'Philosophy, Cosmology, and the Twelfth-Century Renaissance', in P. Dronke, ed., *History of Twelfth-Century Western Philosophy*, Cambridge, 1988.
Whiston, William, *Historical Memoirs of the Life of Dr. Samuel Clarke*, London, 1730.
White, Andrew Dickson, *A History of the Warfare of Science with Theology in Christendom*, 2 vols., New York, 1896.
White, K.D., 'Greek and Roman Technology', in H.H. Scullard, ed., *Aspects of Greek and Roman Life*, Ithaca, N.Y., 1984.
White, Lynn, Jr., 'Cultural Climates and Technological Advance in the Middle Ages', *Viator*, ii (1971): 171-201.
——, *Medieval Religion and Technology: Collected Essays*, Berkeley, 1978.
Whitney, Elspeth, *Paradise Restored: The Mechanical Arts from Antiquity through the Thirteenth Century*, Philadelphia, 1990.
Whitrow, G.J., *Time in History: The Evolution of Our General Awareness of Time and Temporal Perspective*, Oxford, 1988.
Wightman, W.P.D., 'Cullen', *DSB*, iii:494-95.
——, *Science and the Renaissance, Vol. I: An Introduction to the Study of the Emergence of Science in the Sixteenth Century*, Edinburgh, 1962.
——, *Science in a Renaissance Society*, London, 1972.
Wilde, C.B., 'Hutchinsonianism, Natural Philosophy and Religious Controversy in Eighteenth Century Britain', *History of Science*, xviii (1980): 1-24.
——, 'Matter and Spirit as Natural Symbols in Eighteenth-Century British Natural Philosophy', *BJHS*, xv (1982): 99-131.
Wildiers, N. Max, *The Theologian and His Universe: Theology and Cosmology from the Middle Ages to the Present* (*Wereldbeeld en Teologie van de Middelleeuwen tot vandaag*, 1977). Translated by Paul Dunphy, New York, 1982.
Wilken, Robert L., 'Toward a Social Interpretation of Early Christian Apologetics', *CH*, xxxix (1970): 437-58.
Willey, Basil, *The Eighteenth Century Background: Studies in the Idea of Nature in the Thought of the Period*, London, 1940; Boston, 1961.
Williams, Leslie Pearce, 'Faraday', *DSB*, iv:527-40.
——, 'Kant, *Naturphilosophie* and Scientific Method', in R.N. Giere and R.S. Westfall, *Foundations of Scientific Method*, Bloomington, Ind., 1973.
——, *Michael Faraday: A Biography*, London, 1964.
——, 'Oersted', *DSB*, x:182-86.
Williamson, George, *Seventeeth Century Contexts*, London, 1960; Chicago, 1961.
Wilsdorf, Helmut M., 'Agricola', *DSB*, i:77-79.
Wilson, David B., 'Concepts of Physical Nature: John Herschel to Karl Pearson', in U.C. Knoepflmacher and G.B. Tennyson, eds., *Nature and the Victorian Imagination*, Berkeley, 1977.
——, 'Herschel and Whewell's Version of Newtonianism', *JHI*, xxxv (1974): 79-97.
——, 'Kelvin's Scientific Realism: The Theological Context', *The Philosophical Journal*, xi (1974): 41-60.
——, 'Kinetic Atom', *AJP*, xlix (1981): 217-22.
——, 'The Thought of Late Victorian Physicists: Oliver Lodge's Ethereal Body', *Victorian Studies*, xv (1971): 29-48.

Wingren, Gustav, *Luther on Vocation*. Translated from Swedish by Carl C. Rasmussen, Philadelphia, 1957.
Wintle, Justin, ed., *Makers of Nineteenth Century Culture, 1800-1914*, London, 1982.
Wolfson, Harry Austryn, *Repercussions of the Kalam in Jewish Philosophy*, Cambridge, Mass., 1979.
Woolf, Harry, ed., *The Analytic Spirit: Essays in the History of Science in Honor of Henry Guerlac*, Ithaca, N.Y., 1981.
Wormhoudt, Arthur, 'Newton's Natural Philosophy in the Behmenistic Works of William Law', *JHI*, x (1949): 411-29.
Wright, G. Ernest, *Biblical Archeology*, Philadelphia, 1957.
Wright, John Kirkland, *The Geographical Lore of the Time of the Crusades: A Study in the History of Medieval Science and Tradition in Western Europe*, Washington, D.C., 1925; New York, 1965.
Wright, Louis B., *The Alliance Between Piety and Commerce in English Expansion, 1558-1625*, Chapel Hill, N.C., 1943; New York, 1965.
Wrightsman, Bruce, 'Andreas Osiander's Contribution to the Copernican Achievement', in Robert S. Westman, ed., *The Copernican Achievement*, Berkeley, 1975.
Wussing, Hans, ed., *Geschichte der Naturwissenschaften*, Köln, 1983.
Yates, Frances A., *Giordano Bruno and the Hermetic Tradition*, Chicago, 1964.
——, *Lull & Bruno*, Collected Essays, vol. i, London, 1982.
——, *The Occult Philosophy in the Elizabethan Age*, London, 1979.
——, *The Rosicrucian Enlightenment*, London, 1972; Boulder, Colo., 1978.
Young, Davis A., 'The Discovery of Terrestrial History', in H.J. Van Till et al., ed., *Portraits of Creation*, Grand Rapids, Mich., 1990.
Zilsel, Edgar, 'The Genesis of the Concept of Physical Law', *Philosophical Review*, iii (1942): 245-79.
——, 'The Genesis of the Concept of Scientific Progress', *JHI*, vi (1945): 325-49.
Zimmermann, Albert, ed., *Albert der Grosse: Sein Zeit, Sein Werk, Seine Wirkung*, Berlin, 1981.
Zöckler, O., *Geschichte der Beziehungen zwischen Theologie und Naturwissenschaft mit besonderes Rücksicht auf Schöpfungsgeschichte*, 2 vols., Gütersloh, 1877-79.
Zysk, Kenneth C., *Asceticism and Healing in Ancient India: Medicine in the Buddhist Monastery*, New York, 1991.

INDICES

NAMES

Abelard, Peter 50, 53, 103
Abulafia, Abraham 144
Adams, George 334
Adelard of Bath 8n, 26, 27, 51, 54-57, 88, 96n, 99n, 261, 310, 352
Aeschylus 62n, 63n
Aetius 36n
Agricola, George Bauer 156-158, 315
Agrippa of Nettesheim 140, 147, 160, 187, 189, 190-192
Alan of Lille 104n
Albert the Great 8n, 91-92, 95-96, 100, 101, 105, 113, 115, 118-121, 310
Albert of Saxony 103n, 124, 128, 131
Alberti, Leone Battista 138-140, 152
Alcuin of York 49, 77
Alexander of Hales 96n
Alexander of Tralles 74
Alpetragius, see al-Bitruji
Alsted, John Heinrich 207
Ambrose of Milan 41, 45n, 48, 103n
Ampère, André-Marie 377n, 388n
Anaxagoras 27
Anaximander 16, 20
Anaximenes 16
Andreae, Johann 162, 164, 174, 184, 187-188, 201, 207-208
Anselm of Canterbury 54, 95
Anselm of Laon 103n
Anthony the Great 94n
Apian, Peter 103n, 154
Aquinas, Thomas 12, 40, 87, 91, 93, 95-97, 99n, 100-101, 103n, 105, 106, 110-111, 113, 115-116, 118, 123-126, 129-130, 178n, 214, 216, 307, 309
Arbuthnot, John 321
Aristides of Athens 36n
Aristobulus of Alexandria 21n, 24n, 34, 36n
Aristotle, see Aristotelianism
Arnauld, Antoine 216, 221
Arnobius of Sicca 35-37, 78n, 81
Arnold of Bonneval 79
Arnold of Villanova 120-121, 161n, 184
Artapanus 14
Ashmole, Elias 206, 207n
Athanasius 21, 26, 30, 103n, 246
Athenagoras of Athens 29, 36n, 65
Augustine of Hippo 3, 4n, 12, 20n, 22-23, 30, 41-43, 45, 48, 52-58nn, 60, 74, 78, 81, 90, 94, 103n, 108-109, 115, 116, 117n, 146, 163, 203, 256
Avempace, see Ibn Bajjah
Avicenna, see Ibn Sina
Bacon, Francis 138-139, 161-163, 165-166, 173n, 177, 182-184, 188, 191, 193-199, 207-208, 278, 310, 315-316, 327, 329-330, 394, 403
Bacon, Roger 85n, 92, 96n, 103n, 105, 109, 113, 115, 120-121, 126, 129
Baghdadi, al- 40
Bailey, Walter 181
Bardaisan 75
Barrow, Isaac 235, 239-241, 250n
Basil of Caesarea 13, 17-19, 21-22, 25-27, 29-32, 35, 37-41, 45, 48, 49, 57, 60, 68-75, 77-78, 81-83, 101, 103n, 126n, 129, 163, 248, 265, 309, 394, 402, 405
Baxter, Richard 206
Bayle, Pierre 272, 275n
Becher, Johann Joachim 296-297
Becon, Thomas 192
Bede, The Venerable 32, 41, 46, 47, 49, 58, 103-104, 106, 118, 352
Beeckman, Isaac 223
Bellarmine, Robert 187
Benedict of Nursia 47, 74, 76-77, 81-82
Bentley, Richard 242
Berenger of Tours 55
Berkeley, George 247, 248n, 320, 324-327, 334
Bernard of Chartres 51n
Bernard of Clairvaux 53, 79n, 82n, 88, 94n
Bernard Silvestris 104n
Bernouilli, Johann 342n
Bérulle, Pierre de 219
Beza, Theodore 162, 165, 172
Biot, Jean-Baptiste 377n
Biruni, al- 26, 31
Biscop, Benedict 47
Bitruji, al- (Alpetragius) 31, 40, 88, 102
Black, Joseph 300-301
Blake, William 320, 327-331
Boehme, Jakob 205, 210, 239, 327-328
Boerhaave, Herman 277-279, 285, 288-289, 291, 295, 297, 299, 301, 318, 325, 359-360
Boethius 26, 43, 44, 51, 89, 96n, 115, 250n
Bohr, Niels 6, 362, 384, 388-398
Boltzmann, Ludwig von 381, 390n

Bonaventure 12, 91, 93-96, 100, 105-106, 113, 119, 127
Boniface 47-49, 51, 111
Boole, George 381n
Born, Max 373
Boscovich, Roger Joseph 263, 269, 282, 285-286, 288, 303n, 352, 364, 386n
Bossuet, Jacques 285n
Bourquet, Louis 311n
Boyle, Robert 12n, 162, 199, 201-202, 209-210, 214-215, 224, 226, 228-233, 237-238, 240, 243, 255, 264, 266, 271, 278, 326, 327, 328n, 339n, 384
Bradwardine, Thomas 116, 124, 129, 131, 135-136, 141, 171
Brahe, Tycho 173
Brattle, Thomas 272n
Browne, Thomas 225n
Brunelleschi, Filippo 138
Bruno, Eusebius 55n
Bruno, Giordano 165-169, 174, 187, 190, 194, 221, 235n, 336, 340
Bucer, Martin 162, 189, 192
Buckland, William 293, 295
Buckle, Henry 381n
Buffon, Comte de 281, 285, 290, 308-316
Bullein, William 193
Bullinger, Johann Heinrich 162, 180, 192
Buridan, John 8n, 41, 114, 116, 119, 122, 124, 127-131, 135, 149, 171, 310
Butler, Joseph 254n, 259n
Cabanis, Pierre 308
Caelius Aurelianus 45n
Caesarius (brother of Gregory of Nazianzus) 74
Calcagnini, Celio 135-137
Calvin, John 162, 164-166, 168, 172, 177-183, 188-190, 192, 194, 196, 198-199, 257, 265, 294, 318
Campanella, Thomas 165, 208n, 223
Campanus of Novara 103n
Capella, Martianus 59n
Capito, Wolfgang 162
Carleton, Thomas Compton 220n
Casaubon, Isaac 226n
Cassian, John 45, 70n, 76
Cassiodorus, Senator 41, 43-46, 57n, 74, 76-78, 80-82
Cavendish, Henry 305
Cennini, Cennino 154
Chalcidius 56n
Chambers, Robert 369n
Charleton, Walter 224, 226-228, 232-233, 240, 271, 384
Châtelet, Madame du 279n, 283, 311, 314n
Cheyne, George 275n

Chrysippus 38n
Chrysostom, John 71n
Cicero 28n, 38n, 104n, 107, 185n
Clarke, Samuel 247, 254-262, 270, 283, 285, 289, 294-295, 298, 319n, 320, 327, 378
Clausius, Rudolf 374-375, 381
Cleanthes 28
Clement of Alexandria 15, 17, 24n, 25n, 103, 143
Clement of Rome 15, 21
Clodius, Frederick 208
Cocceius, John 220
Colden, Cadwallader 299-300
Coleridge, Samuel Taylor 361n
Collins, Anthony 255
Comenius, Johann Amos 201, 204, 207-210, 214
Condorcet, Marquis de 342-343, 350
Constantine the African 53, 73-74
Conway, Anne 211n, 238
Cooper, Thomas 180-181
Copernicus, Nicholas 134-138, 148-156, 162, 168, 173, 179-180, 185-186, 199, 215, 235n
Cotes, Roger 254, 262, 264, 266-268, 279, 282, 285, 291, 295
Coulomb, Charles Augustin 305, 377n
Croll, Oswald 188, 190-191
Cudworth, Ralph 250n, 273n
Cullen, William 300-301
Cusa, see Nicholas of Cusa
Cyril of Alexandria 46n
D'Ailly, Peter 103n
D'Alembert, Jean 315, 341-344, 353
Dalton, John 334, 352, 360
Damian, Peter 51n-52
Daneau, Lambert 165n
Dante 128n
Davy, Humphry 356-357, 360-363, 366, 371
Dee, John 166, 191
Derham, William 254n, 262, 264 265-266, 270, 274, 276, 279, 291, 295, 302-303, 327, 339n, 378
Desaguliers, John Théophile 262-263, 273n
Descartes, René 40, 201-202, 210-211, 214-224, 226, 228, 231-232, 235-244, 255, 257, 268, 271, 281-283, 286-287, 292, 308, 317, 319n, 324, 326, 327, 338, 339n, 342, 348, 384, 386
Diderot, Denis 201n, 281, 285, 287, 308-309, 314-317, 319, 343
Digby, Kenelm 224-225, 228, 231
Digges, Leonard 336, 340
Digges, Thomas 166-169, 173, 235n
Diodorus 17
Dionysius Exiguus 46
Dioscorides 45n

Dodd, John 196n
Domingo de Soto 136
Duns Scotus, Johannes 111, 124
Dürer, Albrecht 154
Dury, John 209
Edwards, Jonathan 268, 320, 324-327, 333-334, 337
Einstein, Albert 6, 247, 356n, 383-384, 388-398, 406n
Empedocles 16, 36n
Encke, Johann 368
Ephraim of Syria 75-76
Epicurus, see Epicureanism
Epiphanius of Salamis 70
Erastus, Thomas 188
Erigena, John Scotus 58n, 80n
Euclid, see Euclidean geometry
Eudorus of Alexandria 24n, 63n
Eudoxus of Cnidus 292
Euler, Leonhard 277, 279, 287, 295, 342n
Eusebius of Caesaria 14n, 21n, 24n, 30, 36
Eustathius of Sebaste 45, 70,
Ezekiel the Tragedian 28n
Farabi, al- 26, 31
Faraday, Michael 355-357, 362-366, 370-371, 373, 376-377, 380-381, 385, 388, 390
Farel, Guillaume 162
Ferguson, James 335, 337-339, 347
Fermat, Pierre de 281
Ficino, Marsilio 140-156, 159-160, 174, 187, 190-191, 194, 221
Fludd, Robert 191, 221-222, 226-227
Foppl, August 390n
Forbes, J.D. 378n, 379n, 381n
Fourier, Jean Baptiste Joseph 373, 375n, 377n
Fracastoro, Girolamo 136-137, 149, 151-152
Francis of Assisi 94, 119, 139
Francis of Marchia 40, 94, 127, 128n
Frederick II 99
Frederick the Great 279, 311n, 318
Freind, John 262, 264
Fresnel, A.J. 378
Fulbert of Chartres 49-50, 78
Galen, see Galenic medicine
Galileo Galilei 40, 87, 116, 127, 128, 165, 168-169, 177, 187, 199, 220n, 223, 343n
Gassendi, Pierre 201, 210, 215, 221-228, 232, 235-236, 240, 271
Gauden, John 207n
Gellius, Aulus 39n
Geminus 22
Gerbert of Aurillac 49-50
Giese, Tiedemann 155, 162
Gilbert, William 173, 191
Giles of Rome 105n
Goodwin, Thomas 270

Gottschalk 50n
Gregory the Great 77
Gregory IX, Pope 87
Gregory of Nazianzus 21n, 69-78
Gregory of Tours 46n
Gregory, John 382n
Grosseteste, Robert 90, 99n, 104, 115, 116, 118, 126, 131n, 137, 140
Guericke, Otto von 230
Guy de Chauliac 120
Hadrian of Africa 47
Hales, Stephen 298-299, 301, 304
Haller, Albrecht von 278
Hallett, Joseph 260
Halley, Edmond 246
Hamilton, William 378n-380n
Harriot, Thomas 191, 223
Hartley, David 298, 301-305
Hartlib, Samuel 201, 204-205, 207-210, 214, 222, 228-229, 234
Hecataeus 24
Helmholtz, Hermann Ludwig von 369, 377
Helmont, Francis Mercury van 205-206, 210-211, 238n, 239
Helmont, Joan Baptista van 120, 201-207, 209-211, 213, 226-228, 278, 295-296, 315, 327-328
Henry of Ghent 124, 130
Henry of Langenstein 106-107, 111, 116, 119, 128n, 130-131
Heraclitus 20
Herschel, John 368n, 381n
Herschel, William 346, 348
Hertz, Heinrich 386
Hickes, George 320
Hincmar of Reims 50n
Hippocrates, see Hippocratic tradition
Hobbes, Thomas 224-227, 232-233, 235-237, 255
H ffding, Harold 398n
Holbach, Baron d' 285n, 308-309, 316-319, 343
Honorius of Autun 51
Hood, Thomas 180
Hooke, Robert 230
Hooker, Richard 181
Hooper, John 192-193
Horne, George 331
Hugh of St. Victor 56n, 79-81, 265
Hume, David 58, 258, 300-301
Hunayn ibn Ishaq 76
Hutcheson, Francis 302
Hutchinson, John 320, 331-334
Hutton, James 252, 268, 287, 289-295, 297, 301, 311, 340, 352, 359-361, 373
Huyghens, Christian 210-211

Ibn Bajjah (Avempace) 40-41, 126
Ibn Qurra (Thebit) 102
Ibn Sina (Avicenna) 26, 31, 40-41, 88
Innocent III 105n
Innocent VIII 105n
Irenaeus of Lyons 16-17, 22, 24, 49, 65-67, 78n, 111, 394
Isidore of Seville 32, 41, 46, 47, 77, 80n
James of Edessa 40n
Jeans, James 391
Jerome 30, 74, 76, 103n
Jesus ben Sirach 7, 14, 21n, 22, 24, 28n, 29n, 34-36n, 40, 120, 160, 310
Jesus Christ 15, 29, 46, 63-68, 71-72, 75-76, 78, 83, 97, 118, 120, 146, 155-156, 160-162, 167, 176, 184, 190, 194, 216, 247-249, 259-260, 270-271, 277, 303-304, 307, 314, 330, 344, 405
Jewel, John 181
Job of Edessa 40n
John of Damascus 31n
John of Gorze 50n
John of Ripa 101, 124, 166
John of Rupescissa 120, 121
Johnson, Robert 193
Johnson, Samuel 319-323, 334
Johnson, Samuel, of Yale 331
Jones, William 331, 334
Josephus 15n, 45n, 64n, 69n, 103n
Joule, James Prescott 356, 366, 369-378
Jurin, James 324-325
Justin Martyr 15, 65, 66
Kant, Immanuel 7, 309, 335-341, 344, 347-348, 353, 357, 359, 382
Keill, James 254, 262-265, 285-286
Kelvin, William Thomson, Lord 356, 366, 369, 371-387
Kepler, Johannes 49n, 143, 154, 162-165, 168-174, 179, 186-187, 199, 201, 345, 350n, 389, 394, 396, 401
Khunrath, Heinrich 174, 187
Khusro I 73
Kierkegaard, S ren 392
Kilwardby, Robert 126-127
Kindi, al- 26
Knight, Gowin 262, 267-269, 285, 291-292, 299, 360-361, 364
Lactantius 21-22, 29, 30, 49, 66n, 81
Lagrange, Joseph Louis de 335, 341, 344-346, 385
Lamarck, Jean-Baptiste 282, 309
Lamb, Horace 356n
Lang, Andrew 393
Laplace, Pierre Simon de 262, 290, 321, 334-335, 338-341, 343-351, 353, 359, 368, 372, 377n, 381n

Larmor, Joseph 391
Latimer, Hugh 192
Lauterbach, Anthony 185
Lavoisier, Antoine 275n, 297, 303n, 306, 359
Law, William 326n, 332n
Le Fèvre, Nicholas 206
Lefèvre d'Étaples 145-146
Leibniz, Gottfried Wilhelm 199, 201-202, 204-205, 210-214, 234, 246-248, 255-257, 262-267, 276, 279-282, 284, 287, 296-298, 309-311, 314n, 318, 320, 338, 340, 344-345, 349, 370, 372, 375, 394
Leonardo da Vinci 135-137
Libanius 66n, 69
Libavius, Andreas 187-188
Locke, John 250n, 327, 329, 330
Lombard, Peter 54, 97
Lorentz, Hendrik 390n
Lucian 66n
Lucretius 35, 48, 103n, 223
Lull, Raymond 94, 119-120, 139, 201
Luther, Martin 162-164, 168, 171-172, 174, 176-179, 184-189, 192, 194, 257, 265, 294
Lyell, Charles 367, 369, 373, 375
Mach, Ernst 356n
Maclaurin, Colin 287-289, 291, 295, 297
Macrobius 48n, 52n
Maestlin, Michael 49n, 168
Maier, Michael 187
Maillet, Benoit de 312n
Major, John 135-137
Malebranche, Nicolas 221, 277, 319n, 342
Manegold of Lautenbach 52, 53n
Marcion 16
Marlowe, Christopher 195n
Mather, Cotton 254, 269, 271-278, 285, 326
Mather, Increase 272, 273n, 280, 345
Maupertius, Pierre de 277-285, 287, 295, 339n, 342
Maurice, F.D. 380
Maxwell, James Clerk 6, 355-356, 372, 379-391, 396-397
McCosh, James 352n
Melanchthon, Philip 162, 164, 166, 171-172, 177, 179, 184-187192
Mersenne, Marin 215, 221-222, 224-226, 230n
Methodius of Olympus 81
Mettrie, Julien Offray de la 308-309, 317n
Michael Scot 32, 103n, 129
Mill, John Stuart 354
Milton, John 329-330
Minucius Felix 36n
More, Henry 201n, 235-243, 265, 305, 325
Moses of Léon 144
Moses bar Kepha 40n

Napoleon 348-349
Needham, John Tuberville 281, 285, 310n
Newton, Isaac 12, 32, 40, 127, 171, 199, 201, 211-213, 223, 235, 237, 239-270, 274-309, 313n, 316-337, 341, 344-350, 355-356, 363-366, 371-376, 379, 382, 394-397
Nicholas of Autrecourt 101
Nicholas of Cusa 102, 108, 116, 117, 119-120, 125, 130-131, 138-141, 146, 153, 155, 166, 168-169, 171, 174-175, 204, 235n, 315
Nichomachus 43
Nicolas of Amiens 96n
Novara, Domenico 138, 148
Ockham, see William of Ockham
Oecolampadius, John 162, 192
Oersted, Hans Christian 356-362, 364, 366
Oldenburg, Henry 209, 277
Olivi, Peter John 40, 94, 126-128
Oresme, Nicole 101, 103n, 106-107, 116, 129-131
Origen of Alexandria 15, 17, 21, 25, 29, 31, 64n, 68n, 81n, 101, 103-104n, 168, 248, 270-271, 307
Osiander, Andreas 187
Ostwald, Friedrich Wilhelm 356n
Owen, John 190
Pachomius 45, 68, 69, 76
Paley, William 265, 378, 380n
Palladius (monastic historian) 71n
Palladius, Rutilus 80n
Paracelsus 120, 134, 148, 152, 156, 158-162, 166, 184, 187-194, 201, 203, 205, 209, 222n, 239, 265, 296, 299, 315, 327-328, 360, 394
Paraeus, David 207
Pascal, Blaise 230n, 342
Pasteur, Louis 285
Patrizzi, Francesco 174, 223
Paul III, Pope 137
Paul of Aegina 74
Pearson, Karl 356n
Pecham, John 115, 138
Pemberton, Henry 263, 283, 286, 289
Perkins, William 181-182, 194
Peter of Maricourt 117n
Peter of Poitiers 96n
Petrus, Alfonsi 55n, 57n
Petty, William 229, 328n
Peucer, Caspar 172, 179n
Peurbach, Georg 153
Philo of Alexandria 24, 29n, 35, 36n, 64n, 69n, 103n, 143, 304
Philoponus, John 26, 30-31, 39n, 40, 126n
Pico of Mirandola 140-141, 144-147, 152, 160, 190-191, 221
Pierce, James 260

Planck, Max 356n, 394, 406n
Plato 12, 14, 20n, 44, 56, 59n, 63n, 90, 103n, 104n, 115, 165, 199, 271
Plato of Tivoli 85n
Platter, Felix 136
Plattes, Gabriel 208
Playfair, John 293
Pliny 46-47, 66n
Plotinus 236
Pluche, Abbé 282-283n
Poincaré, Jules-Henri 356n
Poisson, Siméon-Denis 377n, 388n
Pomponazzi 225n
Pope, Alexander 321
Porphyry 66n, 89
Posidonius 22
Poynting, H.H. 356n
Priestley, Joseph 258, 260-261, 269, 271, 286, 298, 301-308, 317-318, 328n, 334, 361, 364
Prince, Thomas 254n
Proclus 26
Prout, William 362
Pseudo-Clement 15, 37-38, 49, 58n, 65-67, 78n, 120n
Pseudo-Dionysius 130
Pseudo-Silvanus 21
Ptolemy, see Ptolemaic astronomy
Pym, John 207
Pythagoras, see Pythagoreanism
Qetelet, Adolphe 381
Rabanus Maurus 49
Radbertus 50n, 57n
Raleigh, Sir Walter 191
Rankine, William 377, 382n
Rashi 78n
Ratramnus 50
Ray, John 266, 274, 339n
Raymond of Penafort 92
Raynaud, Theophile 220n
Recorde, Robert 180
Regiomontanus, Johannes 153-154
Reid, Thomas 289, 302, 305, 378, 379n, 382n
Reinhold, Erasmus 179n
Reuchlin, Johann 140, 146-147, 160, 187, 190
Rheticus, George Joachim 155, 165, 169, 174-175, 179n, 185
Richard of Middleton 130
Ridley, Nicholas 192
Ritschl, Albrecht 353, 393n
Robertus Anglicus 103n
Roriczer, Mathias 154
Ross, Alexander 225n
Rothmann, Christoph 173
Rufinus of Aquileia 41
Rutherford, Ernest 362, 390n, 391
Sandeman, Robert 364

Scheele, Carl Wilhelm 306
Schelling, Friedrich 357, 359
Schleiermacher, Friedrich 353
Sebond, Raymond 94
Serverinus, Peter 166
Servetus, Michael 168
Shaftesbury, 3rd Earl of 302
Shaw, Peter 278, 299-301
Shirley, Thomas 206
Siger of Brabant 91, 95, 110, 113
Simplicius 31
Sirach, see Jesus ben Sirach
Smith, Adam 300
Smith, Henry 196n
Smith, John 235-236
Sophocles 152
Sozomen 73n
Spallazani, Lazzaro 285
Spencer, Herbert 393
Spener, Philip Jacob 213
Spinoza, Baruch 255, 394n, 395
Stahl, Georg Ernst 296-301, 306, 311n, 318
Starkey, George 208-209
Starkey, Thomas 192
Stewart, Balfour 375n, 376n
Stewart, Dugald 378n, 379n
Stillingfleet, Edward 250n
Stokes, George Gabriel 378n
Sturm, Johann 162
Suarez, Francisco 214-215, 244
Swift, Jonathan 319, 321-322
Tait, Peter Guthrie 371, 375n-377n
Tartaglia, Niccolo 154-155
Tatian 29
Taylor, C.B. 380n
Taylor, Thomas 328n
Telesio, Bernardino 174, 194
Tempier, Stephen 91, 101, 110
Tertullian 16-17, 21-22, 29, 30, 36n, 49, 111
Thales 16
Thebit, see Ibn Qurra
Theodore of Mopsuestia 30-31, 75
Theodore of Tarsus 46
Theodoric of Freiburg 115
Theophilus of Alexandria 46n
Theophilus of Antioch 24, 65
Thierry of Chartres 50-51, 261
Thomson, J.J. 390n, 391
Thomson, William, see Kelvin
Toland, John 255
Toricelli, Evangelista 230n
Toscanelli, Paolo 138-140, 150
Travers, Walter 181n

Trithemius of Sponheim 147, 192
Turgot, Jacques 254n, 285n
Turner, William 192-193
Turretin, Francis 220-221
Turretin, Jean-Alphonse 221
Tylor, Edward 393
Urban IV 87
Urban V 87
Urban VIII, Pope 177
Valentinus 16
Vaughan, Thomas 208, 229, 239
Vermigli, Peter Martyr 172
Vesalius, Andreas 135-137
Virgil of Salzburg 47-49, 111, 343n
Voltaire 254, 277, 282-286, 316, 319, 339n
Walafrid of Strabo 103n
Wallis, John 211, 229
Walther, Bernard 154
Ward, Seth 229
Watt, James 292, 301
Watts, Isaac 254, 269-271, 276-277, 281, 285, 295, 304, 337
Webster, John 205
Weigel, Valentin 174, 187
Wesley, John 331
Whewell, William 356, 366-369, 372-374, 376, 379n, 380
Whichcote, Benjamin 235
Whiston, William 247, 254-255, 259-262, 266, 270, 274, 276, 281, 285, 295, 310-311, 322
Whitgift, John 181
Wigner, Eugene 406n
Wilkins, John 229
William of Auvergne 178n
William of Conches 8, 32, 51, 53, 56-59, 91, 100-101, 104n, 111, 261
William of Moerbeke 87
William of Ockham 94, 95n, 101, 111, 114, 127-128, 130, 164
William of St. Thierry 32, 53, 88, 111
Witelo of Silesia 115, 138
Wittich, Christoph 220
Wolff, Caspar Friedrich 311n
Wolff, Christian 311, 314n
Wren, Christopher 229
Wright, Thomas 335-336, 338, 340-341
Zachary, Pope 47-48, 343n
Zanchi, Girolamo 172
Zeno, Emperor 73
Zermelo, Ernst 356n
Zöckler, Otto 352n
Zwinger, Theodore 188
Zwingli, Ulrich 162, 177-17

PLACES

Alexandria 26, 74
America 254, 265, 269, 271, 274, 278, 285, 299, 325, 353
Angers 219
Antioch 69
Armenia 70
Athens 16, 69
Augsburg 157
Baghdad 73, 85
Basel 136, 162, 187, 193
Belfast 373
Berlin 279, 284
Berwickshire 289
Birmingham 305
Bohemia 156, 207
Bologna 138, 148-149
Caen 219
Caesarea 17, 70, 72
Calabria 44
Cambridge 182, 192, 201, 205, 235, 237, 239, 260, 266, 273, 366, 373, 378-380, 382, 390n, 391
Canterbury 47
Cappadocia 17
Cassiacum 74
Chartres 32, 47, 50, 56, 78, 100, 115, 118
Cilicia 30
Clairvaux 79
Constantinople 69, 71n, 76
Copenhagen 357, 391
Corinth 87
Cracow 149
Croatia 285
Daventry 261, 304
Edessa 73
Edinburgh 287, 289, 299-300, 379
Egypt 13, 24, 45, 68, 70-71, 76
Ferrara 135
Florence 138-140, 148, 150, 154
Geismar 48
Geneva 162, 167-168, 180, 190, 220, 343
Glasgow 300, 373, 376, 378
Gunde-Shahpur 73
Heidelberg 190, 192, 207
Herborn 207
Iona 46
Jarrow 47
Jerusalem 13, 16, 34
Kulm 155

Langres 315
Leeds 304
Leiden 219, 220, 278, 289, 291n, 297
Liège 220n
Lindisfarne 46
London 181, 193, 208, 261, 284, 358, 360, 367
Louvain 202, 219
Mainz 49, 210
Manchester 390n
Marseilles 76
Mesopotamia 19, 70, 73, 76
Milan 135
Monte Cassino 74
Naples 99n
Neuchâtel 162
Nisibis 73-74
Norwich 331
Novara 138, 148
Nuremberg 153-155
Orvieto 87
Oxford 99n, 135, 205, 229, 262, 264, 331
Padua 135-136, 139
Palestine 35-36, 70
Paris 56, 86, 87, 91, 99n, 101, 110, 121, 136, 219, 221, 224-225
Persia 13, 62, 73
Pontus 70
Prussia 208
Ravenna 52
Rheims 50
Salamanca 214
Salerno 73, 74
Salzburg 48, 1, 49
Saxony 156
Sicily 86
Silesia 115
Sorbonne 312, 343
Strassburg 162, 189-190
Syria 35, 70, 73-74, 75, 90
Thetford 320
Travers 181
Tübingen 167
Utrecht 219-220
Vienna 153
Viterbo 87
Wearmouth 47
Wittenberg 162, 185
Yorkshire 302

Subjects

Abbasid caliphs 73, 85
absolute space 232, 238, 240-241, 247, 256, 259, 263, 295, 308, 324, 327
Acts of John 36n, 65
Acts of Peter 67n
Acts of Paul 65
Acts of Thomas 65
alchemy, 5, 29, 79n, 117, 120, 174, 177, 186-188, 190, 194, 197-198, 203, 208-210, 222, 228, 232n, 240n, 299, 359-360, 361n, 400
Anabaptists 176, 284
angels 14, 23, 25, 27-30, 45, 60-61, 91-92, 104, 106-107, 130-131, 142, 145, 176, 190, 194, 248, 272n, 395
Anglicans 181, 191, 206, 230, 250, 254-255, 258-260, 284, 299, 302, 319-320, 331
Apologists 64-67, 405
Apostolic Constitutions 72n
Apostolic Tradition of Hippolytus 45n, 72n
Arabic culture 26, 40, 50, 54, 73, 74, 76, 79, 84, 85-86, 90, 102, 400
Arianism 247, 249, 260-261, 270, 276, 304, 326
Aristotelianism 10-12, 16-18, 22-31, 38-40, 44, 51, 59n, 64n, 75, 78, 80n, 84, 86, 87-92, 95-137, 141-145, 149-152, 157, 159, 166, 171-183, 191, 198-199, 201, 203-205, 214-215, 218-221, 224, 234, 237, 244, 252, 257, 268, 273n, 283, 292, 321, 325, 389, 400, 407
astrology 14, 46, 154, 179, 192, 194
atheism 59-60, 183, 202, 232n, 233n, 236, 241, 243, 249-250, 263, 285, 298, 316, 331
autonomy of nature 4, 21, 32-60, 82, 94, 98-99, 108, 113-114, 118, 121-122, 156, 162, 175-183, 191, 204, 213, 246-247, 251, 292, 308, 310, 316, 319-320, 330, 351, 353, 361n, 365, 369, 380, 389, 400-405
Averroism 87, 91, 113, 135
Babylonian culture 13-14, 19, 61, 63, 81
Baptists 260-261
Benedictine monasticism 44, 45, 48, 50, 73n, 74, 81-82
Bohemian Brethren 207
Byzantine culture 73-74
Cabalism 140, 141, 144-147, 160, 187, 189, 191n, 194, 199, 210, 238-239
Calvinism 180-183, 188-198, 260, 269, 304, 343
Cambridge Platonists 201, 235-239

Cartesianism 216, 218-221, 236-237, 279-282n, 286, 319-320, 332, 341, 356
Catastrophism 290, 367, 372
Chartres, School of 32, 47, 50, 56, 78, 99n, 100, 115, 118
Christology 249, 260, 271
Cistercians 53, 82n (see also Arianism)
clocks 45, 106-109, 111, 117, 153, 272, 292
Collegium Romanum 285
comets 173, 244-246, 258, 261, 272-275, 278, 290, 294, 311, 322, 344-347, 368
complementarity, principle of 384, 393
comprehensibility of the world 21-27, 32, 49, 59-60, 114, 143, 156, 163-171, 289, 296, 302, 315-316, 319, 330, 345, 348, 351, 365, 369, 380, 398, 400, 404
Condemnation of 1277 87, 101, 106, 110-111, 113-114, 123-124, 129, 151
Congregationalists 254, 261, 269
conservation, principles of 39-41, 120, 203, 211-212, 217, 219, 247, 280, 284, 355, 357, 365, 370-376, 381, 385
Copernicanism 165-166, 169, 175, 177, 186-187
creation *ex nihilo* 59, 62-63, 239, 241
Cynics 66n, 84
deism 35, 94, 249, 253-254, 258, 285, 324, 331
determinism 60, 304-307, 316-317, 350, 382-384
Dissenters 258-261, 270-271, 276-277, 304, 321
Dominicans 95, 111, 113, 120, 126-127, 166
Ecclesiasticus, see Jesus ben Sirach
electricity 243n, 268, 288, 291, 297, 306, 341, 353, 357-360, 363-364, 370-376, 378, 384-386, 389
entropy 246n, 374-376, 381-384
Epicureanism 16, 27, 35-37, 48, 84, 159, 167, 223n, 226-227, 232, 236, 309, 317
ether 103, 171, 242, 267, 278, 286, 288-300, 325, 332, 377, 379-380, 384-385
Euclidean geometry 43, 95-97, 166n
Familists 205-206, 229
Flood, Great 111, 261, 290, 294-295, 311, 372
Franciscans 93, 94, 100, 111, 113, 119-120, 126-127, 137, 139
Galenic medicine 17, 45n, 79n, 136, 153, 159-160, 202, 209
Gnostics 16, 65-66, 75

God
  Clockmaker 102, 106-109, 175
  Designer 234, 292-293
  First Cause 10, 32, 38, 47, 52, 54, 99, 100, 102, 105n, 106, 125-129, 132, 245, 249, 256, 287, 292, 294-295, 316, 367-368
  omnipotence 25, 122, 124, 130, 152, 215, 222, 247, 248, 254, 320, 340
  potentia absoluta 52, 54, 56, 78, 97-99, 104, 109-110, 112, 132n, 172-177, 214, 218, 319, 333n
  potentia ordinata 54-55, 78, 97-98, 104, 109, 112, 132n, 172-178, 181, 214, 218, 233, 235n, 251, 257, 319, 333n
  Prime Mover 99, 102, 106, 122, 234, 263
  Providence 16, 29, 69, 97-98, 105-106, 131-132, 171-174, 178-184, 196-199, 217-218, 227, 233-236, 245, 253-254, 257, 261, 263, 284, 298, 326, 333, 347, 392
  Trinity 92-93, 103, 168169, 173, 239, 250, 255, 258-260, 270-271, 276-277, 320, 326, 331n-333, 342
gravitation 129, 142, 151, 171, 216-217, 235, 237, 240-244, 246, 252-257, 265-268, 280-281, 286, 288, 291, 300, 305-306, 308, 324, 326, 328n, 331-339, 346, 348, 370-371, 397
Hermeticism 29, 104n, 116, 140-147, 150, 152, 166-168, 173n, 174, 186-201, 221-222, 225-226, 252, 278, 314, 389
hexaemeral tradition 17, 18, 25, 30, 38-42, 45n, 81, 128n
Hippocratic tradition 45n, 79n, 84
Huguenots 343n
Hutchinsonians 331-332, 334
impetus 38, 41, 58, 107, 114, 116, 122, 125-132, 135-136, 149, 162, 292
Independents 205, 254, 256, 261, 269, 270, 312, 318, 335, 340
inertia 39-40, 127-128, 131, 211, 213, 266, 270n, 328n, 379
infinity 23, 25-26, 99-101, 123-125, 129-130, 136, 152, 166-171, 199, 223, 227, 238, 240, 247-249, 259, 276, 307, 320, 324, 335-336, 339-340, 375, 387
Islam 31, 73, 76n, 79, 85, 90, 92-93
Jansenists 220n, 221, 343
Jesuits 214, 219-222, 230n, 283, 285, 343, 355
Judaism 6, 13-15, 20, 21n, 23-24n, 28, 33, 34, 36n, 42, 45, 61, 63-64, 71, 74n, 84, 104, 111, 129n, 141, 144, 178n, 238, 322, 391-393, 398, 400, 405
Kabbalah, see Cabalism
Lateran councils 79n
Latitudinarianism 230n, 258, 260, 270, 276-277, 283, 320-323, 331, 355
Levellers 206

Libertines 176
Lutherans 164, 168-169, 171, 173-174, 179, 184-188, 190, 200, 207, 209
magic 64n, 118, 141-142, 145, 167, 190, 192, 198, 222
magnetism 216-217, 237, 240, 243n, 268, 288, 341, 353, 357-360, 364, 371, 373, 376, 378, 384-386, 389
materialism 200, 224-225, 227, 232, 236-237, 251-255, 260, 264-268, 271, 277-282, 285-287, 290, 295-319, 324-326, 332, 334, 343, 354, 361
matter, qualities of 181, 213, 215-216, 218, 222-223, 241-243, 262-263, 265-269, 299, 316, 326-327, 379, 386
mechanical philosophy 10-12, 201, 210, 212-240, 243, 249, 251, 255, 266-270, 278, 286, 292, 295-296, 300, 307-308, 316, 319, 326-327, 334, 337-351, 355-357, 366-379, 381
mechanics 128, 132, 152, 213, 219, 234, 244, 253, 258, 274, 281, 306, 326, 330-331, 335, 341, 344-350, 352-356, 367-368, 372, 389-390
microcosm 143, 159, 174, 203
ministry of healing 5, 21, 60-83, 162, 184-198, 214, 296, 314, 316, 361n, 389, 400, 403, 406
miracles 12, 58, 60, 66-68, 71-72, 75, 91, 104, 161, 167, 173, 175-176, 178, 181-182, 189-191, 213, 217, 225, 255-256, 258, 261, 294, 303, 319, 376, 393
natural theology 92, 95, 98, 252-255, 264, 270, 277, 279-280, 291, 298, 353, 365, 378, 380
naturalism 18-19, 52, 88, 98-99, 109, 111, 118, 132, 140, 148, 173, 177-180, 199, 218, 223, 227, 252, 257, 285n, 331
Neomechanists 254, 258, 262, 290, 303n, 335, 341-351, 355
Neoplatonism 21n, 26, 31-32, 40-42, 44, 50, 51, 90, 109, 114-115, 135, 137-145, 148-150, 152-153, 156, 165-167, 173-175, 191-192, 199, 235, 237, 260, 264, 315, 327, 328n, 389
Neopythagoreanism 21n, 24, 44-45, 139
Nestorians 75-76
Nonconformists 200, 206, 210, 258, 269, 328n, 352
Nonjurors 320
Ockhamists 111
Oxford calculators 96n, 116, 119, 131
Paracelsian tradition 79n, 188, 190, 194, 197, 202-206, 232, 296
perspectivists 115-116, 138-140, 148, 150
physicians 73-74, 78, 79, 120, 166, 179, 184, 187, 193, 206, 209, 278, 283, 323-324

Pietism 213, 296
Platonism 16, 20, 27, 30, 50, 53, 81, 84, 115, 199-202, 225, 234, 235-236, 238-240, 273, 398, 400 (see also Neoplatonism)
Preaching of Peter 24
Presbyterians 182, 197, 260-261, 284, 304
primum mobile 102, 104n, 166, 172-173, 215, 234, 244
Ptolemaic astronomy 17, 22, 43, 136, 138, 149-150, 154, 179
Puritans 173, 180-183, 191, 193, 197, 206, 229-230, 235, 271, 305, 400
Pythagoreanism 14, 16, 20, 27, 84, 141, 147, 151-152, 155, 187, 192, 194, 398, 400 (see also Neopythagoreanism)
Quakers 205-206
quantum theory 362, 397-398, 402n, 404n
Qumran texts 28n, 34n, 36n, 69n, 45
Qur'an 31
rainbow 115
relativity theory 353, 356, 396-398
resurrection 34, 60, 65, 66, 75, 116, 118, 123, 126, 224
Romanticism 199, 294, 361, 391
Rosicrucianism 221-222
Royal Society 207, 209, 230, 244, 304, 320-322, 334, 348n, 367
Royalists 230
Sandemanians 364
secularisation 137, 140, 244, 296, 298, 300, 308, 314-318, 344, 349, 389, 401
seminal principles 42, 141, 203-204, 232-233

Shepherd of Hermas 24
Socinianism 304, 306
spiritualist tradition 29, 199-215, 225-226, 228-229, 230n, 234-235, 239, 243, 273-274, 278, 296-299, 318-319, 327
statistical mechanics 380-384
Stoicism 15-16, 20, 21n, 24, 28-30, 66n, 69, 81, 84, 172, 237, 307, 400
supracelestial waters 58, 103
supramechanical principles 202, 212-213, 233, 238, 240, 242-245, 250, 255-257, 264, 267, 278, 281, 288-289, 292, 324, 326-327, 341, 356, 366
thermodynamics 301, 373-376, 380-384, 387, 389
Thomists 111, 172
Tories 319-320
Trent, Council of 95, 137, 214, 220n
Trinity, see under God
Uniformitarianism 289-290, 293, 295, 311, 369, 373, 375
Unitarianism 261, 254, 260, 270, 276, 304-308, 326, 364n
unity of knowledge 14, 90-94
unity of nature 21, 27-32, 49, 59-60, 147, 156, 171-175, 199, 205, 295-297, 306, 314-315, 319, 333, 351, 357-365, 380, 387-389, 394-397, 400, 404-405
Whigs 260, 269, 320, 328
Wisdom of Solomon 21n, 22n, 24, 29n, 58n, 62n, 114, 139, 147, 153, 329, 380

Studies in the History of Christian Thought

EDITED BY HEIKO A. OBERMAN

1. McNEILL, J. J. *The Blondelian Synthesis.* 1966. Out of print
2. GOERTZ, H.-J. *Innere und äussere Ordnung in der Theologie Thomas Müntzers.* 1967
3. BAUMAN, Cl. *Gewaltlosigkeit im Täufertum.* 1968
4. ROLDANUS, J. *Le Christ et l'Homme dans la Théologie d'Athanase d'Alexandrie.* 2nd ed. 1977
5. MILNER, Jr., B. Ch. *Calvin's Doctrine of the Church.* 1970. Out of print
6. TIERNEY, B. *Origins of Papal Infallibility, 1150-1350.* 2nd ed. 1988
7. OLDFIELD, J. J. *Tolerance in the Writings of Félicité Lamennais 1809-1831.* 1973
8. OBERMAN, H. A. (ed.). *Luther and the Dawn of the Modern Era.* 1974. Out of print
9. HOLECZEK, H. *Humanistische Bibelphilologie bei Erasmus, Thomas More und William Tyndale.* 1975
10. FARR, W. *John Wyclif as Legal Reformer.* 1974
11. PURCELL, M. *Papal Crusading Policy 1244-1291.* 1975
12. BALL, B. W. *A Great Expectation.* Eschatological Thought in English Protestantism. 1975
13. STIEBER, J. W. *Pope Eugenius IV, the Council of Basel, and the Empire.* 1978. Out of print
14. PARTEE, Ch. *Calvin and Classical Philosophy.* 1977
15. MISNER, P. *Papacy and Development.* Newman and the Primacy of the Pope. 1976
16. TAVARD, G. H. *The Seventeenth-Century Tradition.* A Study in Recusant Thought. 1978
17. QUINN, A. *The Confidence of British Philosophers.* An Essay in Historical Narrative. 1977
18. BECK, J. *Le Concil de Basle (1434).* 1979
19. CHURCH, F. F. and GEORGE, T. (ed.). *Continuity and Discontinuity in Church History.* 1979
20. GRAY, P. T. R. *The Defense of Chalcedon in the East (451-553).* 1979
21. NIJENHUIS, W. *Adrianus Saravia (c. 1532-1613).* Dutch Calvinist. 1980
22. PARKER, T. H. L. (ed.). *Iohannis Calvini Commentarius in Epistolam Pauli ad Romanos.* 1981
23. ELLIS, I. *Seven Against Christ.* A Study of 'Essays and Reviews'. 1980
24. BRANN, N. L. *The Abbot Trithemius (1462-1516).* 1981
25. LOCHER, G. W. *Zwingli's Thought.* New Perspectives. 1981
26. GOGAN, B. *The Common Corps of Christendom.* Ecclesiological Themes in Thomas More. 1982
27. STOCK, U. *Die Bedeutung der Sakramente in Luthers Sermonen von 1519.* 1982
28. YARDENI, M. (ed.). *Modernité et nonconformisme en France à travers les âges.* 1983
29. PLATT, J. *Reformed Thought and Scholasticism.* 1982
30. WATTS, P. M. *Nicolaus Cusanus.* A Fifteenth-Century Vision of Man. 1982
31. SPRUNGER, K. L. *Dutch Puritanism.* 1982
32. MEIJERING, E. P. *Melanchthon and Patristic Thought.* 1983
33. STROUP, J. *The Struggle for Identity in the Clerical Estate.* 1984
35. COLISH, M. L. *The Stoic Tradition from Antiquity to the Early Middle Ages.* 1.2. 2nd ed. 1990
36. GUY, B. *Domestic Correspondence of Dominique-Marie Varlet, Bishop of Babylon, 1678-1742.* 1986
37. 38. CLARK, F. *The Pseudo-Gregorian Dialogues.* I. II. 1987
39. PARENTE, Jr. J. A. *Religious Drama and the Humanist Tradition.* 1987
40. POSTHUMUS MEYJES, G. H. M. *Hugo Grotius, Meletius.* 1988
41. FELD, H. *Der Ikonoklasmus des Westens.* 1990
42. REEVE, A. and SCREECH, M. A. (eds.). *Erasmus' Annotations on the New Testament.* Acts — Romans — I and II Corinthians. 1990
43. KIRBY, W. J. T. *Richard Hooker's Doctrine of the Royal Supremacy.* 1990
44. GERSTNER, J. N. *The Thousand Generation Covenant.* Reformed Covenant Theology. 1990
45. CHRISTIANSON, G. and IZBICKI, T. M. (eds.). *Nicholas of Cusa.* 1991
46. GARSTEIN, O. *Rome and the Counter-Reformation in Scandinavia.* 1553-1622. 1992
47. GARSTEIN, O. *Rome and the Counter-Reformation in Scandinavia.* 1622-1656. 1992
48. PERRONE COMPAGNI, V. (ed.). *Cornelius Agrippa, De occulta philosophia Libri tres.* 1992
49. MARTIN, D. D. *Fifteenth-Century Carthusian Reform.* The World of Nicholas Kempf. 1992

50. HOENEN, M. J. F. M. *Marsilius of Inghen*. Divine Knowledge in Late Medieval Thought. 1993
51. O'MALLEY, J. W., IZBICKI, T. M. and CHRISTIANSON, G. (eds.). *Humanity and Divinity in Renaissance and Reformation*. Essays in Honor of Charles Trinkaus. 1993
52. REEVE, A. (ed.) and SCREECH, M. A. (introd.). *Erasmus' Annotations on the New Testament*. Galatians to the Apocalypse. 1993
53. STUMP, Ph. H. *The Reforms of the Council of Constance (1414-1418)*. 1994
54. GIAKALIS, A. *Images of the Divine*. The Theology of Icons at the Seventh Ecumenical Council. With a Foreword by Henry Chadwick. 1994
55. NELLEN, H. J. M. and RABBIE, E. (eds.). *Hugo Grotius – Theologian*. Essays in Honour of G. H. M. Posthumus Meyjes. 1994
56. TRIGG, J. D. *Baptism in the Theology of Martin Luther*. 1994
57. JANSE, W. *Albert Hardenberg als Theologe*. Profil eines Bucer-Schülers. 1994
59. SCHOOR, R.J.M. VAN DE. *The Irenical Theology of Théophile Brachet de La Milletière (1588-1665)*. 1995
60. STREHLE, S. *The Catholic Roots of the Protestant Gospel*. Encounter between the Middle Ages and the Reformation. 1995
61. BROWN, M.L. *Donne and the Politics of Conscience in Early Modern England*. 1995
62. SCREECH, M.A. (ed.). *Richard Mocket, Warden of All Souls College, Oxford, Doctrina et Politia Ecclesiae Anglicanae*. An Anglican Summa. Facsimile with Variants of the Text of 1617. Edited with an Introduction. 1995
63. SNOEK, G.J.C. *Medieval Piety from Relics to the Eucharist*. A Process of Mutual Interaction. 1995
64. PIXTON, P.B. *The German Episcopacy and the Implementation of the Decrees of the Fourth Lateran Council, 1216-1245*. Watchmen on the Tower. 1995
65. DOLNIKOWSKI, E.W. *Thomas Bradwardine: A View of Time and a Vision of Eternity in Fourteenth-Century Thought*. 1995
66. RABBIE, E. (ed.). *Hugo Grotius, Ordinum Hollandiae ac Westfrisiae Pietas (1613)*. Critical Edition with Translation and Commentary. 1995
67. HIRSH, J.C. *The Boundaries of Faith*. The Development and Transmission of Medieval Spirituality. 1996
68. BURNETT, S.G. *From Christian Hebraism to Jewish Studies*. Johannes Buxtorf (1564-1629) and Hebrew Learning in the Seventeenth Century. 1996
69. BOLAND O.P., V. *Ideas in God according to Saint Thomas Aquinas*. Sources and Synthesis. 1996
70. LANGE, M.E. *Telling Tears in the English Renaissance*. 1996
71. CHRISTIANSON, G. and T.M. IZBICKI (eds.). *Nicholas of Cusa on Christ and the Church*. Essays in Memory of Chandler McCuskey Brooks for the American Cusanus Society. 1996
72. MALI, A. *Mystic in the New World*. Marie de l'Incarnation (1599-1672). 1996
73. VISSER, D. *Apocalypse as Utopian Expectation (800-1500)*. The Apocalypse Commentary of Berengaudus of Ferrières and the Relationship between Exegesis, Liturgy and Iconography. 1996
74. O'ROURKE BOYLE, M. *Divine Domesticity*. Augustine of Thagaste to Teresa of Avila. 1997
75. PFIZENMAIER, T.C. *The Trinitarian Theology of Dr. Samuel Clarke (1675-1729)*. Context, Sources, and Controversy. 1997
76. BERKVENS-STEVELINCK, C., J. ISRAEL and G.H.M. POSTHUMUS MEYJES (eds.). *The Emergence of Tolerance in the Dutch Republic*. 1997
77. HAYKIN, M.A.G. (ed.). *The Life and Thought of John Gill (1697-1771)*. A Tercentennial Appreciation. 1997
78. KAISER, C.B. *Creational Theology and the History of Physical Science*. The Creationist Tradition from Basil to Bohr. 1997

Prospectus available on request

KONINKLIJKE BRILL — P.O.B. 9000 — 2300 PA LEIDEN — THE NETHERLANDS